Advances in Experimental Medicine and Biology

Volume 1390

Advances in Experimental Medicine and Biology provides a platform for scientific contributions in the main disciplines of the biomedicine and the life sciences. This series publishes thematic volumes on contemporary research in the areas of microbiology, immunology, neurosciences, biochemistry, biomedical engineering, genetics, physiology, and cancer research. Covering emerging topics and techniques in basic and clinical science, it brings together clinicians and researchers from various fields.

Advances in Experimental Medicine and Biology has been publishing exceptional works in the field for over 40 years, and is indexed in SCOPUS, Medline (PubMed), EMBASE, BIOSIS, Reaxys, EMBiology, the Chemical Abstracts Service (CAS), and Pathway Studio.

2021 Impact Factor: 3.650 (no longer indexed in SCIE as of 2022)

Moray J. Campbell • Charlotte L. Bevan
Editors

Nuclear Receptors in Human Health and Disease

Springer

Editors
Moray J. Campbell
College of Pharmacy
The Ohio State University
Columbus, OH, USA

Charlotte L. Bevan
Department of Surgery & Cancer
Imperial College London
London, UK

ISSN 0065-2598 ISSN 2214-8019 (electronic)
Advances in Experimental Medicine and Biology
ISBN 978-3-031-11835-7 ISBN 978-3-031-11836-4 (eBook)
https://doi.org/10.1007/978-3-031-11836-4

This Springer imprint is published by the registered company Springer Nature Switzerland AG
The registered company address is: Gewerbestrasse 11, 6330 Cham, Switzerland

Introduction

Nuclear Receptors: The Past, the Present and the Future

Nuclear Receptors (NRs) are involved in a multitude of biological pathways and numerous disorders and diseases; it would be difficult to find another family of proteins in the human genome that has such a broad and critical role in both healthy and diseased contexts. Further, they constitute the archetypal proteins for studying some of the most fundamental processes of gene regulation and genomic organization. The first NR was cloned in the 1980s, and we now know that this family is comprised of 48 distinct proteins that share common structural properties. They play essential roles in the development of organs, as evidenced by phenotypic consequences following gene deletion. They are also frequently co-opted or altered in disease states, including cancer and metabolic disorders. One of the special features of the NR superfamily is the fact that they constitute the only class of readily druggable transcription factors. This makes them critical downstream effectors of numerous biological and cellular processes and also the targets of many treatments and therapies. Their vital role in both healthy and pathological contexts likely results from a relatively unique feature of this class of transcription factors: their ligand activated switchable states.

NRs share common molecular features, including the ability to interact with DNA directly, making them potent mediators of gene activation or gene repression. Another unusual feature of NRs is their ability to be activated, either by ligand interaction with the ligand binding domain, or via specific co-factor associations that can also modulate activity. This feature makes them activatable, meaning that their activity can be switched on or off by the presence or absence of a specific ligand or specific co-factors. This combination of ligand/co-factor modulation and direct DNA transcriptional activity provides NRs with a highly unusual, but powerful combination of features that makes them ideal proteins for context-dependent, regulatable activity, since they can be switched on by their cognate ligand when needed (i.e. during development processes) and switched off when their job is done (i.e. when the organ is fully developed). Their ligand-inducible transcriptional activity also makes them ideal proteins for recurring biological changes that need to occur in a time-specific, rapidly-responsive manner, as exemplified by the rapid and substantial changes that occur in relevant organs during the menstrual cycle and pregnancy.

Our understanding of NR function has been largely motivated by their role in specific diseases and disorders. Examples include the key role for Estrogen Receptor alpha (ERα) and Androgen Receptor (AR) in breast and prostate cancer, respectively, or the crucial role of Peroxisome Proliferator Activated Receptors (PPARs) and LXRs in diabetes and other metabolic diseases. New mechanistic insight into how these specific NRs function in these specific contexts have formed the basis for understanding all transcription factors and gene regulatory processes. Fundamental concepts around transcription factor biology, protein complex organization, signalling pathways and gene regulation have been discovered by using NRs as the model system and as such, the impact from studying NRs on our basic understanding of genome regulation and fundamental cellular processes cannot be overstated.

Given their substantial roles in some of the most common and deadly diseases and disorders and their critical role in the development of major organs and physiological processes, it is surprising that so many questions about NRs remain unanswered or are not fully understood. As an example, we know that ligands for one NR can commonly activate related but distinct NRs and that different NRs can vie for similar docking regions on the genome, but this level of cross-activity is poorly characterized. Some of the first co-factors discovered (co-factors being proteins that can influence gene regulation by indirect association with the chromatin) were identified from screens that sought to identify NR-associated proteins and our repertoire of NR-associated co-factors has increased enormously over the years, yet our understanding of the full complement of co-factors, the dynamics between co-factors and their mechanistic roles are not fully defined.

An interesting paradigm in NR-biology is the well-established observation that many NRs have known endogenous ligands, but many other family members lack endogenous ligands or, if endogenous ligands exist, they haven't been discovered yet. These so-called Orphan NRs represent a large class of transcription factors, some of which have been implicated in critical biological processes, although other Orphan NRs are yet to be associated with a function or a biological context. When endogenous NR ligands are known and pharmacological ligands have subsequently been created to alter the structure-function of that NR, they have the potential to change medicine. Some of the most commonly prescribed drugs in the Western world are pharmacological agents that target NRs and they have had profound clinical impact (e.g. as anti-inflammatory agents, cancer treatments or regulators of pregnancy) attesting to the success of understanding and exploiting NR biology. However, the fundamental process of how these drugs work and how they elicit the NR-mediated downstream events are sometimes not fully understood, again highlighting how much we know about NRs and how much we still need to learn.

This book explores the role of NRs in biology, with a focus on these highly unusual, functionally distinct, yet fascinating proteins in human health and disease. The breadth of the topics in this book highlights the diverse and complex nature of NR function, as well as the many contexts wherein they have been implicated. Their roles in fundamental developmental processes are discussed, as are the roles in metabolic systems. Recent advances in our understanding of NR biology in diseases (particularly cancer) are explored and both established and novel therapeutic opportunities in targeting NRs are presented. This book provides a thorough and contemporary discussion of this field and highlights the many physiological and clinical roles for NRs, as described by leaders in the field. The findings and insight will be of relevance to both experts and those with a general interest in this fascinating class of transcriptional regulators.

Cancer Research UK Cambridge Research Institute Jason S. Carroll
Cambridge, UK

Overview

This book, entitled, *Nuclear Receptors in Human Health and Disease*, is designed as an update to an earlier book entitled *Nuclear Receptors, Current Concepts and Future Challenges*, which was published in 2010. As before, the chapters are written by leaders in the field, and are broadly intended to serve as an introduction to the field, to discuss the state-of-the-art, and also to speculate where the field is going to meet the challenges on the horizon.

The first nuclear receptor was cloned over two decades before the first edition in 2010 and provided a rich paradigm for that book. Surprisingly, perhaps, progress from 2010 to the time of writing has in many ways been equally remarkable. This is most evident in the chapter titles and their sectional organization. In the 2010 book, the chapters were largely written around a single nuclear receptor, whereas in the current book chapters are very much more focused on roles of multiple nuclear receptors in shared phenotypes, such as metabolism or reproduction, or how multiple receptors interplay in a single biological function such as in circadian rhythm.

What is also clear is how much general biological insight has been established and underpinned by nuclear receptor research, especially in the fields of epigenetics, genome organization, and transcriptional regulation. Similarly, it is also clear how biological concepts revealed elsewhere are rapidly translated into the nuclear receptor field to profound effect. For example, emerging concepts of the 3D genome, phase separation, and the impact of spatially divergent enhancers are already significantly shaping how nuclear receptor function is understood.

Finally, what is also clear is the explosion of different experimental and analytical approaches being applied to capture nuclear receptor function. A striking illustration of this is that the phrase "ChIP-Seq" did not appear in the 2010 book but is now ubiquitous across chapters. This nuclear receptor field is an early adopter of technologies: variations of next-generation sequencing approaches have been applied to define the nuclear receptor cistrome, epigenome and chromatin accessibility, transcriptome, metabolome, and now, with approaches such as RIME, the proteome. Inevitably, this has required the development and application of integrative analytical approaches to analyze, interpret, and visualize these high-dimensional data sets.

Perhaps chastened by this amazing progress since 2010, the editors quietly dropped the "Future Challenges" from the title of this book.

Contents

Part IV Cancer

Part V New Developments in Transcriptional Control by Nuclear Receptors

Part VI Clinical Translation

Part I

Reproduction and Development

1 Nuclear Receptors in Pregnancy and Outcomes: Clinical Perspective

Luiza Borges Manna and Catherine Williamson

Abstract

Pregnancy is characterised by profound hormonal and metabolic changes in the mother. Both oestrogen and progesterone, along with their respective nuclear receptors, have an important role in maintaining a healthy pregnancy. Equally, other nuclear receptors such as LXR, FXR and the PPARs play important roles in the gradual alterations in metabolism that ensure survival of mother and fetus. Disruptions in nuclear receptor signalling can result in pregnancy disorders such as gestational diabetes mellitus, intrahepatic cholestasis of pregnancy, hypertensive disorders of pregnancy and preterm labour, all of which have both immediate and long-term implications for maternal and fetal health. By reviewing data from human studies and animal models, this chapter will describe the contribution of nuclear receptors to normal pregnancy, their role in gestational disorders and their potential as therapeutic targets.

L. Borges Manna · C. Williamson (✉)
Department of Women and Children's Health, King's College London, London, UK
e-mail: catherine.williamson@kcl.ac.uk

Keywords

Pregnancy · Oestrogen · Progesterone · LXR · FXR · PPAR · Gestational diabetes · Hypertension · Cholestasis

1.1 Introduction

Pregnancy is a unique state in which the maternal organism must undergo a multitude of physiological adaptations to support the growth of a fetus, whilst also maintaining its own health. Numerous cardiovascular, renal, immune and metabolic changes occur in response to rising concentrations of reproductive hormones and the growing conceptus [1]. Not surprisingly, disruptions in the complex regulation of these maternal modifications can result in pregnancy disorders.

In humans, maternal preparations for pregnancy occur in every menstrual cycle regardless of the presence of a conceptus. The uterus and endometrium undergo changes that render them receptive to embryo implantation and placental development [2, 3]. The reproductive hormones oestrogen and progesterone play a key role in this process, along with their respective nuclear receptors (the ERs and PRs). Other nuclear receptors such as peroxisome proliferator-activated receptors (PPARs) and liver X receptors (LXRs) also influence trophoblast development and placental formation. Comprehending

M. J. Campbell, C. L. Bevan (eds.), *Nuclear Receptors in Human Health and Disease*, Advances in Experimental Medicine and Biology 1390, https://doi.org/10.1007/978-3-031-11836-4_1

the mechanisms underlying these early events is not only important for the understanding of early pregnancy pathologies such as recurrent miscarriage and implantation failure, but also later gestational complications. It is known that disruptions in decidualisation, implantation and trophoblast invasion can have a lasting effect on pregnancy, as they can constitute the pathophysiological basis for pre-eclampsia, intrauterine growth restriction and placental abruption [4].

After implantation, maternal metabolism adapts to cater for the increasing energetic demands of the fetus (Fig. 1.1). There are marked alterations in maternal metabolic pathways of uptake, storage and distribution of nutritional fuels to match different stages of fetal development [1]. Early pregnancy is characteristically an anabolic state that guarantees the storage of nutrients in preparation for later stages of gestation. This period is marked by increased insulin sensitivity, lipogenesis and lipid storage [5]. As pregnancy advances, insulin resistance progressively rises towards the third trimester, causing a shift to a catabolic state [5, 6]. Lipolysis is thus stimulated, leading to a state of physiological hyperlipidaemia in the mother [7]. Serum glucose concentrations rise, and glucose is prioritised to the fetus, whilst the mother relies on serum lipids for nutrition [5]. Although the mechanisms behind these changes are not fully under-

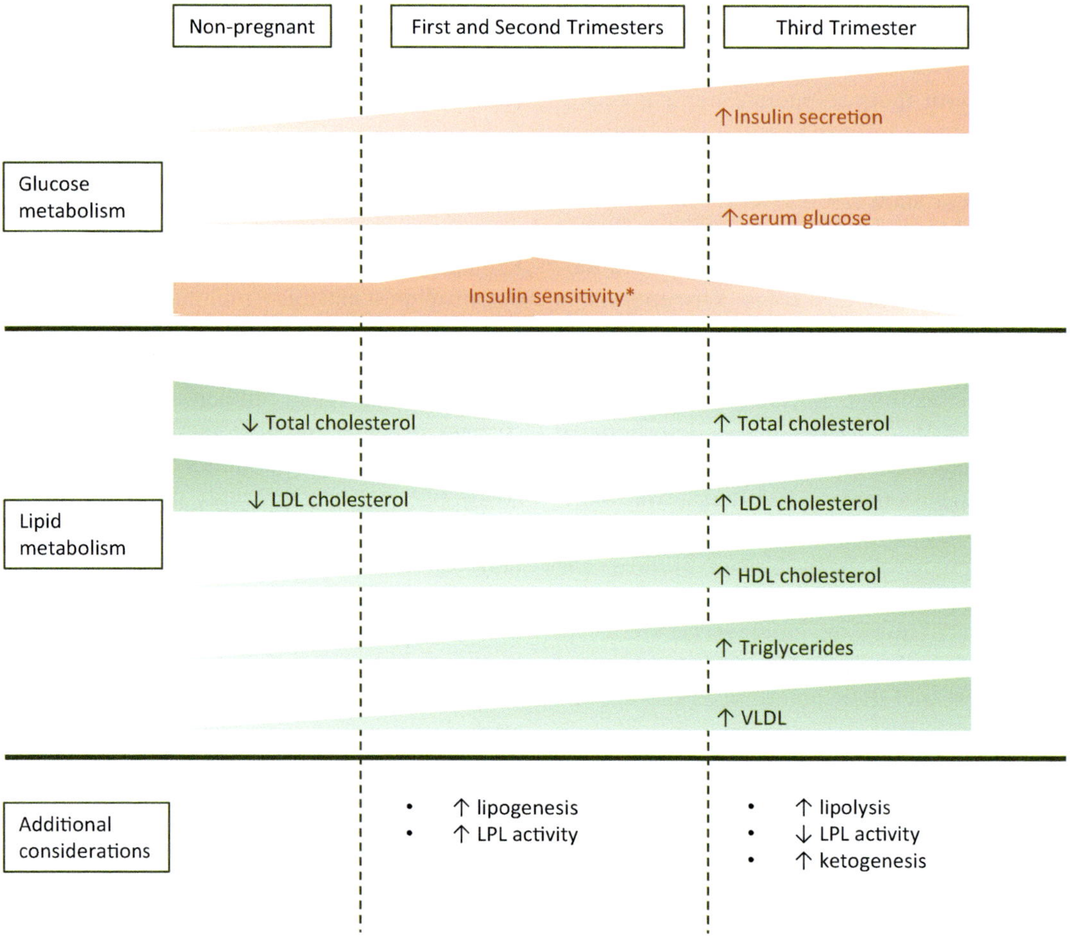

Fig. 1.1 Summary of changes in maternal metabolism during pregnancy. Arrows show direction of change. *: insulin sensitivity increases in the first trimester then progressively declines as the mother enters a catabolic state. *LDL* low-density lipoprotein, *HDL* high-density lipoprotein, *VLDL* very low-density lipoprotein, *LPL* lipoprotein lipase

stood, nuclear receptors have been identified as plausible candidates for their regulation [8].

Close to term, changes in the uterine environment occur to facilitate parturition. The myometrium, previously quiescent, becomes responsive to labour stimuli and undergoes changes that facilitate its contractions. This is a process highly regulated by progesterone and its nuclear receptors.

In this chapter we will explore the contribution of nuclear receptors to the development of a normal pregnancy, focusing on early pregnancy events, maternal metabolic changes and mechanisms behind parturition. We will then describe how nuclear receptors are implicated in disorders such as gestational diabetes, hypertensive disorders of pregnancy, intrahepatic cholestasis of pregnancy and preterm labour.

1.2 The Role of Nuclear Receptors in Maintaining a Healthy Pregnancy

1.2.1 Progesterone Receptors and PPARS in Early Pregnancy

The development of a healthy materno-fetal interface is essential for pregnancy success. A key organ at this interface is the placenta. While maternal uterine receptivity is achieved through the process of endometrial decidualisation, the conceptus is responsible for the development of different trophoblastic lineages that will execute the placental functions of hormonal synthesis, materno-fetal exchange of nutrients and adequate supply to fetal tissues.

The process of decidualisation occurs in the second phase of the endometrial cycle, when progesterone concentrations rise following ovulation. It transforms the oestrogen-primed endometrial stromal cells into specialised secretory cells that facilitate implantation and trophoblast development [3]. Progesterone is a master regulator of this process via stimulation of its nuclear progesterone receptor (PR). Three forms of PRs have been identified in mice and humans: PR-A, PR-B and PR-C, with the first two recognised as the main isoforms present in the uterus [9]. PR can be activated by direct binding of progesterone, as well as through ligand-independent activation [10], illustrating the complexity of its function. Whilst the presence of both PR-A and PR-B is critical for the development of adequate decidual responses in mice, PR-B seems to have a less crucial role. Knockout studies in mice have shown that the absence of PR-B does not induce a markedly abnormal uterine phenotype [11, 12]. A temporal change in the expression of each isoform, as well as their relative expression, is also essential for adequate endometrial proliferation [13].

After fertilisation, the conceptus implants into the decidualised endometrium. Its extraembryonic tissues undergo differentiation into distinct lineages, followed by migration and invasion of maternal tissues to form the placenta. The lineage termed villous trophoblast (VT) forms the chorionic villi, the main materno-fetal exchange surface of the placenta. The extravillous trophoblast (EVT) is the lineage responsible for anchoring the placenta into maternal tissues and remodelling uterine spiral arteries to optimise placental perfusion (Fig. 1.2) [14, 15].

The nuclear peroxisome proliferator-activated receptor (PPAR) has been implicated in this early process of trophoblast differentiation and invasion. All three known PPAR isoforms, PPARα, PPARβ and PPARγ, are expressed in human and rodent placentas [16]. PPARγ and its heterodimer partner RXRα have the most widely reported role in this process. They are expressed in both VT and EVT [17]. Their essential role is illustrated by the fact that PPARγ-null mutations in mice result in early embryo demise secondary to inappropriate placental vascular formation and trophoblast differentiation [18]. *In vitro* experiments with PPARγ agonists showed that PPARγ activation abrogates maternal tissue invasion by the EVT, whilst PPARγ antagonists have the opposite effect [19–22]. There also seems to be an effect of PPARγ agonists on trophoblast differentiation. *In vitro* studies of PPARγ-null trophoblast stem cells showed defects in differentiation of all trophoblast layers [23],

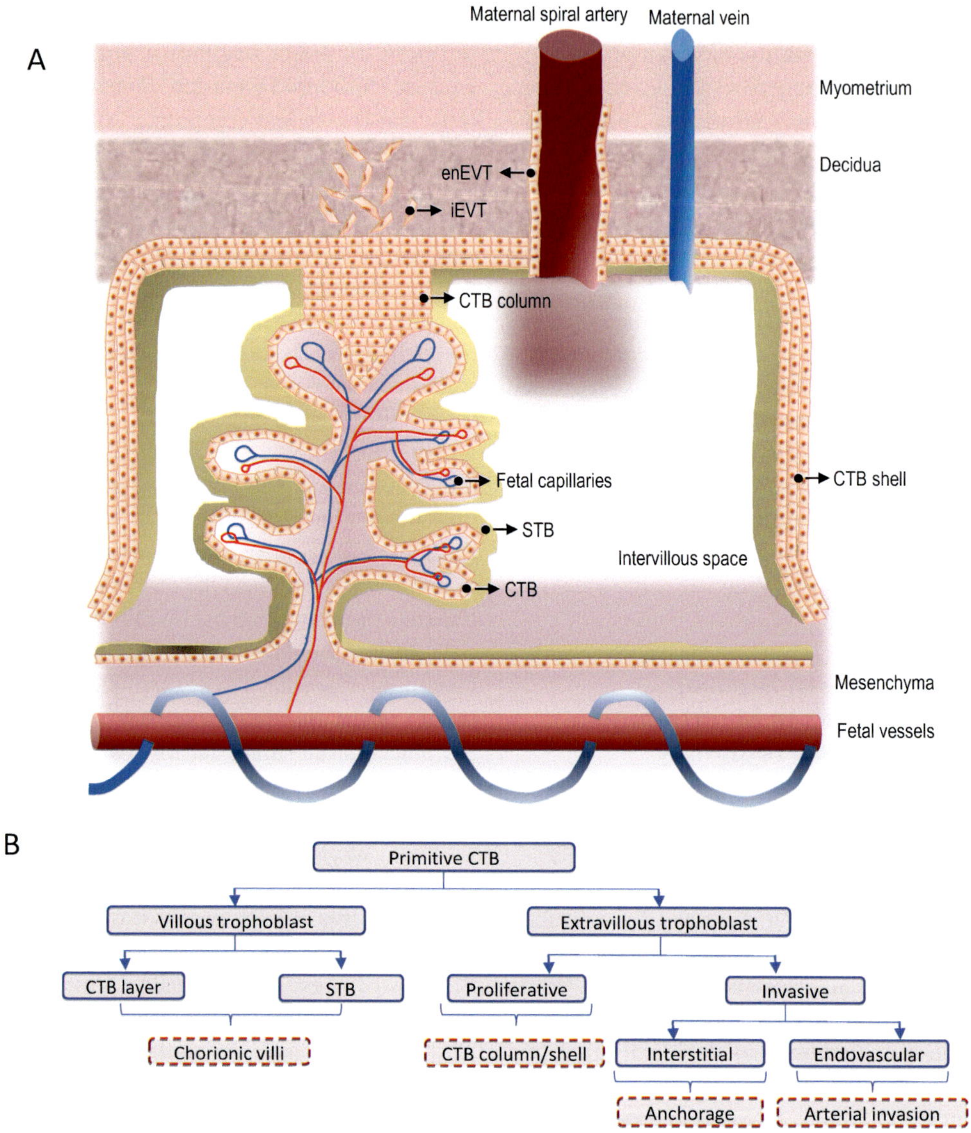

Fig. 1.2 Simplified representation of placental structure (a) and lineages (b). After implantation, the extraembryonic tissues of the blastocyst differentiate into distinct lineages to form the placenta. It first differentiates into the cytotrophoblast (CTB), a single layer of epithelial cells that gives origin to the chorionic villi, the functional units that facilitate feto-maternal exchange. The cytotrophoblast acts as a stem cell layer that generates all other lineages. The fusion of cells creates the multinucleated layer of the syncytiotrophoblast (STB), which is responsible for placental hormone synthesis. Each chorionic villus is made of a mesenchymal chore, fetal capillaries, a layer of cytotrophoblast and a layer of syncytiotrophoblast. The CTB proliferates into columns above the chorionic villi, giving rise to the (EVT), which is responsible for anchoring the placenta into maternal tissues. These columns merge to form a CTB shell, which is a continuous structure only breached by maternal vessels that provide blood to the intervillous space. The EVT then differentiates into the interstitial EVT (iEVT), which invades the maternal decidua, and the endovascular EVT (enEVT), which invades the spiral arteries and replace their smooth muscle to increase placental perfusion. Both LXRs and PPARs are involved in trophoblast differentiation and invasion

although one study showed that this effect might be ligand-dependent [24].

Similarly, liver X receptors (LXR) have been shown to affect trophoblast function. Two subtypes of LXR, LXRα and LXRβ, have been recognised to date. LXRα is highly expressed in tissues with high metabolic activity such as liver and adipose tissue, whereas LXRβ is ubiquitously expressed [25, 26]. Both are expressed in the placenta [27]. LXR is a master regulator of cholesterol metabolism and is activated by endogenous oxysterols [28]. A study in an *in vitro* model of invasive human trophoblast showed that activation of LXRβ by synthetic or endogenous ligands can inhibit trophoblast invasion [29]. LXR activation, by both oxysterols and a synthetic LXR agonist, can also impair trophoblast differentiation [30, 31].

1.2.2 Liver-X-Receptors, Clock Genes and Maternal Metabolic Adaptations in Mid-to-Late Pregnancy

Two groups of nuclear receptors, LXRs and the clock-regulating REV-ERBs, have been shown to influence maternal metabolic adaptations to pregnancy. LXR acts as a cholesterol sensor that prevents cholesterol accumulation in tissues. It is a strong promoter of reverse cholesterol transport, stimulating the transport of cholesterol from the periphery to the liver, whereby it is excreted through the biliary system [25]. In the event of high serum concentrations of cholesterol, LXR induces the expression of transporters ABCA1 and ABCG1, both of which facilitate the transfer of intracellular cholesterol onto apolipoproteins and HDL, and subsequent return of cholesterol to the liver [32, 33]. Despite preventing cholesterol accumulation, LXR has also a seemingly paradoxical role in *de novo* lipogenesis. It upregulates SREBP-1c, ACC, SCD1 and FAS, all of which participate in fatty acid (FA) and triglyceride (TG) synthesis pathways [34]. Thus, LXR stimulation can increase serum concentrations of TG and FAs. By promoting this effect, LXR facilitates cholesterol esterification by FAs, a process that decreases its toxic potential to cells [25]. A summary of the metabolic effects of LXR and its target genes can be found in Fig. 1.3a.

The role of LXR in promoting the marked lipogenic state of early pregnancy has been confirmed in a mouse model [35]. However, LXR did not seem to influence accompanying changes in cholesterol concentrations. The study showed that mouse pregnancy presents the expected findings of increased hepatic concentrations of TG in early stages. A simultaneous upregulation of the LXR targets Fas, Scd-1 and Srebp-1c was also observed. These changes then resolved later in pregnancy, when increased serum concentrations of TG were observed. The same alterations in lipid metabolism were reproduced in non-pregnant females fed LXR agonists, and were disrupted in LXR knockout mice, confirming the role of LXRs in the process.

Data on the contribution of LXR to adaptations in later pregnancy are scarce. LXR expression, along with the expression of other nuclear receptors, was shown to be reduced in the liver of mice in late pregnancy [36]. In a different study, changes in lipid metabolism in late pregnancy occurred in the presence of normal protein levels of both LXRα and LXRβ [35]. However, administration of LXR agonists had little effect on the downstream LXR gene expression profile. It is therefore possible that although LXR expression and protein availability remains constant throughout pregnancy, gestational signals in later stages interfere with its function.

Changes in lipid metabolism in early pregnancy also seem to be associated with disruptions in the body's clock function. Circadian signals are known to influence metabolic pathways [37]. The nuclear receptors REV-ERB-α and REV-ERB-β have been shown to regulate a feedback loop between the body's master clock at the suprachiasmatic nucleus and peripheral organs [38, 39]. A study in mice showed that the expression of the lipogenic genes *Fas, Scd2 and Hmgcr* are increased in early pregnancy in comparison to late pregnancy. This increase seems to be uncoupled from the normal circadian oscillations in *Rev-erb-α* and *Rev-erb-β* expression. In late pregnancy, this synchronicity is restored and becomes similar to that

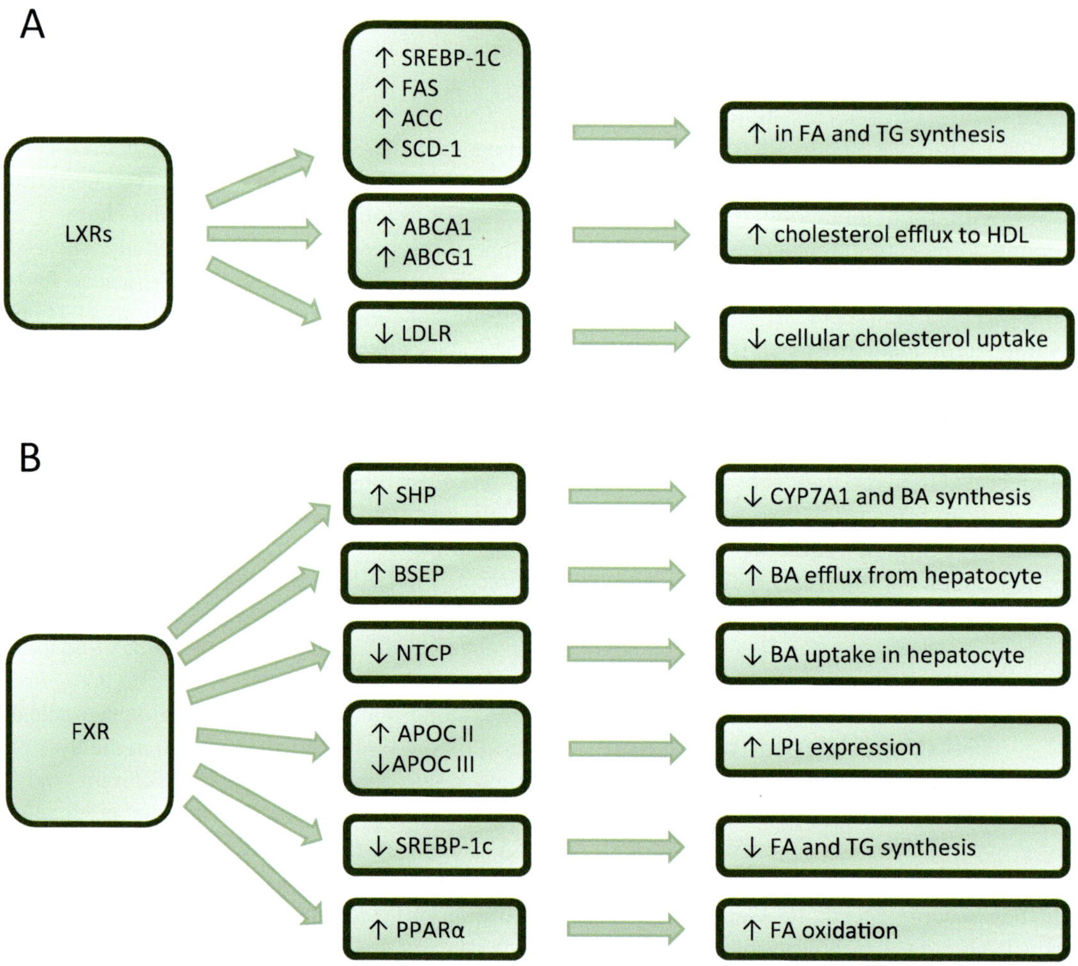

Fig. 1.3 Simplified representation of the metabolic effects of nuclear receptors LXR and FXR. Arrows show the direction of effect on gene targets and physiological processes of (**a**) LXR and (**b**) FXR. *LXR* Liver-X-receptor, *SREBP-1C* Sterol regulatory element-binding protein 1, *FAS* Fatty acid synthase, *ACC* Acetyl-CoA carboxylase, *SCD-1* Stearoyl-CoA desaturase 1, *ABCA1* ATP-binding cassette transporter A1, *ABCG1* ATP-binding cassette transporter G1, *LDLR* Low-density lipoprotein receptor, *FA* fatty acid, *TG* triglyceride, *HDL* high-density lipoprotein, *FXR* farnesoid-Xreceptor, *SHP* small heterodimer partner, *BSEP* bile salt export pump, *NTCP* Sodium-taurocholate co-transporting polypeptide, *APOCII* Apolipoprotein C-II, *APOCIII* apolipoprotein C-III, *PPARα* Peroxisome proliferator-activated receptor alpha, *CYP7A1* 7α-hydroxylase, *BA* bile acid, *LPL* lipoprotein lipase

of non-pregnant females. This shows that, for the anabolic state of early pregnancy to occur, hepatic gene expression becomes independent of the usual hepatic clock system [40].

1.2.3 Parturition

Human labour is a complex event resulting from cervical ripening and myometrial contractions that culminate in the expulsion of the fetus and the placenta. In order to prevent early delivery of the fetus, the uterus remains quiescent throughout gestation until endocrine, pro-inflammatory and mechanical changes occur to trigger myometrial activation [41]. Inflammation is a central feature of human labour (Fig. 1.4), and development of a pro-inflammatory state within the uterus is one of the initial triggers for parturition.

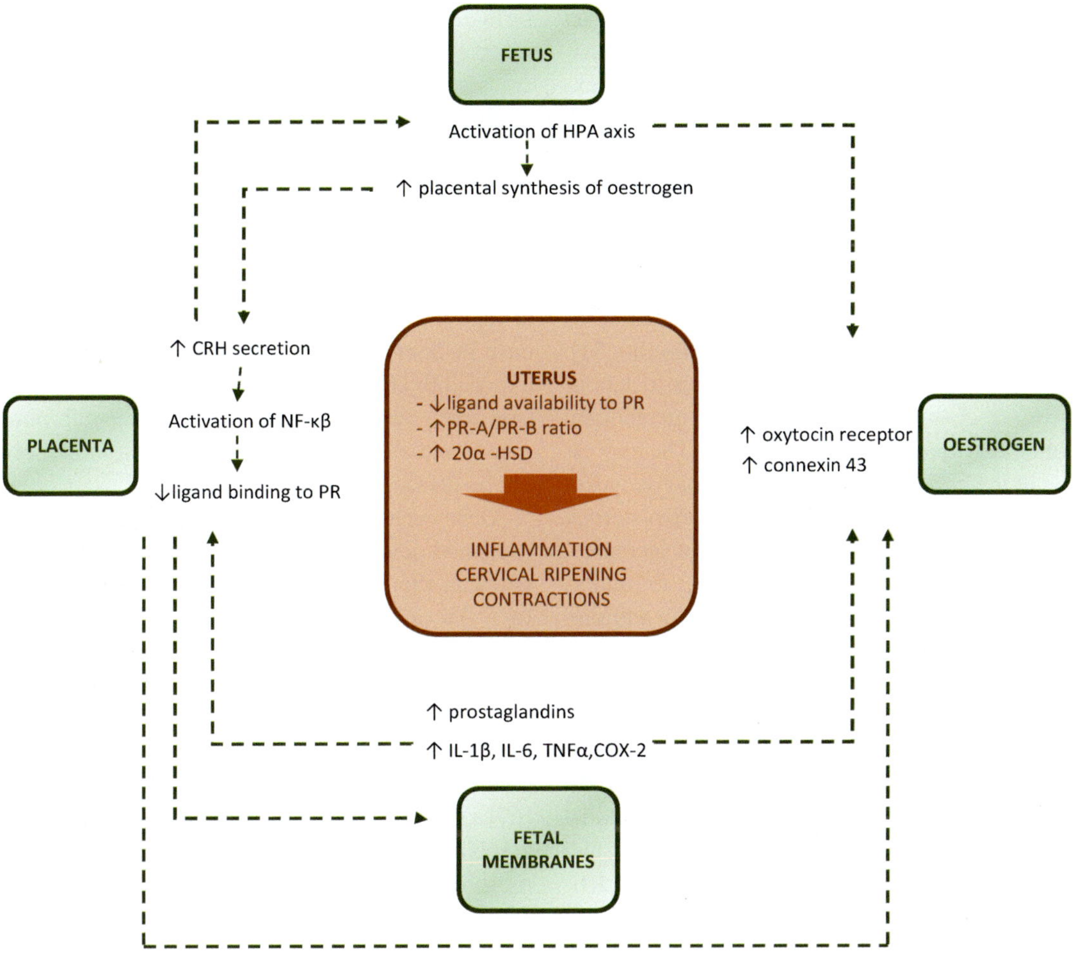

Fig. 1.4 Simplified representation of the mechanisms underlying labour. Dashed arrows represent a positive effect. *PR* progesterone receptor, *HPA axis* hypothalamic-pituitary-adrenal axis, *CRH* corticotropin releasing hormone

Progesterone is a major regulator of uterine quiescence. It provides anti-inflammatory and anti-contractile signals to the myometrium. PR blocks the activation of nuclear factor κβ (NF-κβ), an important initiator of the labour cascade of events, and its downstream inflammatory targets [42, 43]. At the same time, PR upregulates the expression of NF-κβ inhibitor [44]. In the myometrium, activation of PR inhibits the synthesis of connexin 43 (cx43), thus blocking the formation of gap junctions that are responsible for uterine contractions [45]. In addition, by upregulating zinc finger E-box binding homeobox proteins ZEB1 and ZEB2, the PR inhibits the expression of contractile genes, including the oxytocin receptor [46].

In most mammals, the onset of labour is marked by increased inflammatory stimuli in uterine tissues accompanied by a progressive decrease in circulating progesterone concentrations. In human pregnancy, however, serum concentrations of progesterone remain stable throughout gestation. It is thought that labour onset is secondary to a "functional withdrawal" of progesterone, triggered by a change in the relative expression and function of progesterone receptor isoforms [47]. There is substantive evidence to suggest that PR-B is the principal driver of uterine quiescence, whereas PR-A, when not bound to progesterone, has the ability to act as an endogenous repressor of PR-B [48]. A recent study in genetically modified mice has confirmed the distinct roles of PR-A and PR-B in myometrial contractility. Mice that overexpressed the PR-B isoform had an increased length of gestation and poor uterine contractions. Mice overexpressing the PR-A isoform, on the other hand, showed increased uterine contractility. Downstream target genes of both isoforms were also analysed, confirming a stronger anti-contractile role of PR-B [49].

Studies in human myometrium have shown a marked increase in PR-A expression close to term, increasing the PR-A to PR-B ratio [50, 51]. In addition, in the period leading up to labour onset, a change in progesterone metabolism within the myometrium takes place. The expression of the enzyme 20α-hydroxysteroid dehydrogenase (20α-HSD), that converts progesterone into an inactive metabolite, markedly increases, decreasing the ability of progesterone to bind to PR-A [52, 53]. The unliganded PR-A, in addition to repressing PR-B, acts as a transcriptional activator of cx43 [48, 54]. The anti-inflammatory properties of PR-B are then overcome, and unrestrained tissue inflammation perpetuates labour signals [55]. In particular, an increase in IL-1β within the uterus increases NF-κβ activity, whilst at the same time repressing PR-B activity and further perpetuating the cycle of myometrial activation [56].

1.3 Nuclear Receptors and Gestational Disorders

1.3.1 Gestational Diabetes Mellitus

Gestational diabetes mellitus (GDM) is defined by the presence of glucose intolerance that develops, or is first recognised, in pregnancy [57]. The global prevalence of GDM is on the rise, with an estimated 16% of pregnancies affected by some form of hyperglycaemia [58]. This increase is thought to be linked to the equally rising prevalence of obesity amongst reproductive age women, and increase in maternal age [58, 59]. Pregnancies affected by GDM have an increased risk of poor outcomes, with the most prevalent complication being fetal macrosomia and its related birth injuries [57]. Fetal death, preterm birth and neonatal unit admission are also recognised outcomes [60]. Mothers affected by GDM are also more likely to develop pre-eclampsia, adding to the existing maternal and fetal morbidity [60]. The implications of GDM for future health are a much wider public health issue; affected women have an approximately 26% increased risk of developing type 2 diabetes mellitus 15 years after their GDM diagnosis, and are at higher risk of developing cardiovascular disease in later life [61, 62]. Meanwhile, children exposed to GDM in the intrauterine environment can have suboptimal neurodevelopmental outcomes and also increased risk of developing metabolic disease later in life [59, 63, 64].

Oestrogen can influence glucose homeostasis [65], and oestrogen receptors have been investigated in the pathophysiology of diabetes mellitus. An association between the rs1256031 polymorphism in the oestrogen receptor β (Erβ) gene and the development of type 2 diabetes mellitus has been found in a Mexican study [66]. A similar study in a Chinese population did not confirm this association in GDM-affected women [67]. GDM development has, however, been associated with the PVuII single nucleotide polymorphisms in oestrogen receptor α (Erα) [68].

Outside of pregnancy, the development of insulin resistance and diabetes is closely related to disorders in lipid metabolism. Abnormal serum and tissue concentrations of lipids can be both cause and consequence of impaired glucose homeostasis [69–71]. Nuclear receptors involved in lipid regulation have thus been investigated in the context of type 2 diabetes mellitus. LXR agonists have been shown to influence glucose metabolism both *in vitro* and in mice, and are thought to be potent serum glucose-lowering agents [72–74]. However, a concomitant rise in serum triglyceride concentrations with the use of these agents has so far hindered their development as anti-diabetic drugs [74]. The contribution of LXR to GDM pathogenesis and its role in treatment of GDM have not been explored to the same extent. An analysis of gene expression in the adipose tissue of women affected by GDM showed an overall reduced expression of LXR and evidence of abnormal adipose tissue metabolism [75]. Although it is plausible that these changes might contribute to the development of GDM, substantive data are lacking.

Farnesoid X receptors (FXR) are also seen as promising targets for the treatment of glucose disorders [74]. Whilst LXR acts primarily as a cholesterol sensor, FXR is a sensor of the end products of cholesterol metabolism – bile acids (BA). When serum BA concentrations are raised, FXR inhibits further BA synthesis, whilst at the same time promoting BA excretion from the hepatocyte to the biliary system (Fig. 1.5). This is an important step in cholesterol metabolism, as it is excreted in the bile in the form of BAs. Therefore, FXR is also implicated in the control of lipid metabolism (Fig. 1.3b). In addition to modulating cholesterol concentrations, it induces the expression of LPL and downregulates SREBP-1c, generating an overall effect of lowering serum triglyceride concentrations. There also seems to be an impact of FXR on glucose metabolism both directly, via repression of gluconeogenic genes, and indirectly by controlling serum concentrations of TG and free fatty acids (FFAs). Indeed, FXR-null mice show a dyslipidaemic and hyperglycaemic profile with hypertriglyceridemia, high concentrations of circulating FFAs, impaired glucose tolerance and decreased insulin sensitivity [76]. A study in pregnant FXR-null mice also demonstrated new onset of impaired glucose tolerance and insulin resistance in comparison to controls [77]. A randomised controlled trial investigating the effects of the natural FXR agonist obeticholic acid (OCA) showed that it increases insulin sensitivity and improves liver inflammation in adults affected by type 2 diabe-

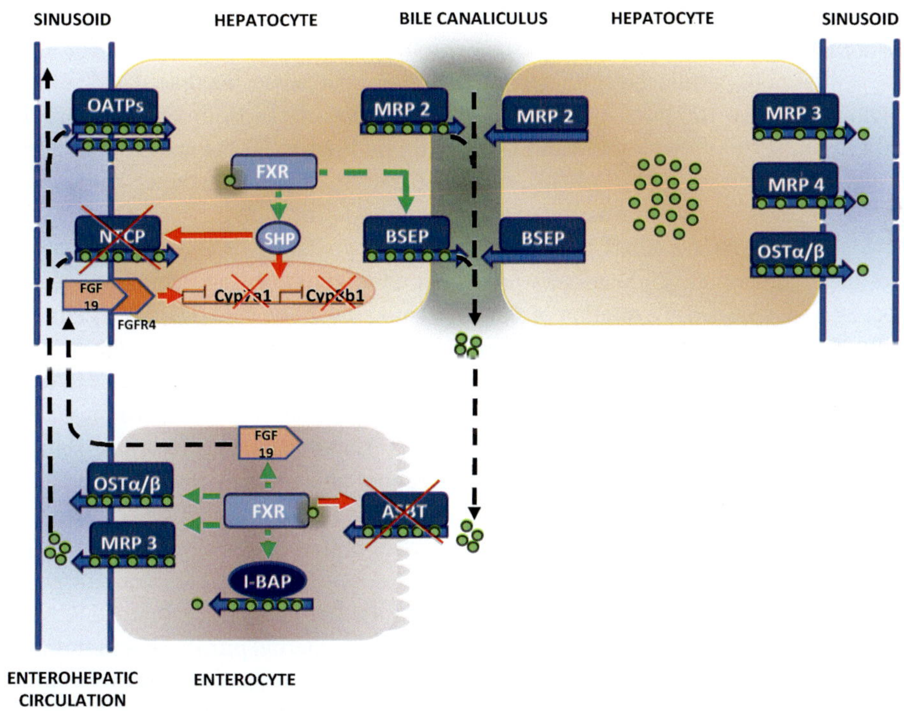

Fig. 1.5 Summary of the main BA transporters in the enterohepatic circulation and FXR effects in the hepatocyte and enterocyte. The hepatocyte on the left represents the effects of FXR activation by bile acids (circles). Dashed green arrows represent transcriptional activation and solid red arrows transcriptional repression. The hepatocyte on the right represents additional bile acid transporters upregulated in the event of cholestasis. Once bile acids reach the intestinal lumen they activate FXR in the enterocyte. FXR then induces the synthesis of FGF19, which reaches the hepatocyte to further repress Cyp7a1 and Cyp8b1 after binding to its receptor, FGFR4. *FXR* Farnesoid X Receptor, *SHP* small heterodimer partner

tes mellitus and non-alcoholic liver disease [78]. Based on these findings, the effects of OCA were also studied in a mouse model of diet-induced GDM [79]. Although a reduction in serum cholesterol concentrations was observed, no changes in glucose tolerance occurred.

1.3.2 Intrahepatic Cholestasis of Pregnancy

Intrahepatic cholestasis of pregnancy (ICP) is a gestational liver disorder that presents with maternal pruritus and increased serum BAs. Its prevalence varies in different ethnicities and around the globe, ranging between 0.2% and 5.6% of pregnancies [80, 81]. Although maternal symptoms tend to resolve soon after delivery, ICP is associated with adverse pregnancy outcomes, which are directly related to serum BA concentrations. Preterm birth and neonatal unit admissions are more likely to occur when serum BAs are above 40 μmolL, whilst the stillbirth rate increases with BAs above 100 μmol/L [82, 83].

ICP has a multifactorial aetiology, with environmental and genetic components [84–87], but FXR and its target genes are a central aspect of the pathophysiology of the disease. FXR is a master controller of the enterohepatic circulation, a process that regulates synthesis and excretion of BAs in the hepatocyte, and their subsequent recycling through the bowel [88] (Fig. 1.5). Its natural ligands consist of both conjugated and unconjugated BAs [89, 90]. When high serum concentrations of BAs are detected, FXR suppresses the enzyme CYP7A1, the rate-limiting step in the synthesis of BAs from cholesterol, whilst at the same time inducing the expression of the transporter BSEP thus downregulating NTCP [89, 91–94]. The overall effect is a reduction in BA synthesis, increase in BA excretion into bile and reduction in BA uptake in the hepatocyte.

There is evidence to suggest that FXR function is blunted in normal murine pregnancy. In fact, both mouse and human pregnancy show increased serum BA concentrations when compared with non-pregnant controls [95]. Microarray followed by Ingenuity Pathway Analysis (IPA) have been performed in FXR-knockout and pregnant mice, showing that the attenuated response to rising BAs is similar in both groups i.e. reduced induction of FXR downstream targets *Shp*, *Bsep*, *Mrp3* and *Mdr1a* [95]. This effect is thought to be mediated by rising concentrations of maternal hormones, as a direct interaction between ERα, sulfated progesterone metabolites and FXR has been reported [86, 95–98]. The exact purpose of this physiological change in FXR function during pregnancy is unknown, but it might play a role in regulating some of the maternal metabolic changes.

In ICP, it is thought that the altered hormonal environment as a consequence of pregnancy unmasks the disease in genetically predisposed women. Sulfated progesterone metabolites are markedly increased in the serum of women affected by ICP when compared to controls [98], and this is thought to interfere with FXR function. Women with ICP also present with dyslipidaemia and are at increased risk of developing GDM [99–101]. Both changes are consistent with findings in FXR knockout mice, confirming the finding of an attenuated FXR response in the condition [76].

The goals of ICP treatment are maternal symptom control and reduction of fetal risks. Ursodeoxycholic acid (UDCA) is commonly prescribed to treat the disease. UDCA is a naturally occurring, relatively hydrophilic BA that makes up approximately 3% of the human BA pool [102]. Its effects occur by transformation of the BA pool into a less hydrophobic, hence less cytotoxic one, and by regulation of hepatic BA transporters both at a transcriptional and protein level [103]. A large 2019 randomised placebo-controlled trial showed that UDCA has some effect on maternal pruritus but in this study it was not effective in reducing adverse perinatal outcomes [104]. However, a more recent individual participant data meta-analysis that included data from a considerably higher number of ICP cases with serum BA concentrations ≥40 μmol/L than in the randomised placebo-controlled trial, showed that UDCA treatment reduces rates of stillbirth and preterm birth when maternal serum

BA concentrations are elevated above this threshold [105].

Similar to GDM, ICP is associated with long-term metabolic consequences for the fetus. A cohort study in affected babies showed that they were likely to develop features of the metabolic syndrome in adolescence. These findings were replicated in a mouse model of gestational cholestasis, and the mechanisms behind these changes are thought to be a disruption of lipid homeostasis in the fetoplacental unit [106]. In mice, UDCA treatment during pregnancy was able to reverse some of these features in the offspring [107].

1.3.3 Pre-eclampsia

Pre-eclampsia is a multisystem disorder of pregnancy characterised by raised maternal blood pressure after 20 weeks of gestation and endothelial dysfunction, and it can result in multiorgan dysfunction [108]. It is one of the leading causes of maternal and fetal morbidity and mortality in low- and middle-income countries [109], causing approximately 14% of maternal deaths worldwide [110].

The placenta seems to be central in the pathophysiology of the disease. Placental dysfunction results in recurrent ischemia-reperfusion injury in the placental bed, triggering an angiogenic imbalance in the mother [111]. The origin of this placental dysfunction is a subject of debate: although conventionally it is thought to be the result of insufficient invasion of spiral arteries by the EVT, new lines of evidence propose that abnormal placental perfusion is secondary to underlying abnormalities in maternal cardiac function that preclude an adequate maternal cardiovascular adaptation to pregnancy [112]. Definitive treatment of pre-eclampsia consists of delivery of the fetus and the placenta; however, this causes a dilemma for clinicians and women when a fetus is preterm. The recommended practice is strict control of maternal blood pressure and planned delivery from 37 weeks of gestation, with the decision to deliver severe cases prior to this taken on a case by case basis [113–115].

Given the influence of PPARγ on trophoblast differentiation and development, there is an increasing interest in its role in the pathogenesis and treatment of pre-eclampsia. The expression of PPARγ in placentas of women affected by pre-eclampsia has been investigated, but no differences have been found in comparison to controls [116, 117]. No associations between polymorphisms of the PPARγ receptor gene and the development or severity of pre-eclampsia have been found either [118]. Administration of PPARγ antagonists in mice induces a phenotype of raised blood pressure, reduced pup weight and endothelial dysfunction, similar to a pre-eclamptic phenotype [119]. In addition, the balance between pro- and anti-angiogenic factors in maternal serum is disrupted in a way similar to the disease in humans, and the studied mice show evidence of impaired trophoblast differentiation. Administration of the PPARγ agonist rosiglitazone reverses the majority of these changes [120, 121]. One study has shown that women who develop pre-eclampsia have decreased serum concentrations of PPARγ activators, which are normally increased in unaffected pregnancies. These findings are present before the onset of disease [122].

LXRs have also been investigated in the context of pre-eclampsia. Their roles in trophoblast development and regulation of placental cholesterol metabolism have been postulated as contributing factors to its pathogenesis [123]. LXRα mRNA expression and LXRβ protein levels have been investigated in placentas from women affected by pre-eclampsia, with variable results [124, 125]. One study showed that expression of both LXRα and its target endoglin, a regulator of trophoblast invasiveness and endothelial function previously implicated in the pathogenesis of pre-eclampsia, were both increased in placentas from affected women [125].

1.3.4 Spontaneous Preterm Labour

Preterm labour (PTL) is defined as the onset of regular uterine contractions and cervical dilatation prior to 37 weeks of pregnancy. An estimated

15 million babies are born premature every year, and the complications of an early birth are the leading cause of mortality in children under 5 years of age [126, 127]. Considering that inflammation is central to the onset of labour, conditions that cause an increase in the inflammatory load of uterine tissues are potential triggers of early labour. Recognised causes are maternal or fetal infection, early activation of the fetal hypothalamic-pituitary-adrenal axis, chorion-decidual haemorrhage, over-distention of the myometrium (e.g. multifetal gestation), changes in the vaginal microbiome and maternal stress [128–130]. However, a significant number of cases of preterm labour do not have an identifiable cause.

So far, no effective treatment for PTL has been found. Pharmacological strategies consist of a reactive approach that aims to delay the onset of parturition for a few days, with the aim of allowing time for fetal lung maturation with exogenous corticosteroids. Progesterone supplementation has been extensively studied as a preventative strategy. The rationale for this approach remains questionable, as it is an established fact that the onset of human labour is not secondary to decreasing progesterone concentrations. Nevertheless, positive results have been found in women at high risk of PTL, such as those with a previous history of PTL, evidence of a short cervix or multifetal pregnancies. The most recent individual participant meta-analysis evaluating randomised clinical trials in this subject has shown that the administration of vaginal progesterone and intramuscular 17-hydroxyprogesterone caproate (17-OHPC) are successful in preventing birth before 34 weeks in high risk singleton pregnancies [131]. This effect seems to be stronger in women with a reduced cervical length.

The challenges in developing strategies for the prevention of preterm birth stem from the fact that it has multiple causative factors, with likely distinct molecular mechanisms. In addition, the background risk of different populations varies, hindering the assessment of interventions. The mechanisms through which progesterone supplementation can prevent PTL are still not fully understood. A study of progesterone supplementation in mice showed no changes in the expression of molecules related to uterine contractility, cervical remodelling or local inflammation [132]. A different study showed that vaginal progesterone, in contrast to intramuscular 17-OHPC, has an influence on the myometrial immune profile and molecules related to cervical ripening [133]. It is also possible that different preparations of progestogens exert distinct effects on PRs and labour mechanisms, or can evade the myometrial changes in progesterone metabolism in different ways [134]. Understanding these mechanisms would allow us to optimise the use of progesterone for prevention of PTL.

1.4 Conclusions

Nuclear receptors are remarkable integrators of hormonal, nutritional and transcriptional pathways that are increasingly recognised as important orchestrators of pregnancy adaptations. They are an essential part of early events of pregnancy, maternal metabolic adaptations and parturition. So far, the prospect of treating gestational disorders with modulators of nuclear receptors has been mainly considered with reference to treatment strategies applied to non-gestational pathologies. A better understanding of the role of nuclear receptors in normal gestation and its specific disorders is necessary to enable consideration of potential new therapeutic strategies.

References

1. Bustamante JJ, Copple BL, Soares MJ, Dai G (2010) Gene profiling of maternal hepatic adaptations to pregnancy. Liver Int 30(3):406–415
2. Brosens JJ, Parker MG, McIndoe A, Pijnenborg R, Brosens IA (2009) A role for menstruation in preconditioning the uterus for successful pregnancy. Am J Obstet Gynecol 200(6):615.e1–615.e6
3. Okada H, Tsuzuki T, Murata H (2018) Decidualization of the human endometrium. Reprod Med Biol 17(3):220
4. Gellersen B, Brosens I, Brosens J (2007) Decidualization of the human endometrium: mechanisms, functions, and clinical perspectives. Semin Reprod Med 25(6):445–453

5. Butte NF (2000) Carbohydrate and lipid metabolism in pregnancy: normal compared with gestational diabetes mellitus. Am J Clin Nutr 71:1256S–1261S. American Society for Nutrition
6. Herrera E, Amusquivar E, López-Soldado I, Ortega H (2006) Maternal lipid metabolism and placental lipid transfer. Horm Res 65:59–64
7. Hadden DR, McLaughlin C (2009) Normal and abnormal maternal metabolism during pregnancy. Semin Fetal Neonatal Med 14(2):66–71
8. Papacleovoulou G, Abu-Hayyeh S, Williamson C (2011) Nuclear receptor-driven alterations in bile acid and lipid metabolic pathways during gestation. Biochim Biophys Acta Mol basis Dis 1812(8):879–887
9. Wu S-P, Li R, Demayo FJ (2018) Progesterone receptor regulation of uterine adaptation for pregnancy. Trends Endocrinol Metab 29(7):481–491
10. Lee K, Jeong J, Tsai M-J, Tsai S, Lydon JP, Demayo FJ (2006) Molecular mechanisms involved in progesterone receptor regulation of uterine function. J Steroid Biochem Mol Biol 102(1–5):41–50
11. Mullac-Jericevic B, Mullinax RA, DeMayo FJ, Lydon JP, Conneely OM (2000) Subgroup of reproductive functions of progesterone mediated by progesterone receptor-B isoform. Science 289(5485):1751–1754
12. Mulac-Jericevic B, Lydon JP, DeMayo FJ, Conneely OM (2003) Defective mammary gland morphogenesis in mice lacking the progesterone receptor B isoform. Proc Natl Acad Sci U S A 100(17):9744
13. Fleisch MC, Chou YC, Cardiff RD, Asaithambi A, Shyamala G (2009) Overexpression of progesterone receptor a isoform in mice leads to endometrial hyperproliferation, hyperplasia and atypia. Mol Hum Reprod 15(4):241–249
14. Aplin J (1991) Implantation, trophoblast differentiation and haemochorial placentation: mechanistic evidence in vivo and in vitro. J Cell Sci 99(4):681–692
15. Ji L, Brkić J, Liu M, Fu G, Peng C, Wang YL (2013) Placental trophoblast cell differentiation: physiological regulation and pathological relevance to preeclampsia. Mol Asp Med 34(5):981–1023
16. Kadam L, Kohan-Ghadr HR, Drewlo S (2015) The balancing act – PPAR-γ's roles at the maternal-fetal interface. Syst Biol Reprod Med 61(2):65–71
17. Fournier T, Handschuh K, Tsatsaris V, Guibourdenche J, Evain-Brion D (2008) Role of nuclear receptors and their ligands in human trophoblast invasion. J Reprod Immunol 77(2):161–170
18. Barak Y, Nelson MC, Ong ES et al (1999) PPARγ is required for placental, cardiac, and adipose tissue development. Mol Cell 4(4):585–595
19. Tarrade A, Schoonjans K, Pavan L et al (2001) PPARγ/RXRα heterodimers control human trophoblast invasion. J Clin Endocrinol Metab 86(10):5017–5024
20. Fournier T, Handschuh K, Tsatsaris V, Evain-Brion D (2007) Involvement of PPARgamma in human trophoblast invasion. Placenta 28(8–9):974–976
21. Fournier T, Thérond P, Handschuh K, Tsatsaris V (2008) Evain-Brion. PPARgamma and early human placental development. Curr Med Chem 15(28):3011–3024
22. Fournier T, Pavan L, Tarrade A et al (2002) The role of PPAR-γ/RXR-α heterodimers in the regulation of human trophoblast invasion. Ann N Y Acad Sci 973:26–30
23. Parast MM, Yu H, Ciric A, Salata MW, Davis V, Milstone DS (2009) PPARγ regulates trophoblast proliferation and promotes labyrinthine Trilineage differentiation. PLoS One 4(11):e8055
24. Schaiff WT, Carlson MG, Smith SD, Levy R, Nelson DM, Sadovsky Y (2000) Peroxisome proliferator-activated receptor-gamma modulates differentiation of human trophoblast in a ligand-specific manner. J Clin Endocrinol Metab 85(10):3874–3881
25. Calkin AC, Tontonoz P (2012) Transcriptional integration of metabolism by the nuclear sterol-activated receptors LXR and FXR. Nat Rev Mol Cell Biol 13(4):213–224
26. Chawta A, Repa JJ, Evans RM, Mangelsdorf DJ (2001) Nuclear receptors and lipid physiology: Opening the x-files. Science (80-) 294(5548):1866–1870
27. Weedon-Fekjaer MS, Duttaroy AK, Nebb HI (2005) Liver X receptors mediate inhibition of hCG secretion in a human placental trophoblast cell line. Placenta 26(10):721–728
28. Janowski BA, Willy PJ, Devi TR, Falck JR, Mangelsdorf DJ (1996) An oxysterol signalling pathway mediated by the nuclear receptor LXRα. Nature 383(6602):728–731
29. Pavan L, Hermouet A, Tsatsaris V et al (2004) Lipids from oxidized low-density lipoprotein modulate human trophoblast invasion: involvement of nuclear liver X receptors. Endocrinology 145(10):4583–4591
30. Aye IL, Waddell BJ, Mark PJ, Keelan JA (2011) Oxysterols inhibit differentiation and fusion of term primary trophoblasts by activating liver X receptors. Placenta 32(2):183–191
31. Larkin JC, Sears SB, Sadovsky Y (2014) The influence of ligand-activated LXR on primary human trophoblasts. Placenta 35(11):919
32. Sabol SL, Brewer HB, Santamarina-Fojo S (2005) The human ABCG1 gene: identification of LXR response elements that modulate expression in macrophages and liver. J Lipid Res 46(10):2151–2167
33. Venkateswaran A, Laffitte BA, Joseph SB et al (2000) Control of cellular cholesterol efflux by the nuclear oxysterol receptor LXRα. Proc Natl Acad Sci U S A 97(22):12097–12102
34. Kalaany NY, Mangelsdorf DJ (2006) LXRS AND FXR: The Yin and Yang of cholesterol and fat metabolism. Annu Rev Physiol 68(1):159–191
35. Nikolova V, Papacleovoulou G, Bellafante E et al (2017) Changes in LXR signaling influence early-pregnancy lipogenesis and protect against dysregu-

lated fetoplacental lipid homeostasis. Am J Physiol Endocrinol Metab 313(4):E463–E472
36. Sweeney TR, Moser AH, Shigenaga JK, Grunfeld C, Feingold KR (2006) Decreased nuclear hormone receptor expression in the livers of mice in late pregnancy. Am J Physiol Metab 290(6):E1313–E1320
37. Bass J, Takahashi JS (2010) Circadian integration of metabolism and energetics. *Science (80-)* 330(6009):1349–1354
38. Cho H, Zhao X, Hatori M et al (2012) Regulation of circadian behaviour and metabolism by REV-ERB-α and REV-ERB-β. Nature 485(7396):123–127
39. Bugge A, Feng D, Everett LJ et al (2012) Rev-erba and Rev-erbb coordinately protect the circadian clock and normal metabolic function. Genes Dev 26(7):657–667
40. Papacleovoulou G, Nikolova V, Oduwole O et al (2017) Gestational disruptions in metabolic rhythmicity of the liver, muscle, and placenta affect fetal size. FASEB J 31(4):1698–1708
41. Shynlova O, Nadeem L, Zhang J, Dunk C, Lye S (2020) Myometrial activation: novel concepts underlying labor. Placenta 92:28–36
42. Hardy DB, Janowski BA, Corey DR, Mendelson CR (2006) Progesterone receptor plays a major Antiinflammatory role in human myometrial cells by antagonism of nuclear factor-κB activation of cyclooxygenase 2 expression. Mol Endocrinol 20(11):2724–2733
43. Kalkhoven E, Wissink S, van der Saag PT, van der Burg B (1996) Negative interaction between the RelA(p65) subunit of NF-kappaB and the progesterone receptor. J Biol Chem 271(11):6217–6224
44. Mendelson CR, Montalbano AP, Gao L, Steroid J, Mol B, Author B (2017) Fetal-to-maternal signaling in the timing of birth HHS public access Author manuscript. J Steroid Biochem Mol Biol 170:19–27
45. Hendrix EM, Myatt L, Sellers S, Russell PT, Larsen WJ (1995) Steroid hormone regulation of rat myometrial gap junction formation: effects on cx43 levels and trafficking. Biol Reprod 52(3):547–560
46. Renthal NE, Chen C-C, Williams KC, Gerard RD, Prange-Kiel J, Mendelson CR (2010) miR-200 family and targets, ZEB1 and ZEB2, modulate uterine quiescence and contractility during pregnancy and labor. Proc Natl Acad Sci 107(48):20828–20833
47. Chai SY, Smith R, Zakar T, Mitchell C, Madsen G (2012) Term myometrium is characterized by increased activating epigenetic modifications at the progesterone receptor-a promoter. Mol Hum Reprod 18(8):401–409
48. Nadeem L, Shynlova O, Matysiak-Zablocki E, Mesiano S, Dong X, Lye S (2016) Molecular evidence of functional progesterone withdrawal in human myometrium. Nat Commun 7:11565
49. Peavey MC, Wu SP, Li R et al (2021) Progesterone receptor isoform B regulates the Oxtr- Plcl2- Trpc3 pathway to suppress uterine contractility. Proc Natl Acad Sci U S A 118(11):e2011643118
50. Merlino AA, Welsh TN, Tan H et al (2007) Nuclear progesterone receptors in the human pregnancy myometrium: evidence that parturition involves functional progesterone withdrawal mediated by increased expression of progesterone receptor-a. J Clin Endocrinol Metab 92(5):1927–1933
51. Pieber D, Allport VC, Hills F, Johnson M, Bennett PR (2001) Interactions between progesterone receptor isoforms in myometrial cells in human labour. Mol Hum Reprod 7(9):875–879
52. Williams KC, Renthal NE, Condon JC, Gerard RD, Mendelson CR (2012) MicroRNA-200a serves a key role in the decline of progesterone receptor function leading to term and preterm labor. Proc Natl Acad Sci U S A 109(19):7529
53. Nadeem L, Balendran R, Dorogin A, Mesiano S, Shynlova O, Lye SJ (2021) Pro-inflammatory signals induce 20α-HSD expression in myometrial cells: a key mechanism for local progesterone withdrawal. J Cell Mol Med 25(14):6773
54. Nadeem L, Shynlova O, Mesiano S, Lye S (2017) Progesterone via its type-a receptor promotes myometrial gap junction coupling. Sci Rep 7(1):13357
55. Lee Y, Sooranna SR, Terzidou V et al (2012) Interactions between inflammatory signals and the progesterone receptor in regulating gene expression in pregnant human uterine myocytes. J Cell Mol Med 16(10):2487
56. Allport VC, Pieber D, Slater DM, Newton R, White JO, Bennett PR (2001) Human labour is associated with nuclear factor-kappaB activity which mediates cyclo-oxygenase-2 expression and is involved with the "functional progesterone withdrawal". Mol Hum Reprod 7(6):581–586
57. Poomalar GK (2015) Changing trends in management of gestational diabetes mellitus. World J Diabetes 6(2):284
58. Cho NH, Shaw JE, Karuranga S et al (2018) IDF diabetes atlas: global estimates of diabetes prevalence for 2017 and projections for 2045. Diabetes Res Clin Pract 138:271–281
59. David Mcintyre H, Kapur A, Divakar H, Hod M (2020) Gestational diabetes mellitus – innovative approach to prediction, diagnosis, management, and prevention of future NCD – mother and offspring. Front Endocrinol (Lausanne) 11:6145333
60. HAPO Study Cooperative Research Group, Metzger BE, Lowe LP et al (2009) Hyperglycemia and Adverse Pregnancy Outcomes. MCN 358(19):1991–2002
61. Lee AJ, Hiscock RJ, Wein P, Walker SP, Permezel M (2007) Gestational diabetes mellitus: clinical predictors and long-term risk of developing type 2 diabetes: a retrospective cohort study using survival analysis. Diabetes Care 30(4):878–883
62. Retnakaran R, Shah BR (2009) Mild glucose intolerance in pregnancy and risk of cardiovascular disease: a population-based cohort study. CMAJ 181(6–7):371–376
63. Ornoy A, Becker M, Weinstein-Fudim L, Ergaz Z (2021) Diabetes during pregnancy: a maternal dis-

ease complicating the course of pregnancy with long-term deleterious effects on the offspring. A clinical review. Int J Mol Sci 22(6):1–38
64. Chu AH, Godfrey KM (2020) Gestational diabetes mellitus and developmental programming. Ann Nutr Metab 76(Suppl 3):4
65. Gregorio KCR, Laurindo CP, Machado UF (2021) Estrogen and glycemic homeostasis: the fundamental role of nuclear estrogen receptors ESR1/ESR2 in glucose transporter GLUT4 regulation. Cells 10(1):90
66. Herrera-Lopez EE, Castelan-Martinez OD, Suarez Sanchez F et al (2018) The rs1256031 of estrogen receptor β gene is associated with type 2 diabetes. Diabetes Metab Syndr 12(5):631–633
67. Li X, Su J, Zheng K et al (2020) Assessment of the association between the polymorphism rs1256031 of the estrogen receptor β gene and GDM susceptibility. Nagoya J Med Sci 82(4):703
68. Li C, Qiao B, Zhou Y, Qi W, Ma C, Zheng L (2020) Asociation of estrogen receptor α gene polymorphism and its expression with gestational diabetes mellitus. Gynecol Obstet Investig 85(1):26–33
69. Hocking S, Samocha-Bonet D, Milner KL, Greenfield JR, Chisholm DJ (2013) Adiposity and insulin resistance in humans: the role of the different tissue and cellular lipid depots. Endocr Rev 34(4):463–500
70. Kitessa SM, Abeywardena MY (2016) Lipid-induced insulin resistance in skeletal muscle: the chase for the culprit Goes from total intramuscular fat to lipid intermediates, and finally to species of lipid intermediates. Nutrients 8(8):466
71. Wilding JP (2007) The importance of free fatty acids in the development of Type 2 diabetes. Diabet Med 24(9):934–945
72. Dong Y, Gao G, Fan H, Li S, Li X, Liu W (2015) Activation of the liver X receptor by Agonist TO901317 improves hepatic insulin resistance via suppressing reactive oxygen species and JNK pathway. PLoS ONE 10(4):e0124778
73. Maczewsky J, Sikimic J, Bauer C et al (2017) The LXR ligand T0901317 acutely inhibits insulin secretion by affecting mitochondrial metabolism. Endocrinology 158(7):2145–2154
74. Ding L, Pang S, Sun Y, Tian Y, Yu L, Dang N (2014) Coordinated actions of FXR and LXR in metabolism: from pathogenesis to pharmacological targets for type 2 diabetes. Int J Endocrinol 2014:7518599
75. Lappas M (2014) Effect of pre-existing maternal obesity, gestational diabetes and adipokines on the expression of genes involved in lipid metabolism in adipose tissue. Metabolism 63(2):250–262
76. Ma K, Saha PK, Chan L, Moore DD (2006) Farnesoid X receptor is essential for normal glucose homeostasis. J Clin Invest 116(4):1102–1109
77. Bellafante E, McIlvride S, Nikolova V et al (2020) Maternal glucose homeostasis is impaired in mouse models of gestational cholestasis. Sci Rep 10(1):11523
78. Mudaliar S, Henry RR, Sanyal AJ et al (2013) Efficacy and safety of the farnesoid x receptor agonist Obeticholic acid in patients with type 2 diabetes and nonalcoholic fatty liver disease. Gastroenterology 145(3):574
79. McIlvride S, Nikolova V, Fan HM et al (2019) Obeticholic acid ameliorates dyslipidemia but not glucose tolerance in mouse model of gestational diabetes. Am J Physiol Endocrinol Metab 317(2):E399–E410
80. Lee RH, Goodwin TM, Greenspoon J, Incerpi M (2006) The prevalence of intrahepatic cholestasis of pregnancy in a primarily Latina Los Angeles population. J Perinatol 26(9):527–532
81. Wood AM, Livingston EG, Hughes BL, Kuller JA (2018) Intrahepatic cholestasis of pregnancy: a review of diagnosis and management. Obstet Gynecol Surv 73(2):103–109
82. Geenes V, Chappell LC, Seed PT, Steer PJ, Knight M, Williamson C (2014) Association of severe intrahepatic cholestasis of pregnancy with adverse pregnancy outcomes: a prospective population-based case-control study. Hepatology 59(4):1482–1491
83. Ovadia C, Seed PT, Sklavounos A et al (2019) Association of adverse perinatal outcomes of intrahepatic cholestasis of pregnancy with biochemical markers: results of aggregate and individual patient data meta-analyses. Lancet 393(10174):899–909
84. Turro E, Astle WJ, Megy K et al (2020) Whole-genome sequencing of patients with rare diseases in a national health system. Nature 583(7814):96–102
85. Dixon PH, Wadsworth CA, Chambers J et al (2014) A comprehensive analysis of common genetic variation around six candidate loci for intrahepatic cholestasis of pregnancy. Am J Gastroenterol 109(1):76–84
86. Abu-Hayyeh S, Martinez-Becerra P, Abdul Kadir SHS et al (2010) Inhibition of Na+–taurocholate co-transporting polypeptide-mediated bile acid transport by cholestatic sulfated progesterone metabolites. J Biol Chem 285(22):16504–16512
87. Sookoian S, Castaño G, Burgueño A, Gianotti TF, Pirola CJ (2008) Association of the multidrug-resistance-associated protein gene (ABCC2) variants with intrahepatic cholestasis of pregnancy. J Hepatol 48(1):125–132
88. Houten SM, Auwerx J (2004) The enterohepatic nuclear receptors are major regulators of the enterohepatic circulation of bile salts. Ann Med 36(7):482–491
89. Makishima M, Okamoto AY, Repa JJ et al (1999) Identification of a nuclear receptor for bite acids. Science (80-) 284(5418):1362–1365
90. Parks DJ, Blanchard SG, Bledsoe RK et al (1999) Bile acids: Natural ligands for an orphan nuclear receptor. Science (80-) 284(5418): 1365–1368
91. Chiang JYL (2009) Bile acids: regulation of synthesis. J Lipid Res 50(10):1955–1966

92. Goodwin B, Jones SA, Price RR et al (2000) A regulatory cascade of the nuclear receptors FXR, SHP-1, and LRH-1 represses bile acid biosynthesis. Mol Cell 6(3):517–526
93. Denson LA, Sturm E, Echevarria W et al (2001) The orphan nuclear receptor, shp, mediates bile acid-induced inhibition of the rat bile acid transporter, ntcp. Gastroenterology 121(1):140–147
94. Müller M, Jansen PLM, Faber KN et al (2002) Farnesoid X receptor and bile salts are involved in transcriptional regulation of the gene encoding the human bile salt export pump. Hepatology 35(3):589–596
95. Milona A, Owen BM, Cobbold JFL et al (2010) Raised hepatic bile acid concentrations during pregnancy in mice are associated with reduced farnesoid X receptor function. Hepatology 52(4):1341–1349
96. Chen Y, Vasilenko A, Song X et al (2015) Estrogen and estrogen receptor-α-mediated Transrepression of bile salt export pump. Mol Endocrinol 29(4):613–626
97. Song X, Vasilenko A, Chen Y et al (2014) Transcriptional dynamics of bile salt export pump during pregnancy: mechanisms and implications in intrahepatic cholestasis of pregnancy. Hepatology 60(6):1993–2007
98. Abu-Hayyeh S, Papacleovoulou G, Lövgren-Sandblom A et al (2013) Intrahepatic cholestasis of pregnancy levels of sulfated progesterone metabolites inhibit farnesoid X receptor resulting in a cholestatic phenotype. Hepatology 57(2):716–726
99. Dann AT, Kenyon AP, Wierzbicki AS, Seed PT, Shennan AH, Tribe RM (2006) Plasma lipid profiles of women with intrahepatic cholestasis of pregnancy. Obstet Gynecol 107(1):106–114
100. Martineau M, Raker C, Powrie R, Williamson C (2014) Intrahepatic cholestasis of pregnancy is associated with an increased risk of gestational diabetes. Eur J Obstet Gynecol Reprod Biol 176(1):80–85
101. Martineau MG, Raker C, Dixon PH et al (2015) The metabolic profile of intrahepatic cholestasis of pregnancy is associated with impaired glucose tolerance, dyslipidemia, and increased fetal growth. Diabetes Care 38(2):243–248
102. Geenes V, Williamson C (2009) Intrahepatic cholestasis of pregnancy. World J Gastroenterol 15(17):2049
103. Trauner M, Wagner M, Fickert P, Zollner G (2005) Molecular regulation of hepatobiliary transport systems: clinical implications for understanding and treating cholestasis. J Clin Gastroenterol 39(4 Suppl 2):S111–S124
104. Chappell LC, Bell JL, Smith A et al (2019) Ursodeoxycholic acid versus placebo in women with intrahepatic cholestasis of pregnancy (PITCHES): a randomised controlled trial. Lancet 394(10201):849–860
105. Ovadia C, Sajous J, Seed PT et al (2021) Ursodeoxycholic acid in intrahepatic cholestasis of pregnancy: a systematic review and individual participant data meta-analysis. Lancet Gastroenterol Hepatol 6(7):547–558
106. Papacleovoulou G, Abu-Hayyeh S, Nikolopoulou E et al (2013) Maternal cholestasis during pregnancy programs metabolic disease in offspring. J Clin Invest 123(7):3172–3181
107. Borges Manna L, Papacleovoulou G, Flaviani F et al (2020) Ursodeoxycholic acid improves feto-placental and offspring metabolic outcomes in hypercholanemic pregnancy. Sci Rep 10(1):10361
108. Chappell LC, Cluver CA, Kingdom J, Tong S (2021) Pre-eclampsia. Lancet 3:341–354
109. Duffy J, Cairns AE, Richards-Doran D et al (2020) A core outcome set for pre-eclampsia research: an international consensus development study. BJOG 127(12):1516–1526
110. WHO recommendations: Policy of interventionist versus expectant management of severe pre-eclampsia before term. WHO Recomm Policy Interv versus Expect Manag Sev pre-eclampsia before term. 2018. https://www.ncbi.nlm.nih.gov/books/NBK535829/. Accessed 15 July 2021
111. Kwiatkowski S, Kwiatkowska E, Torbe A (2019) The role of disordered angiogenesis tissue markers (sflt-1, Plgf) in present day diagnosis of preeclampsia. Ginekol Pol 90(3):173–176
112. Melchiorre K, Giorgione V, Thilaganathan B (2021) The placenta and preeclampsia: villain or victim? Am J Obstet Gynecol 226:S954–S962
113. Magee LA, von Dadelszen P, Rey E et al (2015) Less-tight versus tight control of hypertension in pregnancy. N Engl J Med 372(5):407–417
114. Koopmans CM, Bijlenga D, Groen H et al (2009) Induction of labour versus expectant monitoring for gestational hypertension or mild pre-eclampsia after 36 weeks' gestation (HYPITAT): a multicentre, open-label randomised controlled trial. Lancet 374(9694):979–988
115. Chappell LC, Brocklehurst P, Green ME et al (2019) Planned early delivery or expectant management for late preterm pre-eclampsia (PHOENIX): a randomised controlled trial. Lancet 394(10204):1181–1190
116. Rodie VA, Young A, Jordan F, Sattar N, Greer IA, Freeman DJ (2005) Human placental peroxisome proliferator-activated receptor δ and γ expression in healthy pregnancy and in preeclampsia and intrauterine growth restriction. J Soc Gynecol Investig 12(5):320–329
117. Holdsworth-Carson SJ, Lim R, Mitton A et al (2010) Peroxisome proliferator-activated receptors are altered in pathologies of the human placenta: gestational diabetes mellitus, intrauterine growth restriction and preeclampsia. Placenta 31(3):222–229
118. Laasanen J, Heinonen S, Hiltunen M, Mannermaa A, Laakso M (2002) Polymorphism in the peroxisome proliferator-activated receptor-gamma gene in women with preeclampsia. Early Hum Dev 69(1–2):77–82

119. McCarthy FP, Drewlo S, English FA et al (2011) Evidence implicating peroxisome proliferator-activated receptor-? In the pathogenesis of preeclampsia. Hypertension 58(5):882–887
120. McCarthy FP, Drewlo S, Kingdom J, Johns EJ, Walsh SK, Kenny LC (2011) Peroxisome proliferator-activated receptor-γ as a potential therapeutic target in the treatment of preeclampsia. Hypertension 58(2):280–286
121. Ahham HIG, Masri AAA (2018) The potential therapeutic role of peroxisome ProliferatorActivated receptors agonist in Preeclamptic pregnant rats. J Coll Physicians Surg Pak 28(1):31–35
122. Waite LL, Louie RE, Taylor RN (2005) Circulating activators of peroxisome proliferator-activated receptors are reduced in preeclamptic pregnancy. J Clin Endocrinol Metab 90(2):620–626
123. Plösch T, Gellhaus A, Van Straten EME et al (2010) The liver X receptor (LXR) and its target gene ABCA1 are regulated upon low oxygen in human trophoblast cells: a reason for alterations in preeclampsia? Placenta 31(10):910–918
124. Weedon-Fekjaer MS, Johnsen GM, Anthonisen EH et al (2010) Expression of liver X receptors in pregnancies complicated by preeclampsia. Placenta 31:818–824
125. Wang J, Dong X, Wu H-Y et al (2016) Relationship of liver X receptors α and Endoglin levels in serum and placenta with preeclampsia. PLoS One 11(10):e0163742
126. Preterm birth. https://www.who.int/news-room/fact-sheets/detail/preterm-birth. Accessed 5 Aug 2021
127. Newborns: improving survival and well-being. https://www.who.int/news-room/fact-sheets/detail/newborns-reducing-mortality. Accessed 5 Aug 2021
128. Bayar E, Bennett PR, Chan D, Sykes L, MacIntyre DA (2020) The pregnancy microbiome and preterm birth. Semin Immunopathol 42(4):487–499
129. Rood KM, Buhimschi CS (2017) Genetics, hormonal influences, and preterm birth. Semin Perinatol 41(7):401–408
130. Talati AN, Hackney DN, Mesiano S (2017) Pathophysiology of preterm labor with intact membranes. Semin Perinatol 41(7):420–426
131. Stewart LA, Simmonds M, Duley L et al (2021) Evaluating Progestogens for Preventing Preterm birth International Collaborative (EPPPIC): meta-analysis of individual participant data from randomised controlled trials. Lancet 397(10280):1183–1194
132. Nold C, Maubert M, Anton L, Yellon S, Elovitz MA (2013) Prevention of preterm birth by progestational agents: what are the molecular mechanisms? Am J Obstet Gynecol 208(3):223.e1
133. Furcron A-E, Romero R, Plazyo O et al (2015) Vaginal progesterone, but not 17α-hydroxyprogesterone caproate, has antiinflammatory effects at the murine maternal-fetal interface. Am J Obstet Gynecol 213(6):846.e1
134. Kuon RJ, Shi S-Q, Maul H et al (2010) Pharmacological actions of progestins to inhibit cervical ripening and prevent delivery depend upon their properties, the route of administration and the vehicle. Am J Obstet Gynecol 202(5):455.e1

2 Female Reproductive Systems: Hormone Dependence and Receptor Expression

Kevin K. W. Kuan and Philippa T. K. Saunders

Abstract

The female reproductive system which consists of the ovaries, uterus (myometrium, endometrium), Fallopian tubes, cervix and vagina is exquisitely sensitive to the actions of steroid hormones. The ovaries play a key role in the synthesis of bioactive steroids (oestrogens, androgens, progestins) that act both within the tissue (intracrine/paracrine) as well as on other reproductive organs following release into the blood stream (endocrine action). Sex steroid receptors encoded by the oestrogen (*ESR1, ESR2*), progesterone (*PR*) and androgen (AR) receptor genes, which are members of the superfamily of ligand activated transcription factors are widely expressed within these tissues. These receptors play critical role(s) in regulation of cell proliferation, ovulation, endometrial receptivity, myometrial cell function and inflammatory cell infiltration. Our understanding of their importance has been informed by studies on human tissues and cells, which have employed immunohistochemistry as well as a wide range of molecular and genetic methods to identify which processes are dependent steroid ligand activation. The development of mice with targeted deletions of each of these receptors has provided complementary data that has extended our appreciation of cell-cell interactions in the fine tuning of reproductive tissue function. This large body of work has formed the basis of new and improved therapeutics to treat conditions such as infertility.

Keywords

Ovary · Uterus · Fallopian tube · Cervix · Oestrogen receptors · Androgen receptor · Progesterone receptor

2.1 Introduction

Steroid hormones, acting via their cognate receptors, play a key role in regulation of all organs within the female reproductive system. The ovaries are a major source of endocrine steroid hormones that have body-wide impacts on both reproductive and other tissue systems. In this chapter we will introduce the organs of the female reproductive system, review data on each of the main steroid receptors that bind oestrogens, progestins and androgens, their patterns of expression and impact on the reproductive tissues/cells. The primary focus will be on human tissue function but with some information on model species where this provides complementary information.

K. K. W. Kuan · P. T. K. Saunders (✉)
Centre for Inflammation Research, The University of Edinburgh, Edinburgh, UK
e-mail: P.Saunders@ed.ac.uk

M. J. Campbell, C. L. Bevan (eds.), *Nuclear Receptors in Human Health and Disease*, Advances in Experimental Medicine and Biology 1390, https://doi.org/10.1007/978-3-031-11836-4_2

2.2 Anatomy of the Female Reproductive System

The anatomy of the reproductive system in women is illustrated in Fig. 2.1.

2.2.1 Ovaries

Ovaries are oval-shaped organs located on either side of the uterus composed of both germ cells (developing into oocytes) and somatic cells (granulosa, thecal, stromal). Externally, the ovary is surrounded by simple cuboidal epithelium overlying the cortex where ovarian follicles of various maturities reside. Ovarian follicles typically contain a single oocyte surrounded by granulosa and thecal cells [126]. The innermost layer of the ovary is known as the 'medulla' or 'hilus' mainly comprised of neurovascular structures [52]. Follicular maturation is a complex process that is temporally controlled by interrelated intra- and extra ovarian factors that lead to ovulation of a mature oocyte and transformation of the ruptured follicle into a corpus luteum (Fig. 2.1) [24]. In primates, multiple follicles initiate development during each menstrual cycle but the majority fail to complete the process to become the dominant ovulatory follicle. After ovulation, development of the corpus luteum is associated with extensive angiogenesis and transformation of the follicular cells (granulosa/theca) into luteal cells which are characterised by secretion of progestins [42, 47].

2.2.2 Fallopian Tubes

The Fallopian tubes (oviduct) act as the connection between the ovaries and the uterus facilitating transport of the oocyte following ovulation (Fig. 2.1). They are surrounded by a muscular layer composed of circular and longitudinal smooth muscle fibres and are lined by ciliated cells in the inner mucosal layer [14]. The Fallopian tubes can be divided into four main segments: fimbrae, infundibulum, ampulla, and isthmus. The infundibulum is the widest and most distal section with small, finger-like projections (known as fimbrae) that capture the released ovum in the peritoneal cavity. Once sequestered, the coordinated muscular contractions and beating of cilia on the epithelial cells direct the ovum towards the

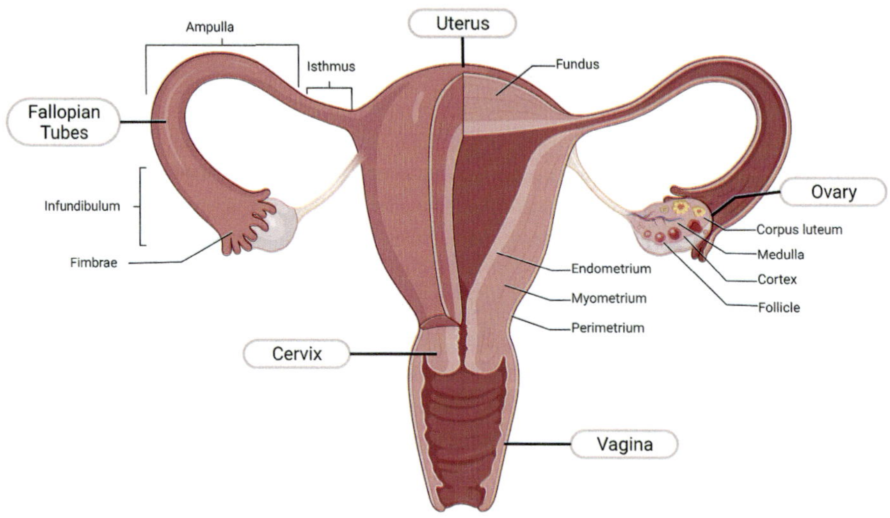

Fig. 2.1 Architecture of the human reproductive organs in women. Note the relationship between the two ovaries found in close association with the fimbrae leading to the Fallopian tubes down which shed oocytes travel en route to the lumen of the uterus. The uterus (womb) has a robust outer layer consisting of muscle cells (myometrium) and an inner luminal layer of endometrium. (Figure prepared by KK using BioRender software)

uterus [148]. Lastly, the isthmus is a short, narrowed segment connecting the ampulla to the uterus.

2.2.3 Uterus

The uterus (womb) is a multi-layered, hollow reproductive organ that plays a key role in nurturing the developing embryo. Like the ovaries, it is suspended by several ligaments with attachments to pelvic structures. The Fallopian tubes join on either side of the superior uterus segment (fundus) and the cervix opens inferiorly from the isthmus [4]. From the outermost to innermost layer, the uterus consists of the perimetrium, myometrium, and endometrium (Fig. 2.1). The perimetrium is a serous layer of epithelial cells lubricating the surface of the organ within the peritoneal cavity. The myometrium is made up of longitudinal and circular smooth muscle layers that enlarge during pregnancy to accommodate the foetus [76].

The endometrium undergoes structural changes throughout the endometrial cycle (Fig. 2.2) as well as contributing to the placenta during pregnancy. The luminal aspect of the endometrium is lined by a simple columnar epithelium that overlies the multicellular stroma containing endometrial stromal fibroblasts/decidual cells, connective tissue, spiral arteries, and glands [35]. The stromal compartment also hosts a complex, dynamic and fluctuating population of immune cells which includes a unique population of CD56 positive natural killer (CD56+ NK) cells, members of the monocyte/macrophage lineage and neutrophils [6, 33, 68]. In women, the inner portion of the endometrium is referred to as the functional layer with the area closest to the myometrium considered as the basal layer: the functional layer is shed at the time of menstruation (Fig. 2.2). One of the most striking alterations in endometrial function occurs following ovulation, when the rapidly rising concentrations of progesterone (P) in the blood bathing the tissue stimulate terminal differentiation (decidualization) of the stromal fibroblasts to create a favourable microenvironment for embryo implantation [115]. Details of the process of decidualization of the endometrium are reviewed in [50]; stromal cell decidualization can be successfully modelled in vitro using cells isolated from endometrium during the proliferative phase [49, 56]. In the absence of pregnancy a rapid fall in the concentration of ovarian-derived progesterone triggers a cascade of events leading to menstruation including an influx of immune cells; increased expression of inflammatory mediators such as prostaglandins increased vessel permeability and increased expression of enzymes that break down extracellular matrix culminating in endometrial tissue breakdown [35]. One of the most remarkable features of menstruation is the piecemeal shedding of the tissue which occurs in parallel with rapid repair and restoration of tissue integrity without forming a scar [48].

2.2.4 Cervix and Vagina

The lower segment of the uterus is known as the cervix and has three distinct functions: maintaining a sterile environment in the upper female reproductive tract, facilitating sperm transport, and retaining the foetus during pregnancy until delivery. These functions are regulated by local and circulating hormones. For example, when oestrogen levels rise, cervical secretion of watery mucous increases, raising the pH, and optimizing the environment for sperm survival. However, when progesterone increases, mucous secretions decrease and become more viscous blocking sperm migration [102]. The cervix opens into the vagina which extends to the vulva forming the vaginal canal. It is a muscular tube covered by stratified squamous epithelial cells. The vagina does not have glands and lubrication is generated from fluid transudate passing through epithelial cells which can be upregulated by sexual stimulation or oestrogen [60]. The fall in circulating concentrations of steroids after menopause can contribute to vaginal dryness and tissue atrophy [87].

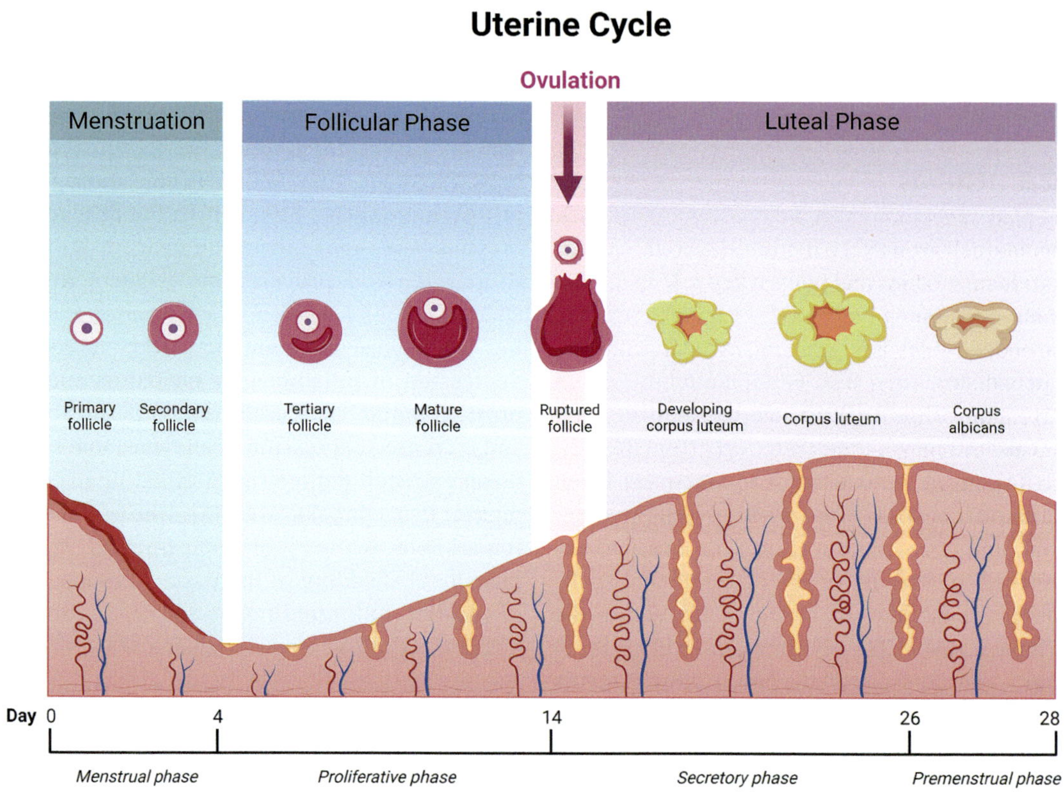

Fig. 2.2 Summary of the different phases of the human menstrual cycle. The upper section of the figure highlights the changes in the ovary illustrating a single follicle as it grows and develops prior to ovulation and thereafter transforms into a corpus luteum which regresses if pregnancy does not occur. The phases of the endometrial cycle that mirror these changes in ovarian function are given below the diagram of the endometrium. During the follicular/proliferative phase, rising concentrations of follicular oestrogen promote cell proliferation and active angiogenesis. After ovulation, the production of progesterone by the corpus luteum promotes functional differentiation (decidualization) of the stromal cells. If pregnancy does not occur, the corpus luteum involutes, circulating levels of progesterone fall rapidly and the inner aspect of the tissue breaks down (menstruation). (Figure prepared by KK adapted from "Uterine Cycle" from BioRender.com 2021)

2.3 Hormone Biosynthesis and Metabolism Within the Female Reproductive System

In women, steroids are synthesised from cholesterol in the ovaries and adrenals via a series of enzymatic conversions which has been extensively reviewed elsewhere [110, 119]. The key enzymes fall into two major classes of proteins: the heme-containing cytochrome P450 (CYP) and the short chain dehydrogenase/reductase (HSD). The review by Hu et al. contains a useful summary of the key locations of these enzymes [72]. For a detailed analysis of the differences between steroid pathways in the ovaries and adrenals, readers are referred to Miller and Auchus [110]. Studies in rodent models have been useful in identifying the role of ovarian steroids but have some limitations because human adrenal glands produce large quantities of the androgens dehydroepiandrosterone (DHEA) and androstenedione but mouse adrenals do not [143].

2.3.1 Endocrine – Ovary

The ovaries are part of the hypothalamus-pituitary-ovarian (HPO) axis, which is a hormone driven regulator of the female reproductive system [38].

In brief, at the hypothalamus, pulsatile secretion of gonadotrophin-releasing hormone (GnRH) acts on the anterior pituitary to stimulate synthesis and secretion of the gonadotrophin hormones luteinising hormone (LH) and follicle-stimulating hormone (FSH). These hormones bind to G-protein coupled receptors expressed in ovarian cells to regulate cell growth and the expression of enzymes that regulate biosynthesis of steroid hormones. Mature ovarian follicles have 4 main cell types: the oocyte which is surrounded by cumulus granulosa cells, the outer mural granulosa cells and the surrounding thecal cells. In the early stages of follicular development the primary effects of LH are on the thecal cells whereas FSH receptors are abundant on granulosa cells [126].

Within the ovary, steroidogenesis is partitioned between the granulosa and theca cells which express different enzymatic components in humans (and mice) summarised as conforming to a 'two cell/two gonadotropin model' [74]. In brief, androgens are synthesized from cholesterol in LH-stimulated theca cells, then converted into oestrogens in FSH-stimulated granulosa cells. The enzyme CYP17, which converts pregnenolone and progesterone to dehydroepiandrosterone (DHEA) and androstenedione, respectively, is expressed primarily in theca cells. Aromatase (CYP19), the protein which plays a critical role in conversion of testosterone to oestradiol, is expressed in mural granulosa cells with levels rising rapidly as the follicles mature [141]. Notably, LH receptors are not exclusively found in theca cells, with expression levels in mural granulosa cells rising in response to FSH just prior to the LH surge [74]. In female mice with targeted deletion of *Cyp19* (Arko) the ovaries contain cells with characteristics of testicular cells including seminiferous tubule-like structures lined with Sertoli cells [16] a phenotype that has clear parallels with that of the ovaries of mice with double knockouts of *Esr1/Esr2* highlighting a role for locally synthesised oestrogens in granulosa cell differentiation/phenotype [45].

Ovulation is a tightly regulated multistep process; following ovulation a rapid reorganization and remodelling of the follicle occurs as the granulosa cells and theca cells luteinize (reviewed in [128]). The capacity to transform cholesterol to progesterone is a universal characteristic of corpora lutea (CL) and involves the mitochondrial P450scc and 3βHSD type 2 located in the endoplasmic reticulum: both enzymes are dramatically upregulated in the CL enabling the organ to produce large quantities of progesterone [138], The CL also produces androgens and oestrogens with the major androgen produced by the ovary being the weak androgen, androstenedione [138].

2.3.2 Intracrine

Studies in mice were the first to highlight an essential role for local expression of aromatase within the decidualized endometrium in regulation of angiogenesis [37]. Subsequent studies using primary endometrial stromal cells decidualized in vitro mirrored these findings [53]. More recently Gibson and colleagues have shown biosynthesis and intracrine metabolism of androgens occurs during decidualization and can influence expression of genes implicated in endometrial receptivity [56]; reductions in the precursor pool of DHEA with age which may contribute to reduced fertility in older women [58]. Taken together all these studies point towards an important role for oestrogen and androgen metabolism in supporting the development of a receptive endometrial tissue microenvironment and establishment of a viable pregnancy. Targets for the actions of the locally generated oestrogens include immune cells [59] and endothelial cells [62]. Intracrine metabolism of steroids has also been investigated in postmenopausal tissues with landmark studies from the Labrie group highlighting the potential use of topical DHEA as a treatment for vaginal atrophy [86].

2.4 Expression and Action of Steroid Receptors Within the Female Reproductive System

2.4.1 Oestrogen Receptors

In the female reproductive system, oestrogen's effects are classically mediated by two nuclear hormone receptors: ERα (encoded by *ESR1*) and ERβ (encoded by *ESR2*). All steroid receptors are believed to have arisen from a common ER-like ancestor with divergence of *ESR1* and *ESR2* sequences occurring following the evolution of jawed vertebrates [139]. Notably, whilst ERα and ERβ exhibit a high degree of homology in their DNA binding domains suggesting they can bind to similar promoter sequences, differences in the amino acids within their ligand binding domains alter their affinity for some ligands (such as phytoestrogens) [36, 100]. These receptors may act via both classical and non-classical pathways with the latter involving rapid signalling [67, 90]. Oestrogens may also bind GPER1 a G-protein coupled receptor that is expressed in human endometrium [82]; much less is known about this receptor than the nuclear receptors and its role is outside the scope of this chapter.

The expression of ERα and ERβ mRNAs and protein vary between different cells and tissues in the reproductive system; the complexity of the system is further complicated by the expression of a number of splice variant isoforms [124, 132, 136]. When ERα and ERβ (or their splice variants) are expressed in the same cell they may form either homo- or heterodimers with varying impacts at regulatory domains acting either by direct binding to EREs (oestrogen response elements) or via tethered binding mechanisms involving additional transcription factors such as AP1, Sp1 and FOXO [62, 117]. Together with the potential for varying ligand affinities, the availability of two different oestrogen receptors may explain why such a variety of oestrogen-dependent responses have been reported in reproductive tissues and reproductive pathologies. We and others have used specific antibodies to reveal cell-specific patterns of expression in human reproductive tissues [28, 34, 131]. Animal models with global or cell specific ablation of *Esr1* and/or *Esr2* have been developed and can be a useful complement to studies on human tissues and cells [45, 84].

Ovary

Analysis of primate ovaries was undertaken using three different antibodies directed against different regions of recombinant ERß protein: protein of appropriate size was detected on Westerns and localized to cellular nuclei in multiple cell types in both marmoset and human ovaries [131]. In this study there was consistent detection of ERβ protein as the predominate ER subtype in the nucleus of granulosa cells (all follicle sizes), thecal cells, corpus lutea, stroma, and epithelium. In the same study (in parallel tissue sections) ERα expression was lower in the stromal/thecal cells and was only expressed in granulosa cells of antral follicles [131]. Splice variant isoforms of the *ESR2* gene have also been identified in the human ovary but their function is unknown [124].

In mice engineered with knockout of *Esr1* (ERαKO), *Esr2* (ERβKO) or both, studies have revealed that the single knockout mice display distinct phenotypes [30, 45]. For example, Dupont et al. reported that the ERαKO females are sterile, whereas ERβKO females are either infertile or exhibit variable degrees of subfertility [45]. Folliculogenesis proceeds normally up to the large antral stage in both ERαKO and ERßKO adults, whereas large antral follicles of ERαβKO adults are markedly deficient in granulosa cells [45]. Couse et al. also reported that steroidogenesis in the ERαKO ovaries was disturbed. For example, Hsd17b3 expression was upregulated with formation of Leydig-like cells in the interstitium [30]. Strikingly, in the ERαβKO, granulosa cells transform into cells displaying junctions that are unique to testicular Sertoli cells and with up regulation of Sox9, a transcription factor involved in differentiation of Sertoli cells in the foetal testis [46]. Notably SOX9 has also been identified as a marker of ER negative luminal cell progenitors in breast cancer [25] and its upregulation has been implicated in resistance to endocrine therapies including administration of the SERM tamoxifen [75].

In adults, ERβKO ovaries exhibit an attenuated response to FSH and reduced expression of *Lhcgr* and *Cyp19a1* both of which are critical for ovarian steroidogenesis [41, 66]. These studies have been complemented and extended by genomic profiling of granulosa cells recovered using laser capture microdissection which confirmed altered expression of genes known to be regulated by FSH (Akap12 and *Runx2*) as well as a extending this relationship to include approximately 300 other genes not previously reported as associated with ERβ regulation of these cells [10]. These studies have been complemented by a large body of literature exploring the function of human granulosa cells in vitro with recent papers highlighting aberrant gene expression and steroid metabolism in cells recovered from the ovaries of women with polycystic ovarian disease [88, 116].

Fallopian tube

Expression of ERα and ERβ is reported to remain relatively constant in the Fallopian tubes although expression of ERα protein was reduced in biopsies from women with ectopic pregnancy [71]. The limited number of studies on human tissues have been complemented by those focused on the oviduct of mice that have reported embryo transport was unaffected in the ERβKO. In contrast when embryos were retrieved from $Wnt7a^{Cre/+}$; $Esr^{f/f}$ (ERαKO) females 3.5 days post coital (dpc), 100% of the embryos were retained in the oviduct whereas they had transited to the uterus in wild types [91]. In another study of conditional knockout mice (cKO) lacking ERα, fertilized eggs failed to survive beyond the 2-cell stage. When antimicrobial activity was measured *in vitro*, significantly higher protease activity was observed in the cKO mice compared to the wild type resulting in disruption of the zona pellucida and altered plasma membrane activity, both detrimental to the survival of embryos [154].

Endometrium

During the menstrual cycle, the fluctuating levels of ovarian-derived oestrogenic hormones results in physiological endometrial changes that are mediated by both ERα and ERβ. The patterns of expression vary between the cell types within the tissue with the prediction that a variety of both homo- and hetero-dimers may be formed. In the functional layer, ERα protein expression is high in glandular epithelial and stromal cells during the proliferative phase but reduced during the mid-late secretory phase [32, 104]. In contrast, ERβ was expressed in the luminal epithelium, endothelial cells and stromal fibroblasts. Whilst protein expression declined in the epithelial cells during the secretory phase, it remained unchanged in the stromal cells [32]. Notably the expression of ERα and ERβ in the cells of the basal compartment did not show such dynamic changes as those in the functional layer.

Expression of ERα is important for both endometrial cell proliferation and expression of PR. Studies in mice have been useful in showing that the impact of E2 on epithelial cell proliferation during the follicular phase is mediated via stromal ERα. Specifically the E2 stimulated ERα-dependent gene expression increases the secretion of insulin growth factor 1 (Igf1) and other proteins (Mad211, Cdkn1a, Cebpb) that stimulate endometrial epithelial cell proliferation [155]. Elevated caspases (pro-apoptotic) were found in uterine epithelial-specific αERKO models, suggesting ERα may directly regulate apoptosis in this cell type [153].

On the contrary, *in vitro* studies suggest that ERβ inhibits endometrial epithelial cell proliferation. When βERKO uteri underwent E_2 stimulation, increased stromal *Igf1* mRNA and decreased epithelial growth factor receptor (Egfr) expression was observed [144, 149]. In mice with selective ERα ablation in luminal and glandular epithelial cells, decidualization was impaired which may be explained by the reduced expression of leukemia inhibitory factor (Lif) as Lif activates the ERK1/2 pathway which induces Indian hedgehog (Ihh) expression in the epithelium and controls stromal decidualization [118]. Apart from regulating cellular proliferation and decidualization, ERs also modulate endometrial vascularisation and expression of angiogeneic factors such as vascular endothelial growth factor (Vegf) [113]. Cell based studies using endothelial cells from different vascular beds including endometrium and myometrium reported that these cells contained ERβ mRNA and protein but no ERα (consistent with their phenotype in intact tissue and that the impact of oestrogens on these cells was via a tethered receptor binding mechanism involving Sp1 [62].

Oestrogens play a key role in regulating the function of endometrial immune cells including uterine natural killer (uNK) cells [55, 69, 135] and mast cells [39]. When CD56+ uNK cells were retrieved from human endometrial tissue, mRNA for ERβ could be detected and cells were immunopositive for ERβ throughout all stages of the menstrual cycle, suggesting that ERβ homodimers form in these cells [69]. Gibson et al. found that oestrogen stimulation increased uNK cell migration and chemokine ligand 2 secretion which promotes endothelial angiogenesis and modulates vascular functioning [55] and that the cells expressed a variant of ERα (ERα46) on their cell membranes which might mediate rapid E2 dependent signalling and cell mobility [59]. Expression of this variant in other immune cells is yet to be explored. Aberrant expression of endometrial ERs has been implicated in a range of uterine disorders including endometriosis, adenomyosis, and endometrial cancer the discussion of which is beyond the scope of this chapter [129, 156].

Myometrium

Within the myometrium, ER plays a crucial role in pregnancy. Quantitative RT-PCR of human myometrial tissue collected at term mainly identified ERα mRNA expression while ERβ mRNA was negligible [150]. During labour, the spike in E2 levels increases uterine contractility by several mechanisms including an accentuated oxytocin response [150], increased connexin-43 gap junction protein, increased prostaglandin-E2 and F2a [85], and inhibition of myometrial K^+ channels [81]. As oestrogen levels continuously rise throughout pregnancy, it has been suggested that the ratio of spliced ERα variants, ERΔ7 and ERα46, may play a role in preventing premature uterine contractions [5]. In a recent study transcriptomic analysis was used to unravel the complex co-regulation of genes that is involved in the transformation of myometrial cells into a contractile phenotype revealing an important role for long non coding RNAs and microRNAs with ESR1 identified as one of 3 master regulators opening up new avenues for research into the role of this receptor subtype in regulation of the myometrium [142]

Vagina

Oestrogen receptors are important to normal functioning of the vagina: only the basal layer undergoes mitogenic activity whereas suprabasal cells are keratinized with a squamous appearance. In post-menopausal women, there is a reduction in both ERα and ERβ in the vaginal mucosa [22] and treatments with selective ER-modulators such as Ospemifine are currently being explored to increase receptor expression and alleviate the symptoms of vaginal atrophy [89]. Ayehunie and colleagues developed a Hormone-Responsive Organotypic Human Vaginal Tissue Model and used this to explore the expression of receptors as well as the impact of E2 and progesterone (P) on gene expression demonstrating a significant upregulation in immune regulating genes in response to E2 [8].

In ERαKO mice, the absence of ERα led to a reduction of keratin receptors *Krt6a* and *Krt10* and failure of epithelial cells to undergo keratinized differentiation [92, 112]. Studies in mice suggest ERα is also responsible for preventing vaginal epithelial atrophy and maintaining cellular integrity [92]. Oestrogen receptors may also play a role in regulation of leukocyte activity in the vagina. The ERαKO mice showed excessive neutrophil infiltration throughout the estrous cycle which can damage tissue integrity by heightening neutrophil elastase and matrix metalloproteinase activity [92].

2.4.2 Progesterone Receptors

There is a single PR gene in human and rodent but this encodes two isoforms of the protein (PRA, PRB) [29, 40]. The human PR isoforms (hPRA, 94 kDa; hPRB, 114 kDa) are transcribed from distinct, oestrogen-inducible promoters and the only difference between them is that the first 164 amino acids of hPRB are absent in hPRA [51]. The classical actions of hPR involve binding to DNA at progesterone response elements within the promoter or distant enhancer of a target gene [9]. In common with the ERs discussed above, PR can also modulate gene expression through a pathway involving PR tethering to transcription factors such as AP1, SP1, NFkB, and signal transducer activator of transcription 3 (Stat3) [64]. Following ligand binding, redistribution of PR into discrete subnuclear foci occurs in endometrial cells that is dependent upon binding to the nuclear matrix and is associated with transcriptional activity: mutants lacking the ability to interact with the matrix have been identified [61]. In addition to transcription factors involved in the tethered response, several other co-factors important for PR dependent responses have been identified in endometrial tissues. One example is FOXO1A which plays an important role in regulation of genes involved in decidualization such as IGFBP1 [79]. An alternative signalling pathway is mediated via membrane-bound GPCR and membrane spanning receptors that are beyond the scope of this chapter but were recently comprehensively reviewed by Medina-Laver et al. [108].

Since the discovery of the two isoforms of PR, there has been an effort to determine their relative contributions to progesterone dependent impacts on cell function. In vitro studies have demonstrated PRA inhibits PRB action via the inhibitory domain (ID) present in its extra amino acid domain and this decreases the effects of progesterone on target cells [122]. In addition, our understanding of the key role(s) played by PR in reproductive function took a major step forward with studies on female mice with targeted deletion of the entire *Pr* gene [PrKO] which revealed an inability to ovulate, uterine hyperplasia and inflammation as well as major impacts on mammary gland development [96]. The selective ablation of the PRA (PRAKO) and PRB (PRBKO) isoforms confirmed PRA is the isoform most important for ovarian and uterine function as its ablation leads to female infertility [29].

Ovary

The development of antibodies to the PR protein in the 1980's led to a number of landmark immunohistochemical studies reporting its localisation to the different cell types in the human and primate ovary. For example, Press and Greene detected expression in ovarian surface epithelium, stroma and luteal cells [125]. Detailed studies in primates, where access to ovarian tissue at different stages of the cycle is easier than in humans, has reported PR positive staining of theca cells of both healthy and atretic follicles at all stages of the cycle with some granulosa cells of primordial and primary follicles being immunopositive but only the granulosa layer of large preovulatory follicles associated with of luteinization after the LH surge having staining equivalent the theca [70]. Duffy et al. showed that the ratios of PRA to PRB changes in the monkey corpus luteum (CL) during the luteal phase with PRA levels decreasing while PRB levels were unchanged [44]. In humans, PR expression is maintained in the active corpus luteum, but it ceases in the late corpus luteum. In their immunohistochemical study on human CL, Maybin et al. detected PR in all steroidogenic cells and stromal fibroblasts but endothelial cells, pericytes, macrophages and fibroblasts within the central CL clot were immunonegative [105].

Notably, studies in mice suggest PR plays a key role in ovulation with the PRKO females forming corpora lutea with retained oocytes. The failure to release the oocyte in the PRKO is associated with reduced biosynthesis of proteases including cathepsin L [127].

Fallopian tube

In a study using human FT explants Horne et al. [71] reported PRAB and PRB mRNAs were decreased in midluteal phase compared to follicular phase and down-regulated in human FT treated in vitro with progestogen. Progesterone also controls gamete transport in the oviduct by regulating muscular contraction (by upregulating endothelin 1) and ciliary beat frequency (membrane isoforms mPRγ and mPRβ have been found on the apical cell membrane and cilia respectively) [21, 114]. Although the exact localization of PR in the oviduct is inconclusive, focal slowing at the ampullary region may facilitate fertilization [20].

Endometrium

In an early study, Wang et al. used antibodies specific to the PRB isoform to examine its distribution in human endometrium across the menstrual cycle and to compare this to the pattern of protein expression detected using an antibody that recognised PRA + B [147]. They reported that that both PR subtypes were present in glands and stroma in the proliferative phase and by comparing the A + B pattern with that of B alone they inferred that PRA was most strongly expressed in stroma during the secretory phase. Subsequent analysis of full thickness sections highlighted the parallels between ERα and PRA staining in the proliferative phase and persistence of PRA in stromal cells during the secretory phase [145]. Studies in knockout mice have also showed specific ablation of PRA alone was sufficient to induce infertility associated with failure of decidualization and implantation [29].

The most well studied genomic impact of progesterone, acting via PR, on endometrial cells is the transformation (decidualization) of stromal fibroblasts, the process of which can be reliably and reproducibly induced in vitro using primary human cells [49, 50]. Wide ranging studies have identified progesterone-dependent patterns of gene expression including impacts on cell survival and senescence [18] and induction of factors that play a key role in uterine receptivity to the blastocyst [1, 40, 93] and there are extensive genomic datasets available to those interested in this aspect of steroid hormone action. Factors induced in response to P-induced decidualization include the transcription factors HAND2 and FOXO1 as well as interleukin 15 (IL15) [15, 18]. The importance of progesterone in induction of IL15 has been confirmed following analysis of endometrium from women treated with a progesterone receptor modulator [152]. The production of IL15 plays a key role in recruitment and differentiation of uNK cells that are involved in regulation of angiogenesis which is important for successful implantation [83]. Notably, a critical role for uNK cells downstream of progesterone-induced changes in tissue function has been supported by evidence that disfunction in, or aberrant recruitment of, uNKs has been implicated as a cause of recurrent miscarriage [123]. A suboptimal response to progesterone, so called 'progesterone resistance', has also been proposed as a contributing factor in the aetiology of endometriosis, a condition associated with sub/infertility [2].

The endometrium of mice with ablation of *Pr* exhibit an exaggerated response to exogenous E2 [96]. Studies in women and primates have also highlighted the impact of progesterone receptor antagonists on endometrial cell proliferation [12] with this property exploited as a therapy for a range of endometrial disorders [35]. Further studies exploring the impact of new classes of selective progesterone receptor modulators (SPRMs) has revealed specific impacts on PR mediated gene expression including upregulation of AR [151].

Myometrium

Two aspects of the impact of progesterone on the myometrium have attracted particular attention: the role of progesterone/PR activity in maintaining myometrial quiescence during pregnancy and on the proliferative activity of cells found in fibroids with a considerable literature attached to both. Lye and colleagues recently reviewed the mechanisms that can stimulate myometrial contractions at the end of pregnancy with 'functional progesterone withdrawal' being considered as the prime factor for myometrial activation and labour induction [133]. Notably, prior to labour in the human myometrium there is an increase in local activity of the enzyme 20α-HSD which metabolizes bioactive progesterone to an inactive metabolite, 20α-dihydroprogesterone, providing another example of the importance of intracrine regulation in reproductive tissue function [120]. Progesterone induces the growth of fibroids (benign myometrial growths) by regulating key genes that control proliferation and apoptosis with recent studies focused on the impact of the steroid on stem/progenitor cells [19, 80].

Vagina

Immunoreactivity of PRB in the vagina is highest during the luteal phase which correlates with the rise in blood levels of progesterone produced by the CL. There has been limited studies on the expression of PR in the human vaginal with clinical trials in postmenopausal women reporting poor induction of PR in response to topical oestrogens [111]. Studies in mice have been helpful in revealing a role for epithelial PR in regulating apoptosis and differentiation of the vagina [109].

2.4.3 Androgen Receptor

Androgens are synthesised and secreted by both ovary and adrenals in women. The actions of bioactive androgens (testosterone and dihydrotestosterone) are mediated by AR which is expressed on the X chromosome [94]. In contrast to ER and PR proteins, the AR has a very large N terminal domain which contains important sites for post translational modifications including phosphorylation which can have a significant impact on the activity of the receptor [27]. The large size of the receptor also enables interactions between the N and C terminal domains which can modulate binding to DNA domains on androgen responsive genes (AREs) [17].

Ovary

Our understanding of the role(s) played by androgens and its receptor in ovarian function has been informed both by detailed immunohistochemical studies using human tissues and those of animal models including primates and rodents [107, 131, 146]. A particular focus of many of the investigations has been on the role played by androgens, such as testosterone, in the development and clinical consequences of polycystic ovarian syndrome (PCOS) which have been recently reviewed [137]. In the human ovary, AR can been detected by immunohistochemistry in granulosa cells of all follicle sizes as well as in thecal cells, the stromal fibroblasts and surface epithelium [131]. Mcewan et al. conducted studies in a primate model, the Common marmoset, to explore the phosphorylation status of ovarian AR [107]. Using phosphorylation-specific antibodies combined with ovarian tissue sections they were able to detect AR+phosphoserines 81, 308, and 650 in the granulosa cells of developing follicles, the surface epithelium, and vessel endothelial cells suggesting AR was active in these cells [107].

Mouse models with targeted deletion of *Ar* from different ovarian cell types as well as models in which excess androgens are administered have done much to refine our understanding of the importance of androgens in normal follicle development and ovulation and these have been extensively reviewed in recent papers [7, 137, 146].

Fallopian tube

The Fallopian tube epithelium exists as a continuum of the endometrium but studies by Mclean et al. using explants recovered from fertile women showed lower proliferation and higher expression of epithelial AR than endometrial samples [97]. In other studies designed to evaluate whether the high levels of androgens in women with PCOS might have an impact on Fallopian tube function, the authors isolated epithelial cells from 12 women then treated them for 14 days in low testosterone (0.8 nM) or a PCOS-like, testosterone concentration (2 nM) using both static and dynamic conditions in microfluidic devices [73]. Whilst this was a small study the novel findings included evidence that treatment with high testosterone slowed ciliary beat and reduced response to oestrogen which may have some implications for understanding the reduced fertility experienced by some PCOS patients [73].

Endometrium

The most prominent expression of AR in the human endometrium is in the nuclei of stromal fibroblasts. Saunders and colleagues published data from full thickness sections of endometrium highlighting intense expression in the basal compartment throughout the cycle but variable expression in the functional layer with evidence of upregulation in epithelial cells in the mid/late secretory phased [57, 101]. They used an in silico strategy to identify putative androgen-regulated genes and reported evidence that in vitro treatment of primary AR-positive endometrial stromal cells with the potent androgen DHT could reduce apoptosis and cell migration [101]. These findings are in agreement with those highlighting upregulation in epithelial cell AR expression in response to treatment with anti-progestins such as RU486 (mifepristone) which have implicated AR in reduced epithelial cell proliferation [13]. AR are also expressed in epithelial cells of endometrial cancers and the perivascular myoid cells surrounding blood vessels in the endometrium [31, 54] and they may play a role in the aetiology of both malignancy and benign disorders such as endometriosis and heavy menstrual bleeding [134].

Several studies have explored the impact of androgens on stromal cell decidualization using in vitro models. For example, Cloke et al. [26] reported that AR and PR regulate the expression of distinct decidual gene networks. Notably AR-induced genes were involved in cytoskeletal organization and cell motility, whereas analysis of AR-repressed genes suggested involvement in cell cycle regulation. In contrast, PR depletion perturbed a number of signalling intermediates and knockdown of PR, but not AR, compromised activation of WNT/ß catenin which plays an important role in endometrial tissue function [26]. In follow up studies treatment of cells with the potent AR receptor ligand DHT stimulated cytoplasmic expansion, lipid droplet formation, the production of an abundant extracellular matrix, and gap junction formation in decidualized primary stromal cells [78] and enhanced their resistance to oxidative stress [77].

Myometrium

The cells of the muscular myometrium contain abundant nuclear AR during reproductive life: interest in the role(s) of androgens in myometrial function have included studies on their impact on cell proliferation and in myometrial contractility with particular emphasis on parturition [98, 99]. Whilst there have been limited studies in women compared to animal models, there is evidence that circulating concentrations of testosterone, and its precursor androstenedione, are significantly higher during pregnancy than in the non-pregnant state, and their values increase throughout pregnancy [106] leading to suggestions they may complement the impacts of progestins and the maintenance of myometrial quiescence [121]. Studies exploring the impact of androgens on contractile activity of the myometrium have made use of tissue recovered from women undergoing elective caesarean [99, 121]. These studies that have reported the relaxation response to androgens had a very rapid time course and appears to work through non-genomic pathways involving calcium which were not abolished by AR knockdown [99]. In contrast, the impact of androgens on myometrial endothelial cells in vitro did appear AR-dependent although the doses of T used in the study were in the pharmacological range [43].

Vagina

In a recent study both vaginal tissues and smooth muscle cells (hvSMC) isolated from the vagina were evaluated for AR mRNA expression. Whilst AR mRNA was significantly lower than ERα, in isolated hvSMCs, its mRNA expression was higher than PR and both ERs [23]. In addition to these studies focused on the smooth muscle cells, the complex interplay between different receptors in the vaginal mucosa also needs to be expanded to consider the immune cells that populate the tissue and this area of research should also complement that in endometrium [65]. These studies have translational potential for example in smooth mucle cells activation of AR by DHT can reduce their potential to be involved in the initiation and maintaining of inflammation [103] enforcing the proposed beneficial effects of topical androgen administration after menopause [86].

2.5 Summary and Future Prospects

In this brief review we have highlighted the evidence that spatial, temporal and cell-specific patterns of expression of the sex steroid receptors (ESR1, ESR2, PR and AR) within the organs of the female reproductive system are essential for the fine-tuning of tissue function required for normal functioning and for fertility. Malfunctions or mal-adaptations of sex steroid biosynthesis (or action) via these receptors contributes to common reproductive disorders including recurrent miscarriage, heavy menstrual bleeding and endometriosis [35, 95, 156]. The role(s) played by these receptors have been extensively studied both in human cells/tissues as well as in model species, most notably mice [29, 35, 63, 84, 130].

A number of recent technical developments including improved isolation and characterisation of stem cells from endometrial tissue and menstrual fluid [11] and refinement of in vitro models such as organoids and organ on a chip [3, 140] all offer new opportunities to rapidly increase our understanding of the impact of sex steroid receptor dependent signalling in the reproductive system and to develop new smarter therapies for reproductive disorders.

Acknowledgements Studies exploring the role(s) of steroid hormone receptors on reproductive function conducted in the senior author's laboratory have been funded by grants from the UK Medical Research Council (MR/N024524/1; G1100356/1).

References

1. Aghajanova L, Hamilton AE, Giudice LC (2008) Uterine receptivity to human embryonic implantation: histology, biomarkers, and transcriptomics. Semin Cell Dev Biol 19:204–211
2. Aghajanova L, Velarde MC, Giudice LC (2010) Altered gene expression profiling in endometrium: evidence for progesterone resistance. Semin Reprod Med 28:51–58
3. Ahn J, Yoon MJ, Hong SH, Cha H, Lee D, Koo HS, Ko JE, Lee J, Oh S, Jeon NL, Kang YJ (2021) Three-dimensional microengineered vascularised endometrium-on-a-chip. Hum Reprod 36:2720–2731
4. Ameer MA, Fagan SE, Sosa-Stanley JN, Peterson DC (2021) Anatomy, abdomen and pelvis, uterus. StatPearls. Treasure Island (FL): StatPearls Publishing Copyright © 2021, StatPearls Publishing LLC
5. Anamthathmakula P, Kyathanahalli C, Ingles J, Hassan SS, Condon JC, Jeyasuria P (2019) Estrogen receptor alpha isoform ERdelta7 in myometrium modulates uterine quiescence during pregnancy. EBioMedicine 39:520–530
6. Armstrong GM, Maybin JA, Murray AA, Nicol M, Walker C, Saunders PTK, Rossi AG, Critchley HOD (2017) Endometrial apoptosis and neutrophil infiltration during menstruation exhibits spatial and temporal dynamics that are recapitulated in a mouse model. Sci Rep 7:17416
7. Astapova O, Minor BMN, Hammes SR (2019) Physiological and pathological androgen actions in the ovary. Endocrinology 160:1166–1174
8. Ayehunie S, Islam A, Cannon C, Landry T, Pudney J, Klausner M, Anderson DJ (2015) Characterization of a hormone-responsive Organotypic human vaginal tissue model: morphologic and immunologic effects. Reprod Sci 22:980–990
9. Beato M, Wright RHG, Dily FL (2020) 90 years of progesterone: molecular mechanisms of progesterone receptor action on the breast cancer genome. J Mol Endocrinol 65:T65–T79
10. Binder AK, Rodriguez KF, Hamilton KJ, Stockton PS, Reed CE, Korach KS (2013) The absence of ER-beta results in altered gene expression in ovarian granulosa cells isolated from in vivo preovulatory follicles. Endocrinology 154:2174–2187

11. Bozorgmehr M, Gurung S, Darzi S, Nikoo S, Kazemnejad S, Zarnani AH, Gargett CE (2020) Endometrial and menstrual blood mesenchymal stem/stromal cells: biological properties and clinical application. Front Cell Dev Biol 8:497
12. Brenner RM, Slayden OD (2005) Progesterone receptor antagonists and the endometrial antiproliferative effect. Semin Reprod Med 23:74–81
13. Brenner RM, Slayden OD, Critchley HO (2002) Anti-proliferative effects of progesterone antagonists in the primate endometrium: a potential role for the androgen receptor. Reproduction 124:167–172
14. Briceag I, Costache A, Purcarea VL, Cergan R, Dumitru M, Briceag I, Sajin M, Ispas AT (2015) Fallopian tubes–literature review of anatomy and etiology in female infertility. J Med Life 8:129–131
15. Brighton PJ, Maruyama Y, Fishwick K, Vrljicak P, Tewary S, Fujihara R, Muter J, Lucas ES, Yamada T, Woods L, Lucciola R, Hou Lee Y, Takeda S, Ott S, Hemberger M, Quenby S, Brosens JJ (2017) Clearance of senescent decidual cells by uterine natural killer cells in cycling human endometrium. elife 6:e31274
16. Britt KL, Kerr J, O'donnell L, Jones ME, Drummond AE, Davis SR, Simpson ER, Findlay JK (2002) Estrogen regulates development of the somatic cell phenotype in the eutherian ovary. FASEB J 16:1389–1397
17. Brodie J, Mcewan IJ (2005) Intra-domain communication between the N-terminal and DNA-binding domains of the androgen receptor: modulation of androgen response element DNA binding. J Mol Endocrinol 34:603–615
18. Brosens JJ, Lam EW (2013) Progesterone and FOXO1 signaling: harnessing cellular senescence for the treatment of ovarian cancer. Cell Cycle 12:1660–1661
19. Bulun SE, Moravek MB, Yin P, Ono M, Coon JST, Dyson MT, Navarro A, Marsh EE, Zhao H, Maruyama T, Chakravarti D, Kim JJ, Wei JJ (2015) Uterine leiomyoma stem cells: linking progesterone to growth. Semin Reprod Med 33:357–365
20. Bylander A, Lind K, Goksör M, Billig H, Larsson DGJ (2013) The classical progesterone receptor mediates the rapid reduction of fallopian tube ciliary beat frequency by progesterone. Reprod Biol Endocrinol 11:33–33
21. Bylander A, Gunnarsson L, Shao R, Billig H, Larsson DGJ (2015) Progesterone-mediated effects on gene expression and oocyte-cumulus complex transport in the mouse fallopian tube. Reprod Biol Endocrinol 13:40
22. Cavallini A, Dinaro E, Giocolano A, Caringella AM, Ferreri R, Tutino V, Loverro G (2008) Estrogen receptor (ER) and ER-related receptor expression in normal and atrophic human vagina. Maturitas 59:219–225
23. Cellai I, DI Stasi V, Comeglio P, Maseroli E, Todisco T, Corno C, Filippi S, Cipriani S, Sorbi F, Fambrini M, Petraglia F, Scavello I, Rastrelli G, Acciai G, Villanelli F, Danza G, Sarchielli E, Guarnieri G, Morelli A, Maggi M, Vignozzi L (2021) Insight on the Intracrinology of menopause: androgen production within the human vagina. Endocrinology 162:bqaa219
24. Channing CP, Schaerf FW, Anderson LD, Tsafriri A (1980) Ovarian follicular and luteal physiology. Int Rev Physiol 22:117–201
25. Christin JR, Wang C, Chung CY, Liu Y, Dravis C, Tang W, Oktay MH, Wahl GM, Guo W (2020) Stem cell determinant SOX9 promotes lineage plasticity and progression in basal-like breast cancer. Cell Rep 31:107742
26. Cloke B, Huhtinen K, Fusi L, Kajihara T, Yliheikkila M, Ho KK, Teklenburg G, Lavery S, Jones MC, Trew G, Kim JJ, Lam EW, Cartwright JE, Poutanen M, Brosens JJ (2008) The androgen and progesterone receptors regulate distinct gene networks and cellular functions in decidualizing endometrium. Endocrinology 149:4462–4474
27. Coffey K, Robson CN (2012) Regulation of the androgen receptor by post-translational modifications. J Endocrinol 215:221–237
28. Collins F, Macpherson S, Brown P, Bombail V, Williams AR, Anderson RA, Jabbour HN, Saunders PT (2009) Expression of oestrogen receptors, ERalpha, ERbeta, and ERbeta variants, in endometrial cancers and evidence that prostaglandin F may play a role in regulating expression of ERalpha. BMC Cancer 9:330
29. Conneely OM, Mulac-Jericevic B, Demayo F, Lydon JP, O'malley BW (2002) Reproductive functions of progesterone receptors. Recent Prog Horm Res 57:339–355
30. Couse JF, Yates MM, Rodriguez KF, Johnson JA, Poirier D, Korach KS (2006) The Intraovarian actions of estrogen receptor-{alpha} (ER{alpha}) are necessary to repress the formation of morphological and functional Leydig-like cells in the female gonad. Endocrinology 147:3666–3678
31. Critchley HO, Saunders PT (2009) Hormone receptor dynamics in a receptive human endometrium. Reprod Sci 16:191–199
32. Critchley HO, Brenner RM, Henderson TA, Williams K, Nayak NR, Slayden OD, Millar MR, Saunders PT (2001a) Estrogen receptor beta, but not estrogen receptor alpha, is present in the vascular endothelium of the human and nonhuman primate endometrium. J Clin Endocrinol Metab 86:1370–1378
33. Critchley HO, Kelly RW, Brenner RM, Baird DT (2001b) The endocrinology of menstruation–a role for the immune system. Clin Endocrinol 55:701–710
34. Critchley HO, Henderson TA, Kelly RW, Scobie GS, Evans LR, Groome NP, Saunders PT (2002) Wild-type estrogen receptor (ERbeta1) and the splice variant (ERbetacx/beta2) are both expressed within the human endometrium throughout the normal menstrual cycle. J Clin Endocrinol Metab 87:5265–5273

35. Critchley HOD, Maybin JA, Armstrong GM, Williams ARW (2020) Physiology of the endometrium and regulation of menstruation. Physiol Rev 100:1149–1179
36. Damdimopoulos AE, Spyrou G, Gustafsson JA (2008) Ligands differentially modify the nuclear mobility of estrogen receptors alpha and beta. Endocrinology 149:339–345
37. Das A, Mantena SR, Kannan A, Evans DB, Bagchi MK, Bagchi IC (2009) De novo synthesis of estrogen in pregnant uterus is critical for stromal decidualization and angiogenesis. Proc Natl Acad Sci U S A 106:12542–12547
38. Davis HC, Hackney AC (2017) The hypothalamic–pituitary–ovarian Axis and Oral contraceptives: regulation and function. In: Hackney AC (ed) Sex hormones, exercise and women: scientific and clinical aspects. Springer International Publishing, Cham
39. DE Leo B, Esnal-Zufiaurre A, Collins F, Critchley HOD, Saunders PTK (2017) Immunoprofiling of human uterine mast cells identifies three phenotypes and expression of ERbeta and glucocorticoid receptor. F1000Res 6:667
40. Demayo FJ, Lydon JP (2020) 90 years of progesterone: new insights into progesterone receptor signaling in the endometrium required for embryo implantation. J Mol Endocrinol 65:T1–T14
41. Deroo BJ, Rodriguez KF, Couse JF, Hamilton KJ, Collins JB, Grissom SF, Korach KS (2009) Estrogen receptor beta is required for optimal cAMP production in mouse granulosa cells. Mol Endocrinol (Baltimore, Md.) 23:955–965
42. Devoto L, Vega M, Kohen P, Castro O, Carvallo P, Palomino A (2002) Molecular regulation of progesterone secretion by the human corpus luteum throughout the menstrual cycle. J Reprod Immunol 55:11–20
43. Dietrich W, Gaba A, Zhegu Z, Bieglmayer C, Mairhofer M, Mikula M, Tschugguel W, Yotova I (2011) Testosterone dependent androgen receptor stabilization and activation of cell proliferation in primary human myometrial microvascular endothelial cells. Fertil Steril 95:1247–1255.e2
44. Duffy DM, Wells TR, Haluska GJ, Stouffer RL (1997) The ratio of progesterone receptor isoforms changes in the monkey corpus luteum during the luteal phase of the menstrual cycle. Biol Reprod 57:693–699
45. Dupont S, Krust A, Gansmuller A, Dierich A, Chambon P, Mark M (2000) Effect of single and compound knockouts of estrogen receptors alpha (ERalpha) and beta (ERbeta) on mouse reproductive phenotypes. Development 127:4277–4291
46. Dupont S, Dennefeld C, Krust A, Chambon P, Mark M (2003) Expression of Sox9 in granulosa cells lacking the estrogen receptors, ERalpha and ERbeta. Dev Dyn 226:103–106
47. Fraser HM, Bell J, Wilson H, Taylor PD, Morgan K, Anderson RA, Duncan WC (2005) Localization and quantification of cyclic changes in the expression of endocrine gland vascular endothelial growth factor in the human corpus luteum. J Clin Endocrinol Metab 90:427–434
48. Garry R, Hart R, Karthigasu KA, Burke C (2009) A re-appraisal of the morphological changes within the endometrium during menstruation: a hysteroscopic, histological and scanning electron microscopic study. Hum Reprod 24:1393–1401
49. Gellersen B, Brosens J (2003) Cyclic AMP and progesterone receptor cross-talk in human endometrium: a decidualizing affair. J Endocrinol 178:357–372
50. Gellersen B, Brosens JJ (2014) Cyclic decidualization of the human endometrium in reproductive health and failure. Endocr Rev 35:851–905
51. Giangrande PH, Mcdonnell DP (1999) The A and B isoforms of the human progesterone receptor: two functionally different transcription factors encoded by a single gene. Recent Prog Horm Res 54:291–313. discussion 313-4
52. Gibson E, Mahdy H (2021) Anatomy, abdomen and pelvis, ovary. StatPearls. Treasure Island (FL): StatPearls Publishing, Copyright © 2021, StatPearls Publishing LLC
53. Gibson DA, Mcinnes KJ, Critchley HO, Saunders PT (2013) Endometrial Intracrinology--generation of an estrogen-dominated microenvironment in the secretory phase of women. J Clin Endocrinol Metab 98:E1802–E1806
54. Gibson DA, Simitsidellis I, Collins F, Saunders PT (2014) Evidence of androgen action in endometrial and ovarian cancers. Endocr Relat Cancer 21:T203–T218
55. Gibson DA, Greaves E, Critchley HO, Saunders PT (2015) Estrogen-dependent regulation of human uterine natural killer cells promotes vascular remodelling via secretion of CCL2. Hum Reprod 30:1290–1301
56. Gibson DA, Simitsidellis I, Cousins FL, Critchley HO, Saunders PT (2016) Intracrine androgens enhance Decidualization and modulate expression of human endometrial receptivity genes. Sci Rep 6:19970
57. Gibson DA, Saunders PTK, Mcewan IJ (2018a) Androgens and androgen receptor: above and beyond. Mol Cell Endocrinol 465:1–3
58. Gibson DA, Simitsidellis I, Kelepouri O, Critchley HOD, Saunders PTK (2018b) Dehydroepiandrosterone enhances decidualization in women of advanced reproductive age. Fertil Steril 109:728–734 e2
59. Gibson DA, Esnal-Zufiaurre A, Bajo-Santos C, Collins F, Critchley HOD, Saunders PTK (2020) Profiling the expression and function of oestrogen receptor isoform ER46 in human endometrial tissues and uterine natural killer cells. Hum Reprod 35:641–651

60. Gold JM, Shrimanker I (2021) Physiology, vaginal. StatPearls. Treasure Island (FL): StatPearls Publishing, Copyright © 2021, StatPearls Publishing LLC
61. Graham JD, Hanson AR, Croft AJ, Fox AH, Clarke CL (2009) Nuclear matrix binding is critical for progesterone receptor movement into nuclear foci. FASEB J 23:546–556
62. Greaves E, Collins F, Critchley HO, Saunders PT (2013) ERbeta-dependent effects on uterine endothelial cells are cell specific and mediated via Sp1. Hum Reprod 28:2490–2501
63. Greaves E, Critchley HOD, Horne AW, Saunders PTK (2017) Relevant human tissue resources and laboratory models for use in endometriosis research. Acta Obstet Gynecol Scand 96:644–658
64. Grimm SL, Hartig SM, Edwards DP (2016) Progesterone Receptor Signaling Mechanisms. J Mol Biol 428:3831–3849
65. Gu C, Duluc D, Wiest M, Xue Y, Yi J, Gorvel JP, Joo H, Oh S (2021) Cell type-specific expression of estrogen and progesterone receptors in the human vaginal mucosa. Clin Immunol 232:108874
66. Gulappa T, Menon B, Menon KMJ (2017) LHCGR expression during follicle stimulating hormone-induced follicle growth is negatively regulated by eukaryotic initiation factor 5A. Endocrinology 158:2672–2679
67. Hall JM, Couse JF, Korach KS (2001) The multifaceted mechanisms of estradiol and estrogen receptor signaling. J Biol Chem 276:36869–36872
68. Henderson TA, Saunders PT, Moffett-King A, Groome NP, Critchley HO (2003a) Steroid receptor expression in uterine natural killer cells. J Clin Endocrinol Metab 88:440–449
69. Henderson TA, Saunders PTK, Moffett-King A, Groome NP, Critchley HOD (2003b) Steroid receptor expression in uterine natural killer cells. J Clin Endocrinol Metabol 88:440–449
70. Hild-Petito S, Stouffer RL, Brenner RM (1988) Immunocytochemical localization of estradiol and progesterone receptors in the monkey ovary throughout the menstrual cycle. Endocrinology 123:2896–2905
71. Horne AW, King AE, Shaw E, Mcdonald SE, Williams AR, Saunders PT, Critchley HO (2009a) Attenuated sex steroid receptor expression in fallopian tube of women with ectopic pregnancy. J Clin Endocrinol Metab 94:5146–5154
72. Hu J, Zhang Z, Shen WJ, Azhar S (2010) Cellular cholesterol delivery, intracellular processing and utilization for biosynthesis of steroid hormones. Nutr Metab (Lond) 7:47
73. Jackson-Bey T, Colina J, Isenberg BC, Coppeta J, Urbanek M, Kim JJ, Woodruff TK, Burdette JE, Russo A (2020) Exposure of human fallopian tube epithelium to elevated testosterone results in alteration of cilia gene expression and beating. Hum Reprod 35:2086–2096
74. Jamnongjit M, Hammes SR (2006) Ovarian steroids: the good, the bad, and the signals that raise them. Cell Cycle 5:1178–1183
75. Jeselsohn R, Cornwell M, Pun M, Buchwalter G, Nguyen M, Bango C, Huang Y, Kuang Y, Paweletz C, Fu X, Nardone A, De Angelis C, Detre S, Dodson A, Mohammed H, Carroll JS, Bowden M, Rao P, Long HW, Li F, Dowsett M, Schiff R, Brown M (2017) Embryonic transcription factor SOX9 drives breast cancer endocrine resistance. Proc Natl Acad Sci U S A 114:E4482–E4491
76. Kagami K, Ono M, Iizuka T, Matsumoto T, Hosono T, Sekizuka-Kagami N, Shinmyo Y, Kawasaki H, Fujiwara H (2020) A novel third mesh-like myometrial layer connects the longitudinal and circular muscle fibers–a potential stratum to coordinate uterine contractions. Sci Rep 10:8274
77. Kajihara T, Tochigi H, Prechapanich J, Uchino S, Itakura A, Brosens JJ, Ishihara O (2012) Androgen signaling in decidualizing human endometrial stromal cells enhances resistance to oxidative stress. Fertil Steril 97:185–191
78. Kajihara T, Tanaka K, Oguro T, Tochigi H, Prechapanich J, Uchino S, Itakura A, Sucurovic S, Murakami K, Brosens JJ, Ishihara O (2014) Androgens modulate the morphological characteristics of human endometrial stromal cells decidualized in vitro. Reprod Sci 21:372–380
79. Kim JJ, Buzzio OL, Li S, Lu Z (2005) Role of FOXO1A in the regulation of insulin-like growth factor-binding protein-1 in human endometrial cells: interaction with progesterone receptor. Biol Reprod 73:833–839
80. Kim JJ, Kurita T, Bulun SE (2013) Progesterone action in endometrial cancer, endometriosis, uterine fibroids, and breast cancer. Endocr Rev 34:130–162
81. Knock GA, Tribe RM, Hassoni AA, Aaronson PI (2001) Modulation of potassium current characteristics in human myometrial smooth muscle by 17β-estradiol and Progesterone1. Biol Reprod 64:1526–1534
82. Kolkova Z, Noskova V, Ehinger A, Hansson S, Casslen B (2010) G protein-coupled estrogen receptor 1 (GPER, GPR 30) in normal human endometrium and early pregnancy decidua. Mol Hum Reprod 16:743–751
83. Kong CS, Ordonez AA, Turner S, Tremaine T, Muter J, Lucas ES, Salisbury E, Vassena R, Tiscornia G, Fouladi-Nashta AA, Hartshorne G, Brosens JJ, Brighton PJ (2021) Embryo biosensing by uterine natural killer cells determines endometrial fate decisions at implantation. FASEB J 35:e21336
84. Korach KS, Emmen JM, Walker VR, Hewitt SC, Yates M, Hall JM, Swope DL, Harrell JC, Couse JF (2003) Update on animal models developed for analyses of estrogen receptor biological activity. J Steroid Biochem Mol Biol 86:387–391
85. Kota SK, Gayatri K, Jammula S, Kota SK, Krishna SVS, Meher LK, Modi KD (2013)

Endocrinology of parturition. Indian J Endocrinol Metab 17:50–59
86. Labrie F (2018) Intracrinology and menopause: the science describing the cell-specific intracellular formation of estrogens and androgens from DHEA and their strictly local action and inactivation in peripheral tissues. Menopause 26:220–224
87. Labrie F, Martel C, Pelletier G (2017) Is vulvovaginal atrophy due to a lack of both estrogens and androgens? Menopause 24:452–461
88. Lerner A, Owens LA, Coates M, Simpson C, Poole G, Velupillai J, Liyanage M, Christopoulos G, Lavery S, Hardy K, Franks S (2019) Expression of genes controlling steroid metabolism and action in granulosa-lutein cells of women with polycystic ovaries. Mol Cell Endocrinol 486:47–54
89. Lethaby A, Ayeleke RO, Roberts H (2016) Local oestrogen for vaginal atrophy in postmenopausal women. Cochrane Database Syst Rev 2016:Cd001500
90. Levin ER (2008) Rapid signaling by steroid receptors. Am J Physiol Regul Integr Comp Physiol 295:R1425–R1430
91. Li S, O'neill SRS, Zhang Y, Holtzman MJ, Takemaru K-I, Korach KS, Winuthayanon W (2017) Estrogen receptor α is required for oviductal transport of embryos. FASEB J 31:1595–1607
92. Li S, Herrera GG, Tam KK, Lizarraga JS, Beedle M-T, Winuthayanon W (2018) Estrogen action in the epithelial cells of the mouse vagina regulates neutrophil infiltration and vaginal tissue integrity. Sci Rep 8:11247
93. Lim H, Ma L, Ma WG, Maas RL, Dey SK (1999) Hoxa-10 regulates uterine stromal cell responsiveness to progesterone during implantation and decidualization in the mouse. Mol Endocrinol 13:1005–1017
94. Lubahn DB, Brown TR, Simental JA, Higgs HN, Migeon CJ, Wilson EM, French FS (1989) Sequence of the intron/exon junctions of the coding region of the human androgen receptor gene and identification of a point mutation in a family with complete androgen insensitivity. Proc Natl Acad Sci U S A 86:9534–9538
95. Lucas ES, Vrljicak P, Muter J, Diniz-Da-Costa MM, Brighton PJ, Kong CS, Lipecki J, Fishwick KJ, Odendaal J, Ewington LJ, Quenby S, Ott S, Brosens JJ (2020) Recurrent pregnancy loss is associated with a pro-senescent decidual response during the peri-implantation window. Commun Biol 3:37
96. Lydon JP, Demayo FJ, Conneely OM, O'malley BW (1996) Reproductive phenotpes of the progesterone receptor null mutant mouse. J Steroid Biochem Mol Biol 56:67–77
97. Maclean A, Bunni E, Makrydima S, Withington A, Kamal AM, Valentijn AJ, Hapangama DK (2020) Fallopian tube epithelial cells express androgen receptor and have a distinct hormonal responsiveness when compared with endometrial epithelium. Hum Reprod 35:2097–2106
98. Makieva S, Saunders PT, Norman JE (2014) Androgens in pregnancy: roles in parturition. Hum Reprod Update 20:542–559
99. Makieva S, Hutchinson LJ, Rajagopal SP, Rinaldi SF, Brown P, Saunders PT, Norman JE (2016) Androgen-induced relaxation of uterine myocytes is mediated by blockade of both Ca flux and MLC phosphorylation. J Clin Endocrinol Metab 101:jc20152851
100. Mangelsdorf DJ, Thummel C, Beato M, Herrlich P, Schutz G, Umesono K, Blumberg B, Kastner P, Mark M, Chambon P, Evans R (1995) The nuclear receptor superfamily: the second decade. Cell 83:835–839
101. Marshall E, Lowrey J, Macpherson S, Maybin JA, Collins F, Critchley HO, Saunders PT (2011) In silico analysis identifies a novel role for androgens in the regulation of human endometrial apoptosis. J Clin Endocrinol Metab 96:E1746–E1755
102. Martyn F, Mcauliffe FM, Wingfield M (2014) The role of the cervix in fertility: is it time for a reappraisal? Hum Reprod 29:2092–2098
103. Maseroli E, Cellai I, Filippi S, Comeglio P, Cipriani S, Rastrelli G, Rosi M, Sorbi F, Fambrini M, Petraglia F, Amoriello R, Ballerini C, Lombardelli L, Piccinni MP, Sarchielli E, Guarnieri G, Morelli A, Maggi M, Vignozzi L (2020) Anti-inflammatory effects of androgens in the human vagina. J Mol Endocrinol 65:109–124
104. Matsuzaki S, Fukaya T, Suzuki T, Murakami T, Sasano H, Yajima A (1999) Oestrogen receptor α and β mRNA expression in human endometrium throughout the menstrual cycle. Mol Hum Reprod 5:559–564
105. Maybin JA, Duncan WC (2004) The human corpus luteum: which cells have progesterone receptors? Reproduction 128:423–431
106. Mcclamrock HD, Adashi EY (1992) Gestational hyperandrogenism. Fertil Steril 57:257–274
107. Mcewan IJ, Mcguinness D, Hay CW, Millar RP, Saunders PT, Fraser HM (2010) Identification of androgen receptor phosphorylation in the primate ovary in vivo. Reproduction 140:93–104
108. Medina-Laver Y, Rodriguez-Varela C, Salsano S, Labarta E, Dominguez F (2021) What do we know about classical and non-classical progesterone receptors in the human female reproductive tract? A review. Int J Mol Sci 22:11278
109. Mehta FF, Son J, Hewitt SC, Jang E, Lydon JP, Korach KS, Chung S-H (2016) Distinct functions and regulation of epithelial progesterone receptor in the mouse cervix, vagina, and uterus. Oncotarget 7:17455–17467
110. Miller WL, Auchus RJ (2011) The molecular biology, biochemistry, and physiology of human steroidogenesis and its disorders. Endocr Rev 32:81–151
111. Mirkin S, Simon JA, Liu JH, Archer DF, Castro PD, Graham S, Bernick B, Komm B (2021) Evaluation of endometrial progesterone receptor expression after 12 weeks of exposure to a low-dose vaginal estradiol insert. Menopause 28:998–1003

112. Miyagawa S, Iguchi T (2015) Epithelial estrogen receptor 1 intrinsically mediates squamous differentiation in the mouse vagina. Proc Natl Acad Sci 112:12986–12991
113. Mueller MD, Vigne JL, Minchenko A, Lebovic DI, Leitman DC, Taylor RN (2000) Regulation of vascular endothelial growth factor (VEGF) gene transcription by estrogen receptors alpha and beta. Proc Natl Acad Sci U S A 97:10972–10977
114. Nutu M, Weijdegård B, Thomas P, Thurin-Kjellberg A, Billig H, Larsson DGJ (2009) Distribution and hormonal regulation of membrane progesterone receptors β and γ in ciliated epithelial cells of mouse and human fallopian tubes. Reprod Biol Endocrinol 7:89
115. Okada H, Tsuzuki T, Murata H (2018) Decidualization of the human endometrium. Reprod Med Biol 17:220–227
116. Owens LA, Kristensen SG, Lerner A, Christopoulos G, Lavery S, Hanyaloglu AC, Hardy K, Yding Andersen C, Franks S (2019) Gene expression in granulosa cells from small antral follicles from women with or without polycystic ovaries. J Clin Endocrinol Metab 104:6182–6192
117. Paech K, Webb P, Kuiper GG, Nilsson S, Gustafsson J, Kushner PJ, Scanlan TS (1997) Differential ligand activation of estrogen receptors ERalpha and ERbeta at AP1 sites. Science 277:1508–1510
118. Pawar S, Laws MJ, Bagchi IC, Bagchi MK (2015) Uterine epithelial estrogen receptor-α controls Decidualization via a paracrine mechanism. Mol Endocrinol (Baltimore, Md.) 29:1362–1374
119. Payne AH, Hales DB (2004) Overview of steroidogenic enzymes in the pathway from cholesterol to active steroid hormones. Endocr Rev 25:947–970
120. Penning TM, Drury JE (2007) Human aldo-keto reductases: function, gene regulation, and single nucleotide polymorphisms. Arch Biochem Biophys 464:241–250
121. Perusquia M, Navarrete E, Jasso-Kamel J, Montano LM (2005) Androgens induce relaxation of contractile activity in pregnant human myometrium at term: a nongenomic action on L-type calcium channels. Biol Reprod 73:214–221
122. Pieber D, Allport VC, Bennett PR (2001) Progesterone receptor isoform A inhibits isoform B-mediated transactivation in human amnion. Eur J Pharmacol 427:7–11
123. Polanski LT, Barbosa MA, Martins WP, Baumgarten MN, Campbell B, Brosens J, Quenby S, Raine-Fenning N (2014) Interventions to improve reproductive outcomes in women with elevated natural killer cells undergoing assisted reproduction techniques: a systematic review of literature. Hum Reprod 29:65–75
124. Poola I, Abraham J, Baldwin K, Saunders A, Bhatnagar R (2005) Estrogen receptors beta4 and beta5 are full length functionally distinct ERbeta isoforms: cloning from human ovary and functional characterization. Endocrine 27:227–238
125. Press MF, Greene GL (1988) Localization of progesterone receptor with monoclonal antibodies to the human progestin receptor. Endocrinology 122:1165–1175
126. Richards JS, Pangas SA (2010) The ovary: basic biology and clinical implications. J Clin Invest 120:963–972
127. Robker RL, Russell DL, Espey LL, Lydon JP, O'malley BW, Richards JS (2000a) Progesterone-regulated genes in the ovulation process: ADAMTS-1 and cathepsin L proteases. Proc Natl Acad Sci U S A 97:4689–4694
128. Robker RL, Russell DL, Yoshioka S, Sharma SC, Lydon JP, O'malley BW, Espey LL, Richards JS (2000b) Ovulation: a multi-gene, multi-step process. Steroids 65:559–570
129. Sanderson PA, Critchley HO, Williams AR, Arends MJ, Saunders PT (2017) New concepts for an old problem: the diagnosis of endometrial hyperplasia. Hum Reprod Update 23:232–254
130. Saunders PTK (2020) What have we learned from animal models of endometriosis and how can we use the knowledge gained to improve treatment of patients? Adv Anat Embryol Cell Biol 232:99–111
131. Saunders PT, Millar MR, Williams K, Macpherson S, Harkiss D, Anderson RA, Orr B, Groome NP, Scobie G, Fraser HM (2000) Differential expression of estrogen receptor-alpha and -beta and androgen receptor in the ovaries of marmosets and humans. Biol Reprod 63:1098–1105
132. Shaaban AM, Green AR, Karthik S, Alizadeh Y, Hughes TA, Harkins L, Ellis IO, Robertson JF, Paish EC, Saunders PT, Groome NP, Speirs V (2008) Nuclear and cytoplasmic expression of ERbeta1, ERbeta2, and ERbeta5 identifies distinct prognostic outcome for breast cancer patients. Clin Cancer Res 14:5228–5235
133. Shynlova O, Nadeem L, Zhang J, Dunk C, Lye S (2020) Myometrial activation: novel concepts underlying labor. Placenta 92:28–36
134. Simitsidellis I, Saunders PTK, Gibson DA (2018) Androgens and endometrium: new insights and new targets. Mol Cell Endocrinol 465:48–60
135. Sojka DK, Yang L, Yokoyama WM (2019) Uterine natural killer cells. Front Immunol 10:960–960
136. Springwald A, Lattrich C, Skrzypczak M, Goerse R, Ortmann O, Treeck O (2010) Identification of novel transcript variants of estrogen receptor alpha, beta and progesterone receptor gene in human endometrium. Endocrine 37:415–424
137. Stener-Victorin E, Padmanabhan V, Walters KA, Campbell RE, Benrick A, Giacobini P, Dumesic DA, Abbott DH (2020) Animal models to understand the etiology and pathophysiology of polycystic ovary syndrome. Endocr Rev 41:bnaa010
138. Stocco C, Telleria C, Gibori G (2007) The molecular control of corpus luteum formation, function, and regression. Endocr Rev 28:117–149
139. Thornton JW (2001) Evolution of vertebrate steroid receptors from an ancestral estrogen receptor by

ligand exploitation and serial genome expansions. Proc Natl Acad Sci U S A 98:5671–5676
140. Turco MY, Gardner L, Hughes J, Cindrova-Davies T, Gomez MJ, Farrell L, Hollinshead M, Marsh SGE, Brosens JJ, Critchley HO, Simons BD, Hemberger M, Koo BK, Moffett A, Burton GJ (2017) Long-term, hormone-responsive organoid cultures of human endometrium in a chemically defined medium. Nat Cell Biol 19:568–577
141. Turner KJ, Macpherson S, Millar MR, Mcneilly AS, Williams K, Cranfield M, Groome NP, Sharpe RM, Fraser HM, Saunders PT (2002) Development and validation of a new monoclonal antibody to mammalian aromatase. J Endocrinol 172:21–30
142. Tyagi S, Chan EC, Barker D, Mcelduff P, Taylor KA, Riveros C, Singh E, Smith R (2022) Transcriptomic analysis reveals myometrial topologically associated domains linked to onset of human term labor. Mol Hum Reprod 28(3):gaac003
143. VAN Weerden WM, Bierings HG, VAN Steenbrugge GJ, DE Jong FH, Schroder FH (1992) Adrenal glands of mouse and rat do not synthesize androgens. Life Sci 50:857–861
144. Wada-Hiraike O, Hiraike H, Okinaga H, Imamov O, Barros RP, Morani A, Omoto Y, Warner M, Gustafsson JA (2006) Role of estrogen receptor beta in uterine stroma and epithelium: insights from estrogen receptor beta-/- mice. Proc Natl Acad Sci U S A 103:18350–18355
145. Wagenfeld A, Saunders PT, Whitaker L, Critchley HO (2016) Selective progesterone receptor modulators (SPRMs): progesterone receptor action, mode of action on the endometrium and treatment options in gynecological therapies. Expert Opin Ther Targets 20:1045–1054
146. Walters KA, Handelsman DJ (2018) Role of androgens in the ovary. Mol Cell Endocrinol 465:36–47
147. Wang H, Critchley HO, Kelly RW, Shen D, Baird DT (1998) Progesterone receptor subtype B is differentially regulated in human endometrial stroma. Mol Hum Reprod 4:407–412
148. Wånggren K, Stavreus-Evers A, Olsson C, Andersson E, Gemzell-Danielsson K (2008) Regulation of muscular contractions in the human Fallopian tube through prostaglandins and progestagens. Hum Reprod 23:2359–2368
149. Weihua Z, Saji S, Mäkinen S, Cheng G, Jensen EV, Warner M, Gustafsson JA (2000) Estrogen receptor (ER) beta, a modulator of ERalpha in the uterus. Proc Natl Acad Sci U S A 3:5936–5941
150. Welsh T, Johnson M, Yi L, Tan H, Rahman R, Merlino A, Zakar T, Mesiano S (2012) Estrogen receptor (ER) expression and function in the pregnant human myometrium: estradiol via ERα activates ERK1/2 signaling in term myometrium. J Endocrinol 212:227–238
151. Whitaker LH, Murray AA, Matthews R, Shaw G, Williams AR, Saunders PT, Critchley HO (2017) Selective progesterone receptor modulator (SPRM) ulipristal acetate (UPA) and its effects on the human endometrium. Hum Reprod 32:531–543
152. Wilkens J, Male V, Ghazal P, Forster T, Gibson DA, Williams AR, Brito-Mutunayagam SL, Craigon M, Lourenco P, Cameron IT, Chwalisz K, Moffett A, Critchley HO (2013) Uterine NK cells regulate endometrial bleeding in women and are suppressed by the progesterone receptor modulator asoprisnil. J Immunol 191:2226–2235
153. Winuthayanon W, Hewitt SC, Orvis GD, Behringer RR, Korach KS (2010) Uterine epithelial estrogen receptor α is dispensable for proliferation but essential for complete biological and biochemical responses. Proc Natl Acad Sci U S A 107:19272–19277
154. Winuthayanon W, Bernhardt ML, Padilla-Banks E, Myers PH, Edin ML, Lih FB, Hewitt SC, Korach KS, Williams CJ (2015) Oviductal estrogen receptor α signaling prevents protease-mediated embryo death. elife 4:e10453–e10453
155. Winuthayanon W, Lierz SL, Delarosa KC, Sampels SR, Donoghue LJ, Hewitt SC, Korach KS (2017) Juxtacrine activity of estrogen receptor α in uterine stromal cells is necessary for estrogen-induced epithelial cell proliferation. Sci Rep 7:8377–8377
156. Yilmaz BD, Bulun SE (2019) Endometriosis and nuclear receptors. Hum Reprod Update 25:473–485

Nuclear Receptors in Ovarian Function

3

Doan Thao Dinh and Darryl Lyndon Russell

Abstract

The ovary undergoes cycles of hormone production that regulate physiological changes necessary for folliculogenesis, ovulation and luteinisation, ultimately contributing to female reproductive success. Crucial to these biological processes is stage-specific nuclear receptor signalling. While the transcriptional regulatory roles of steroid receptors in female fertility and especially ovarian functions have long been documented, non-steroid receptors also play an important part in regulating gene expression at various stages of ovarian development. The recent application of high-throughput genomic and transcriptomic technologies has begun to shed light on the molecular mechanisms underlying ovarian nuclear receptor actions and pointed to a complex interplay between highly specific transcription co-regulators as well as between nuclear receptors in mediating mutual as well as unique target genes. Interrelationships between nuclear receptors as well as the involvement of context-specific protein and non-protein co-regulators are likely keys to the precise and specific nuclear receptor action in the ovary. Leveraging such knowledge on the nuclear receptor network is especially valuable in the development of novel fertility treatments as well as female contraceptives.

Keywords

Reproductive biology · Steroid receptor · Nuclear receptor · Ovary · Ovulation · Female reproduction

3.1 Introduction

The ovary is responsible for ensuring female reproductive success through the generation of viable oocytes for fertilisation and development as well as the production of hormones that coordinate the reproductive cycle and support pregnancy and lactation. Critical ovarian functions include follicle development (folliculogenesis), oocyte maturation, ovulation and luteinisation. Crucial to the precise regulation of all ovarian functions is the involvement of reproductive hormones; in particular, reproductive steroids and their receptors are the archetypal hormone network. Ligand-activated receptors provide an elegant mechanism for communication between different organs or cell types to control and coordinate the many critical reproductive processes. In the ovary multiple nuclear hormone receptors, including steroid and non-steroid receptors, are activated at specific stages and regulate a com-

D. T. Dinh · D. L. Russell (✉)
Robinson Research Institute, School of Biomedicine, Faculty of Health & Medical Sciences, The University of Adelaide, Adelaide, SA, Australia
e-mail: Darryl.russell@adelaide.edu.au

M. J. Campbell, C. L. Bevan (eds.), *Nuclear Receptors in Human Health and Disease*, Advances in Experimental Medicine and Biology 1390, https://doi.org/10.1007/978-3-031-11836-4_3

plex network of signalling pathways via their target genes.

3.2 Hormonal Control of Dynamic Physiological Change in the Ovary

(a) Folliculogenesis:

The ovarian follicle is composed of an oocyte surrounded by somatic cells – mural granulosa cells, cumulus cells and theca cells. The complex interactions between each of these compartments, often involving steroid hormone signals, are vital for ovarian functions. A typical ovarian follicular cycle is illustrated in Fig. 3.1. Folliculogenesis initiates prior to birth and in the developing ovary, oocytes enter meiosis and germ cell division is arrested early at meiosis I prophase in prenatal development [1]. At, or shortly after birth, meiotically arrested oocytes are assembled into primordial follicles in which they are surrounded by a layer of flat, un-differentiated pre-granulosa cells [2]. Follicle growth and development (folliculogenesis) is sporadically initiated each day in a small number of primordial follicles. During the early stage of follicle development, granulosa cells also display morphological changes, becoming cuboidal and proliferative [3]. During early follicle growth granulosa cells are not steroidogenic, but as the follicle grows, specialised stromal cells called theca cells are recruited from a progenitor pool in the ovarian stroma, then proliferate and differentiate to form the theca layer surrounding the exterior of the follicle. Theca cells express steroidogenic enzymes that are necessary to convert cholesterol to testosterone (T) under the control of luteinising hormone (LH) [4]. This testosterone is secreted and taken up by granulosa cells, which convert it to estrogen (E2) via the P450 Aromatase enzyme encoded by the ***Cyp19a1*** gene, regulated by follicle stimulating hormone (FSH) from the pituitary. This regulation of two independent cell types by two distinct gonadotrophins, known as the two-cell two-gonadotropin theory, provides exquisite control of the regular female hormone cycle driven by the developmental status of the ovarian follicles. E2 produced by growing follicles acts on the pituitary to repress FSH production, while also stimulating GnRH synthesis and release by the hypothalamus, thus promoting the release of LH pulses. As a result, granulosa cells of dominant follicles acquire FSH-independent growth and development, while rising LH levels further stimulate theca cells and begin to also act on granulosa cells, promoting their differentiation to preovulatory stage. During folliculogenesis, granulosa cell specification and the formation of the fluid-filled antral space also lead to the differentiation between cumulus cells, which immediately surround the oocyte and are important in promoting oocyte growth and developmental competence, and mural granulosa cells which are involved in steroid and protein hormone production in response to FSH and LH [5].

(b) Ovulation:

Continued rising E2 from preovulatory follicles causes larger and more frequent pulses of LH release from the pituitary until the pulses merge to become the mid-cycle LH-surge. Preovulatory ovarian follicles respond to the LH surge, resulting in a number of dynamic morphological, molecular and biochemical events in preparation for the release of the mature oocyte into the oviduct and potential fertilisation, embryo development and implantation [6]. A multifaceted interplay between different components of the pre-ovulatory follicle, including oocytes and their surrounding somatic cells, has to be coordinated to achieve ovulation. In oocytes, meiotic resumption occurs leading to the extrusion of the first polar body, which carries half of the genetic material, and the second meiotic arrest at MII stage. At the same time, the surrounding cumulus cell layers produce a specialised extracellular matrix (ECM), causing the cumulus oocyte complex (COC) to expand and increase in volume, as well as gaining additional migratory and invasive properties which are necessary for ovulation [7]. The COC, containing a mature oocyte, is then released into the oviduct from the peri-ovulatory

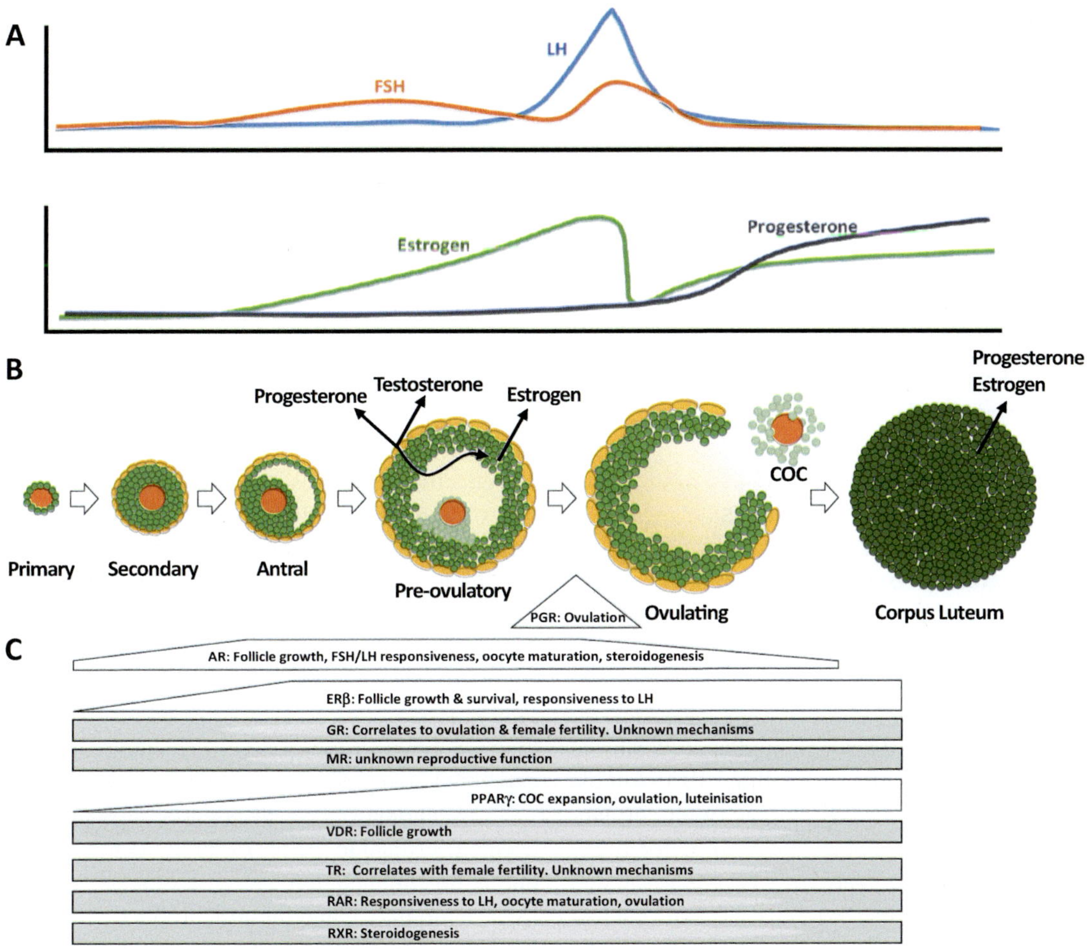

Fig. 3.1 Follicle development and nuclear receptor action in the ovary. (**a**) Circulating levels of gonadotropins; follicle stimulating hormone (FSH) and luteinising hormone (LH) and steroids associated with stages of follicle development (folliculogenesis). (**b**) During folliculogenesis, follicles grow and differentiate, form the antrum cavity, granulosa cells (green) diverge into mural granulosa and cumulus cells. Steroidogenic theca cells (yellow) produce progesterone, most of which is converted to testosterone. Testosterone diffuses to the granulosa cell layers, where the aromatase enzyme converts it to estrogen. The rise in circulating estrogen stimulates the hypothalamus and pituitary, prompting the LH surge which triggers release of the mature oocyte from the follicle (ovulation), while residual granulosa cells luteinise, forming the highly steroidogenic corpus luteum which secretes progesterone to support implantation and pregnancy. (**c**) Expression and role of nuclear receptors at different stages of folliculogenesis. Position and size of boxes reflect the temporal expression of each nuclear receptor in granulosa cells. Nuclear receptors that are present in granulosa cells but do not have a well-described temporal expression pattern are greyed

follicle at the follicle apex. For this to occur, the physical cellular barrier of the follicle, composed of multiple layers of ECM as well as granulosa, theca and surface epithelial cells, needs to be thinned and broken down through tissue remodelling. This involves many concurrent processes, including proteolytic degradation of ECM layers, surface epithelial cell apoptosis, immune cell recruitment and theca cell migration [6]. Aside from tissue remodelling, precisely-timed muscle contraction as well as vasocontraction at the apex are also required for the release of the oocyte. The ovulatory surge of LH also induces terminal differentiation of granulosa cells into highly steroidogenic luteal cells, which synthesise cholesterol and convert it to progesterone that acts on the uterus to promote implantation and gestation. Another important part of the tissue remodelling process is the generation of new vasculature around the periovulatory follicle, which is necessary for the formation of the corpus luteum (CL) from the ovulated follicle by providing nutrients and hormones to the developing CL, and providing ready access for highly active hormone secretion from the CL to reach circulation.

3.3 Physiological Effects of Nuclear Hormone Receptors on Ovarian Functions

Nuclear hormone receptors are a family of ligand-dependent transcription factors that are usually activated through binding with steroid hormones or other signalling lipid-soluble molecules and directly interact with chromatin. Despite the name, the ligands for many nuclear receptors are as yet unknown and these are thus referred to as 'orphan' nuclear receptors. In addition to genomic actions, many are also known to have non-genomic roles in various contexts [8, 9]. While several orphan receptors have important ovarian roles, in particular SF1 for early ovarian development and LRH1 for folliculogenesis and ovulation, for the purpose of this review, only hormone receptors with well-described ligands and their genomic actions will be considered, with orphan receptors having been reviewed elsewhere [10]. The ligand activated receptors are grouped into two classes:

(a) Steroid receptors (SR), which are steroid hormone-binding transcription factors (NR3 family) including progesterone receptor (PGR), estrogen receptor (ER), androgen receptor (AR), glucocorticoid receptor (GR) and mineralocorticoid receptor (MR).

(b) Non-steroid receptors, loosely including transcription factors not in the NR3 family that bind and are regulated by ligands that are lipid permeable compounds, such as vitamins, lipid metabolites or retinoids. These include peroxisome proliferator-activated receptor (PPAR), thyroid hormone receptor (TR), vitamin D receptor (VDR), retinoic acid receptor (RAR) and retinoid X receptor (RXR).

(a) Steroid receptors in the ovary:

The steroid hormones are classical regulators of reproductive processes. Highly regulated secretion of hormones and expression of their receptors enable communication and exquisite coordination of the functions of the different reproductive organs in preparation for fertilisation and pregnancy. The ovary is the primary source of estrogen, androgens and progesterone in females and ovaries are themselves responsive to these signals through steroid receptors that are expressed at key developmental stages.

Progesterone (P4) is an essential reproductive hormone critical in the ovary for ovulation, in the uterus for implantation and gestation, as well as in the mammary gland for milk production. In the final stages of ovarian follicle maturation, differentiated granulosa cells of preovulatory follicles respond to the LH surge and begin to express *Cyp11a1,* which encodes the P450 side chain cleavage enzyme that is rate-limiting for P4 production [11]. This results in steadily increasing P4 secretion by the ovarian follicular granulosa cells immediately prior to ovulation, which continues to rise as the follicle luteinises and remains high throughout gestation. P4 mainly functions through the direct binding and activation of its cognate receptor PGR, a nuclear steroid receptor that has profound importance in the regulation and maintenance of normal female reproductive physiology. In different female reproductive tissues, PGR responding to rising P4 secretion from the ovary shows distinct functions that are highly dependent on each tissue context, revealed in studies on PGR knockout (KO) mouse models

[12, 13]. In the pre-ovulatory ovary, PGR is expressed exclusively in granulosa cells and is highly induced in response to the ovulatory LH-surge [14]. PGRKO female mice are infertile due to complete anovulation [12]. Likewise, treatment with PGR antagonist results in ovulation suppression in rodents and humans [15–17], and silencing of ovarian PGR expression results in ovulation disruption in macaques [18]. Luteinisation of follicles is unaffected in PGRKO or PGR-antagonist treated models with the resulting CLs containing entrapped oocytes, indicating that ovulation is specifically dependent on PGR action in the ovary [12]. PGR is a key regulator of a number of ovulatory genes that are involved in tissue remodelling (*Adamts1*), cumulus expansion (*Areg, Ereg*) and also acts upstream to other ovulatory transcription factors (*Pparg, Hif1a*) [19]. PGR includes two main isoforms, PGR-A and PGR-B, both of which are present in most PGR-positive cells. Even though both isoforms are expressed in granulosa cells of pre-ovulatory follicles, PGR-A is credited as the more essential isoform in ovulation, as determined from studies on null mouse models that are specific to each PGR isoform [20, 21]. Female mice that have a mutation which prevents production of functional PGR-A exhibit a specific failure of follicle rupture, but not luteinisation, even after gonadotropin stimulation. However, female mice lacking PGR-B have normal ovulation and fertility. Analysis of total and isoform-specific knockout granulosa transcriptomes indicates that such phenotypic properties are a result of broad differences in gene expression patterns that are driven by PGR-A and not PGR-B, in which PGR-B deletion had very limited impact on gene expression in LH-stimulated ovaries, while PGR-A deletion caused very similar gene expression changes to the total PGRKO [22]. The role of PGR on oocyte development, however, is less clear. Oocytes from total PGRKO mice that are extracted from preovulatory ovaries and subjected to *in vitro* maturation are capable of COC expansion, fertilisation and developing into normal pups [23]. Furthermore, while there is *in vitro* evidence that PGR antagonist treatment has detrimental effects on cumulus expansion in pigs [24], there is no evidence for a role for PGR in human cumulus cells or oocyte maturation.

A direct intraovarian role for estrogen to promote FSH-independent survival, proliferation and differentiation of granulosa cells is well known [25]. In granulosa cells, estrogen receptor β (ERβ) is expressed at all stages of development, from the secondary follicle stage onwards to CL [26, 27]. ERα, however, is not found in granulosa cells but rather in theca and interstitial cells. Correspondingly, it has been shown through a number of mouse models that ERβ is the more important form in ovulation. Knockout of ERβ in female mice results in reduced cumulus expansion, ovulation and corpus luteum formation and hence reduced litter size, which cannot be rescued through gonadotrophin stimulation [28]. A number of FSH-regulated genes, including the LH receptor-encoding gene *Lhcgr* and LH-regulated downstream target genes, show disrupted expression in ERβKO granulosa cells [29]. Thus, in response to E2, ERβ mediates a gene expression profile that is required for granulosa cell differentiation to the fully LH-responsive preovulatory stage. ERβ also has a role in supporting the emergence of dominant follicles and their progression to become preovulatory follicles [30]. This is in contrast to the ERαKO model, in which anovulation can be ameliorated through exogenous gonadotrophin stimulation, indicating that the key role for ERα is in the regulation of gonadotropin release from the pituitary [31]. The involvement of non-classical ERα actions in fertility has also been investigated in separate transgenic mouse models carrying point mutations in the LBD or AF-2 region of ERα respectively, resulting in disrupted ERα ligand binding function and plasma membrane association. Both of these mouse models showed a similar reversible anovulation phenotype due to defects in survival and proliferation of granulosa cells and theca cells. This suggested that ERα can have extranuclear and ligand-independent ovarian functions [32, 33]. Furthermore, theca-specific KO of ERα leads to a less severe reproductive phenotype, where aberrant oestrus cycling pattern results in more pronounced fertility decline in older female mice, further indicating that ERα has only a minor role in the theca, and is most important in regulating gonadotropin release to mediate reproduction [34]. In breast cells and in the endometrium, ER has been shown to be immediately upstream of PGR expression through direct binding of ER to response elements within the PGR promoter [35]. Such sites are dispensable for PGR expression in granulosa cells [36]. Rather, the effect of ERβ on PGR expression in this context is more likely indirect, through the mediation of LH receptor expression, which is required for LH-induced PGR induction as shown through transcriptomic analysis of ERβKO vs WT granulosa cells [29, 37]. A similar pathway is likely the mechanism by which ERβ regulates other ovulatory transcription factors, including RUNX1 and RUNX2. Rather than having a direct role in ovulation, transcription analysis of ERβ KO vs WT in pre-ovulatory follicles indicates that ERβ is required for growth and development of follicles, in particular steroidogenesis and the PKA-cAMP signalling pathway that is responsive to FSH.

Androgen receptor (AR) is a nuclear hormone receptor closely related to PGR, with very similar protein structure and DNA binding sequence specificity [38]. In the ovary, AR is expressed in the oocyte, cumulus, granulosa and theca cells at most stages throughout folliculogenesis [39]. Androgens, the key ligands of AR, are synthesised in the ovarian theca cells which express the rate limiting steroidogenic enzyme *Cyp17a1* under the control of LH [40]. Treatment with the AR ligand T or the non-aromatisable AR ligand dihydrotestosterone (DHT) promotes follicle growth *in vitro* [41, 42]. T is also required for *Fshr* and *Lhcgr* expression and hence for the induction of PGR in cultured granulosa cells [43]. *In vivo* treatment with non-aromatisable ligand DHT also stimulates the expression of LH-responsive ovulatory genes, indicating this is a direct effect of androgen, not its conversion (via aromarisation) to E2 [40]. Global KO of AR in mice results in overall poorer female fertility, with a reduction in antral follicle count, impaired oocyte maturation and reduced expression of steroidogenesis genes [44]. When AR is knocked out specifically in granulosa cells, defective folliculogenesis is again observed as well as disruption in steroidogenesis and the estrus cycle [45, 46]. However, in young mice ovulation can be rescued with exogenous gonadotropin, suggesting that AR also has non-ovarian reproductive functions. Knockout of AR in other ovarian cell types, including the oocyte and theca cells, has no effect on female fertility [45, 47], indicating that only AR action in granulosa cells is compulsory for female reproduction. Another key physiological focus on androgen action in the ovary is in the aetiology of polycystic ovary syndrome (PCOS), which is linked to elevated androgen exposure during development and affects androgen levels, metabolism, insulin sensitivity, fat deposition, risk of cardiovascular disease and many other diseases in adults. In the ovary, elevated T levels cause arrested follicle growth at the antral stage, leading to an accumulation of immature cystic follicle structures – which gave the condition its name – and resulting in failure to ovulate, hence sub-fertility [48]. Ablation of AR in neuronal cells can ameliorate many of the features of PCOS, indicating that the effects of androgen excess are multifactorial and includes effects on the central nervous system [49]. Thus the balance of AR signalling appears to be important for fertility regulation, with either too low or too high stimulation being detrimental [50].

Corticosteroid receptors include glucocorticoid receptor (GR) and mineralocorticoid receptor (MR), which are activated by the adrenal hormones cortisol and aldosterone. GR is expressed in oocytes and granulosa cells [51] and is shown in macaques to be LH-induced [52]. Due to the lethal effect of GR knockout in mouse, little is known about the role of GR in the context of reproduction. The recent generation of a viable GRKO model in zebrafish has begun to indicate a role of GR in female fertility, as GRKO female fish display reduced ovulation and fertilisation rate [53]. However, it is unknown whether this is a specific consequence of ovarian GR ablation or whether it is due to systemic lack of GR. MR is reported to have different expression patterns in the ovary depending on the species [52, 54], however its roles remain unknown.

In summary, the ovary is the primary source of female reproductive hormones, while specialised spatio-temporal expression of corresponding SR are also mediated through hormonally controlled mechanisms. As a consequence, ovarian functions are tightly governed by steroid hormones and corresponding SR, resulting in the highly-coordinated regulation of folliculogenesis, ovulation, oocyte maturation and luteinisation. While the roles of of ERβ, PGR and AR in female fertility have been described extensively, the roles of GR and MR remain largely unexplored.

(b) Non-steroid nuclear receptors:

Several families of nuclear receptors that are structurally related to the steroid receptor family but are regulated by ligands linked to cell homeostasis and metabolism, such as lipid derivatives, fatty acids or vitamins, are also expressed in the ovary. These ligand-receptor interactions play important paracrine roles in regulation of folliculogenesis and ovulation.

The peroxisome proliferator activated receptor (PPAR) family, consisting of PPARα, PPARδ and PPARγ, has fatty acids and prostaglandins as its activating ligands. PPARα and PPARδ are present in theca and stromal cells [55], while PPARγ is expressed in mouse granulosa cells at most stages of follicular development [55] and is induced through a PGR-dependent mechanism after the ovulatory LH-surge [56]. Consequentially, PPARγ has been shown to be critical for ovarian functions. A granulosa-specific PPARγ KO mouse model has dramatically impaired ovulation as well as reduced CL formation and progesterone production [56]. Evidence suggests there are species differences in the role of PPARγ during ovulation, since the expression of PPARγ mRNA has been shown to be reduced after ovulation induction in macaque [57] and rat granulosa cells [55]. While the pattern of regulation in human is yet to be demonstrated, pharmacological activators of PPARγ have been shown to improve ovulation in women with PCOS [58]. In mice, treatment in the peri-ovulatory stage with agonists of PPARγ have also been shown to improve the developmental competence of oocytes impacted by metabolic disturbance [59]. PPARα and PPARδ have not been found to participate in the regulation of reproduction in genetic ablation models, with PPARα KO mice being fertile [60] and PPARδ KO being embryonically lethal [61]. A number of PGR-regulated genes are now recognised to be downstream of PPARγ during ovulation in mice, including *Edn2* and *Il6*, which are important in smooth muscle contraction and cumulus expansion [56]. PPARα and PPARγ are also present in ovarian macrophages, where expression of the inflammatory mediator *Nos2* is regulated by PPAR agonist [59].

The nuclear receptor for vitamin D (VDR) is expressed in granulosa cells [62] and associated with follicle growth and granulosa cell proliferation [63]. The ablation of VDR in mice thus results in female infertility due to impaired folliculogenesis [64]. In some reports, this reproductive phenotype can be ameliorated through a calcium-supplemented diet, however other data contradict this suggestion [63]. In a pathology context, vitamin D signalling has also been linked to PCOS, and vitamin D supplement has been shown to be beneficial in PCOS patients in improving glucose and lipid metabolism, testosterone level, insulin resistance and ovarian follicle development [65–69].

The thyroid hormone receptor (TR) family consists of isoforms TRα and TRβ, both of which are expressed in oocytes, granulosa cells and theca cells at different stages of follicle development [70]. For TRα, an alternative splicing isoform (TRα-2) is more important for female reproduction, as shown through impaired fertility in TRα-2 KO female mice [71]. Mice that have either TRα-1 (the canonical TRα isoform) or TRβ ablated are reported to have normal fertility [72, 73], however double KO of both transcription factors results in reduced fertility rate [74], alluding to the existence of a shared mechanism of TRα-1 and TRβ in regulating female reproduction which until now has remained unexplored. Additionally, a recent report has suggested a correlation between TRα in human granulosa cells and fertility, in which TRα-2 mRNA level is higher in infertile women and TRα expression is negatively correlated to *Has2* and *Ptgs2* [75].

The three subtypes of retinoic acid receptor (RAR) – RARα/RARβ/RARγ – are expressed in the ovary, specifically in granulosa cells and oocytes [76], and the role of RA in folliculogenesis and granulosa cell functions has been indicated in a number of studies. In granulosa cells, treatment with RA promotes the expression of LHR through inducing *Lhcgr* promoter demethylation in a granulosa cell-specific manner [77]. Mice given a vitamin A-deficient diet or treated with an inhibitor of alcohol dehydrogenase (required for RA conversion) show reduced ovulation and oocyte maturation rate [78]. Using a lacZ reporter mouse model, it has been shown that RA acts through the activation of RAR, but this did not differentiate between different RAR isoforms. Conversely, triple KO of all three RAR in granulosa cells does not affect fertility [79], thus necessitating further studies into the mechanism through which RA regulates ovarian functions. Another nuclear receptor with little known reproductive function is RXR. Activated by a number of retinoid molecules, there has been little to no research on the involvement of the three RXR proteins (RXRα, RXRβ, RXRγ) in ovarian functions, although RXRβ and RXRγ KO mice reportedly reproduce normally [80]. In granulosa cells, PPARγ and RXR have been shown to regulate the ovary specific promoter of *Cyp19a1*, thus modulating E2 production [81].

Together with steroid hormones, non-steroid ligands and their nuclear receptors play diverse roles in follicle development and ovulation, as summarised in Fig. 3.2. However, in contrast to SR which have been the focus of reproduction biology for many years, details on the importance and mechanism of non-steroid receptors in female fertility are largely absent from the literature, apart from more recent works on the PPAR family.

3.4 Signalling Mechanism of Nuclear Receptors in the Ovary

(a) Steroid receptor genome interactions in the ovary

As discussed, most steroid ligands are produced in the ovary, hence local concentrations of steroids are elevated at certain stages of the reproductive cycle, with their receptors being expressed under the control of reproductive hormones including steroids and gonadotropins. Upon activation by ligand binding, steroid receptors dimerise and, if cytoplasmic, translocate from the cytoplasm to the nucleus [82]. As in all target tissues, SR in the ovary mainly exert their effects through directly binding DNA at specific hormone nuclear receptor response element sequences (HRE), leading to the transcriptional induction or repression of specific genes. The canonical response elements bound by PGR, AR, GR and MR are minor variations on a highly similar core motif (5′-GnACAnnnTGTnC-3′), whereas ER utilises a different motif (ERE, 5′-AGGTCAnnnTGACCT-3′). Further characterisation of specific binding motif preference found that sequences flanking the core HRE motif are important for GR and AR specific interaction [83, 84], but this has not been elaborated for PGR. These motifs are present in the regulatory regions (promoter or enhancer) of many target genes and bound by specific activated receptors to regulate transcription. The influence of SR is not only restricted to genes with full consensus HRE, as SR are also recruited to regions with HRE half-sites [85–87], or can be tethered to chromatin through interaction with other DNA-binding transcription factors [88, 89].

Growing evidence that SR can interact with target chromatin sites through tethering at non-canonical motifs adds complexity to the mechanisms of the transcriptional regulation by SR. In

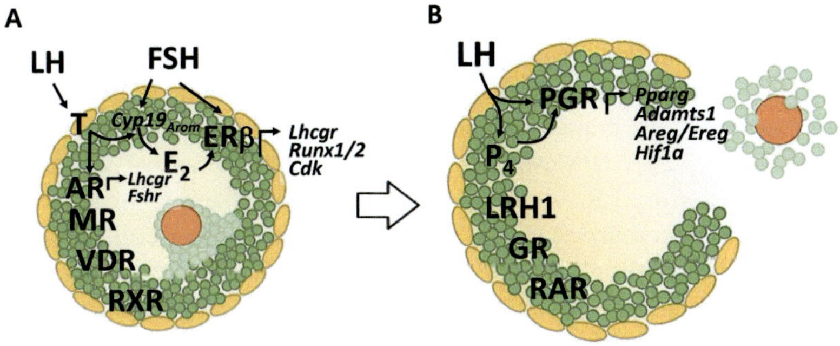

Fig. 3.2 Hormone regulation of gene expression during folliculogenesis and ovulation. (**a**) During folliculogenesis gonadotrophins LH and FSH promote production of testosterone (T) and its conversion by Cyp19a1 (aromatase) to estradiol (E2) in theca and granulosa cells respectively. Testosterone and E2 in turn act through their nuclear receptors, androgen receptor (AR) and estrogen receptor beta (ERβ), promoting expression of FSH and LH receptors (*Fshr, Lhcgr*), cell proliferation genes such as cyclin dependent kinases (*Cdk*), and others. Mineralocorticoid receptor (MR), vitamin D receptor (VDR) and Retinoid X receptor (RXR) also contribute to control of folliculogenesis. (**b**) The ovulation activating LH-surge induces high expression of progesterone receptor (PGR) which stimulates a cascade of ovulation genes including peroxisome proliferating receptor gamma (*Pparg*), A disintegrin and metalloproteinase-1 (*Adamts1*), Amphiregulin and Epireguliln (*Areg, Ereg*) and Hypoxia induced factor-1 alpha (*Hif1a*). The orphan receptor LRH1, as well as glucocorticoid receptor (GR) and retinoic acid receptor (RAR) are also important regulators of ovulatory gene expression

such cases, the cooperation between SR and other DNA-binding transcription factors in a context-specific manner is crucial, as is the role for each transcription factor in the recruitment of the transcriptional machinery. Despite the importance of steroid hormones and their receptors in female fertility, the unique molecular mechanisms that distinguish their ovarian functions from other hormone-responsive organs remain largely unexplored. Recently, the unique roles of PGR in different female reproductive tissues have been identified at the cistromic and transcriptomic levels. PGR displays specific preferences for DNA binding in each tissue, which results in tissue-specific gene regulation patterns. A study comparing PGR cistromes between T47D breast cancer cell line versus primary leiomyoma found less than 15% overlap in PGR-binding sites [90]. Similarly, less than 10% of PGR binding sites were found to be shared between progesterone-responsive granulosa cells and uterine tissue, which leads to the regulation of distinct sets of genes in different female reproductive tissues with little overlap [14]. Further exploration into the chromatin binding patterns of PGR in each context discovered a strong preference for proximal promoter regions (within 3 kb of transcription start sites) in granulosa cells but not in the uterus. Additionally, a predilection for interaction with distinct non-canonical motifs was also indicated in granulosa cells, suggesting direct interaction of PGR with AP1 and RUNX transcription factors in an ovarian-specific context [14]. Apart from regulating gene expression through promoter binding, PGR also shows the potential to mediate enhancer action through binding non-promoter regions. For example, in granulosa cells PGR binds a number of chromatin sites within *Zbtb16* intronic bodies, including sites previously shown to have enhancer action that promotes the expression of *Zbtb16* [91]. PGR chromatin binding is highly associated with chromatin accessibility, as demonstrated through ATAC-seq and H3K27ac ChIP-seq of mouse peri-ovulatory granulosa cells [14, 22]. Importantly, PGR shows an active role in driving chromatin accessibility and does not only take advantage of pre-accessible chromatin sites, suggesting PGR-chromatin binding is not dependent on a pioneer factor [22]. Although several studies have focused on AR and GR chromatin binding [92–94], none has been performed in the context of the ovary. Given the stark differences in SR action in different tissue contexts, investigation into the ovarian cistromic action of these SR will be required to fully understand their importance in ovarian functions.

ERβ plays a critical role in granulosa cell specification and function [29]. However, the activation of gene expression by ERβ in granulosa cells is dependent on the presence of FOXL2, another granulosa cell specification factor [95, 96]. This guidance of chromatin binding by cell-specific co-factors could explain the mechanism for unique transcriptional activity of ERβ in granulosa cells, however to date there has been no systematic comparison of ERβ and FOXL2 binding sites. Similarly, FOXL2 has also been implicated in AR action in granulosa cells [95]. The exact mechanism for such involvement remains unknown, however in prostate and mammary gland the related transcription factor FOXA2 has been shown to play a vital pioneer function for ER and AR [97].

(b) Steroid receptor isoforms:

Many SR are expressed in various different isoforms as a result of diverse translation initiation sites or alternative transcript splicing from a single gene. The two main PGR isoforms, A and B, generated from different translational start codons, have long been the focus of attention due to their discrete roles in different reproductive tissues. The longer PGR-B isoform includes the additional activation function-3 (AF-3) transactivation sequence in the N-terminal region, which mediates different co-regulator interactions [98]. This results in a higher transactivation capacity for PGR-B compared to PGR-A, and specific transcriptomes governed by each isoform. Not only does each PGR isoform exhibit discrete tissue-specific functions, the interplay between the isoforms can be highly complex and is precisely regulated in a spatiotemporal pattern and tissue-specific manner. In the ovary, both PGR-A and PGR-B are present and induced in response to the LH surge, with PGR-A being slightly predominant [20]. In the context of cancer, the balance of PGR-A:PGR-B ratio is important for cellular responses and the elevation of tumour development [99]. Interestingly, elevated PGR-A abundance can cause trans-repression of not only PGR-B but also other SR including GR and ER, without affecting their expression level [100]. Attempts have been made to elucidate the nature of such trans-repressive function, however the exact nature of the inhibitory process, such as the involvement of other co-repressors or the effect on PGR-B stability, is still poorly understood. In the uterus, this auto-inhibitory function plays an important role during parturition, in which uterine progesterone withdrawal induces PGR-A trans-repression of PGR-B function, leading to an upregulation in contraction and inflammation genes and consequently to the onset of labour [101, 102]. Whether such a mechanism also influences PGR action in granulosa cells remains unknown.

AR, GR and MR can also be translated in multiple isoforms, with the two main isoforms of AR and MR generated through separate translation start sites. For AR, it has been shown in the human ovary that the abundance of the full-length AR-B outweighs that of the slightly more truncated AR-A [103]. Like PGR, AR-A and AR-B are also shown to be functionally diverse [104]. Less is known about MR isoforms and isoform-specific expression pattern in the ovary; however it has been shown that MR-A possesses stronger transactivation action than MR-B [105]. The main isoforms of GR are GRα and GRβ, generated through alternative splicing events, and within each isoform multiple variants can arise based on different translation start sites. GRβ can act as a dominant negative inhibitor of GRα at glucocorticoid-responsive target genes [106]. The composition and dynamics of GR in the ovary and whether different GR isoforms are involved in the mediation of GR action in the ovary remains a mystery. The ERα and ERβ isoforms are expressed from separate genes and are less commonly found in the same cell types. In the ovary in particular, *Esr2*, which encodes ERβ, plays the predominant role in granulosa cells mediating folliculogenesis, while *Esr1* encoding ERα is more predominantly expressed in theca cells [26, 27], thus it is less likely that the two isoforms are directly functionally linked in the ovarian context.

(c) Steroid receptor protein interactions:

A wide range of coactivators and corepressors has been associated with SR in various biological contexts. A classic coactivator family is the aptly named SR coactivators (SRC), whose members, especially the earliest known coactivators SRC-1, SRC-2 and SRC-3, were identified through their ability to bind SR upon ligand activation and mediate SR transcriptional activation [107, 108]. This ability to promote SR transactivation is explained by the histone acetyltransferase activity of SRC [109]; furthermore, SRC can also interact with other histone modifiers, thus promoting additional chromosomal modifications in preparation for transcription. One key example is CBP/p300, which can act in synergy with SRC-1 to promote PGR and ER activation of gene expression *in vitro* [110]. Recent work on the synergy of the ER/SRC/CBP interaction has further elucidated the relationship between different components of the nuclear receptor-related transcription complex, in which SRC-3 proteins act as linkage between ER and CBP/p300 that in turn acetylates nearby histones and facilitates chromatin accessibility and gene transcription [111]. The expression of SRC1–3 as well as SRA and the corepressors NCOR and SMRT has been demonstrated in the ovary as well as in granulosa tumor cells [112]. However, to date there is insufficient study on the expression and actions of the SRC family during ovarian folliculogenesis, thus this aspect of steroid action remains not fully understood.

Aside from recruiting chromatin remodellers and components of the basal transcription complex, SR can also interact with members of other DNA-binding transcription factor families, which can enable tethering to non-canonical motifs or cooperative mechanisms that lead to the targeting of an expanded range of genes without the HRE motif. In the ovary, the identification of specific PGR binding partners at PGR-bound chromatin sites in individual PGR-regulated genes led to the suggestion that PGR interacts with SP1 related transcription factors [88]. Further genome-wide assays identified enrichment of AP1 and RUNX motifs at PGR bound sites [14]. Such studies have indicated a specific suite of transcription factors that are likely to be involved in PGR regulation of ovarian function. This PGR-RUNX interaction has to date only been identified in granulosa cells, suggesting that this may be a tissue-specific mechanism of hormone action. PGR colocalisation with both RUNX1 and RUNX2 in response to ovulatory cues was demonstrated through proximity ligation assay and comparative ChIP-seq analysis showed that PGR and RUNX1 chromatin binding regions closely overlapped, sharing a high number of mutual chromatin binding sites as well as downstream target genes. These findings illustrate physical and functional interactions of PGR and RUNX1/2. At the same time, PGR was also shown to interact with members of the JUN/FOS and NR5A families, members of which are also expressed in ovarian granulosa cells and play a role in ovulation. Whether all of these proteins assemble into one mutual transcription complex or whether each exhibits unique interacting dynamics with PGR remains to be explored.

Together, these findings on the interactions of SR indicate that the precise co-expression pattern of the different SR members, as well as other transcription factor families, can influence the hormone response, providing a potential mechanism for cell-specific regulatory action. As SR members can share binding partner repertoires, in granulosa cells where PGR, AR, ERβ and GR are known to be co-expressed, deciphering the individual and mutual interactomes of these SR will be complex.

(d) Interaction with non-protein co-regulators:

SR action can also be modulated by RNA components, which are often overlooked due to their low abundance. The classic RNA regulator of SR is *Sra1*, a long non-coding RNA (lncRNA) that forms a physical interaction with and promotes SR transactivation [113, 114]. Curiously, *Sra1* can also exhibit SR regulatory function in the form of an encoded protein, named SRAP [115]. The *Sra1* lncRNA seems to generically bind SRs and non-steroid nuclear receptors and can mediate their interaction with other protein co-regulators [116], while the mechanism for SRAP action remains unknown [117]. While the roles of *Sra1* and SRAP are mainly examined in the context of tumorigenesis [118, 119], transgenic mice with overexpressed *Sra1* are subfertile and the presence of *Sra1* and its protein counterpart has been linked to reproductive disorders that affect the ovary and uterus [120–122]. The spatial and temporal patterns of *Sra1* expression during folliculogenesis and ovulation have not been explored in depth, however our unpublished data shows an induction in *Sra* transcription and associated interaction with PGR post-LH surge. Another lncRNA that has been attributed to SR regulation is *Gas5*. The genomic structure of *Gas5* is complex and generates various isoforms due to alternative splicing and intronic retention [123]. Furthermore, small nucleolar RNA (snoRNA) encoded in the *Gas5* introns are also functional regulators of protein methylation, in particular, the methylation of ribosomal subunits that regulates their stability and translational activity [124]. Unlike *Sra1* where a functional protein has been identified, so far there has been no protein product found for *Gas5*. Originally linked to cellular response to stress conditions, *Gas5* also plays prominent roles in the modulation of SR activity. This has been particularly demonstrated for GR, but *Gas5* also interacts with all members of the NR3C steroid receptor family [125, 126]. In this context, *Gas5* secondary RNA structure mimics the HRE chromatin folding structure and acts as a decoy, forming a physical interaction with the DNA binding domain of GR and competing with target DNA for GR occupancy, and inhibits GR transactivation functions. Evidence has shown that *Gas5* in cumulus cells is associated with pregnancy outcomes [127] and other studies have indicated the presence of *Gas5* in oocytes and granulosa cells [128], as well as an association with stem cell renewal and pluripotency [129]. Both lncRNA and other short ncRNA including miRNA have been shown to play various roles in ovarian functions, such as oocyte development and ovulation [130].

3.5 Conclusions

Nuclear receptors have long been linked to the physiology of female reproductive cycles and fertility success, indeed steroid receptor regulation of reproductive processes are among the earliest known hormone actions. These steroid hormones and their receptors have unique as well as shared roles within the ovary, suggesting that there are interrelationships between nuclear receptors in regulating transcription networks that are important for various aspects of ovarian functions, specifically in guiding the progress of folliculogenesis, ovulation and luteinisation. Given that nuclear receptor action is highly dependent on tissue context, it is also likely that nuclear receptor ovarian functions are a result of a unique combination of transcription modulators as well as specific interactions between each hormone receptor and their co-regulators or other transcription factors. Evidence is emerging from investigations into these ovarian interactomes as well as non-protein cofactor partners that supports the formation of ovary-specific transcriptional complexes. The identification and characterisation of the complex regulatory network that governs various aspects of ovarian function is crucial in our understanding of female fertility. This is especially important in the development of infertility treatment as well as novel targets for female contraceptives.

References

Bury L, Coelho PA, Glover DM (2016) In: DePamphilis ML (ed) Current topics in developmental biology, vol 120. Academic, pp 125–171

Rimon-Dahari N, Yerushalmi-Heinemann L, Alyagor L, Dekel N (2016) In: Piprek RP (ed) Molecular mechanisms of cell differentiation in gonad development. Springer, pp 167–190

Hirshfield AN (1991) In: Jeon KW, Friedlander M (eds) International review of cytology, vol 124. Academic, pp 43–101

Wood JR, Strauss JF (2002) Multiple signal transduction pathways regulate ovarian steroidogenesis. Rev Endocr Metab Disord 3:33–46

Patel S, Zhou C, Rattan S, Flaws JA (2015) Effects of endocrine-disrupting chemicals on the ovary1. Biol Reprod 93. https://doi.org/10.1095/biolreprod.115.130336

Russell DL, Robker RL (2019) In: Leung PCK, Adashi EY (eds) The ovary, 3rd edn. Academic, pp 217–234

Akison LK, Alvino ER, Dunning KR, Robker RL, Russell DL (2012) Transient invasive migration in mouse cumulus oocyte complexes induced at ovulation by luteinizing hormone1. Biol Reprod 86. https://doi.org/10.1095/biolreprod.111.097345

Boonyaratanakornkit V, McGowan E, Sherman L, Mancini MA, Cheskis BJ, Edwards DP (2007) The role of extranuclear signaling actions of progesterone receptor in mediating progesterone regulation of gene expression and the cell cycle. Mol Endocrinol 21:359–375. https://doi.org/10.1210/me.2006-0337

Samarasinghe RA, Di Maio R, Volonte D, Galbiati F, Lewis M, Romero G, DeFranco DB (2011) Nongenomic glucocorticoid receptor action regulates gap junction intercellular communication and neural progenitor cell proliferation. Proc Natl Acad Sci 108:16657–16662. https://doi.org/10.1073/pnas.1102821108

Guzmán A, Hughes CHK, Murphy BD (2021) Orphan nuclear receptors in angiogenesis and follicular development. Reproduction 162:R35–R54. https://doi.org/10.1530/rep-21-0118

Okada M, Lee L, Maekawa R, Sato S, Kajimura T, Shinagawa M, Tamura I, Taketani T, Asada H, Tamura H, Sugino N (2016) Epigenetic changes of the Cyp11a1 promoter region in granulosa cells undergoing luteinization during ovulation in female rats. Endocrinology 157:3344–3354. https://doi.org/10.1210/en.2016-1264

Lydon JP, DeMayo FJ, Funk CR, Mani SK, Hughes AR, Montgomery CA Jr, Shyamala G, Conneely OM, O'Malley BW (1995) Mice lacking progesterone receptor exhibit pleiotropic reproductive abnormalities. Genes Dev 9:2266–2278

Park CJ, Lin P-C, Zhou S, Barakat R, Bashir ST, Choi JM, Cacioppo JA, Oakley OR, Duffy DM, Lydon JP, Ko CJ (2020) Progesterone receptor serves the ovary as a trigger of ovulation and a terminator of inflammation. Cell Rep 31:107496. https://doi.org/10.1016/j.celrep.2020.03.060

Dinh DT, Breen J, Akison LK, DeMayo FJ, Brown HM, Robker RL, Russell DL (2019) Tissue-specific progesterone receptor-chromatin binding and the regulation of progesterone-dependent gene expression. Sci Rep 9:11966–11966. https://doi.org/10.1038/s41598-019-48333-8

Gaytan F, Bellido C, Gaytan M, Morales C, Sanchez-Criado JE (2003) Differential effects of RU486 and indomethacin on follicle rupture during the ovulatory process in the rat. Biol Reprod 69:99–105. https://doi.org/10.1095/biolreprod.102.013755

Loutradis D, Bletsa R, Aravantinos L, Kallianidis K, Michalas S, Psychoyos A (1991) Preovulatory effects of the progesterone antagonist mifepristone (RU486) in mice. Hum Reprod 6:1238–1240. https://doi.org/10.1093/oxfordjournals.humrep.a137519

Gemzell-Danielsson K, Berger C, Lalitkumar PGL (2013) Emergency contraception – mechanisms of action. Contraception 87:300–308. https://doi.org/10.1016/j.contraception.2012.08.021

Bishop CV, Hennebold JD, Kahl CA, Stouffer RL (2016) Knockdown of progesterone receptor (PGR) in macaque granulosa cells disrupts ovulation and progesterone production. Biol Reprod 94:109. https://doi.org/10.1095/biolreprod.115.134981

Akison L, Robker R (2012) The critical roles of progesterone receptor (PGR) in ovulation, oocyte developmental competence and oviductal transport in mammalian reproduction. Reprod Domest Anim 47:288–296. https://doi.org/10.1111/j.1439-0531.2012.02088.x

Mulac-Jericevic B, Mullinax RA, DeMayo FJ, Lydon JP, Conneely OM (2000) Subgroup of reproductive functions of progesterone mediated by progesterone receptor-B isoform. Science 289:1751–1754. https://doi.org/10.1126/science.289.5485.1751

Mulac-Jericevic B, Lydon JP, DeMayo FJ, Conneely OM (2003) Defective mammary gland morphogenesis in mice lacking the progesterone receptor B isoform. Proc Natl Acad Sci 100:9744–9749. https://doi.org/10.1073/pnas.1732707100

Dinh D, Breen J, Nicol B, Smith K, Nicholls M, Emery A, Wong Y, Barry S, Yao H, Robker R, Russell D (2021) Progesterone receptor-A isoform interaction with RUNX transcription factors controls chromatin remodelling at promoters during ovulation. bioRxiv 202120062017448908. https://doi.org/10.1101/2021.06.17.448908

Robker RL, Richards JS (2000) Ovulation. Springer, pp 121–129

Shimada M, Yamashita Y, Ito J, Okazaki T, Kawahata K, Nishibori M (2004) Expression of two progesterone receptor isoforms in cumulus cells and their roles during meiotic resumption of porcine oocytes. J Mol Endocrinol 33:209–225. https://doi.org/10.1677/jme.0.0330209

Robker RL, Richards JS (1998) Hormone-induced proliferation and differentiation of granulosa cells: a coordinated balance of the cell cycle regulators cyclin D2 and p27Kip1. Mol Endocrinol 12:924–940. https://doi.org/10.1210/mend.12.7.0138

Sar M, Welsch F (1999) Differential expression of estrogen receptor-β and estrogen receptor-α in the rat ovary. Endocrinology 140:963–971. https://doi.org/10.1210/endo.140.2.6533

Duffy DM, Chaffin CL, Stouffer RL (2000) Expression of estrogen receptor α and β in the rhesus monkey corpus luteum during the menstrual cycle: regulation by luteinizing hormone and progesterone*. Endocrinology 141:1711–1717. https://doi.org/10.1210/endo.141.5.7477

Krege JH, Hodgin JB, Couse JF, Enmark E, Warner M, Mahler JF, Sar M, Korach KS, Gustafsson JA, Smithies O (1998) Generation and reproductive phenotypes of mice lacking estrogen receptor beta. Proc

Natl Acad Sci U S A 95:15677–15682. https://doi.org/10.1073/pnas.95.26.15677
Binder AK, Rodriguez KF, Hamilton KJ, Stockton PS, Reed CE, Korach KS (2013) The absence of ER-β results in altered gene expression in ovarian granulosa cells isolated from in vivo preovulatory follicles. Endocrinology 154:2174–2187. https://doi.org/10.1210/en.2012-2256
Chakravarthi VP, Ratri A, Masumi S, Borosha S, Ghosh S, Christenson LK, Roby KF, Wolfe MW, Rumi MAK (2021) Granulosa cell genes that regulate ovarian follicle development beyond the antral stage: the role of estrogen receptor β. Mol Cell Endocrinol 528:111212. https://doi.org/10.1016/j.mce.2021.111212
Rosenfeld CS, Murray AA, Simmer G, Hufford MG, Smith MF, Spears N, Lubahn DB (2000) Gonadotropin induction of ovulation and corpus luteum formation in young estrogen receptor-α knockout mice1. Biol Reprod 62:599–605. https://doi.org/10.1095/biolreprod62.3.599
Sinkevicius KW, Burdette JE, Woloszyn K, Hewitt SC, Hamilton K, Sugg SL, Temple KA, Wondisford FE, Korach KS, Woodruff TK, Greene GL (2008) An estrogen receptor-α knock-in mutation provides evidence of ligand-independent signaling and allows modulation of ligand-induced pathways in vivo. Endocrinology 149:2970–2979. https://doi.org/10.1210/en.2007-1526
Adlanmerini M, Solinhac R, Abot A, Fabre A, Raymond-Letron I, Guihot A-L, Boudou F, Sautier L, Vessières E, Kim SH, Lière P, Fontaine C, Krust A, Chambon P, Katzenellenbogen JA, Gourdy P, Shaul PW, Henrion D, Arnal J-F, Lenfant F (2014) Mutation of the palmitoylation site of estrogen receptor α in vivo reveals tissue-specific roles for membrane versus nuclear actions. Proc Natl Acad Sci 111:E283–E290. https://doi.org/10.1073/pnas.1322057111
Lee S, Kang D-W, Hudgins-Spivey S, Krust A, Lee E-Y, Koo Y, Cheon Y, Gye MC, Chambon P, Ko C (2009) Theca-specific estrogen receptor-α knockout mice lose fertility prematurely. Endocrinology 150:3855–3862. https://doi.org/10.1210/en.2008-1774
Diep CH, Ahrendt H, Lange CA (2016) Progesterone induces progesterone receptor gene (PGR) expression via rapid activation of protein kinase pathways required for cooperative estrogen receptor alpha (ER) and progesterone receptor (PR) genomic action at ER/PR target genes. Steroids 114:48–58. https://doi.org/10.1016/j.steroids.2016.09.004
Sriraman V, Sharma SC, Richards JS (2003) Transactivation of the progesterone receptor gene in granulosa cells: evidence that Sp1/Sp3 binding sites in the proximal promoter play a key role in luteinizing hormone inducibility. Mol Endocrinol 17:436–449. https://doi.org/10.1210/me.2002-0252
Rodriguez KF, Couse JF, Jayes FL, Hamilton KJ, Burns KA, Taniguchi F, Korach KS (2010) Insufficient luteinizing hormone-induced intracellular signaling disrupts ovulation in preovulatory follicles lacking estrogen receptor-β. Endocrinology 151:2826–2834
Denayer S, Helsen C, Thorrez L, Haelens A, Claessens F (2010) The rules of DNA recognition by the androgen receptor. Mol Endocrinol 24:898–913. https://doi.org/10.1210/me.2009-0310
Horie K, Takakura K, Fujiwara H, Suginami H, Liao S, Mori T (1992) Immunohistochemical localization of androgen receptor in the human ovary throughout the menstrual cycle in relation to oestrogen and progesterone receptor expression. Hum Reprod 7:184–190. https://doi.org/10.1093/oxfordjournals.humrep.a137614
Yazawa T, Kawabe S, Kanno M, Mizutani T, Imamichi Y, Ju Y, Matsumura T, Yamazaki Y, Usami Y, Kuribayashi M, Shimada M, Kitano T, Umezawa A, Miyamoto K (2013) Androgen/androgen receptor pathway regulates expression of the genes for cyclooxygenase-2 and amphiregulin in periovulatory granulosa cells. Mol Cell Endocrinol 369:42–51. https://doi.org/10.1016/j.mce.2013.02.004
Laird M, Thomson K, Fenwick M, Mora J, Franks S, Hardy K (2017) Androgen stimulates growth of mouse preantral follicles in vitro: interaction with follicle-stimulating hormone and with growth factors of the TGFβ superfamily. Endocrinology 158:920–935. https://doi.org/10.1210/en.2016-1538
Tarumi W, Itoh MT, Suzuki N (2014) Effects of 5α-dihydrotestosterone and 17β-estradiol on the mouse ovarian follicle development and oocyte maturation. PLoS One 9:e99423. https://doi.org/10.1371/journal.pone.0099423
Liu T, Cui Y-Q, Zhao H, Liu H-B, Zhao S-D, Gao Y, Mu X-L, Gao F, Chen Z-J (2015) High levels of testosterone inhibit ovarian follicle development by repressing the FSH signaling pathway. J Huazhong Univ Sci Technolog Med Sci 35:723–729. https://doi.org/10.1007/s11596-015-1497-z
Wang R-S, Chang H-Y, Kao S-H, Kao C-H, Wu Y-C, Yeh S, Tzeng C-R, Chang C (2015) Abnormal mitochondrial function and impaired granulosa cell differentiation in androgen receptor knockout mice. Int J Mol Sci 16:9831–9849
Sen A, Hammes SR (2010) Granulosa cell-specific androgen receptors are critical regulators of ovarian development and function. Mol Endocrinol 24:1393–1403. https://doi.org/10.1210/me.2010-0006
Walters KA, Middleton LJ, Joseph SR, Hazra R, Jimenez M, Simanainen U, Allan CM, Handelsman DJ (2012) Targeted loss of androgen receptor signaling in murine granulosa cells of preantral and antral follicles causes female subfertility1. Biol Reprod 87. https://doi.org/10.1095/biolreprod.112.102012
Ma Y, Andrisse S, Chen Y, Childress S, Xue P, Wang Z, Jones D, Ko C, Divall S, Wu S (2016) Androgen receptor in the ovary theca cells plays a critical role in androgen-induced reproductive dysfunction. Endocrinology 158:98–108. https://doi.org/10.1210/en.2016-1608

Goodarzi MO, Dumesic DA, Chazenbalk G, Azziz R (2011) Polycystic ovary syndrome: etiology, pathogenesis and diagnosis. Nat Rev Endocrinol 7:219–231. https://doi.org/10.1038/nrendo.2010.217

Cox MJ, Edwards MC, Rodriguez Paris V, Aflatounian A, Ledger WL, Gilchrist RB, Padmanabhan V, Handelsman DJ, Walters KA (2020) Androgen action in adipose tissue and the brain are key mediators in the development of pcos traits in a mouse model. Endocrinology 161. https://doi.org/10.1210/endocr/bqaa061

Walters KA, Rodriguez Paris V, Aflatounian A, Handelsman DJ (2019) Androgens and ovarian function: translation from basic discovery research to clinical impact. J Endocrinol 242:R23–R50. https://doi.org/10.1530/joe-19-0096

Pontes JT, Maside C, Lima LF, Magalhães-Padilha DM, Padilha RT, Matos MHT, Figueiredo JR, Campello CC (2019) Immunolocalization for glucocorticoid receptor and effect of cortisol on in vitro development of preantral follicles. Vet Anim Sci 7:100060. https://doi.org/10.1016/j.vas.2019.100060

Fru KN, VandeVoort CA, Chaffin CL (2006) Mineralocorticoid synthesis during the periovulatory interval in macaques1. Biol Reprod 75:568–574. https://doi.org/10.1095/biolreprod.106.053470

Maradonna F, Gioacchini G, Notarstefano V, Fontana CM, Citton F, Dalla Valle L, Giorgini E, Carnevali O (2020) Knockout of the glucocorticoid receptor impairs reproduction in female zebrafish. Int J Mol Sci 21:9073

Mukangwa M, Takizawa K, Aoki Y, Hamano S, Tetsuka M (2019) Expression of genes encoding mineralocorticoid biosynthetic enzymes and the mineralocorticoid receptor, and levels of mineralocorticoids in the bovine follicle and corpus luteum. J Reprod Dev advpub. https://doi.org/10.1262/jrd.2019-127

Komar CM, Braissant O, Wahli W, Curry TE Jr (2001) Expression and localization of PPARs in the rat ovary during follicular development and the periovulatory period. Endocrinology 142:4831–4838. https://doi.org/10.1210/endo.142.11.8429

Kim J, Sato M, Li Q, Lydon JP, DeMayo FJ, Bagchi IC, Bagchi MK (2008) Peroxisome proliferator-activated receptor γ is a target of progesterone regulation in the preovulatory follicles and controls ovulation in mice. Mol Cell Biol 28:1770–1782. https://doi.org/10.1128/mcb.01556-07

Puttabyatappa M, VandeVoort CA, Chaffin CL (2010) hCG-induced down-regulation of PPARγ and liver X receptors promotes periovulatory progesterone synthesis by macaque granulosa cells. Endocrinology 151:5865–5872. https://doi.org/10.1210/en.2010-0698

Azziz R, Ehrmann D, Legro RS, Whitcomb RW, Hanley R, Fereshetian AG, O'Keefe M, Ghazzi MN (2001) Troglitazone improves ovulation and hirsutism in the polycystic ovary syndrome: a multicenter, double blind, placebo-controlled trial1. J Clin Endocrinol Metab 86:1626–1632. https://doi.org/10.1210/jcem.86.4.7375

Minge CE, Ryan NK, Hoek KHVD, Robker RL, Norman RJ (2006) Troglitazone regulates peroxisome proliferator-activated receptors and inducible nitric oxide synthase in murine ovarian macrophages1. Biol Reprod 74:153–160. https://doi.org/10.1095/biolreprod.105.043729

Lee SS, Pineau T, Drago J, Lee EJ, Owens JW, Kroetz DL, Fernandez-Salguero PM, Westphal H, Gonzalez FJ (1995) Targeted disruption of the alpha isoform of the peroxisome proliferator-activated receptor gene in mice results in abolishment of the pleiotropic effects of peroxisome proliferators. Mol Cell Biol 15:3012–3022. https://doi.org/10.1128/MCB.15.6.3012

Nadra K, Anghel SI, Joye E, Tan NS, Basu-Modak S, Trono D, Wahli W, Desvergne B (2006) Differentiation of trophoblast giant cells and their metabolic functions are dependent on peroxisome proliferator-activated receptor beta/delta. Mol Cell Biol 26:3266–3281. https://doi.org/10.1128/MCB.26.8.3266-3281.2006

Thill M, Becker S, Fischer D, Cordes T, Hornemann A, Diedrich K, Salehin D, Friedrich M (2009) Expression of prostaglandin metabolising enzymes COX-2 and 15-PGDH and VDR in human granulosa cells. Anticancer Res 29:3611–3618

Xu F, Wolf S, Green OR, Xu J (2021) Vitamin D in follicular development and oocyte maturation. Reproduction 161:R129–R137. https://doi.org/10.1530/rep-20-0608

Yoshizawa T, Handa Y, Uematsu Y, Takeda S, Sekine K, Yoshihara Y, Kawakami T, Arioka K, Sato H, Uchiyama Y, Masushige S, Fukamizu A, Matsumoto T, Kato S (1997) Mice lacking the vitamin D receptor exhibit impaired bone formation, uterine hypoplasia and growth retardation after weaning. Nat Genet 16:391–396. https://doi.org/10.1038/ng0897-391

Mu Y, Cheng D, Yin T-L, Yang J (2021) Vitamin D and polycystic ovary syndrome: a narrative review. Reprod Sci 28:2110–2117. https://doi.org/10.1007/s43032-020-00369-2

Miao CY, Fang XJ, Chen Y, Zhang Q (2020) Effect of vitamin D supplementation on polycystic ovary syndrome: a meta-analysis. Exp Ther Med 19:2641–2649. https://doi.org/10.3892/etm.2020.8525

Fang F, Ni K, Cai Y, Shang J, Zhang X, Xiong C (2017) Effect of vitamin D supplementation on polycystic ovary syndrome: a systematic review and meta-analysis of randomized controlled trials. Complement Ther Clin Pract 26:53–60. https://doi.org/10.1016/j.ctcp.2016.11.008

Jamilian M, Foroozanfard F, Rahmani E, Talebi M, Bahmani F, Asemi Z (2017) Effect of two different doses of vitamin D supplementation on metabolic profiles of insulin-resistant patients with polycystic ovary syndrome. Nutrients 9:1280

Maktabi M, Chamani M, Asemi Z (2017) The effects of vitamin D supplementation on metabolic status of patients with polycystic ovary syndrome: a randomized, double-blind, placebo-controlled trial. Horm Metab Res 49:493–498

Zhang SS, Carrillo AJ, Darling DS (1997) Expression of multiple thyroid hormone receptor mRNAs in human oocytes, cumulus cells, and granulosa cells. Mol Hum Reprod 3:555–562. https://doi.org/10.1093/molehr/3.7.555

Saltó C, Kindblom JM, Johansson C, Wang Z, Gullberg H, Nordström K, Mansén A, Ohlsson C, Thorén P, Forrest D, Vennström BR (2001) Ablation of TRα2 and a concomitant overexpression of α1 yields a mixed hypo- and hyperthyroid phenotype in mice. Mol Endocrinol 15:2115–2128. https://doi.org/10.1210/mend.15.12.0750

Quignodon L, Vincent SV, Winter H, Samarut J, Flamant FDR (2007) A point mutation in the activation function 2 domain of thyroid hormone receptor α1 expressed after CRE-mediated recombination partially recapitulates hypothyroidism. Mol Endocrinol 21:2350–2360. https://doi.org/10.1210/me.2007-0176

Forrest D, Hanebuth E, Smeyne RJ, Everds N, Stewart CL, Wehner JM, Curran T (1996) Recessive resistance to thyroid hormone in mice lacking thyroid hormone receptor beta: evidence for tissue-specific modulation of receptor function. EMBO J 15:3006–3015. https://doi.org/10.1002/j.1460-2075.1996.tb00664.x

Göthe S, Wang Z, Ng L, Kindblom JM, Barros AC, Ohlsson C, Vennström B, Forrest D (1999) Mice devoid of all known thyroid hormone receptors are viable but exhibit disorders of the pituitary–thyroid axis, growth, and bone maturation. Genes Dev 13:1329–1341

López Navarro E, Ortega FJ, Francisco-Busquets E, Sabater-Masdeu M, Álvarez-Castaño E, Ricart W, Fernández-Real JM (2016) Thyroid hormone receptors are differentially expressed in granulosa and cervical cells of infertile women. Thyroid 26:466–473. https://doi.org/10.1089/thy.2015.0416

Atikuzzaman M, Koo OJ, Kang JT, Kwon DK, Park SJ, Kim SJ, Gomez MNL, Oh HJ, Hong SG, Jang G, Lee B-C (2011) The 9-cis retinoic acid signaling pathway and its regulation of prostaglandin-endoperoxide synthase 2 during in vitro maturation of pig cumulus cell-oocyte complexes and effects on parthenogenetic embryo production1. Biol Reprod 84:1272–1281. https://doi.org/10.1095/biolreprod.110.086595

Kawai T, Richards JS, Shimada M (2018) The cell type–specific expression of Lhcgr in mouse ovarian cells: evidence for a DNA-demethylation–dependent mechanism. Endocrinology 159:2062–2074. https://doi.org/10.1210/en.2018-00117

Kawai T, Yanaka N, Richards JS, De Shimada M (2016) Novo-synthesized retinoic acid in ovarian antral follicles enhances FSH-mediated ovarian follicular cell differentiation and female fertility. Endocrinology 157:2160–2172. https://doi.org/10.1210/en.2015-2064

Minkina A, Lindeman RE, Gearhart MD, Chassot A-A, Chaboissier M-C, Ghyselinck NB, Bardwell VJ, Zarkower D (2017) Retinoic acid signaling is dispensable for somatic development and function in the mammalian ovary. Dev Biol 424:208–220. https://doi.org/10.1016/j.ydbio.2017.02.015

Krezel W, Dupé V, Mark M, Dierich A, Kastner P, Chambon P (1996) RXR gamma null mice are apparently normal and compound RXR alpha +/−/RXR beta −/−/RXR gamma −/− mutant mice are viable. Proc Natl Acad Sci U S A 93:9010–9014. https://doi.org/10.1073/pnas.93.17.9010

Fan W, Yanase T, Morinaga H, Mu Y-M, Nomura M, Okabe T, Goto K, Harada N, Nawata H (2005) Activation of peroxisome proliferator-activated receptor-γ and retinoid X receptor inhibits aromatase transcription via nuclear factor-κB. Endocrinology 146:85–92. https://doi.org/10.1210/en.2004-1046

O'Malley BW, Tsai M-J (1992) Molecular pathways of steroid receptor action. Biol Reprod 46:163–167. https://doi.org/10.1095/biolreprod46.2.163

Zhang L, Martini GD, Rube HT, Kribelbauer JF, Rastogi C, FitzPatrick VD, Houtman JC, Bussemaker HJ, Pufall MA (2018) SelexGLM differentiates androgen and glucocorticoid receptor DNA-binding preference over an extended binding site. Genome Res 28:111–121. https://doi.org/10.1101/gr.222844.117

Schöne S, Jurk M, Helabad MB, Dror I, Lebars I, Kieffer B, Imhof P, Rohs R, Vingron M, Thomas-Chollier M, Meijsing SH (2016) Sequences flanking the core-binding site modulate glucocorticoid receptor structure and activity. Nat Commun 7:12621. https://doi.org/10.1038/ncomms12621

Johnson TA, Paakinaho V, Kim S, Hager GL, Presman DM (2021) Genome-wide binding potential and regulatory activity of the glucocorticoid receptor's monomeric and dimeric forms. Nat Commun 12:1987. https://doi.org/10.1038/s41467-021-22234-9

Buser AC, Obr AE, Kabotyanski EB, Grimm SL, Rosen JM, Edwards DP (2011) Progesterone receptor directly inhibits β-casein gene transcription in mammary epithelial cells through promoting promoter and enhancer repressive chromatin modifications. Mol Endocrinol 25:955–968. https://doi.org/10.1210/me.2011-0064

Massie CE, Adryan B, Barbosa-Morais NL, Lynch AG, Tran MG, Neal DE, Mills IG (2007) New androgen receptor genomic targets show an interaction with the ETS1 transcription factor. EMBO Rep 8:871–878. https://doi.org/10.1038/sj.embor.7401046

Doyle KMH, Russell DL, Sriraman V, Richards JS (2004) Coordinate transcription of the ADAMTS-1 gene by luteinizing hormone and progesterone receptor. Mol Endocrinol 18:2463–2478. https://doi.org/10.1210/me.2003-0380

Starick SR, Ibn-Salem J, Jurk M, Hernandez C, Love MI, Chung H-R, Vingron M, Thomas-Chollier M, Meijsing SH (2015) ChIP-exo signal associated with DNA-binding motifs provides insight into the genomic binding of the glucocorticoid receptor and cooperating transcription factors. Genome Res 25:825–835. https://doi.org/10.1101/gr.185157.114

Yin P, Roqueiro D, Huang L, Owen JK, Xie A, Navarro A, Monsivais D, Coon VJS, Kim JJ, Dai Y, Bulun SE (2012) Genome-wide progesterone receptor binding: cell type-specific and shared mechanisms in T47D breast cancer cells and primary leiomyoma cells. PLoS One 7:e29021. https://doi.org/10.1371/journal.pone.0029021

Mao A-P, Ishizuka IE, Kasal DN, Mandal M, Bendelac A (2017) A shared Runx1-bound Zbtb16 enhancer directs innate and innate-like lymphoid lineage development. Nat Commun 8:863. https://doi.org/10.1038/s41467-017-00882-0

Hickey TE, Selth LA, Chia KM, Laven-Law G, Milioli HH, Roden D, Jindal S, Hui M, Finlay-Schultz J, Ebrahimie E, Birrell SN, Stelloo S, Iggo R, Alexandrou S, Caldon CE, Abdel-Fatah TM, Ellis IO, Zwart W, Palmieri C, Sartorius CA, Swarbrick A, Lim E, Carroll JS, Tilley WD (2021) The androgen receptor is a tumor suppressor in estrogen receptor-positive breast cancer. Nat Med 27:310–320. https://doi.org/10.1038/s41591-020-01168-7

Ogara MF, Rodríguez-Seguí SA, Marini M, Nacht AS, Stortz M, Levi V, Presman DM, Vicent GP, Pecci A (2019) The glucocorticoid receptor interferes with progesterone receptor-dependent genomic regulation in breast cancer cells. Nucleic Acids Res 47:10645–10661. https://doi.org/10.1093/nar/gkz857

Pihlajamaa P, Sahu B, Lyly L, Aittomäki V, Hautaniemi S, Jänne OA (2014) Tissue-specific pioneer factors associate with androgen receptor cistromes and transcription programs. EMBO J 33:312–326. https://doi.org/10.1002/embj.201385895

Georges A, L'Hôte D, Todeschini AL, Auguste A, Legois B, Zider A, Veitia RA (2014) The transcription factor FOXL2 mobilizes estrogen signaling to maintain the identity of ovarian granulosa cells. elife 3:e04207. https://doi.org/10.7554/eLife.04207

Herman L, Legois B, Todeschini A-L, Veitia RA (2021) Genomic exploration of the targets of FOXL2 and ESR2 unveils their implication in cell migration, invasion, and adhesion. FASEB J 35:e21355. https://doi.org/10.1096/fj.202002444R

Li Z, Tuteja G, Schug J, Kaestner KH (2012) Foxa1 and Foxa2 are essential for sexual dimorphism in liver cancer. Cell 148:72–83. https://doi.org/10.1016/j.cell.2011.11.026

Tung L, Shen T, Abel MG, Powell RL, Takimoto GS, Sartorius CA, Horwitz KB (2001) Mapping the unique activation function 3 in the progesterone B-receptor upstream segment: two LXXLL motifs and A tryptophan residue are required for activity. J Biol Chem 276:39843–39851. https://doi.org/10.1074/jbc.M106843200

Singhal H, Greene ME, Zarnke AL, Laine M, Al Abosy R, Chang Y-F, Dembo AG, Schoenfelt K, Vadhi R, Qiu X, Rao P, Santhamma B, Nair HB, Nickisch KJ, Long HW, Becker L, Brown M, Greene GL (2017) Progesterone receptor isoforms, agonists and antagonists differentially reprogram estrogen signaling. Oncotarget 9:4282–4300. https://doi.org/10.18632/oncotarget.21378

Vegeto E, Shahbaz MM, Wen DX, Goldman ME, O'Malley BW, McDonnell DP (1993) Human progesterone receptor A form is a cell- and promoter-specific repressor of human progesterone receptor B function. Mol Endocrinol 7:1244–1255. https://doi.org/10.1210/mend.7.10.8264658

Merlino AA, Welsh TN, Tan H, Yi LJ, Cannon V, Mercer BM, Mesiano S (2007) Nuclear progesterone receptors in the human pregnancy myometrium: evidence that parturition involves functional progesterone withdrawal mediated by increased expression of progesterone receptor-A. J Clin Endocrinol Metab 92:1927–1933. https://doi.org/10.1210/jc.2007-0077

Nadeem L, Shynlova O, Matysiak-Zablocki E, Mesiano S, Dong X, Lye S (2016) Molecular evidence of functional progesterone withdrawal in human myometrium. Nat Commun 7:11565. https://doi.org/10.1038/ncomms11565

Wilson CM, McPhaul MJ (1996) A and B forms of the androgen receptor are expressed in a variety of human tissues. Mol Cell Endocrinol 120:51–57. https://doi.org/10.1016/0303-7207(96)03819-1

Liegibel UM, Sommer U, Boercsoek I, Hilscher U, Bierhaus A, Schweikert HU, Nawroth P, Kasperk C (2003) Androgen receptor isoforms AR-A and AR-B display functional differences in cultured human bone cells and genital skin fibroblasts. Steroids 68:1179–1187. https://doi.org/10.1016/j.steroids.2003.08.016

Pascual-Le Tallec L, Demange C, Lombès M (2004) Human mineralocorticoid receptor A and B protein forms produced by alternative translation sites display different transcriptional activities. Eur J Endocrinol 150:585–590. https://doi.org/10.1530/eje.0.1500585

Kino T, Su YA, Chrousos GP (2009) Human glucocorticoid receptor isoform β: recent understanding of its potential implications in physiology and pathophysiology. Cell Mol Life Sci 66:3435–3448. https://doi.org/10.1007/s00018-009-0098-z

York B, O'Malley BW (2010) Steroid receptor coactivator (SRC) family: masters of systems biology. J Biol Chem 285:38743–38750. https://doi.org/10.1074/jbc.R110.193367

Kollara A, Brown TJ (2012) Expression and function of nuclear receptor co-activator 4: evidence of a potential role independent of co-activator activity. Cell Mol Life Sci 69:3895–3909. https://doi.org/10.1007/s00018-012-1000-y

Spencer TE, Jenster G, Burcin MM, Allis CD, Zhou J, Mizzen CA, McKenna NJ, Onate SA, Tsai SY, Tsai M-J, O'Malley BW (1997) Steroid receptor coactivator-1 is a histone acetyltransferase. Nature 389:194–198. https://doi.org/10.1038/38304

Li X, Wong J, Tsai SY, Tsai M-J, O'Malley BW (2003) Progesterone and glucocorticoid receptors recruit distinct coactivator complexes and promote distinct patterns of local chromatin modification. Mol Cell Biol 23:3763–3773. https://doi.org/10.1128/mcb.23.11.3763-3773.2003
Yi P, Wang Z, Feng Q, Chou CK, Pintilie GD, Shen H, Foulds CE, Fan G, Serysheva I, Ludtke SJ, Schmid MF, Hung MC, Chiu W, O'Malley BW (2017) Structural and functional impacts of ER coactivator sequential recruitment. Mol Cell 67:733–743 e734. https://doi.org/10.1016/j.molcel.2017.07.026
Hussein-Fikret S, Fuller PJ (2005) Expression of nuclear receptor coregulators in ovarian stromal and epithelial tumours. Mol Cell Endocrinol 229:149–160. https://doi.org/10.1016/j.mce.2004.08.005
Lanz RB, Razani B, Goldberg AD, O'Malley BW (2002) Distinct RNA motifs are important for coactivation of steroid hormone receptors by steroid receptor RNA activator (SRA). Proc Natl Acad Sci 99:16081–16086. https://doi.org/10.1073/pnas.192571399
Agoulnik IU, Weigel NL (2009) Coactivator selective regulation of androgen receptor activity. Steroids 74:669–674. https://doi.org/10.1016/j.steroids.2009.02.007
Cooper C, Vincett D, Yan Y, Hamedani MK, Myal Y, Leygue E (2011) Steroid receptor RNA activator bi-faceted genetic system: heads or tails? Biochimie 93:1973–1980. https://doi.org/10.1016/j.biochi.2011.07.002
Liu C, Wu H-T, Zhu N, Shi Y-N, Liu Z, Ao B-X, Liao D-F, Zheng X-L, Qin L (2016) Steroid receptor RNA activator: biologic function and role in disease. Clin Chim Acta 459:137–146. https://doi.org/10.1016/j.cca.2016.06.004
McKay DB, Xi L, Barthel KKB, Cech TR (2014) Structure and function of steroid receptor RNA activator protein, the proposed partner of SRA noncoding RNA. J Mol Biol 426:1766–1785. https://doi.org/10.1016/j.jmb.2014.01.006
Yan Y, Cooper C, Hamedani MK, Guppy B, Xu W, Tsuyuki D, Zhang C, Nugent Z, Blanchard A, Davie JR, McManus K, Murphy LC, Myal Y, Leygue E (2015) The steroid receptor RNA activator protein (SRAP) controls cancer cell migration/motility. FEBS Lett 589:4010–4018. https://doi.org/10.1016/j.febslet.2015.11.007
Lanz RB, Chua SS, Barron N, Söder BM, DeMayo F, O'Malley BW (2003) Steroid receptor RNA activator stimulates proliferation as well as apoptosis in vivo. Mol Cell Biol 23:7163–7176
Lin K, Zhan H, Ma J, Xu K, Wu R, Zhou C, Lin J (2017) Silencing of SRA1 regulates ER expression and attenuates the growth of stromal cells in ovarian endometriosis. Reprod Sci 24:836–843. https://doi.org/10.1177/1933719116670036
Li Y, Zhao W, Wang H, Chen C, Zhou D, Li S, Zhang X, Zhao H, Zhou D, Chen B (2019) Silencing of LncRNA steroid receptor RNA activator attenuates polycystic ovary syndrome in mice. Biochimie 157:48–56. https://doi.org/10.1016/j.biochi.2018.10.021
Eoh KJ, Paek J, Kim SW, Kim HJ, Lee HY, Lee SK, Kim YT (2017) Long non-coding RNA, steroid receptor RNA activator (SRA), induces tumor proliferation and invasion through the NOTCH pathway in cervical cancer cell lines. Oncol Rep 38:3481–3488. https://doi.org/10.3892/or.2017.6023
Pickard MR, Williams GT (2015) Molecular and cellular mechanisms of action of tumour suppressor GAS5 LncRNA. Gene 6:484–499
Smith CM, Steitz JA (1998) Classification of gas5 as a multi-small-nucleolar-RNA (snoRNA) host gene and a member of the 5′-terminal oligopyrimidine gene family reveals common features of snoRNA host genes. Mol Cell Biol 18:6897–6909
Hudson WH, Pickard MR, de Vera IM, Kuiper EG, Mourtada-Maarabouni M, Conn GL, Kojetin DJ, Williams GT, Ortlund EA (2014) Conserved sequence-specific lincRNA-steroid receptor interactions drive transcriptional repression and direct cell fate. Nat Commun 5:5395. https://doi.org/10.1038/ncomms6395
Kino T, Hurt DE, Ichijo T, Nader N, Chrousos GP (2010) Noncoding RNA gas5 is a growth arrest- and starvation-associated repressor of the glucocorticoid receptor. Sci Signal 3:2000568
Gebhardt KM, Feil DK, Dunning KR, Lane M, Russell DL (2011) Human cumulus cell gene expression as a biomarker of pregnancy outcome after single embryo transfer. Fertil Steril 96:47–52.e42. https://doi.org/10.1016/j.fertnstert.2011.04.033
Cheng Y, Kim J, Li XX, Hsueh AJ (2015) Promotion of ovarian follicle growth following mTOR activation: synergistic effects of AKT stimulators. PLoS One 10:e0117769. https://doi.org/10.1371/journal.pone.0117769
Tu J, Tian G, Cheung H-H, Wei W, Lee T-L (2018) Gas5 is an essential lncRNA regulator for self-renewal and pluripotency of mouse embryonic stem cells and induced pluripotent stem cells. Stem Cell Res Ther 9:71–71. https://doi.org/10.1186/s13287-018-0813-5
Robles V, Valcarce DG, Riesco MF (2019) Non-coding RNA regulation in reproduction: their potential use as biomarkers. Non-coding RNA Res 4:54–62. https://doi.org/10.1016/j.ncrna.2019.04.001

Part II

Metabolism

Nuclear Receptors in Energy Metabolism

4

Alina A. Walth-Hummel, Stephan Herzig, and Maria Rohm

Abstract

Nuclear receptors are master regulators of energy metabolism through the conversion of extracellular signals into gene expression signatures. The function of the respective nuclear receptor is tissue specific, signal and co-factor dependent. While normal nuclear receptor function is central to metabolic physiology, aberrant nuclear receptor signaling is linked to various metabolic diseases such as type 2 diabetes mellitus, obesity, or hepatic steatosis. Thus, the tissue specific manipulation of nuclear receptors is a major field in biomedical research and represents a treatment approach for metabolic syndrome. This chapter focuses on key nuclear receptors involved in regulating the metabolic function of liver, adipose tissue, skeletal muscle, and pancreatic β-cells. It also addresses the importance of nuclear co-factors for fine-tuning of nuclear receptor function. The mode of action, role in energy metabolism, and therapeutic potential of prominent nuclear receptors is outlined.

Keywords

Energy homeostasis · Glucose and lipid metabolism · Nuclear receptor-based therapies · Metabolic syndrome · Transcriptional co-factors

4.1 Introduction and Outline

Nuclear receptors govern multiple essential functions in metabolism. In the current chapter, we aim to introduce the most important nuclear receptor-related functions and factors in the organ-specific regulation of glucose and lipid

A. A. Walth-Hummel · M. Rohm (✉)
Institute for Diabetes and Cancer, Helmholtz Center Munich, Neuherberg, Germany

Joint Heidelberg-IDC Translational Diabetes Program, Inner Medicine 1, Heidelberg University Hospital, Heidelberg, Germany

German Center for Diabetes Research (DZD), Neuherberg, Germany
e-mail: maria.rohm@helmholtz-muenchen.de

S. Herzig
Institute for Diabetes and Cancer, Helmholtz Center Munich, Neuherberg, Germany

Joint Heidelberg-IDC Translational Diabetes Program, Inner Medicine 1, Heidelberg University Hospital, Heidelberg, Germany

German Center for Diabetes Research (DZD), Neuherberg, Germany

Chair Molecular Metabolic Control, Technical University Munich,
Munich, Germany

M. J. Campbell, C. L. Bevan (eds.), *Nuclear Receptors in Human Health and Disease*, Advances in Experimental Medicine and Biology 1390, https://doi.org/10.1007/978-3-031-11836-4_4

metabolism, as well as diseases associated with their malfunction, and novel approaches to target them. Notably, there is a multitude of additional nuclear receptors, classified either as orphan receptors including the estrogen related receptor (ERR) and the retinoic acid related receptor (ROR), or with known ligands such as the estrogen receptor (ER), the androgen receptor (AR), and the retinoic acid receptor (RAR). These receptors are also involved to some extent in the regulation of metabolism, but cannot be covered exhaustively within the scope of this chapter. Our increasing understanding of the organ- and context-specific regulation of nuclear receptors and their co-factors (Box 4.1) has already led to the development of promising therapeutics for common diseases and will likely yield novel treatment approaches for metabolic diseases in the future. We here discuss the roles of the most prominent nuclear receptors in metabolism, PPARs, LXR, FXR, and GR, in the major metabolic organs and summarize the current state of play as regards therapeutic targeting of these receptors in metabolic diseases.

4.2 Liver

Constant food accessibility and an overly sedentary lifestyle have led to an obesity pandemic. The imbalance of energy availability and expenditure is detrimental especially for the liver, which is one of the central organs for metabolism. Consequently, conditions such as non-alcoholic fatty liver disease (NAFLD) and non-alcoholic steatohepatitis (NASH) are on the rise. To date, there are no efficient treatments for these diseases. Increasing knowledge of the diverse functions of nuclear receptors has led to growing interest in pharmacological compounds that can manipulate their activity. Tissue- or pathway-specific manipulation of nuclear receptors could represent novel treatment possibilities.

4.2.1 PPARα Is the Key to Liver Lipid Metabolism

Peroxisome proliferator-activated receptor alpha (PPARα) is a nuclear receptor highly abundant in liver and other tissues with high rates of fatty acid oxidation such as cardiac muscle, brown adipose tissue, and kidney [1–3]. PPARα was initially identified as an activator of peroxisome proliferation induced by hepatocarcinogens [4], but was since established as a master regulator of liver lipid metabolism. PPARα expression levels were found to be reduced in NAFLD patients, but increased in parallel with NAFLD histological improvements secondary to lifestyle intervention or bariatric surgery [5]. In line with this, hepatocyte specific disruption of PPARα in mice resulted in steatosis and steatohepatitis indicating an essential role in lipid utilization [6]. Indeed, PPARα expression is increased during suckling [7] and fasting [8], both states in which fat serves as the primary energy source. During fasting, fatty acids released from adipose tissue serve as endogenous ligands for PPARα and promote the activation of the majority of pathways involved in lipid catabolism, including lipid uptake, intracellular lipid trafficking, peroxisomal and mitochondrial β-oxidation, and ketone body synthesis. Cellular lipid uptake, which is the first step in lipid catabolism, is facilitated by the fatty acid transporters fatty acid transporter (FAT/CD36) and fatty acid transporter protein (FATP). Both are direct target genes of PPARα [9] highlighting the importance of PPARα function not only for hepatic cellular metabolism but also for fatty acid clearance from the periphery. Apart from lipid uptake, PPARα regulates medium-chain acyl-coenzyme A (CoA) dehydrogenase [10, 11] and acyl-CoA oxidase 1 [11] which are the rate limiting enzymes of mitochondrial and peroxisomal fatty acid β-oxidation (Fig. 4.1). These miscellaneous regulatory functions of PPARα in lipid metabolism sparked interest in developing compounds based on PPARα target gene products to counteract abnormalities and disorders associated with the metabolic syn-

Box 4.1: Transcriptional Co-factors as Tissue- and Context-Specific Regulators of Nuclear Receptor Function

The transcriptional control of biological processes requires tight regulation and the ability to rapidly adapt in response to metabolic changes. NRs facilitate the transcription of target genes ligand dependently, however also rely on the recruitment of co-factors. These co-factors either induce or suppress transcription and are referred to as co-activators or co-repressors, respectively. Co-factors are not exclusive to the NRs, however in general, unliganded NRs preferentially interact with co-repressors and thereby inhibit transcription while ligand binding promotes NR:co-activator interaction which facilitates transcription. Interestingly the transcriptional co-factors transducin β-like protein 1 (TBL1), and TBL-related 1 (TBLR1) were reported to act as so called "nuclear exchange factors", which regulate gene repression and expression by exchanging co-repressors and co-activators [147]. Co-factors modulate NR activity by determination of cellular localization, regulation of NR stability, or posttranslational modification of the NR itself or the chromatin [148–150]. One of the better studied repressor complexes is the NR co-repressor (NCoR) and silencing mediator of retinoic acid and thyroid hormone receptors (SMRT) co-repressor complex (NCoR/SMRT complex). This repressor complex is comprised of the core subunits NCoR or SMRT as well as histone deacetylase 3 (HDAC3), G-protein pathway suppressor 2 (GPS2), TBL1, and TBLR1 [151, 152]. Repression through the NCoR/SMRT complex is partly achieved by HDAC3 mediated deacetylation of chromatin, which disables transcription. HDAC3 is a negative regulator of white adipose tissue (WAT) browning [153], and a modulator of liver lipid metabolism [154] and pancreatic β-cell function [155, 156], indicating its prominent and tissue-specific function in molecular metabolic regulation. GPS2 regulates liver lipid metabolism [157] as well as adipose tissue metabolism and its endocrine function [158, 159]. Interestingly, while hepatocyte specific loss of GPS2 ameliorates NASH [157], TBL1 and TBLR1 loss of function in the liver promotes steatosis development [160], indicating that the complex core components regulate transcriptional events independently of NCoR and SMRT and/or through recruitment of additional regulatory units. In addition, TBL1 and TBLR1 were shown to directly interact with NRs and facilitate diverse tissue specific metabolic events including proliferation in pancreatic cancer cells [161, 162] and adipose tissue lipid metabolism [163]. It is currently estimated that the group of co-regulators includes around 150–400 proteins in humans [164], providing a unique tissue- and context-specific targeting opportunity in future research. As disruption of co-factor function results in various metabolic diseases, insights into the mechanistic action of these co-factors are essential. The diverse features of these co-factors and their ability to function as complexes but also independently highlight their importance for the maintenance of a normal and healthy metabolism and their potential in the development of novel therapeutic drugs.

drome. A recent prominent example which gained interest as a pharmacological target is fibroblast growth factor (FGF) 21 [12]. FGF21 administration reduced body weight, blood glucose levels, circulating plasma insulin, and hepatic gluconeogenesis in diet-induced and genetic mouse models of obesity, as well as in non-human primates [13–16]. Additionally, while FGF21 ablation resulted in severe hepatic steatosis and inflammation [17, 18], elevation of circulating FGF21 levels reversed fatty liver and NASH [18, 19]. Due to poor pharmacokinetic properties of natural FGF21, several modified FGF21 analogues were developed, which are currently undergoing clinical trials.

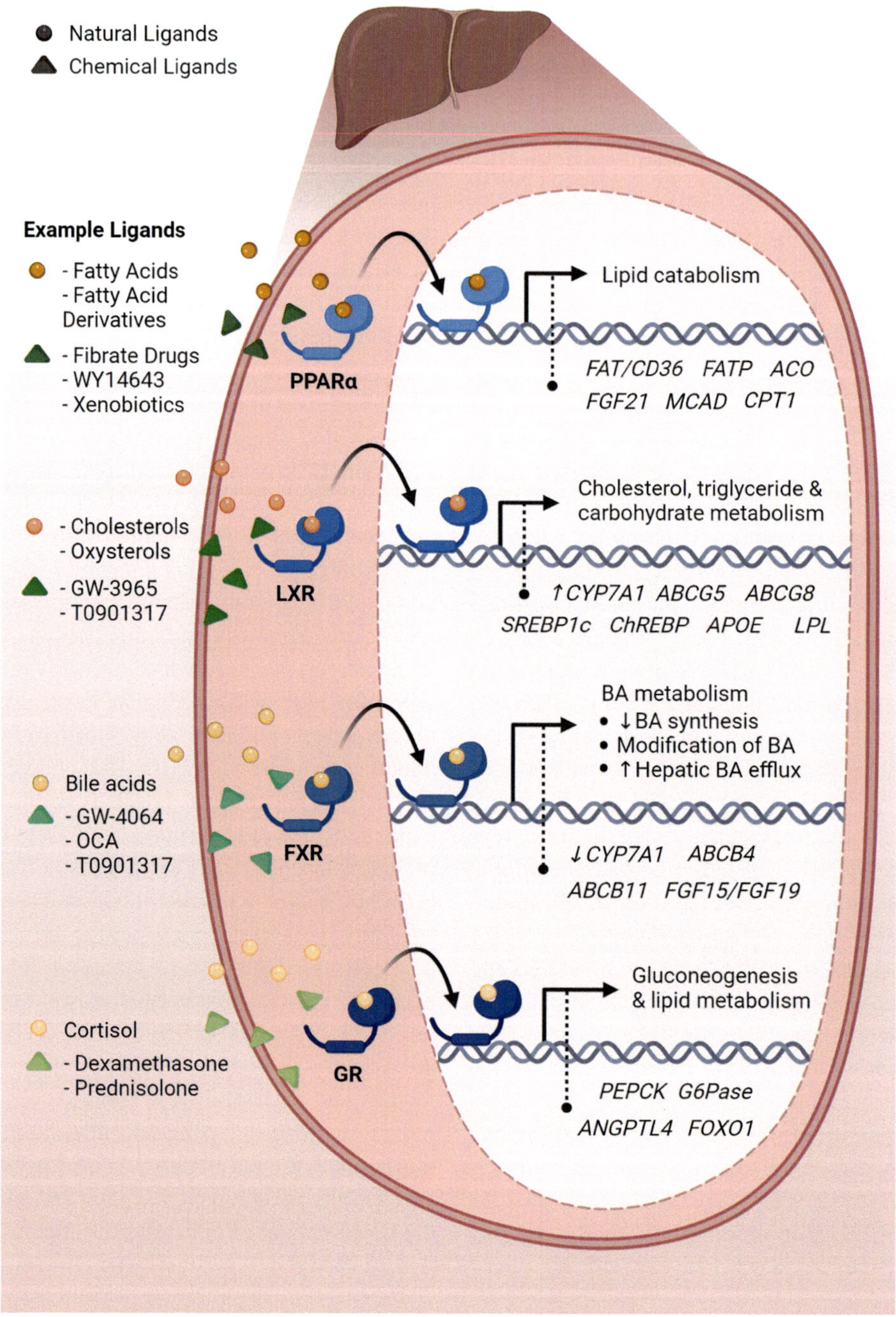

Fig. 4.1 Nuclear receptors involved in regulating liver metabolism. Schematic representation of the main functions of nuclear receptors in the liver and their ligands, main mode of action, and key target genes. (**a**) Upon ligand binding, PPARα induces lipid catabolism through direct induction of genes involved in cellular lipid uptake (FAT/CD36, FATP), fatty acid β-oxidation (ACO, MCAD, CPT1), and ketone body synthesis. (**b**) Apart from cholesterol metabolism (CYP7A1, ABCG5, ABCG8), LXR controls genes involved in carbohydrate (ChREBP) and triglyceride (SREBP1c, APOE, LPL) metabolism (**c**) FXR controls hepatic bile acid metabolism and is a known LXR counterplayer. (**d**) Glucocorticoid induced GR activity promotes gluconeogenesis (PEPCK, G6Pase) and lipid metabolism (ANGPTL4, FOXO1) in the hepatocytes. *FAT/CD36* fatty acid transporter, *FATP* fatty acid transport protein, *ACO* acyl-CoA oxidase, *MACD* medium-chain acyl-CoA dehydrogenase, *FGF21* fibroblast growth factor

Recently completed phase 2 studies assessed the efficacy and safety of FGF21 covalently conjugated to polyethylene glycol (PEGylation). PEGylated FGF21 significantly reduced hepatic fat, liver stiffness, fibrosis markers, and markers of liver damage [20, 21]. In contrast to recombinant FGF21 and other FGF21 analogues, PEGylated FGF21 did not induce bone loss or increased blood pressure in the pre-clinical setting [22–24], suggesting a low risk for chronic treatment in patients. Currently two phase 2b studies are ongoing, which evaluate the effects of PEGylated FGF21 in NASH patients with severe fibrosis (ClinicalTrials.gov Identifiers: NCT03486899 and NCT03486912). Another long-acting FGF21 fusion protein (Fc-FGF21) was shown to improve glucose metabolism and plasma lipid levels across species, including humans [25, 26]. Completion of a phase 2 clinical study, which investigated the effects of Fc-FGF21 in NASH patients, reported reduction of the hepatic fat fraction and markers of liver damage, while improving glycemic control [27], indicating the potential of this compound as a novel treatment possibility for NASH and NAFLD. The current lack of approved treatments for these conditions underlines the significance of the pharmacological improvements in obesity related morbidities after FGF21 analogue administration. As a direct PPARα target gene, this indicates how crucial the physiological function of PPARα in the liver is, and how promising manipulations of these pathways are to develop novel treatment possibilities for the metabolic syndrome and its manifestation as NAFLD and NASH.

4.2.2 LXR and FXR Are Regulators of Cholesterol Metabolism

The two liver X receptor (LXR) isoforms, α and β, are key regulators of cholesterol, triglyceride, and carbohydrate metabolism in the liver [28–30]. LXRα was initially discovered in the liver where it is highly abundant [31], whereas LXRβ is ubiquitously expressed [32]. Cholesterol, cholesterol derivatives, and cholesterol precursors were identified as natural LXR ligands indicating a central role of LXR in cholesterol metabolism [33]. Indeed, LXR is an intracellular cholesterol sensor and modulator by directly regulating genes involved in reverse cholesterol transport (RCT), conversion of cholesterol into bile acids, and intestinal excretion of cholesterol. In rodents, LXR induced cholesterol 7 α-hydrolase (CYP7A1) expression upon ligand binding, which is the first step and the rate-limiting enzyme for bile acid synthesis [28]. Interestingly, binding of bile acids to the farnesoid X receptor (FXR), another nuclear receptor highly abundant in the liver, downregulated CYP7A1 expression [34], identifying FXR as a LXR counterplayer. The downregulation of CYP7A1 was in part facilitated by the FXR target gene FGF15 in mice and its orthologue FGF19 in humans [35]. In addition to bile acid synthesis inhibition, FXR promoted the modification of bile acids into less toxic molecules [36] and hepatic bile acid efflux via ATP-binding cassettes ABCB11 and ABCB4 [37–39], while LXRα and β regulated cholesterol efflux from the liver into the bile via ABCG5 and ABCG8 [40] (Fig. 4.1).

Despite sharing 78% similarity in their amino acid sequence [41], LXRα and β do not possess identical functions in metabolism [42]. In mice lacking LXRα only, cholesterol removal from the body was severely impaired [28, 42], while LXRβ knockout (KO) mice were protected from such a phenotype [42]. Interestingly, LXRs were also identified to regulate glucose metabolism through energy utilization in brown fat [43], pancreatic insulin secretion [44], and direct up-regulation of the glucose transporter GLUT4 in adipose tissue and muscle [30, 45, 46]. Additionally, ligand activated LXR activity inhibited the gluconeogenic program through

Fig. 4.1 (continued) 21, *CPT1* carnitine palmitoyltransferase 1, *CYP7A1* cholesterol 7 α-hydrolase, *ABCG5* ATP-binding cassette G 5, *SREBP1c* sterol regulatory element-binding transcription factor 1, *ChREBP* carbohydrate-responsive element-binding protein, *APOE* apolipoprotein E, *LPL* lipoprotein lipase, *BA* bile acid, *PEPCK* phosphoenolpyruvate carboxykinase, *G6Pase* glucose-6-phosphatase, *ANGPTL4* angiopoietin-like 4, *FOXO1* forkhead box protein O1

down-regulation of peroxisome proliferator-activated receptor-γ coactivator 1-α (PGC-1), phosphoenolpyruvate carboxykinase (PEPCK), and glucose-6-phosphatase (G6Pase) expression in the liver (Fig. 4.1). Suppression of gluconeogenesis was accompanied by increased glucokinase expression, which promoted glucose utilization and blood glucose clearance [30, 45, 47]. Similar to LXR, FXR also regulates glucose metabolism through fine tuning of gluconeogenesis. In the fed state FXR inhibited gluconeogenesis through repression of the key enzymes PEPCK and G6Pase [48, 49], while at fasted state FXR promoted the gluconeogenic program [49]. Accordingly, mice lacking FXR are prone to glucose intolerance and insulin resistance [50].

Due to the primarily beneficial action of LXR on glucose metabolism, LXR agonists were initially thought to be ideal therapeutic agents to treat hepatic steatosis and hyperglycemia. Indeed, LXR agonist administration suppressed the gluconeogenic program and thereby reduced blood glucose levels [45]. Some of the promising effects of LXR agonists on glucose metabolism were, however, shown to be rodent specific [51]. Additionally, synthetic LXR agonists promoted hepatic lipogenesis and steatosis, via transcriptional activation of the triglyceride master regulator sterol regulatory binding transcription factor (SREBP) 1 [29, 52]. Reduced lipogenesis was observed in LXRα KO mice in comparison to wild type mice [28]. However, LXR agonist administration in LXRα KO mice increased SREBP1 gene expression [29] indicating that both LXRα and LXRβ regulate lipogenesis and triglyceride synthesis in the liver. The species-specific effects as well as adverse effects of synthetic LXR agonists raised concerns as to the suitability of LXR agonists for lipid metabolism-associated disorders. In recent studies inverse LXR agonists gained interest for NASH/NAFLD treatment. Similar to agonists, inverse agonists bind to the same receptor, however they exert the opposite effect on the target cells. The inverse LXR agonist 10rr was found to inhibit lipogenesis by downregulating the expression of SREBP1, acetyl-CoA carboxylase (ACC), fatty acid synthase (FAS) and stearoyl-CoA desaturase-1 (SCD-1) [53]. Interestingly, the liver specific inverse LXR agonist SR9238 exerts antifibrotic and anti-inflammatory effects on NASH [54, 55] and suppresses hepatic steatosis [56]. Surprisingly, in contrast to what was seen in LXR KO mice, reduced plasma cholesterol levels were observed, partly through downregulation of HMG CoA reductase (HMGCR) [56], a key regulator of cholesterol synthesis. So far, no studies were performed testing inverse LXR agonists in humans. However, administration or combination of such inverse agonists could provide novel implications and treatment possibilities in fatty liver associated diseases. In contrast, several FXR agonists already underwent clinical studies especially for the treatment of NASH. The FXR agonist obeticholic acid (OCA) reduced bile acid synthesis by repression of CYP7A1 gene expression. Moreover, OCA protected from NASH induced apoptosis of hepatocytes through suppression of p53 [57]. In NASH patients, treatment with OCA improved liver histology, however induced pruritus (itching) [58], which was also observed in treatment with other FXR agonists [59]. Currently a long term phase 3 study (NCT02548351) is ongoing to evaluate the effects of OCA on mortality, liver-related clinical outcomes, and long-term safety [60]. Cilofexor, another FXR agonist, was reported to improve hypertension and liver fibrosis in rats [61]. Additionally, in a phase 2 clinical trial NASH patients receiving cilofexor showed improved hepatic steatosis and liver transaminases [59]. In conclusion, FXR agonists represent novel and attractive candidates for NASH treatment. The current ongoing studies to determine safety and efficiency highlight the potential of such FXR agonists.

4.2.3 GR – Linking Inflammation and Metabolism

The glucocorticoid receptor (GR) is a hormone-dependent nuclear receptor which regulates a wide range of metabolic processes including inflammation, lipid and glucose metabolism. More than 50 genes are under direct GR control in the liver alone [62]. GR directly regulates the expression of the gluconeogenic key enzymes PEPCK and G6Pase, suggesting an essential role of GR in gluconeogenesis. Thus, suppression of GR activity in the liver improved hyperglycemia and dyslipidemia in genetic and inducible models of diabetes through down-regulation of PEPCK and G6Pase [63–65] (Fig. 4.1). Apart from gluconeogenesis, GR was shown to control lipid metabolism by regulating enzymes involved in lipogenesis, triglyceride (TG) uptake and fatty acid β-oxidation, resulting in hepatic lipid accumulation upon GR activity [66, 67]. Conversely, upon liver specific GR dysfunction, hepatic steatosis in db/db mice was ameliorated mainly through the induction of hairy enhancer of split 1 (HES1) gene expression [67]. Additionally, hepatocyte-specific GR KO impaired systemic bile acid distribution, hepatic bile acid uptake, and increased the susceptibility to develop cholesterol gallstones [68]. This indicates the importance of functional and balanced GR signaling in the liver. Previous studies identified regulatory factors such as microRNAs, transcription factors, or co-factors that directly interact with GR and thereby fine tune and balance its actions in the liver and other tissues [69–72]. Pharmacological activation of GR through cortisone or dexamethasone has immune-suppressive properties and is therefore commonly used in inflammatory or auto-immune diseases [73], however the use is overshadowed due to its severe metabolic side effects. Chronic glucocorticoid (GC) administration and the induced GR activity can result in metabolic abnormalities including hyperglycemia, insulin resistance, hepatic dyslipidemia, and hypertension [74]. Despite these severe negative effects that come with chronic GR activation, GCs are still widely prescribed. Interestingly, it was previously shown that due to alternative splicing, GR is expressed as two isoforms, GRα and GRβ [75]. Both isoforms are ubiquitously expressed in most tissues, with GRβ to a lesser extent than GRα. Additionally, in contrast to GRα, GRβ lacks a binding pocket for GCs [75, 76]. GRβ was initially shown to act as a GRα antagonist by binding to glucocorticoid responsive elements without inducing gene expression and also through heterodimerization with GRα [75, 77–79]. However, recent studies suggest that GRβ has, apart from the GRα antagonizing property, distinct regulatory functions on gene expression. Animal studies revealed that GRβ is the main regulator of hepatic gluconeogenesis and lipid storage. GRβ gene expression was elevated in the liver upon diet induced obesity [80]. Moreover, upon liver specific GRβ overexpression, hepatic and serum TG levels were significantly elevated. Additionally, GRβ overexpression resulted in hyperglycemia without alterations in circulating insulin levels suggesting increased gluconeogenesis or reduced hepatic insulin signaling [80]. Interestingly, short term GC administration induced lipolysis in adipose tissue through transcription of hormone-sensitive lipase (HSL) and adipose triglyceride lipase (ATGL) [81]. An explanation for this observation might be that short-term exposure to GC specifically activates GRα, which facilitates lipolysis [81]. GRα action however is antagonized by GRβ upon chronic GC administration [75, 77, 78] which in turn induces gluconeogenesis and lipogenesis [80], notably through interaction with PPARα pathways, finally resulting in metabolic disruptions such as hepatic dyslipidemia and hyperglycemia. Given the antagonizing effect of GRβ on GRα and the direct regulation of hepatic glucose and lipid metabolism by GRβ, differentiation between the isoforms is essential in future studies. Specific regulation of either of the GR isoforms thus might provide novel treatment possibilities or reduce the severity of GC-associated side-effects.

4.3 Adipose Tissue

The white adipose tissue (WAT) is the central organ for energy storage. Excess energy is converted into TG and stored in lipid droplets in adipocytes. In nutrient deprived states, fatty acids are released and serve as energy source for other tissues such as liver or skeletal muscle. Additionally, adipose tissue was identified as an endocrine organ and releases a wide range of adipokines which regulate immune responses, control blood pressure, or modify glucose homeostasis. Modern lifestyle but also genetic predisposition account for excessive fat storage in WAT, which is accompanied by hyperglycemia and dyslipidemia. In contrast, brown adipose tissue (BAT) is specialized to dissipate chemical energy in the form of non-shivering thermogenesis. The ability of white adipocytes to convert into brown-like or beige adipocytes, named browning and partly regulated by nuclear receptors (see also Box 4.1), represents an intensively studied field to target lifestyle induced obesity.

4.3.1 PPARγ Is the Master Regulator of Adipose Tissue Function

The master regulator of adipose tissue function is PPARγ. Its expression is rapidly induced during adipogenesis [82], regulating hundreds of genes central to adipocyte function, including lipid transport (FABP4), fatty acid uptake (FATP, LPL), recycling of fatty acids (PEPCK), and lipolysis (GPR81) [83–87] (Fig. 4.2). Indeed, PPARγ KO in pre-adipocytes completely inhibited adipocyte formation [88], proving the vital importance of this nuclear receptor for adipogenesis. In patients, a mutation in the ligand-binding domain of PPARγ led to partial lipodystrophy and insulin resistance [89], and mice lacking PPARγ in the adipose tissue displayed the same phenotype [90]. Aside from the effects on adipogenesis, PPARγ is also important for insulin sensitivity as it regulates adiponectin and resistin expression [91, 92]. Synthetic PPARγ agonists, in particular thiazolidinediones, ameliorate insulin resistance and are widely used in type 2 diabetes mellitus (T2DM) treatment [93]. Fat accumulation in insulin sensitive tissues such as liver and skeletal muscle has been shown to promote insulin resistance. PPARγ agonist-induced upregulation of genes involved in fatty acid uptake and storage in the adipose tissue promotes redistribution, and could thereby prevent ectopic fat accumulation in liver or skeletal muscle. Weight gain, liver damage, and cardiovascular events are, however, common side-effects upon chronic PPARγ activation [94, 95]. Interestingly, in vitro exposure of white adipocytes and in vivo exposure of mice to synthetic PPARγ agonist induced expression of brown fat marker genes including uncoupling protein 1 (UCP1), PR

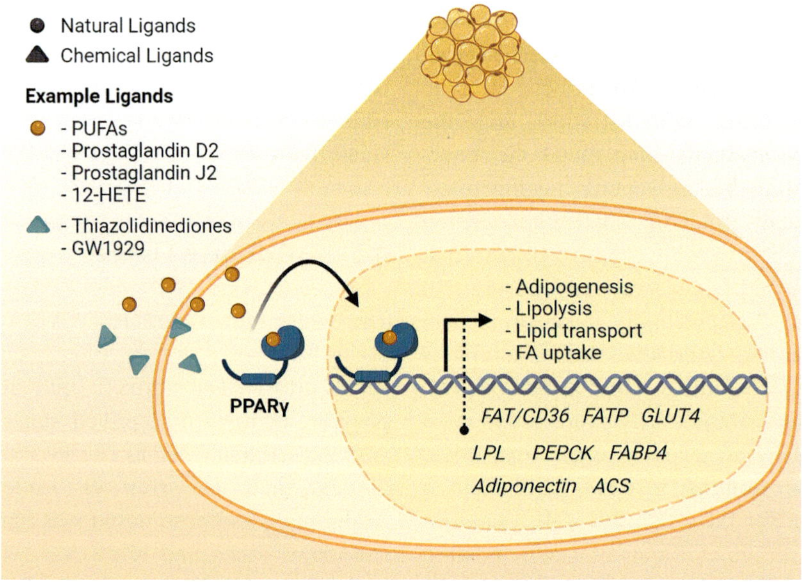

Fig. 4.2 Regulation of adipocyte function by PPARγ. Upon ligand binding PPARγ induces genes involved in adipogenesis, lipid transport (FABP4), fatty acid uptake (FAT/CD36 FATP), lipid recycling (PEPCK, GLUT4), and lipolysis (LPL). *PUFA* polyunsaturated fatty acid, *12-HETE* 12-hydroxyeicosatetraenoic acid, *FAT/CD36* fatty acid transporter, *FATP* fatty acid transport protein, *GLUT4* glucose transporter type 4, *LPL* lipoprotein lipase, *PEPCK* phosphoenolpyruvate carboxykinase, *ACS* acetyl-CoA synthetase, *FABP4* fatty acid-binding protein 4

domain containing 16 (PRDM16), and Cell Death-Inducing DFFA-Like Effector A (CIDEA) [96–98]. The potential of increased energy expenditure by promoting browning in obesity treatment has been previously reviewed [99]. However, although PPARγ agonist administration induced browning of white adipocytes, PPARγ overexpression had no such effect [93] Previous studies have shown that administration of dual agonists, compounds activating two targets simultaneously, surpass effects that are reached by conventional agonists [100]. Interestingly, in vivo simultaneous activation of PPARα and PPARγ through dual agonists synergistically induced browning of white adipocytes [101]. Moreover, combinatorial PPARα and γ activation reduced body weight and ameliorated insulin resistance in diet induced obesity, mainly through FGF21 signaling. Although PPARα is the key regulator of FGF21 expression [12], its effects on browning rely on pharmacological PPARγ activation [101].

Not only white adipocyte metabolism and the white-to-brown transition of adipocytes are regulated by PPARγ, but also BAT relies on normal PPARγ function. In BAT, PPARγ is essential for adipogenesis, adipocyte differentiation, survival, and functionality [90, 102–106]. Similar to WAT, PPARγ ablation in BAT inhibited adipocyte formation [90, 102]. In addition, mature brown adipocytes lost their ability to induce non-shivering thermogenesis as PPARγ directly regulated the key thermogenic proteins UCP1 and PRDM16 [104, 107]. Accordingly, chronic treatment of mice with the PPARγ agonist rosiglitazone increased UCP1 levels in brown adipocytes, and thereby thermogenesis [103]. Interestingly, PGC1α, a PPARγ target gene itself [108], directly interacted with PPARγ to enhance UCP1 gene expression [109]. Conversely, mice lacking PGC1α failed to induce thermogenesis in response to cold exposure [110]. The interaction between PPARγ and PGC1α is highly tissue- and target gene-specific. For example, PGC1α was differentially expressed between BAT and WAT [109], suggesting a BAT-specific function. Moreover, PPARγ-controlled FABP4 expression was PGC1α independent [109], indicating that PCG1α selectively facilitated PPARγ mediated thermogenesis in BAT. Apart from UCP1 expression, rosiglitazone administration upregulated triacylglyceride (TAG) synthesis [103], underlining the importance of PPARγ in BAT, as it regulates expression of key proteins but also the formation of substrates for non-shivering thermogenesis.

Taken together, PPARγ regulates many aspects of white and brown adipose tissue metabolism, which reveals PPARγ as highly promising target for metabolism associated abnormalities. Browning of adipocytes especially, but also increased combusting of energy through BAT, represent two intriguing possibilities to counteract obesity and its related morbidities. To date, the investigation of novel drugs for the treatment of obesity in humans by the induction of browning of white adipocytes has proven difficult.

4.4 Muscle

Skeletal muscle is the largest metabolically active organ in the human body. It is the major site of insulin dependent glucose uptake, glycogen storage, and fatty acid oxidation. Under metabolic disorders such as obesity and diabetes, severe changes occur in the skeletal muscle amongst which the switch from type 1 slow-twitch to type 2 fast-twitch fibers has profound consequences. The transition of the fiber types promotes insulin resistance, further driving the vicious cycle of the metabolic syndrome. Increase of the favorable type 1 slow-twitch muscles will ameliorate the metabolic syndrome by retrieving insulin sensitivity.

4.4.1 PPARβ/δ – Regulator of Skeletal Muscle

PPARβ/δ is the key transcription factor in skeletal muscle function and metabolism (Fig. 4.3). In the skeletal muscle PPARβ/δ regulates glucose and fatty acid metabolism [111, 112], myogenesis [113] and the transition from fast glycolytic 2b to slow/fast oxidative 1/2a fibers [114–116]. In mice and humans, PPARβ/δ expression is

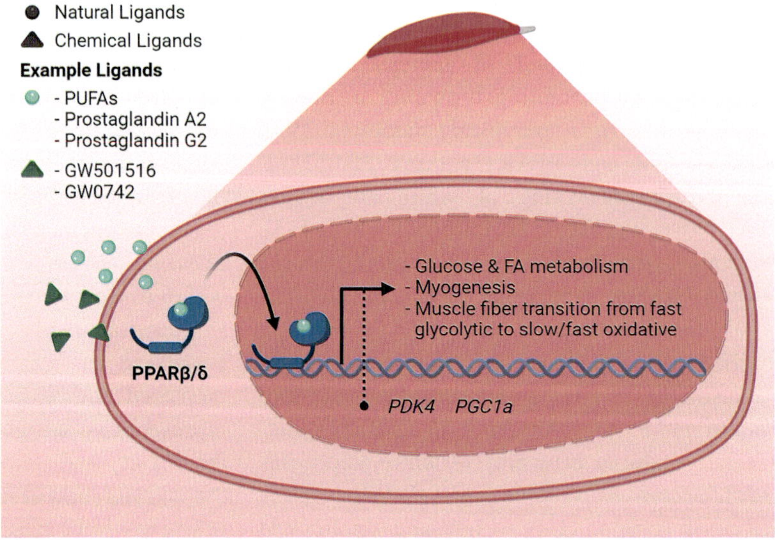

Fig. 4.3 PPARβ/δ and its main function in muscle metabolism. In the muscle, PPARβ/δ regulates genes involved in glucose and fatty acid metabolism. Moreover, ligand induced PPARβ/δ activation favours the muscle fiber transition from fast glycolytic to slow/fast oxidative. *PUFA* polyunsaturated fatty acid, *FA* fatty acid, *PDK4* pyruvate dehydrogenase kinase 4, *PGC1α* peroxisome proliferator-activated receptor γ coactivator 1-α

higher in slow/oxidative muscle types in comparison to fast/glycolytic muscles [117, 118] indicating a significant role in muscle type transition. Indeed, during endurance training, which promotes slow/fast oxidative fiber formation, PPARβ/δ expression was elevated [119]. The increase in the number of slow/oxidative muscle fibers upon PPARβ/δ induction resulted from increased muscle progenitor cell abundance, partly through antagonism of myostatin activity, a potent myokine inhibiting muscle growth [113, 120, 121]. Moreover, induction of overexpression of PPARβ/δ improved wheel-running performance, favored the number of slow/fast oxidative 1/2a fibers, and decreased body fat mass in rodents, partly regulated by the PPARβ/δ target gene PGC1α [114, 116, 119]. Interestingly, as observed above PGC1α was also identified as a PPARβ/δ coactivator [109], leading to a feed-forward loop which ensures constant PGC1α expression and thereby maintenance of slow/oxidative fibers [122, 123]. Accordingly, muscle specific overexpression of PGC1α phenocopies PPARβ/δ overexpression suggesting that both PGC1α and PPARβ/δ facilitate skeletal muscle metabolism and function [123]. Moreover, PPARβ/δ modulates fatty acid metabolism through direct transcriptional control of enzymes involved in lipolysis, lipid uptake, and fatty acid β-oxidation [111, 115]. Regulation of β-oxidation in the muscle by PPARβ/δ is facilitated by its direct target gene pyruvate dehydrogenase kinase 4 (PDK4) [111]. PDK4 inactivates the pyruvate dehydrogenase complex (PDH), which is rate limiting for carbohydrate metabolism, leading to the up-regulation of fatty acid β-oxidation [124]. Accordingly, mice lacking muscle-specific PPARβ/δ suffer from dyslipidemia [115]. In diet-induced and genetic mouse models of obesity, PPARβ/δ agonist administration increased fatty acid β-oxidation and thereby improved dyslipidemia [111]. Moreover, PPARβ/δ agonist treatment improved insulin resistance, elevated proliferation of mitochondria, and reduced lipid droplets in skeletal muscle [111, 116], highlighting the therapeutic potential of PPARβ/δ agonists in the metabolic syndrome. Although agonist-activated PPARβ/δ was shown to oppose T2DM and obesity progression, and mimiced endurance training, none of the PPARβ/δ agonists has reached human application yet. This is largely because, apart from the overall positive effects on skeletal muscle metabolism, PPARβ/δ agonist administration was linked to liver fibrosis and hepatic carcinoma [125, 126]. In a recent study, novel and highly muscle-specific PPARβ/δ agonists were synthesized [127], yet remain to be tested for efficiency and side effects. The development of tissue-specific PPARβ/δ agonists is crucial in order to bypass the severe side effects in other tissues which limit their potential to improve skeletal muscle function.

4.5 Pancreas

Pancreatic β-cells are the main regulators of glucose homeostasis as they secrete insulin in response to glucose, enabling glucose uptake into peripheral tissues. Chronic elevation of blood glucose levels results in β-cell hypertrophy, exhaustion, and dysfunction. Hallmarks of β-cell dysfunction are loss of identity, apoptosis, and insufficient insulin secretion, promoting the progression of T2DM. Counteracting β-cell dysfunction by nuclear receptor manipulation might prevent progression of diabetes.

4.5.1 The NR4A Family of Orphan Nuclear Receptors as Regulators of β-Cell Physiology

The NR4A family of orphan nuclear receptors comprises of three members: nuclear receptor 4 A1 (NR4A1), NR4A2, and NR4A3. All three were identified as important regulators of apoptosis, inflammation, and metabolism. While no function in β-cells for NR4A2 was reported thus far, the NR4A members NR4A1 and NR4A3 have gained substantial interest in pancreatic β-cell research and their roles are summarized in Fig. 4.4. NR4A1 expression was induced by glucose and fatty acids in β-cells [128, 129], indicating a significant role of NR4A1 in β-cell function and metabolism. Indeed, the NR4A1 promoter was hypomethylated in pancreatic islets from T2DM patients and mouse models of T2DM and in turn, induction of NR4A1 expression decreased blood glucose levels [130]. Moreover, NR4A1 deletion in insulin-secreting INS1 832/13 cells inhibited glucose stimulated insulin secretion through impaired mitochondrial respiration and tricarboxylic acid cycle [131, 132]. Additionally, NR4A1 was characterized as a direct NK homeobox 6.1 (NKX6.1) target and thereby induced β-cell proliferation in rat pancreatic islets through up-regulation of cell cycle activating genes [133]. Surprisingly, in MIN6 cells – a murine insulinoma β-cell line capable of insulin secretion in response to glucose stimulation [134] – fatty acid induced NR4A1 expression impaired insulin biosynthesis and insulin secretion through direct protein-protein interaction with forkhead box protein O1 (FOXO1) and down-regulation of pancreatic and duodenal homeobox 1 (PDX-1), MAF BZIP transcription factor A (MAFA), and neurogenic differentiation 1 (NEUROD1), essential transcription factors regulating β-cell identity and function [135]. Apart from insulin secretion, NR4A1 directly regulates endoplasmic reticulum (ER) stress induced apoptosis. ER stress, a result of sustained hyperglycemia and dyslipidemia, is a driver for T2DM progression by inducing apoptosis in pancreatic β-cells [136]. Interestingly, NR4A1 expression positively correlated with the induction of ER stress in vitro and ex vivo, while overexpression of NR4A1 ameliorated ER stress induced apoptosis [136, 137]. NR4A1 was identified as mitogen-activated protein kinase phosphatase 7 (MKP7) transcription factor, which counteracted c-Jun N-terminal kinase (JNK) activity and thereby apoptosis by dephosphorylation of JNK [138]. The lack of suitable and specific NR4A1 ligands has hampered detailed research on NR4A1 function in β-cells. Although cytosporone B (Csn-B) was identified as one of the first naturally occurring agonists for NR4A1, its effects on β-cells remain to be investigated [139]. Systemically, Csn-B administration resulted in increased blood glucose levels partly induced by upregulation of gluconeogenic genes in the liver. Additionally, Csn-B induced apoptosis in tumor cells to inhibit xenograft tumor growth [139], highlighting its promising properties to treat hypoglycemia and cancer.

Similar to NR4A1, findings on NR4A3 function in β-cells are contradictory. Initially, NR4A3 was proposed as novel candidate gene for β-cell function, as common genetic variations within the NR4A3 locus were associated with improved insulin secretion [140]. Glucose, fatty acids, and pro-inflammatory cytokines promote NR4A3

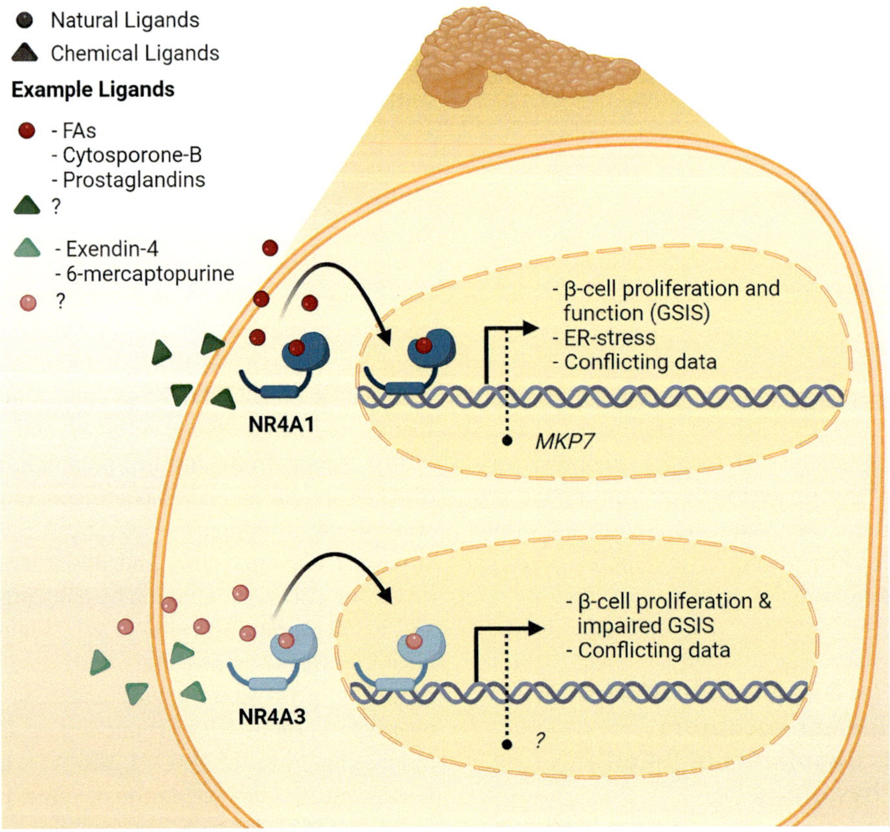

Fig. 4.4 The orphan nuclear receptors NR4A1 and NR4A3 and their function in β-cells. Ligands, main mode of action, and key target genes of NR4A1 and NR4A3 in the pancreatic β-cells. For both nuclear receptors, studies are contradictory. (**a**) NR4A1 expression promotes β-cell proliferation and ameliorates ER stress induced apoptosis through MKP7. However, NR4A1 expression was also associated with impaired β-cell functionality. (**b**) NR4A3 was described as positive regulator of β-cell proliferation. However, NR4A3 expression was linked to impaired β-cell function. The underlying conflicting results might be explained by the different models used in the respective studies. *FA* fatty acid, *MKP7* Mitogen-activated protein kinase phosphatase 7

gene expression [141, 142]. Accordingly, NR4A3 expression was elevated in human islets from T2DM patients in comparison to healthy controls, while global NR4A3 deletion in mice resulted in increased β-cell mass, enhanced β-cell proliferation, and improved glucose tolerance [141]. Additionally, in vitro NR4A3 overexpression negatively correlated with insulin gene expression and secretion [142], suggesting NR4A3 activity impaired β-cell function (Fig. 4.4). Unexpectedly, NKX6.1 was also characterized as a direct regulator of NR4A3 expression, and NR4A3 overexpression promoted β-cell proliferation [133]. Various compounds directly regulate NR4A3 activity. 6-Mercaptopurine was identified as specific NR4A3 agonist in skeletal muscle [143, 144]. Moreover, NR4A3 expression was induced by β-adrenergic receptor agonists, indicating a role in lipid metabolism [145]. Further, exendin-4 was shown to attenuate NR4A3 expression in vascular smooth muscle cells [146]. However, the effects and functionality on β-cells remain to be shown.

In summary, current data support a direct regulatory function of the orphan nuclear receptors NR4A1 and NR4A3 in β-cell functionality and metabolism. The conflicting results on NR4A1 and NR4A3 regulated insulin gene expression and secretion may be explained by the different models used in the respective studies. The conduction of in vivo studies with β-cell specific

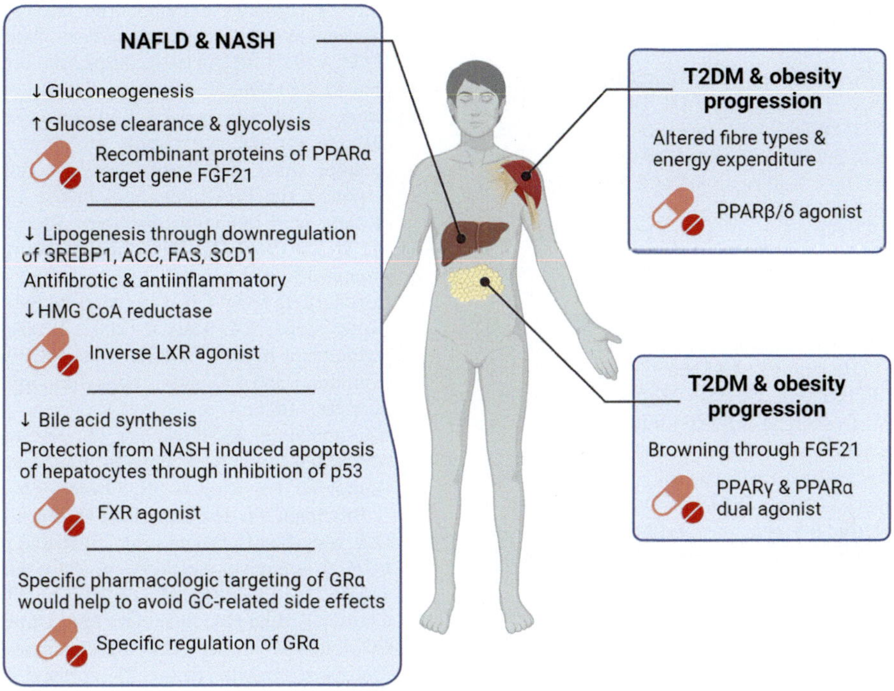

Fig. 4.5 Potential treatment strategies in metabolic diseases. Exemplary treatment strategies for metabolic diseases by pharmacologically targeting nuclear receptors. (**a**) NAFLD & NASH treatment: Recombinant FGF21 proteins downregulate hepatic gluconeogenesis and promote glucose clearance and glycolysis. Inverse LXR agonists reduce lipogenesis and reduce plasma cholesterol levels though inhibition of HMG CoA reductase. FXR agonists reduce bile acid synthesis and thereby protect from NASH induced apoptosis in hepatocytes. Specific GRα manipulation would avoid glucocorticoid induced side effects, which are likely induced by GRβ. (**b**) T2DM and obesity progression: PPARβ/δ activity alters fiber types and increases energy expenditure. A dual PPARα and PPARγ agonist induces browning of white adipocytes and thereby increases energy expenditure. *FGF21* fibroblast growth factor 21, *SREBP* sterol regulatory element-binding transcription factor 1, *ACC* acetyl-CoA carboxylase, *FAS* fatty acid synthase, *SCD1* stearoyl-CoA desaturase

NR4A manipulation will help to further understand the function of NR4A1 and NR4A3 in β-cells. Additionally, the continuous search for novel specific agonists is essential for clinical applications such as in the treatment of β-cell dysfunction in T2DM.

4.6 Conclusion

Nuclear receptors are key regulators of metabolism and their function is indispensable for metabolic health. Together with co-factors and other co-regulators, they govern a wide range of tissue- and context-specific functions influencing lipid and glucose metabolism. As a result, manipulation and therapeutic targeting of nuclear receptor function has been intensively studied and continues to produce novel and promising drug candidates for metabolic diseases including T2DM, NAFLD, and NASH (Fig. 4.5).

Acknowledgements We thank Dr. Luke Harrison for figure design and editorial support. Figures were created with Biorender.com.

Funding A.W. and M.R. are funded through Helmholtz Association – Initiative and Networking Fund. M.R. is funded by the European Association for the Study of Diabetes (EASD), and the Helmholtz Future Topic Aging and Metabolic Programming (AMPro, ZT-0026). S.H. receives funding by the Deutsche Forschungsgemeinschaft (DFG; SFB 1123 and SFB1118) and the Else Kröner-Fresenius Foundation.

References

1. Kliewer SA, Forman BM, Blumberg B, Ong ES, Borgmeyer U, Mangelsdorf DJ, Umesono K, Evans RM (1994) Differential expression and activation of a family of murine peroxisome proliferator-activated receptors. Proc Natl Acad Sci U S A 91:7355–7359. https://doi.org/10.1073/pnas.91.15.7355
2. Braissant O, Foufelle F, Scotto C, Dauça M, Wahli W (1996) Differential expression of peroxisome proliferator-activated receptors (PPARs): tissue distribution of PPAR-alpha, -beta, and -gamma in the adult rat. Endocrinology 137:354–366. https://doi.org/10.1210/endo.137.1.8536636
3. Escher P, Braissant O, Basu-Modak S, Michalik L, Wahli W, Desvergne B (2001) Rat PPARs: quantitative analysis in adult rat tissues and regulation in fasting and refeeding. Endocrinology 142:4195–4202
4. Issemann I, Green S (1990) Activation of a member of the steroid hormone receptor superfamily by peroxisome proliferators. Nature 347:645–650. https://doi.org/10.1038/347645a0
5. Francque S, an Verrijken SC, Prawitt J, Paumelle R, Derudas B, Lefebvre P, Taskinen M-R, van Hul W, Mertens I, Hubens G, van Marck E, Michielsen P, van Gaal L, Staels B (2015) PPARα gene expression correlates with severity and histological treatment response in patients with non-alcoholic steatohepatitis. J Hepatol 63:164–173. https://doi.org/10.1016/j.jhep.2015.02.019
6. Montagner A, Polizzi A, Fouché E, Ducheix S, Lippi Y, Lasserre F, Barquissau V, Régnier M, Lukowicz C, Benhamed F, Iroz A, Bertrand-Michel J, Al Saati T, Cano P, Mselli-Lakhal L, Mithieux G, Rajas F, Lagarrigue S, Pineau T, Loiseau N, Postic C, Langin D, Wahli W, Guillou H (2016) Liver PPARα is crucial for whole-body fatty acid homeostasis and is protective against NAFLD. Gut 65:1202–1214. https://doi.org/10.1136/gutjnl-2015-310798
7. Panadero M, Herrera E, Bocos C (2005) Different sensitivity of PPARalpha gene expression to nutritional changes in liver of suckling and adult rats. Life Sci 76:1061–1072. https://doi.org/10.1016/j.lfs.2004.10.018
8. Kersten S, Seydoux J, Peters JM, Gonzalez FJ, Desvergne B, Wahli W (1999) Peroxisome proliferator-activated receptor alpha mediates the adaptive response to fasting. J Clin Invest 103:1489–1498. https://doi.org/10.1172/JCI6223
9. Motojima K, Passilly P, Peters JM, Gonzalez FJ, Latruffe N (1998) Expression of putative fatty acid transporter genes are regulated by peroxisome proliferator-activated receptor alpha and gamma activators in a tissue- and inducer-specific manner. J Biol Chem 273:16710–16714. https://doi.org/10.1074/jbc.273.27.16710
10. Gulick T, Cresci S, Caira T, Moore DD, Kelly DP (1994) The peroxisome proliferator-activated receptor regulates mitochondrial fatty acid oxidative enzyme gene expression. Proc Natl Acad Sci U S A 91:11012–11016. https://doi.org/10.1073/pnas.91.23.11012
11. Leone TC, Weinheimer CJ, Kelly DP (1999) A critical role for the peroxisome proliferator-activated receptor alpha (PPARalpha) in the cellular fasting response: the PPARalpha-null mouse as a model of fatty acid oxidation disorders. Proc Natl Acad Sci U S A 96:7473–7478. https://doi.org/10.1073/pnas.96.13.7473
12. Lundåsen T, Mary CH, Lisa-Mari N, Sabyasachi S, Bo A, Stefan EA, Mats R (2007) PPARα is a key regulator of hepatic FGF21. Biochem Biophys Res Commun 360:437–440. https://doi.org/10.1016/j.bbrc.2007.06.068
13. Kharitonenkov A, Shiyanova TL, Koester A, Ford AM, Micanovic R, Galbreath EJ, Sandusky GE, Hammond LJ, Moyers JS, Owens RA, Gromada J, Brozinick JT, Hawkins ED, Wroblewski VJ, Li D-S, Mehrbod F, Jaskunas SR, Shanafelt AB (2005) FGF-21 as a novel metabolic regulator. J Clin Invest 115:1627–1635. https://doi.org/10.1172/JCI23606
14. Coskun T, Bina HA, Schneider MA, Dunbar JD, Hu CC, Chen Y, Moller DE, Kharitonenkov A (2008) Fibroblast growth factor 21 corrects obesity in mice. Endocrinology 149:6018–6027. https://doi.org/10.1210/en.2008-0816
15. Berglund ED, Li CY, Bina HA, Lynes SE, Michael MD, Shanafelt AB, Kharitonenkov A, Wasserman DH (2009) Fibroblast growth factor 21 controls glycemia via regulation of hepatic glucose flux and insulin sensitivity. Endocrinology 150:4084–4093. https://doi.org/10.1210/en.2009-0221
16. Kharitonenkov A, Wroblewski VJ, Koester A, Chen Y-F, Clutinger CK, Tigno XT, Hansen BC, Shanafelt AB, Etgen GJ (2007) The metabolic state of diabetic monkeys is regulated by fibroblast growth factor-21. Endocrinology 148:774–781
17. Tanaka N, Takahashi S, Zhang Y, Krausz KW, Smith PB, Patterson AD, Gonzalez FJ (2015) Role of fibroblast growth factor 21 in the early stage of NASH induced by methionine- and choline-deficient diet. Biochim Biophys Acta (BBA) – Mol Basis Dis 1852:1242–1252. https://doi.org/10.1016/j.bbadis.2015.02.012
18. Fisher FM, Chui PC, Nasser IA, Popov Y, Cunniff JC, Lundasen T, Kharitonenkov A, Schuppan D, Flier JS, Maratos-Flier E (2014) Fibroblast growth factor 21 limits lipotoxicity by promoting hepatic fatty acid activation in mice on methionine and choline-deficient diets. Gastroenterology 147:1073–83.e6. https://doi.org/10.1053/j.gastro.2014.07.044
19. Xu J, Lloyd DJ, Hale C, Stanislaus S, Chen M, Sivits G, Vonderfecht S, Hecht R, Li Y-S, Lindberg RA, Chen J-L, Jung DY, Zhang Z, Ko H-J, Kim JK, Véniant MM (2009) Fibroblast growth factor 21 reverses hepatic steatosis, increases energy expenditure, and improves insulin sensitivity in diet-induced

obese mice. Diabetes 58:250–259. https://doi.org/10.2337/db08-0392
20. Sanyal A, Charles ED, Neuschwander-Tetri BA, Loomba R, Harrison SA, Abdelmalek MF, Lawitz EJ, Halegoua-DeMarzio D, Kundu S, Noviello S, Luo Y, Christian R (2019) Pegbelfermin (BMS-986036), a PEGylated fibroblast growth factor 21 analogue, in patients with non-alcoholic steatohepatitis: a randomised, double-blind, placebo-controlled, phase 2a trial. Lancet 392:2705–2717. https://doi.org/10.1016/S0140-6736(18)31785-9
21. Charles ED, Neuschwander-Tetri BA, Pablo Frias J, Kundu S, Luo Y, Tirucherai GS, Christian R (2019) Pegbelfermin (BMS-986036), PEGylated FGF21, in patients with obesity and type 2 diabetes: results from a randomized phase 2 study. Obesity (Silver Spring) 27:41–49. https://doi.org/10.1002/oby.22344
22. Wei W, Dutchak PA, Wang X, Ding X, Wang X, Bookout AL, Goetz R, Mohammadi M, Gerard RD, Dechow PC, Mangelsdorf DJ, Kliewer SA, Wan Y (2012) Fibroblast growth factor 21 promotes bone loss by potentiating the effects of peroxisome proliferator-activated receptor γ. Proc Natl Acad Sci U S A 109:3143–3148. https://doi.org/10.1073/pnas.1200797109
23. Thompson KE, Guillot M, Graziano MJ, Mangipudy RS, Chadwick KD (2021) Pegbelfermin, a PEGylated FGF21 analogue, has pharmacology without bone toxicity after 1-year dosing in skeletally-mature monkeys. Toxicol Appl Pharmacol 428:115673. https://doi.org/10.1016/j.taap.2021.115673
24. Kim AM, Somayaji VR, Dong JQ, Rolph TP, Weng Y, Chabot JR, Gropp KE, Talukdar S, Calle RA (2017) Once-weekly administration of a long-acting fibroblast growth factor 21 analogue modulates lipids, bone turnover markers, blood pressure and body weight differently in obese people with hypertriglyceridaemia and in non-human primates. Diabetes Obes Metab 19:1762–1772. https://doi.org/10.1111/dom.13023
25. Stanislaus S, Hecht R, Yie J, Hager T, Hall M, Spahr C, Wang W, Weiszmann J, Li Y, Deng L, Winters D, Smith S, Zhou L, Li Y, Véniant MM, Xu J (2017) A novel Fc-FGF21 with improved resistance to proteolysis, increased affinity toward β-Klotho, and enhanced efficacy in mice and cynomolgus monkeys. Endocrinology 158:1314–1327. https://doi.org/10.1210/en.2016-1917
26. Kaufman A, Abuqayyas L, Denney WS, Tillman EJ, Rolph T (2020) AKR-001, an Fc-FGF21 analog, showed sustained pharmacodynamic effects on insulin sensitivity and lipid metabolism in type 2 diabetes patients. Cell Rep Med 1:100057. https://doi.org/10.1016/j.xcrm.2020.100057
27. Harrison SA, Ruane PJ, Freilich BL, Neff G, Patil R, Behling CA, Hu C, Fong E, de Temple B, Tillman EJ, Rolph TP, Cheng A, Yale K (2021) Efruxifermin in non-alcoholic steatohepatitis: a randomized, double-blind, placebo-controlled, phase 2a trial. Nat Med 27:1262–1271. https://doi.org/10.1038/s41591-021-01425-3
28. Peet DJ, Turley SD, Ma W, Janowski BA, Lobaccaro JM, Hammer RE, Mangelsdorf DJ (1998) Cholesterol and bile acid metabolism are impaired in mice lacking the nuclear oxysterol receptor LXR alpha. Cell 93:693–704. https://doi.org/10.1016/s0092-8674(00)81432-4
29. Repa JJ, Liang G, Ou J, Bashmakov Y, Lobaccaro JM, Shimomura I, Shan B, Brown MS, Goldstein JL, Mangelsdorf DJ (2000) Regulation of mouse sterol regulatory element-binding protein-1c gene (SREBP-1c) by oxysterol receptors, LXRalpha and LXRbeta. Genes Dev 14:2819–2830. https://doi.org/10.1101/gad.844900
30. Commerford SR, Vargas L, Dorfman SE, Mitro N, Rocheford EC, Mak PA, Li X, Kennedy P, Mullarkey TL, Saez E (2007) Dissection of the insulin-sensitizing effect of liver X receptor ligands. Mol Endocrinol 21:3002–3012. https://doi.org/10.1210/me.2007-0156
31. Willy PJ, Umesono K, Ong ES, Evans RM, Heyman RA, Mangelsdorf DJ (1995) LXR, a nuclear receptor that defines a distinct retinoid response pathway. Genes Dev 9:1033–1045. https://doi.org/10.1101/gad.9.9.1033
32. Teboul M, Enmark E, Li Q, Wikström AC, Pelto-Huikko M, Gustafsson JA (1995) OR-1, a member of the nuclear receptor superfamily that interacts with the 9-cis-retinoic acid receptor. Proc Natl Acad Sci U S A 92:2096–2100. https://doi.org/10.1073/pnas.92.6.2096
33. Janowski BA, Willy PJ, Devi TR, Falck JR, Mangelsdorf DJ (1996) An oxysterol signalling pathway mediated by the nuclear receptor LXRα. Nature 383:728–731. https://doi.org/10.1038/383728a0
34. Chiang JY, Kimmel R, Weinberger C, Stroup D (2000) Farnesoid X receptor responds to bile acids and represses cholesterol 7alpha-hydroxylase gene (CYP7A1) transcription. J Biol Chem 275:10918–10924. https://doi.org/10.1074/jbc.275.15.10918
35. Inagaki T, Choi M, Moschetta A, Peng L, Cummins CL, McDonald JG, Luo G, Jones SA, Goodwin B, Richardson JA, Gerard RD, Repa JJ, Mangelsdorf DJ, Kliewer SA (2005) Fibroblast growth factor 15 functions as an enterohepatic signal to regulate bile acid homeostasis. Cell Metab 2:217–225. https://doi.org/10.1016/j.cmet.2005.09.001
36. Guo GL, Lambert G, Negishi M, Ward JM, Brewer HB, Kliewer SA, Gonzalez FJ, Sinal CJ (2003) Complementary roles of farnesoid X receptor, pregnane X receptor, and constitutive androstane receptor in protection against bile acid toxicity. J Biol Chem 278:45062–45071
37. Huang L, Zhao A, Lew J-L, Zhang T, Hrywna Y, Thompson JR, de Pedro N, Royo I, Blevins RA, Peláez F, Wright SD, Cui J (2003) Farnesoid X

receptor activates transcription of the phospholipid pump MDR3. J Biol Chem 278:51085–51090. https://doi.org/10.1074/jbc.M308321200
38. Plass JRM, Mol O, Heegsma J, Geuken M, Faber KN, Jansen PLM, Müller M (2002) Farnesoid X receptor and bile salts are involved in transcriptional regulation of the gene encoding the human bile salt export pump. Hepatology 35:589–596. https://doi.org/10.1053/jhep.2002.31724
39. Ananthanarayanan M, Balasubramanian N, Makishima M, Mangelsdorf DJ, Suchy FJ (2001) Human bile salt export pump promoter is transactivated by the farnesoid X receptor/bile acid receptor. J Biol Chem 276:28857–28865. https://doi.org/10.1074/jbc.M011610200
40. Repa JJ, Berge KE, Pomajzl C, Richardson JA, Hobbs H, Mangelsdorf DJ (2002) Regulation of ATP-binding cassette sterol transporters ABCG5 and ABCG8 by the liver X receptors alpha and beta. J Biol Chem 277:18793–18800. https://doi.org/10.1074/jbc.M109927200
41. Alberti S, Steffensen KR, Gustafsson JA (2000) Structural characterisation of the mouse nuclear oxysterol receptor genes LXRalpha and LXRbeta. Gene 243:93–103. https://doi.org/10.1016/s0378-1119(99)00555-7
42. Alberti S, Schuster G, Parini P, Feltkamp D, Diczfalusy U, Rudling M, Angelin B, Björkhem I, Pettersson S, Gustafsson JA (2001) Hepatic cholesterol metabolism and resistance to dietary cholesterol in LXRbeta-deficient mice. J Clin Invest 107:565–573. https://doi.org/10.1172/JCI9794
43. Korach-André M, Archer A, Barros RP, Parini P, Gustafsson J-Å (2011) Both liver-X receptor (LXR) isoforms control energy expenditure by regulating brown adipose tissue activity. Proc Natl Acad Sci U S A 108:403–408
44. Ogihara T, Chuang J-C, Vestermark GL, Garmey JC, Ketchum RJ, Huang X, Brayman KL, Thorner MO, Repa JJ, Mirmira RG, Evans-Molina C (2010) Liver X receptor agonists augment human islet function through activation of anaplerotic pathways and glycerolipid/free fatty acid cycling. J Biol Chem 285:5392–5404. https://doi.org/10.1074/jbc.M109.064659
45. Laffitte BA, Chao LC, Li J, Walczak R, Hummasti S, Joseph SB, Castrillo A, Wilpitz DC, Mangelsdorf DJ, Collins JL, Saez E, Tontonoz P (2003) Activation of liver X receptor improves glucose tolerance through coordinate regulation of glucose metabolism in liver and adipose tissue. Proc Natl Acad Sci U S A 100:5419–5424. https://doi.org/10.1073/pnas.0830671100
46. Baranowski M, Zabielski P, Błachnio-Zabielska AU, Harasim E, Chabowski A, Górski J (2014) Insulin-sensitizing effect of LXR agonist T0901317 in high-fat fed rats is associateddoi: with restored muscle GLUT4 expression and insulin-stimulated AS160 phosphorylation. Cell Physiol Biochem 33:1047–1057. https://doi.org/10.1159/000358675
47. Cao G, Liang Y, Broderick CL, Oldham BA, Beyer TP, Schmidt RJ, Zhang Y, Stayrook KR, Suen C, Otto KA, Miller AR, Dai J, Foxworthy P, Gao H, Ryan TP, Jiang X-C, Burris TP, Eacho PI, Etgen GJ (2003) Antidiabetic action of a liver x receptor agonist mediated by inhibition of hepatic gluconeogenesis. J Biol Chem 278:1131–1136. https://doi.org/10.1074/jbc.M210208200
48. Zhang Y, Lee FY, Barrera G, Lee H, Vales C, Gonzalez FJ, Willson TM, Edwards PA (2006) Activation of the nuclear receptor FXR improves hyperglycemia and hyperlipidemia in diabetic mice. Proc Natl Acad Sci U S A 103:1006–1011. https://doi.org/10.1073/pnas.0506982103
49. Renga B, Mencarelli A, D'Amore C, Cipriani S, Baldelli F, Zampella A, Distrutti E, Fiorucci S (2012) Glucocorticoid receptor mediates the gluconeogenic activity of the farnesoid X receptor in the fasting condition. FASEB J 26:3021–3031. https://doi.org/10.1096/fj.11-195701
50. Cariou B, van Harmelen K, Duran-Sandoval D, van Dijk TH, Grefhorst A, Abdelkarim M, Caron S, Torpier G, Fruchart J-C, Gonzalez FJ, Kuipers F, Staels B (2006) The farnesoid X receptor modulates adiposity and peripheral insulin sensitivity in mice. J Biol Chem 281:11039–11049. https://doi.org/10.1074/jbc.M510258200
51. Goodwin B, Watson MA, Kim H, Miao J, Kemper JK, Kliewer SA (2003) Differential regulation of rat and human CYP7A1 by the nuclear oxysterol receptor liver X receptor-alpha. Mol Endocrinol 17:386–394. https://doi.org/10.1210/me.2002-0246
52. Schultz JR, Tu H, Luk A, Repa JJ, Medina JC, Li L, Schwendner S, Wang S, Thoolen M, Mangelsdorf DJ, Lustig KD, Shan B (2000) Role of LXRs in control of lipogenesis. Genes Dev 14:2831–2838. https://doi.org/10.1101/gad.850400
53. Chen Z, Chen H, Zhang Z, Ding P, Yan X, Li Y, Zhang S, Gu Q, Zhou H, Xu J (2020) Discovery of novel liver X receptor inverse agonists as lipogenesis inhibitors. Eur J Med Chem 206:112793. https://doi.org/10.1016/j.ejmech.2020.112793
54. Huang P, Kaluba B, Jiang X, Chang S, Tang X, Mao L, Zhang Z, Huang F, Zhai L (2018) Liver X receptor inverse agonist SR9243 suppresses nonalcoholic steatohepatitis intrahepatic inflammation and fibrosis. Biomed Res Int 2018:8071093. https://doi.org/10.1155/2018/8071093
55. Griffett K, Welch RD, Flaveny CA, Kolar GR, Neuschwander-Tetri BA, Burris TP (2015) The LXR inverse agonist SR9238 suppresses fibrosis in a model of non-alcoholic steatohepatitis. Mol Metab 4:353–357. https://doi.org/10.1016/j.molmet.2015.01.009
56. Griffett K, Solt LA, El-Gendy BE-DM, Kamenecka TM, Burris TP (2013) A liver-selective LXR inverse

agonist that suppresses hepatic steatosis. ACS Chem Biol 8:559–567. https://doi.org/10.1021/cb300541g
57. Goto T, Itoh M, Suganami T, Kanai S, Shirakawa I, Sakai T, Asakawa M, Yoneyama T, Kai T, Ogawa Y (2018) Obeticholic acid protects against hepatocyte death and liver fibrosis in a murine model of nonalcoholic steatohepatitis. Sci Rep 8:8157. https://doi.org/10.1038/s41598-018-26383-8
58. Neuschwander-Tetri BA, Loomba R, Sanyal AJ, Lavine JE, van Natta ML, Abdelmalek MF, Chalasani N, Dasarathy S, Diehl AM, Hameed B, Kowdley KV, McCullough A, Terrault N, Clark JM, Tonascia J, Brunt EM, Kleiner DE, Doo E (2015) Farnesoid X nuclear receptor ligand obeticholic acid for non-cirrhotic, non-alcoholic steatohepatitis (FLINT): a multicentre, randomised, placebo-controlled trial. Lancet 385:956–965. https://doi.org/10.1016/S0140-6736(14)61933-4
59. Patel K, Harrison SA, Elkhashab M, Trotter JF, Herring R, Rojter SE, Kayali Z, Wong VW-S, Greenbloom S, Jayakumar S, Shiffman ML, Freilich B, Lawitz EJ, Gane EJ, Harting E, Xu J, Billin AN, Chung C, Djedjos CS, Subramanian GM, Myers RP, Middleton MS, Rinella M, Noureddin M (2020) Cilofexor, a nonsteroidal FXR agonist, in patients with noncirrhotic NASH: a phase 2 randomized controlled trial. Hepatology 72:58–71. https://doi.org/10.1002/hep.31205
60. Ratziu V, Sanyal AJ, Loomba R, Rinella M, Harrison S, Anstee QM, Goodman Z, Bedossa P, MacConell L, Shringarpure R, Shah A, Younossi Z (2019) REGENERATE: design of a pivotal, randomised, phase 3 study evaluating the safety and efficacy of obeticholic acid in patients with fibrosis due to nonalcoholic steatohepatitis. Contemp Clin Trials 84:105803. https://doi.org/10.1016/j.cct.2019.06.017
61. Schwabl P, Hambruch E, Budaş GR, Supper P, Burnet M, Liles JT, Birkel M, Brusilovskaya K, Königshofer P, Peck-Radosavljevic M, Watkins WJ, Trauner M, Breckenridge DG, Kremoser C, Reiberger T (2021) The non-steroidal FXR agonist cilofexor improves portal hypertension and reduces hepatic fibrosis in a rat NASH model. Biomedicine 9. https://doi.org/10.3390/biomedicines9010060
62. Phuc Le P, Friedman JR, Schug J, Brestelli JE, Parker JB, Bochkis IM, Kaestner KH (2005) Glucocorticoid receptor-dependent gene regulatory networks. PLoS Genet 1:e16. https://doi.org/10.1371/journal.pgen.0010016
63. Opherk C, Tronche F, Kellendonk C, Kohlmüller D, Schulze A, Schmid W, Schütz G (2004) Inactivation of the glucocorticoid receptor in hepatocytes leads to fasting hypoglycemia and ameliorates hyperglycemia in streptozotocin-induced diabetes mellitus. Mol Endocrinol 18:1346–1353. https://doi.org/10.1210/me.2003-0283
64. Liang Y, Osborne MC, Monia BP, Bhanot S, Watts LM, She P, DeCarlo SO, Chen X, Demarest K (2005) Antisense oligonucleotides targeted against glucocorticoid receptor reduce hepatic glucose production and ameliorate hyperglycemia in diabetic mice. Metabolism 54:848–855. https://doi.org/10.1016/j.metabol.2005.01.030
65. Watts LM, Manchem VP, Leedom TA, Rivard AL, McKay RA, Bao D, Neroladakis T, Monia BP, Bodenmiller DM, Cao JX-C, Zhang HY, Cox AL, Jacobs SJ, Michael MD, Sloop KW, Bhanot S (2005) Reduction of hepatic and adipose tissue glucocorticoid receptor expression with antisense oligonucleotides improves hyperglycemia and hyperlipidemia in diabetic rodents without causing systemic glucocorticoid antagonism. Diabetes 54:1846–1853. https://doi.org/10.2337/diabetes.54.6.1846
66. Cole TG, Wilcox HG, Heimberg M (1982) Effects of adrenalectomy and dexamethasone on hepatic lipid metabolism. J Lipid Res 23:81–91. https://doi.org/10.1016/S0022-2275(20)38176-1
67. Lemke U, Krones-Herzig A, Diaz MB, Narvekar P, Ziegler A, Vegiopoulos A, Cato AC, Bohl S, Klingmüller U, Screaton RA, Müller-Decker K, Kersten S, Herzig S (2008) The glucocorticoid receptor controls hepatic dyslipidemia through Hes1. Cell Metab 8:212–223. https://doi.org/10.1016/j.cmet.2008.08.001
68. Rose AJ, Díaz MB, Reimann A, Klement J, Walcher T, Krones-Herzig A, Strobel O, Werner J, Peters A, Kleyman A, Tuckermann JP, Vegiopoulos A, Herzig S (2011) Molecular control of systemic bile acid homeostasis by the liver glucocorticoid receptor. Cell Metab 14:123–130. https://doi.org/10.1016/j.cmet.2011.04.010
69. de Guia RM, Rose AJ, Sommerfeld A, Seibert O, Strzoda D, Zota A, Feuchter Y, Krones-Herzig A, Sijmonsma T, Kirilov M, Sticht C, Gretz N, Dallinga Thie G, Diederichs S, Klöting N, Blüher M, Berriel Diaz M, Herzig S (2015) microRNA-379 couples glucocorticoid hormones to dysfunctional lipid homeostasis. EMBO J 34:344–360. https://doi.org/10.15252/embj.201490464
70. Glantschnig C, Koenen M, Gil-Lozano M, Karbiener M, Pickrahn I, Williams-Dautovich J, Patel R, Cummins CL, Giroud M, Hartleben G, Vogl E, Blüher M, Tuckermann J, Uhlenhaut H, Herzig S, Scheideler M (2019) A miR-29a-driven negative feedback loop regulates peripheral glucocorticoid receptor signaling. FASEB J 33:5924–5941. https://doi.org/10.1096/fj.201801385RR
71. Hemmer MC, Wierer M, Schachtrup K, Downes M, Hübner N, Evans RM, Uhlenhaut NH (2019) E47 modulates hepatic glucocorticoid action. Nat Commun 10:306. https://doi.org/10.1038/s41467-018-08196-5
72. Jones CL, Bhatla T, Blum R, Wang J, Paugh SW, Wen X, Bourgeois W, Bitterman DS, Raetz EA, Morrison

DJ, Teachey DT, Evans WE, Garabedian MJ, Carroll WL (2014) Loss of TBL1XR1 disrupts glucocorticoid receptor recruitment to chromatin and results in glucocorticoid resistance in a B-lymphoblastic leukemia model. J Biol Chem 289:20502–20515. https://doi.org/10.1074/jbc.M114.569889
73. Greulich F, Wierer M, Mechtidou A, Gonzalez-Garcia O, Uhlenhaut NH (2021) The glucocorticoid receptor recruits the COMPASS complex to regulate inflammatory transcription at macrophage enhancers. Cell Rep 34:108742. https://doi.org/10.1016/j.celrep.2021.108742
74. Schäcke H, Döcke WD, Asadullah K (2002) Mechanisms involved in the side effects of glucocorticoids. Pharmacol Ther 96:23–43. https://doi.org/10.1016/s0163-7258(02)00297-8
75. Hinds TD Jr, Ramakrishnan S, Cash HA, Stechschulte LA, Heinrich G, Najjar SM, Sanchez ER (2010) Discovery of glucocorticoid receptor-beta in mice with a role in metabolism. Mol Endocrinol 24:1715–1727. https://doi.org/10.1210/me.2009-0411
76. Oakley RH, Webster JC, Sar M, Parker CR Jr, Cidlowski JA (1997) Expression and subcellular distribution of the beta-isoform of the human glucocorticoid receptor. Endocrinology 138:5028–5038. https://doi.org/10.1210/endo.138.11.5501
77. Bamberger CM, Bamberger AM, de Castro M, Chrousos GP (1995) Glucocorticoid receptor beta, a potential endogenous inhibitor of glucocorticoid action in humans. J Clin Invest 95:2435–2441. https://doi.org/10.1172/JCI117943
78. Chatzopoulou A, Roy U, Meijer AH, Alia A, Spaink HP, Schaaf MJM (2015) Transcriptional and metabolic effects of glucocorticoid receptor α and β signaling in zebrafish. Endocrinology 156:1757–1769. https://doi.org/10.1210/en.2014-1941
79. Oakley RH, Jewell CM, Yudt MR, Bofetiado DM, Cidlowski JA (1999) The dominant negative activity of the human glucocorticoid receptor β isoform: specificity and mechanisms of action*. J Biol Chem 274:27857–27866. https://doi.org/10.1074/jbc.274.39.27857
80. Marino JS, Stechschulte LA, Stec DE, Nestor-Kalinoski A, Coleman S, Hinds TD Jr (2016) Glucocorticoid receptor β induces hepatic steatosis by augmenting inflammation and inhibition of the peroxisome proliferator-activated receptor (PPAR) α. J Biol Chem 291:25776–25788. https://doi.org/10.1074/jbc.M116.752311
81. Xu C, He J, Jiang H, Zu L, Zhai W, Pu S, Xu G (2009) Direct effect of glucocorticoids on lipolysis in adipocytes. Mol Endocrinol 23:1161–1170. https://doi.org/10.1210/me.2008-0464
82. Chawla A, Schwarz EJ, Dimaculangan DD, Lazar MA (1994) Peroxisome proliferator-activated receptor (PPAR) gamma: adipose-predominant expression and induction early in adipocyte differentiation. Endocrinology 135:798–800. https://doi.org/10.1210/endo.135.2.8033830
83. Tontonoz P, Graves RA, Budavari AI, Erdjument-Bromage H, Lui M, Hu E, Tempst P, Spiegelman BM (1994) Adipocyte-specific transcription factor ARF6 is a heterodimeric complex of two nuclear hormone receptors, PPAR gamma and RXR alpha. Nucleic Acids Res 22:5628–5634. https://doi.org/10.1093/nar/22.25.5628
84. Frohnert BI, Hui TY, Bernlohr DA (1999) Identification of a functional peroxisome proliferator-responsive element in the murine fatty acid transport protein gene. J Biol Chem 274:3970–3977. https://doi.org/10.1074/jbc.274.7.3970
85. Jeninga EH, Bugge A, Nielsen R, Kersten S, Hamers N, Dani C, Wabitsch M, Berger R, Stunnenberg HG, Mandrup S, Kalkhoven E (2009) Peroxisome proliferator-activated receptor gamma regulates expression of the anti-lipolytic G-protein-coupled receptor 81 (GPR81/Gpr81). J Biol Chem 284:26385–26393. https://doi.org/10.1074/jbc.M109.040741
86. Tontonoz P, Hu E, Devine J, Beale EG, Spiegelman BM (1995) PPAR gamma 2 regulates adipose expression of the phosphoenolpyruvate carboxykinase gene. Mol Cell Biol 15:351–357. https://doi.org/10.1128/MCB.15.1.351
87. Schoonjans K, Peinado-Onsurbe J, Lefebvre AM, Heyman RA, Briggs M, Deeb S, Staels B, Auwerx J (1996) PPARalpha and PPARgamma activators direct a distinct tissue-specific transcriptional response via a PPRE in the lipoprotein lipase gene. EMBO J 15:5336–5348
88. Rosen ED, Sarraf P, Troy AE, Bradwin G, Moore K, Milstone DS, Spiegelman BM, Mortensen RM (1999) PPARγ is required for the differentiation of adipose tissue in vivo and in vitro. Mol Cell 4:611–617. https://doi.org/10.1016/S1097-2765(00)80211-7
89. Savage DB, Tan GD, Acerini CL, Jebb SA, Agostini M, Gurnell M, Williams RL, Umpleby AM, Thomas EL, Bell JD, Dixon AK, Dunne F, Boiani R, Cinti S, Vidal-Puig A, Karpe F, Chatterjee VKK, O'Rahilly S (2003) Human metabolic syndrome resulting from dominant-negative mutations in the nuclear receptor peroxisome proliferator-activated receptor-gamma. Diabetes 52:910–917. https://doi.org/10.2337/diabetes.52.4.910
90. He W, Barak Y, Hevener A, Olson P, Liao D, Le J, Nelson M, Ong E, Olefsky JM, Evans RM (2003) Adipose-specific peroxisome proliferator-activated receptor gamma knockout causes insulin resistance in fat and liver but not in muscle. Proc Natl Acad Sci U S A 100:15712–15717. https://doi.org/10.1073/pnas.2536828100
91. Iwaki M, Matsuda M, Maeda N, Funahashi T, Matsuzawa Y, Makishima M, Shimomura I (2003) Induction of adiponectin, a fat-derived antidiabetic

and antiatherogenic factor, by nuclear receptors. Diabetes 52:1655–1663. https://doi.org/10.2337/diabetes.52.7.1655
92. Tomaru T, Steger DJ, Lefterova MI, Schupp M, Lazar MA (2009) Adipocyte-specific expression of murine resistin is mediated by synergism between peroxisome proliferator-activated receptor gamma and CCAAT/enhancer-binding proteins. J Biol Chem 284:6116–6125. https://doi.org/10.1074/jbc.M808407200
93. Sugii S, Olson P, Sears DD, Saberi M, Atkins AR, Barish GD, Hong S-H, Castro GL, Yin Y-Q, Nelson MC, Hsiao G, Greaves DR, Downes M, Yu RT, Olefsky JM, Evans RM (2009) PPARgamma activation in adipocytes is sufficient for systemic insulin sensitization. Proc Natl Acad Sci U S A 106:22504–22509. https://doi.org/10.1073/pnas.0912487106
94. Graham DJ, Ouellet-Hellstrom R, MaCurdy TE, Ali F, Sholley C, Worrall C, Kelman JA (2010) Risk of acute myocardial infarction, stroke, heart failure, and death in elderly Medicare patients treated with rosiglitazone or pioglitazone. JAMA 304:411–418. https://doi.org/10.1001/jama.2010.920
95. Floyd JS, Barbehenn E, Lurie P, Wolfe SM (2009) Case series of liver failure associated with rosiglitazone and pioglitazone. Pharmacoepidemiol Drug Saf 18:1238–1243. https://doi.org/10.1002/pds.1804
96. Fukui Y, Masui S, Osada S, Umesono K, Motojima K (2000) A new thiazolidinedione, NC-2100, which is a weak PPAR-gamma activator, exhibits potent antidiabetic effects and induces uncoupling protein 1 in white adipose tissue of KKAy obese mice. Diabetes 49:759–767. https://doi.org/10.2337/diabetes.49.5.759
97. Petrovic N, Walden TB, Shabalina IG, Timmons JA, Cannon B, Nedergaard J (2010) Chronic peroxisome proliferator-activated receptor gamma (PPARgamma) activation of epididymally derived white adipocyte cultures reveals a population of thermogenically competent, UCP1-containing adipocytes molecularly distinct from classic brown adipocytes. J Biol Chem 285:7153–7164. https://doi.org/10.1074/jbc.M109.053942
98. Ohno H, Shinoda K, Spiegelman BM, Kajimura S (2012) PPARγ agonists induce a white-to-brown fat conversion through stabilization of PRDM16 protein. Cell Metab 15:395–404. https://doi.org/10.1016/j.cmet.2012.01.019
99. Jeremic N, Chaturvedi P, Tyagi SC (2017) Browning of white fat: novel insight into factors, mechanisms, and therapeutics. J Cell Physiol 232:61–68. https://doi.org/10.1002/jcp.25450
100. Day JW, Ottaway N, Patterson JT, Gelfanov V, Smiley D, Gidda J, Findeisen H, Bruemmer D, Drucker DJ, Chaudhary N, Holland J, Hembree J, Abplanalp W, Grant E, Ruehl J, Wilson H, Kirchner H, Lockie SH, Hofmann S, Woods SC, Nogueiras R, Pfluger PT, Perez-Tilve D, DiMarchi R, Tschöp MH (2009) A new glucagon and GLP-1 co-agonist eliminates obesity in rodents. Nat Chem Biol 5:749–757. https://doi.org/10.1038/nchembio.209
101. Kroon T, Harms M, Maurer S, Bonnet L, Alexandersson I, Lindblom A, Ahnmark A, Nilsson D, Gennemark P, O'Mahony G, Osinski V, McNamara C, Boucher J (2020) PPARγ and PPARα synergize to induce robust browning of white fat in vivo. Mol Metab 36:100964. https://doi.org/10.1016/j.molmet.2020.02.007
102. Jones JR, Barrick C, Kim K-A, Lindner J, Blondeau B, Fujimoto Y, Shiota M, Kesterson RA, Kahn BB, Magnuson MA (2005) Deletion of PPARgamma in adipose tissues of mice protects against high fat diet-induced obesity and insulin resistance. Proc Natl Acad Sci U S A 102:6207–6212. https://doi.org/10.1073/pnas.0306743102
103. Festuccia WT, Blanchard P-G, Richard D, Deshaies Y (2010) Basal adrenergic tone is required for maximal stimulation of rat brown adipose tissue UCP1 expression by chronic PPAR-gamma activation. Am J Phys Regul Integr Comp Phys 299:R159–R167. https://doi.org/10.1152/ajpregu.00821.2009
104. Lasar D, Rosenwald M, Kiehlmann E, Balaz M, Tall B, Opitz L, Lidell ME, Zamboni N, Krznar P, Sun W, Varga L, Stefanicka P, Ukropec J, Nuutila P, Virtanen K, Amri E-Z, Enerbäck S, Wahli W, Wolfrum C (2018) Peroxisome proliferator activated receptor gamma controls mature brown adipocyte inducibility through glycerol kinase. Cell Rep 22:760–773. https://doi.org/10.1016/j.celrep.2017.12.067
105. Imai T, Takakuwa R, Marchand S, Dentz E, Bornert J-M, Messaddeq N, Wendling O, Mark M, Desvergne B, Wahli W, Chambon P, Metzger D (2004) Peroxisome proliferator-activated receptor gamma is required in mature white and brown adipocytes for their survival in the mouse. Proc Natl Acad Sci U S A 101:4543–4547. https://doi.org/10.1073/pnas.0400356101
106. Tai T-AC, Jennermann C, Brown KK, Oliver BB, MacGinnitie MA, Wilkison WO, Brown HR, Lehmann JM, Kliewer SA, Morris DC, Graves RA (1996) Activation of the nuclear receptor peroxisome proliferator-activated receptor γ promotes brown adipocyte differentiation*. J Biol Chem 271:29909–29914. https://doi.org/10.1074/jbc.271.47.29909
107. Xiong W, Zhao X, Villacorta L, Rom O, Garcia-Barrio MT, Guo Y, Fan Y, Zhu T, Zhang J, Zeng R, Chen YE, Jiang Z, Chang L (2018) Brown adipocyte-specific PPARγ (peroxisome proliferator-activated receptor γ) deletion impairs perivascular adipose tissue development and enhances atherosclerosis in mice. Arterioscler Thromb Vasc Biol 38:1738–1747. https://doi.org/10.1161/ATVBAHA.118.311367
108. Hondares E, Mora O, Yubero P, de la Concepción R, Marisa R, Iglesias M, Giralt FV (2006) Thiazolidinediones and rexinoids induce peroxisome proliferator-activated receptor-coactivator (PGC)-

1alpha gene transcription: an autoregulatory loop controls PGC-1alpha expression in adipocytes via peroxisome proliferator-activated receptor-gamma coactivation. Endocrinology 147:2829–2838. https://doi.org/10.1210/en.2006-0070

109. Puigserver P, Wu Z, Park CW, Graves R, Wright M, Spiegelman BM (1998) A cold-inducible coactivator of nuclear receptors linked to adaptive thermogenesis. Cell 92:829–839. https://doi.org/10.1016/S0092-8674(00)81410-5
110. Lin J, Wu P-H, Tarr PT, Lindenberg KS, St-Pierre J, Zhang C-Y, Mootha VK, Jäger S, Vianna CR, Reznick RM, Cui L, Manieri M, Donovan MX, Wu Z, Cooper MP, Fan MC, Rohas LM, Zavacki AM, Cinti S, Shulman GI, Lowell BB, Krainc D, Spiegelman BM (2004) Defects in adaptive energy metabolism with CNS-linked hyperactivity in PGC-1α null mice. Cell 119:121–135. https://doi.org/10.1016/j.cell.2004.09.013
111. Tanaka T, Yamamoto J, Iwasaki S, Asaba H, Hamura H, Ikeda Y, Watanabe M, Magoori K, Ioka RX, Tachibana K, Watanabe Y, Uchiyama Y, Sumi K, Iguchi H, Ito S, Doi T, Hamakubo T, Naito M, Auwerx J, Yanagisawa M, Kodama T, Sakai J (2003) Activation of peroxisome proliferator-activated receptor delta induces fatty acid beta-oxidation in skeletal muscle and attenuates metabolic syndrome. Proc Natl Acad Sci U S A 100:15924–15929. https://doi.org/10.1073/pnas.0306981100
112. Nahlé Z, Hsieh M, Pietka T, Coburn CT, Grimaldi PA, Zhang MQ, Das D, Abumrad NA (2008) CD36-dependent regulation of muscle FoxO1 and PDK4 in the PPAR delta/beta-mediated adaptation to metabolic stress. J Biol Chem 283:14317–14326. https://doi.org/10.1074/jbc.M706478200
113. Angione AR, Jiang C, Pan D, Wang Y-X, Kuang S (2011) PPARδ regulates satellite cell proliferation and skeletal muscle regeneration. Skelet Muscle 1:33. https://doi.org/10.1186/2044-5040-1-33
114. Chen W, Gao R, Xie X, Zheng Z, Li H, Li S, Dong F, Wang L (2015) A metabolomic study of the PPARδ agonist GW501516 for enhancing running endurance in Kunming mice. Sci Rep 5:9884. https://doi.org/10.1038/srep09884
115. Schuler M, Ali F, Chambon C, Duteil D, Bornert J-M, Tardivel A, Desvergne B, Wahli W, Chambon P, Metzger D (2006) PGC1α expression is controlled in skeletal muscles by PPARβ, whose ablation results in fiber-type switching, obesity, and type 2 diabetes. Cell Metab 4:407–414. https://doi.org/10.1016/j.cmet.2006.10.003
116. Wang Y-X, Zhang C-L, Yu RT, Cho HK, Nelson MC, Bayuga-Ocampo CR, Ham J, Kang H, Evans RM (2004) Regulation of muscle fiber type and running endurance by PPARdelta. PLoS Biol 2:e294. https://doi.org/10.1371/journal.pbio.0020294
117. Lunde IG, Ekmark M, Rana ZA, Buonanno A, Gundersen K (2007) PPARdelta expression is influenced by muscle activity and induces slow muscle properties in adult rat muscles after somatic gene transfer. J Physiol 582:1277–1287. https://doi.org/10.1113/jphysiol.2007.133025
118. Krämer DK, Ahlsén M, Norrbom J, Jansson E, Hjeltnes N, Gustafsson T, Krook A (2006) Human skeletal muscle fibre type variations correlate with PPAR alpha, PPAR delta and PGC-1 alpha mRNA. Acta Physiol 188:207–216. https://doi.org/10.1111/j.1748-1716.2006.01620.x
119. Luquet S, Lopez-Soriano J, Holst D, Fredenrich A, Melki J, Rassoulzadegan M, Grimaldi PA (2003) Peroxisome proliferator-activated receptor delta controls muscle development and oxidative capability. FASEB J 17:2299–2301. https://doi.org/10.1096/fj.03-0269fje
120. Chandrashekar P, Manickam R, Ge X, Bonala S, McFarlane C, Sharma M, Wahli W, Kambadur R (2015) Inactivation of PPARβ/δ adversely affects satellite cells and reduces postnatal myogenesis. Am J Physiol Endocrinol Metab 309:E122–E131. https://doi.org/10.1152/ajpendo.00586.2014
121. McPherron AC, Lawler AM, Lee SJ (1997) Regulation of skeletal muscle mass in mice by a new TGF-beta superfamily member. Nature 387:83–90. https://doi.org/10.1038/387083a0
122. Wang Y-X, Lee C-H, Tiep S, Yu RT, Ham J, Kang H, Evans RM (2003) Peroxisome-proliferator-activated receptor δ activates fat metabolism to prevent obesity. Cell 113:159–170. https://doi.org/10.1016/S0092-8674(03)00269-1
123. Lin J, Wu H, Tarr PT, Zhang C-Y, Wu Z, Boss O, Michael LF, Puigserver P, Isotani E, Olson EN, Lowell BB, Bassel-Duby R, Spiegelman BM (2002) Transcriptional co-activator PGC-1α drives the formation of slow-twitch muscle fibres. Nature 418:797–801. https://doi.org/10.1038/nature00904
124. Pettersen IKN, Tusubira D, Ashrafi H, Dyrstad SE, Hansen L, Liu X-Z, Nilsson LIH, Løvsletten NG, Berge K, Wergedahl H, Bjørndal B, Fluge Ø, Bruland O, Rustan AC, Halberg N, Røsland GV, Berge RK, Tronstad KJ (2019) Upregulated PDK4 expression is a sensitive marker of increased fatty acid oxidation. Mitochondrion 49:97–110. https://doi.org/10.1016/j.mito.2019.07.009
125. Glinghammar B, Skogsberg J, Hamsten A, Ehrenborg E (2003) PPARδ activation induces COX-2 gene expression and cell proliferation in human hepatocellular carcinoma cells. Biochem Biophys Res Commun 308:361–368. https://doi.org/10.1016/S0006-291X(03)01384-6
126. Kostadinova R, Montagner A, Gouranton E, Fleury S, Guillou H, Dombrowicz D, Desreumaux P, Wahli W (2012) GW501516-activated PPARβ/δ promotes liver fibrosis via p38-JNK MAPK-induced hepatic stellate cell proliferation. Cell Biosci 2:34. https://doi.org/10.1186/2045-3701-2-34
127. Da'adoosh B, Marcus D, Rayan A, King F, Che J, Goldblum A (2019) Discovering highly selective and diverse PPAR-delta agonists by ligand based machine learning and structural modeling. Sci Rep 9:1106. https://doi.org/10.1038/s41598-019-38508-8

128. Susini S, Roche E, Prentki M, Schlegel W (1998) Glucose and glucoincretin peptides synergize to induce c-fos, c-jun, junB, zif-268, and nur-77 gene expression in pancreatic β (INS-1) cells. FASEB J 12:1173–1182
129. Roche E, Buteau J, Aniento I, Reig JA, Soria B, Prentki M (1999) Palmitate and oleate induce the immediate-early response genes c-fos and nur-77 in the pancreatic beta-cell line INS-1. Diabetes 48:2007–2014. https://doi.org/10.2337/diabetes.48.10.2007
130. Chen Y-T, Liao J-W, Tsai Y-C, Tsai F-J (2016) Inhibition of DNA methyltransferase 1 increases nuclear receptor subfamily 4 group A member 1 expression and decreases blood glucose in type 2 diabetes. Oncotarget 7:39162–39170. https://doi.org/10.18632/oncotarget.10043
131. Reynolds MS, Hancock CR, Ray JD, Kener KB, Draney C, Garland K, Hardman J, Bikman BT, Tessem JS (2016) β-Cell deletion of Nr4a1 and Nr4a3 nuclear receptors impedes mitochondrial respiration and insulin secretion. Am J Physiol Endocrinol Metab 311:E186–E201
132. Close A-F, Dadheech N, Lemieux H, Wang Q, Buteau J (2020) Disruption of beta-cell mitochondrial networks by the orphan nuclear receptor Nor1/Nr4a3. Cell 9:168. https://doi.org/10.3390/cells9010168
133. Tessem JS, Moss LG, Chao LC, Arlotto M, Lu D, Jensen MV, Stephens SB, Tontonoz P, Hohmeier HE, Newgard CB (2014) Nkx6.1 regulates islet β-cell proliferation via Nr4a1 and Nr4a3 nuclear receptors. Proc Natl Acad Sci U S A 111:5242–5247. https://doi.org/10.1073/pnas.1320953111
134. Ishihara H, Asano T, Tsukuda K, Katagiri H, Inukai K, Anai M, Kikuchi M, Yazaki Y, Miyazaki J-I, Oka Y (1993) Pancreatic beta cell line MIN6 exhibits characteristics of glucose metabolism and glucose-stimulated insulin secretion similar to those of normal islets. Diabetologia 36:1139–1145. https://doi.org/10.1007/BF00401058
135. Briand O, Helleboid-Chapman A, Ploton M, Hennuyer N, Carpentier R, Pattou F, Vandewalle B, Moerman E, Gmyr V, Kerr-Conte J, Eeckhoute J, Staels B, Lefebvre P (2012) The nuclear orphan receptor Nur77 is a lipotoxicity sensor regulating glucose-induced insulin secretion in pancreatic β-cells. Mol Endocrinol 26:399–413. https://doi.org/10.1210/me.2011-1317
136. Yu C, Cui S, Zong C, Gao W, Xu T, Gao P, Chen J, Qin D, Guan Q, Liu Y (2015) The orphan nuclear receptor NR4A1 protects pancreatic β-cells from endoplasmic reticulum (ER) stress-mediated apoptosis. J Biol Chem 290:20687–20699
137. Pu Z, Liu D, Mouguegue L, Patherny HP, Jin C, Sadiq E, Qin D, Yu T, Zong C, Chen J, Zhao R (2020) NR4A1 counteracts JNK activation incurred by ER stress or ROS in pancreatic β-cells for protection. J Cell Mol Med 24:14171–14183
138. Pu Z, Yu T, Liu D, Jin C, Sadiq E, Qiao X, Li X, Chen Y, Zhang J, Tian M, Li S, Zhao R, Wang X (2021) NR4A1 enhances MKP7 expression to diminish JNK activation induced by ROS or ER-stress in pancreatic β cells for surviving. Cell Death Dis 7:133. https://doi.org/10.1038/s41420-021-00521-0
139. Zhan Y, Du X, Chen H, Liu J, Zhao B, Huang D, Li G, Xu Q, Zhang M, Weimer BC, Chen D, Cheng Z, Zhang L, Li Q, Li S, Zheng Z, Song S, Huang Y, Ye Z, Su W, Lin S-C, Shen Y, Wu Q (2008) Cytosporone B is an agonist for nuclear orphan receptor Nur77. Nat Chem Biol 4:548–556. https://doi.org/10.1038/nchembio.106
140. Weyrich P, Staiger H, Stancáková A, Schäfer SA, Kirchhoff K, Ullrich S, Ranta F, Gallwitz B, Stefan N, Machicao F, Kuusisto J, Laakso M, Fritsche A, Häring H-U (2009) Common polymorphisms within the NR4A3 locus, encoding the orphan nuclear receptor Nor-1, are associated with enhanced beta-cell function in non-diabetic subjects. BMC Med Genet 10:77. https://doi.org/10.1186/1471-2350-10-77
141. Close A-F, Dadheech N, Villela BS, Rouillard C, Buteau J (2019) The orphan nuclear receptor Nor1/Nr4a3 is a negative regulator of β-cell mass. J Biol Chem 294:4889–4897. https://doi.org/10.1074/jbc.RA118.005135
142. Gao W, Fu Y, Yu C, Wang S, Zhang Y, Zong C, Xu T, Liu Y, Li X, Wang X (2014) Elevation of NR4A3 expression and its possible role in modulating insulin expression in the pancreatic beta cell. PLoS One 9:e91462. https://doi.org/10.1371/journal.pone.0091462
143. Wansa KDSA, Harris JM, Yan G, Ordentlich P, Muscat GEO (2003) The AF-1 domain of the orphan nuclear receptor NOR-1 mediates trans-activation, coactivator recruitment, and activation by the purine anti-metabolite 6-mercaptopurine. J Biol Chem 278:24776–24790. https://doi.org/10.1074/jbc.M300088200
144. Liu Q, Zhu X, Xu L, Fu Y, Garvey WT (2013) 6-Mercaptopurine augments glucose transport activity in skeletal muscle cells in part via a mechanism dependent upon orphan nuclear receptor NR4A3. Am J Physiol Endocrinol Metab 305:E1081–E1092. https://doi.org/10.1152/ajpendo.00169.2013
145. Pearen MA, Ryall JG, Maxwell MA, Ohkura N, Lynch GS, Muscat GEO (2006) The orphan nuclear receptor, NOR-1, is a target of beta-adrenergic signaling in skeletal muscle. Endocrinology 147:5217–5227. https://doi.org/10.1210/en.2006-0447
146. Takahashi H, Nomiyama T, Terawaki Y, Kawanami T, Hamaguchi Y, Tanaka T, Tanabe M, Bruemmer D, Yanase T (2019) GLP-1 receptor agonist exendin-4 attenuates NR4A orphan nuclear receptor NOR1 expression in vascular smooth muscle cells. J Atheroscler Thromb 26:183–197. https://doi.org/10.5551/jat.43414
147. Perissi V, Aggarwal A, Glass CK, Rose DW, Rosenfeld MG (2004) A corepressor/coactivator exchange complex required for transcriptional

activation by nuclear receptors and other regulated transcription factors. Cell 116:511–526. https://doi.org/10.1016/s0092-8674(04)00133-3
148. Jeyakumar M, Liu X-F, Erdjument-Bromage H, Tempst P, Bagchi MK (2007) Phosphorylation of thyroid hormone receptor-associated nuclear receptor corepressor holocomplex by the DNA-dependent protein kinase enhances its histone deacetylase activity. J Biol Chem 282:9312–9322. https://doi.org/10.1074/jbc.M609009200
149. Perissi V, Scafoglio C, Zhang J, Ohgi KA, Rose DW, Glass CK, Rosenfeld MG (2008) TBL1 and TBLR1 phosphorylation on regulated gene promoters overcomes dual CtBP and NCoR/SMRT transcriptional repression checkpoints. Mol Cell 29:755–766. https://doi.org/10.1016/j.molcel.2008.01.020
150. Li J, Wang C-Y (2008) TBL1–TBLR1 and β-catenin recruit each other to Wnt target-gene promoter for transcription activation and oncogenesis. Nat Cell Biol 10:160–169. https://doi.org/10.1038/ncb1684
151. Guenther MG, Lane WS, Fischle W, Verdin E, Lazar MA, Shiekhattar R (2000) A core SMRT corepressor complex containing HDAC3 and TBL1, a WD40-repeat protein linked to deafness. Genes Dev 14:1048–1057
152. Yoon H-G, Chan DW, Huang Z-Q, Li J, Fondell JD, Qin J, Wong J (2003) Purification and functional characterization of the human N-CoR complex: the roles of HDAC3, TBL1 and TBLR1. EMBO J 22:1336–1346
153. Ferrari A, Longo R, Fiorino E, Silva R, Mitro N, Cermenati G, Gilardi F, Desvergne B, Andolfo A, Magagnotti C, Caruso D, de Fabiani E, Hiebert SW, Crestani M (2017) HDAC3 is a molecular brake of the metabolic switch supporting white adipose tissue browning. Nat Commun 8:93. https://doi.org/10.1038/s41467-017-00182-7
154. Sun Z, Feng D, Fang B, Mullican SE, You S-H, Lim H-W, Everett LJ, Nabel CS, Li Y, Selvakumaran V, Won K-J, Lazar MA (2013) Deacetylase-independent function of HDAC3 in transcription and metabolism requires nuclear receptor corepressor. Mol Cell 52:769–782. https://doi.org/10.1016/j.molcel.2013.10.022
155. Chen W-B, Gao L, Wang J, Wang Y-G, Dong Z, Zhao J, Mi Q-S, Zhou L (2016) Conditional ablation of HDAC3 in islet beta cells results in glucose intolerance and enhanced susceptibility to STZ-induced diabetes. Oncotarget 7:57485–57497. https://doi.org/10.18632/oncotarget.11295
156. Remsberg JR, Ediger BN, Ho WY, Damle M, Li Z, Teng C, Lanzillotta C, Stoffers DA, Lazar MA (2017) Deletion of histone deacetylase 3 in adult beta cells improves glucose tolerance via increased insulin secretion. Mol Metab 6:30–37. https://doi.org/10.1016/j.molmet.2016.11.007
157. Liang N, Damdimopoulos A, Goñi S, Huang Z, Vedin L-L, Jakobsson T, Giudici M, Ahmed O, Pedrelli M, Barilla S, Alzaid F, Mendoza A, Schröder T, Kuiper R, Parini P, Hollenberg A, Lefebvre P, Francque S, van Gaal L, Staels B, Venteclef N, Treuter E, Fan R (2019) Hepatocyte-specific loss of GPS2 in mice reduces non-alcoholic steatohepatitis via activation of PPARα. Nat Commun 10:1684. https://doi.org/10.1038/s41467-019-09524-z
158. Drareni K, Ballaire R, Barilla S, Mathew MJ, Toubal A, Fan R, Liang N, Chollet C, Huang Z, Kondili M, Foufelle F, Soprani A, Roussel R, Gautier J-F, Alzaid F, Treuter E, Venteclef N (2018) GPS2 deficiency triggers maladaptive white adipose tissue expansion in obesity via HIF1A activation. Cell Rep 24:2957–2971.e6. https://doi.org/10.1016/j.celrep.2018.08.032
159. Drareni K, Ballaire R, Alzaid F, Goncalves A, Chollet C, Barilla S, Nguewa J-L, Dias K, Lemoine S, Riveline J-P, Roussel R, Dalmas E, Velho G, Treuter E, Gautier J-F, Venteclef N (2020) Adipocyte reprogramming by the transcriptional coregulator GPS2 impacts beta cell insulin secretion. Cell Rep 32:108141. https://doi.org/10.1016/j.celrep.2020.108141
160. Kulozik P, Jones A, Mattijssen F, Rose AJ, Reimann A, Strzoda D, Kleinsorg S, Raupp C, Kleinschmidt J, Müller-Decker K, Wahli W, Sticht C, Gretz N, von Loeffelholz C, Stockmann M, Pfeiffer A, Stöhr S, Dallinga-Thie GM, Nawroth PP, Diaz MB, Herzig S (2011) Hepatic deficiency in transcriptional cofactor TBL1 promotes liver steatosis and hypertriglyceridemia. Cell Metab 13:389–400. https://doi.org/10.1016/j.cmet.2011.02.011
161. Stoy C, Sundaram A, Rios Garcia M, Wang X, Seibert O, Zota A, Wendler S, Männle D, Hinz U, Sticht C, Muciek M, Gretz N, Rose AJ, Greiner V, Hofmann TG, Bauer A, Hoheisel J, Berriel Diaz M, Gaida MM, Werner J, Schafmeier T, Strobel O, Herzig S (2015) Transcriptional co-factor Transducin beta-like (TBL) 1 acts as a checkpoint in pancreatic cancer malignancy. EMBO Mol Med 7:1048–1062. https://doi.org/10.15252/emmm.201404837
162. Gu J-F, Fu W, Qian H-X, Gu W-X, Zong Y, Chen Q, Lu L (2020) TBL1XR1 induces cell proliferation and inhibit cell apoptosis by the PI3K/AKT pathway in pancreatic ductal adenocarcinoma. World J Gastroenterol 26:3586–3602. https://doi.org/10.3748/wjg.v26.i25.3586
163. Rohm M, Sommerfeld A, Strzoda D, Jones A, Sijmonsma TP, Rudofsky G, Wolfrum C, Sticht C, Gretz N, Zeyda M, Leitner L, Nawroth PP, Stulnig TM, Diaz MB, Vegiopoulos A, Herzig S (2013) Transcriptional cofactor TBLR1 controls lipid mobilization in white adipose tissue. Cell Metab 17:575–585. https://doi.org/10.1016/j.cmet.2013.02.010
164. Schaefer U, Schmeier S, Bajic VB (2011) TcoF-DB: dragon database for human transcription co-factors and transcription factor interacting proteins. Nucleic Acids Res 39:D106–D110. https://doi.org/10.1093/nar/gkq945

Nuclear Receptors and Lipid Sensing

5

James L. Thorne and Giorgia Cioccoloni

Abstract

Fluctuations in concentration of diverse lipid classes occur in response to diet and metabolism. These changes are managed and mediated by a cell network of enzymes, pumps, and carriers under the control of the lipid responsive nuclear receptors. The understanding of how dysregulation of lipid metabolism are causes and indicators of disease beyond the cardiovascular system has developed in the last decade. A particular emphasis on the role of lipids and lipid-sensing nuclear receptors has emerged in the fields of cancer and the immune system's interaction with cancer. The range of known lipid-based ligands has also expanded. Lipids are not just signalling molecules, but also play structural roles in cells and tissues, for example as major constituents of the lipid bilayer – positioning them as integrators and mediators of signaling. This chapter will discuss the major groups of lipid-sensing nuclear receptors focusing on the liver x receptors, farnesoid x receptor, and the peroxisome proliferator-activated receptors. Initially the reader is presented with information on how these receptors behave and function at the molecular biology level, the range of selective modulation of function by endogenous ligands, and examples of how activity is fine-tuned by mechanisms such as miRNA regulation and post-translational modification of the proteins. We then explore the advances in understanding that have positioned these receptors as therapeutic targets in cancer and immuno-oncology. Finally, the chapter explains the gaps in understanding and experimental challenges that should be prioritized in the coming decade.

Keywords

Lipids · Oxysterols · Cancer · Immuno-oncology · Transcription · Splicing · Selective modulation

5.1 The Molecular Biology of Lipid-Sensing Nuclear Receptors

Nuclear receptors (NR) sense lipids from broad subclasses that include fatty acids, phospholipids, sphingolipids, and sterols. Cholesterol, integral to the plasma membrane's barrier function, is the precursor for an array of ligands including hormones, seco-steroids, oxysterols, and bile acids, which drive NR activity. Cholesterol constitutes around 40% of the mammalian cell membrane and mediates signal transduction pathways that originate from liquid ordered nanodomains

J. L. Thorne (✉) · G. Cioccoloni
Nutrition, Epigenetics and Cancer Research Group, School of Food Science and Nutrition, University of Leeds, Leeds, UK
e-mail: j.l.thorne@leeds.ac.uk

M. J. Campbell, C. L. Bevan (eds.), *Nuclear Receptors in Human Health and Disease*, Advances in Experimental Medicine and Biology 1390, https://doi.org/10.1007/978-3-031-11836-4_5

within this barrier to the external milieu. Fatty acids and phospholipids also influence plasma membrane fluidity and function and act as ligands for the peroxisome proliferation activated receptors (NR1C1-3/PPARs) and liver receptor homologue (NR5A2/LRH1) respectively. Lipid-NR pathways therefore link the cell's external barrier and its metabolic state to transcriptional activity and cell fate processes.

5.1.1 Liver x Receptors (NR1H2, NR1H3)

Cellular and tissue regulation of cholesterol is controlled by liver x receptors alpha (LXRα) and beta (LXRβ). LXRα and LXRβ are expressed in the liver and a variety of extra-hepatic tissues including brain, reproductive organs, gut-axis, bone, and vital organs. The two paralogues target a battery of genes involved in flux (e.g. ATP binding cassette transporters), metabolism (e.g. CYP450s, hydroxylases), and transport (apolipoproteins) of sterols. The LXRs therefore underpin cholesterol's physiological roles in the liver, integrity of the blood brain barrier, neuronal function, amyloid pathology, cellular proliferation and migration, inflammation and immune cell differentiation and function, xenobiotic efflux, and autophagy.

5.1.1.1 Structure of the LXRs

The LXRs are classically considered to be bound to gene regulatory regions that contain a direct repeat with a four-nucleotide spacer (DR4: AGGTCA). LXRs are activated by hydroxylated cholesterol metabolites that dock to the broad affinity ligand binding domain (LBD). This binding event induces allosteric change in the protein structure and co-repressors are exchanged for co-activators. The presence of ligand increases the amount of LXR at binding sites [54], either via auto-regulated induction of expression or stabilization of the transcription factor-DNA interaction. The LXRs contain an activation function domain (AF1) at the N-terminus and a flexible hinge region that links the LBD and DBD. In the C-terminal region of the LBD is a second activation function 2 domain (AF2) [168]. The hinge region of LXR is subject to post-translational modifications that modifies the response to ligand and co-factor exchange [13, 14, 191]. Alternative splicing and alternative promoter usage lead to differential expression of these protein domains patterns for LXRα (Fig. 5.1a) and to a lesser extend LXRβ (Fig. 5.1b) as well. Five LXRα splice variants have been experimentally validated to date [31, 51, 112] but as many as 62 differentially spliced transcripts are predicted to exist [6]. LXRα2 and LXRα5 have shortened ligand binding domains and LXRα4 is generated from an alternative transcriptional start site altering the length of the AF1 domain. Contrarily, only a single LXRβ splice variant appears to have been reported as expressed at the protein level to date (LXRβ4) and contains a shortened LBD [112]. The function of this variant has not been established, and is expressed in whole breast tumour tissues, but not epithelial cell lines, so may be expressed under some pathological conditions or in non-tumour cells of the cancer microenvironment.

5.1.1.2 Endogenous Selective Modulators of LXR

Oxysterols were discovered as ligands for the LXRs in the 1990s [81] and can be synthesized by a range of enzymes or through auto-oxidative routes (Fig. 5.2). As a diverse class of cholesterol derived lipids, over the last decade the range of known oxysterol based LXR ligands has increased, as has our understanding of the role that the oxysterol:LXR axis plays in health and disease [69, 117]. Although historically considered as intermediates in the metabolism of cholesterol to bile acids or steroid hormones, oxysterols are now established as potent signaling molecules in their own right, with an array of cell biology functions elucidated.

The diversity of oxysterols allows categorization based on the chemical moiety(s) that distinguishes them from cholesterol. Hydroxy-, epoxy-, or keto- modifications to the cholesterol backbone leads to distinct oxysterol classes with different transactivation potential for LXR. In physiological systems a pool of these LXR

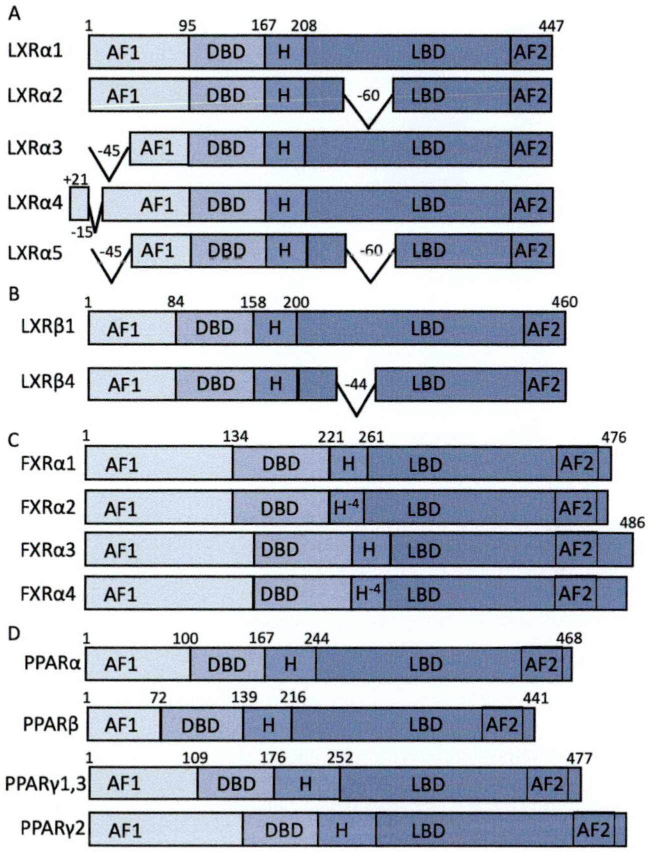

Fig. 5.1 **Structure of lipid-sensing nuclear receptors.** The main transcripts and splice variants for LXRα (**a**) LXRβ (**b**) FXRα (**c**) and PPARs (**d**). *AF* activation function domain, *DBD* DNA binding domain, *H* hinge region, *LBD* ligand binding domain. Numbers above protein indicate feature position by amino acid number. Numbers with plus/minus sign indicate number of additional or lost amino acids relative to first variant. (Data for figure acquired by Dr Alex Websdale and Dr Priscilia Lianto)

ligands would typically be present, which differ between tissues. The oxysterol constituents of this pool are not present at equimolar concentrations and have varying EC_{50} values in their capacity to activate LXR. For example, when comparing the side-chain hydroxycholesterols (scOHC): 22OHC, 24OHC, 25OHC, 27OHC and the epoxycholesterol 24,25EC, there are distinct differences in both circulating concentrations [163] and LXR transactivation potential [77, 78, 82]. For example, 27OHC is by far the most abundant of these oxysterols in the circulation (by >50-fold when compared with 22-OHC or 24,25EC) and in some tumours such as breast [159] but is a significantly weaker LXR agonist [78, 82]. In different tissues other oxysterols are the majority species. In the brain 27OHC is rare and 24S-OHC (cerebrosterol) is the dominant OHC. These differences in the concentration and ability to transactivate the LXRs provide tissue specific modulation of the receptors at a level beyond expression of the NR protein. Selective modulation of LXR has also been found in epoxy class of oxysterols mediated by conjugation to histamine or via metabolism by 11βHSD2, and generates cross-talk between LXR and other NRs. When 5,6β-EC is metabolized by 11βHSD2 to 6-oxo-cholestan-3β,5α-diol (OCDO) it binds to GR (KD = 10 μM) and LXRβ (KD = 12.5 μM) with similar affinity, and promotes proliferation in a GR dependent manner [183]. However, if the stereoisomer 5,6α-EC is instead conjugated to histamine it generates dendrogenin A, a compound that is reduced in tumour tissue and that can drive lethal autophagy in cancer via LXR activation [154]. OCDO is considered a competitive inhibitor of cortisol given that it decreases cortisol affinity to GR at 1 μM [183]. The synthesis of OCDO and dendrogenin A and their roles as LXR ligands are comprehensively reviewed [144, 145].

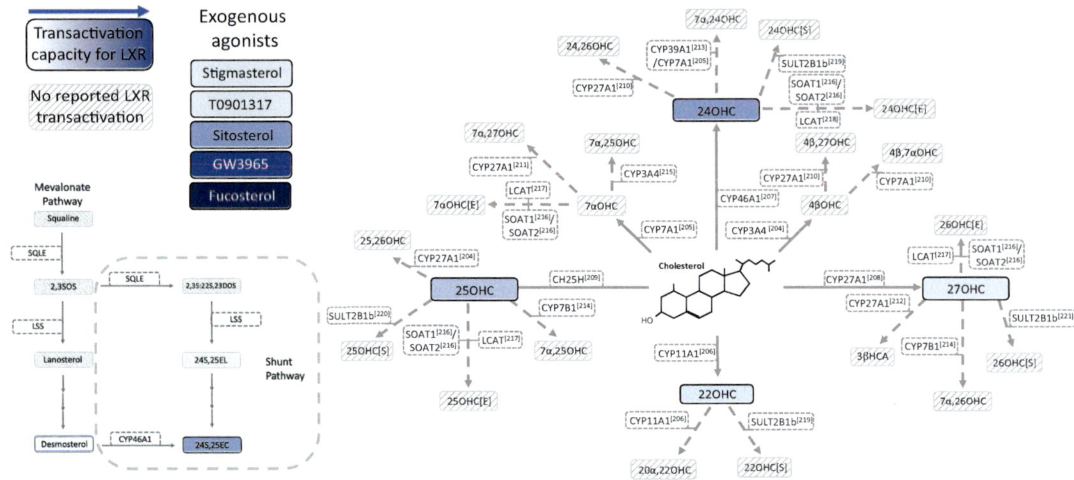

Fig. 5.2 Complexity of the endogenous liver x receptor ligand pool. Sulfonated and esterified oxysterols are represented with [S] or [E] following their name. A sequence of three arrows in a row indicates a series of enzymatic functions leading to the synthesis of the next product. LXR ligands are shown in blue boxes, depth of colour indicates LXR transactivation potential according to [77, 78, 81, 82, 143]. Compounds with unreported ability to induce a response from LXR shown in thatched boxes. The initial hydroxylation of cholesterol by cytochrome p450 enzymes; CYP3A4, CYP7A1, CYP11A1, CYP46A1, CYP27A1 and endoplasmic reticulum oxidoreductase, CH25H, leads to the formation of 4βOHC [204], 7αOHC [205], 22OHC [206], 24OHC [207], 27OHC [208] and 25OHC [209] respectively. CYP27A1 can further metabolise 4βOHC, 7αOHC, 24OHC and 27OHC to 4β,27OHC [210], 7α,27OHC [211], 24,26OHC [210] and 3b-hydroxycholest-5-en-(25R)26-oic acid (3b-HCA) [212], respectively. Either CYP7A1 [205] or CYP39A1 [213] can hydroxylate 24OHC at C7 to produce 7α,24OHC. CYP7A1 also hydroxylates 4βOHC to produce 4β,7αOHC [210]. CYP7B1 hydroxylates 25OHC and 27OHC at C7 to produce 7α,25OHC and 7α,26OHC, respectively [214]. CYP3A4 hydroxylates 7αOHC to produce 7α,25OHC [215]. CYP11A1 hydroxylates 22OHC to 20α,20OHC [215]. Lecithin-cholesterol acyltransferase (LCAT) and Sterol O-acyltransferase (SOAT) 1 and 2 have been reported to esterify 7αOHC, 25OHC and 27OHC [216, 217]. 24OHC can also be esterified by LCAT , SOAT1 and SOAT2 [216] Esterification of 22OHC has not been reported yet. SULT2B1b is capable of sulphating 22OHC [218], 24OHC [219], 25OHC [220], and 27OHC [221]. Figure generated by Dr Alex Websdale

The plethora of endogenous LXR ligands with their differing ability to transactivate LXR that is typically found in human serum and tissue therefore add nuance and adaptability to the oxysterol:LXR axis. Selective modulation of LXR activity by OHCs is a function of the local oxysterol pool; although a relatively simple ligand-receptor interaction at the molecular level, extensive specialization is conferred to the pathway at the tissue level. Genome-wide analyses have been reported for LXR knockdown in mouse [16, 19], the synthetic ligand T0901317 [54, 140] and GW3965 [133], but formal genome wide-comparisons of multiple endogenous ligands at the RNA- or ChIP-Seq level are yet to be reported. Such comparisons would add valuable information regarding selective modulation by ligands.

5.1.1.3 Fine-Tuning of LXR Signaling

As for most NR, LXR activity is fine-tuned at the cell and tissue level by several mechanisms. Differences in expression of splice variants, ligand concentration, and co-factors act to alter the transcriptional output from LXR. Interestingly, LXRβ is predicted to bind far more co-repressors than LXRα, yet LXRα is predicted to bind to a wider range of co-activators [21]. Experimental validation of such predictions is challenging and are currently not yet evidenced in the literature. Given the changes to protein structure generated by alternative splicing, it is plausible that co-factors may form distinct complexes depending on the LXR variant to which they are binding. Such a variety of overlapping but diverging transcriptional complexes allows for subtle fine-tuning for the response of LXR to ligand. Another important factor that alters LXR's activity is post-translational modification of the Hinge region. Phosphorylation of S198 confers specificity to LXRα transcriptional activity, with expression of some but not all target genes differentially modulated in the presence of this post-translational modification [176]. An adjacent modification on LXRα (pS196) has been linked to severity of liver disease and to activation of a subset of LXR target genes [14]. Activation of LXR target genes is also modified by interactions with miRNAs. LXR induces ABCA1 through promoter binding, but also down-regulates expression of miR-26, a miRNA that binds and degrades the *ABCA1* transcript [164]. A coherent type-IV feed forward loop that simultaneously activates expression of the target gene while transrepressing it's miRNA inhibitor is therefore established, which allows for rapid and massive induction of *ABCA1* transcript.

5.1.2 Farnesoid x Receptor Alpha (NR1H4)

FXR function is linked to bile acid (BA) metabolism and cholesterol bioavailability for BA synthesis. Although the initial steps of cholesterol metabolism towards the BA route are initiated by the LXRs. In non-primate mammals FXRβ is activated by the cholesterol precursor lanosterol, so expression of this gene results in an alternative pathway for cholesterol metabolism in all other mammals [135]. Several excellent reviews are available regarding FXR's gene/protein structure and ligand binding repertoire [86], and function [59]. Aberrant FXR activity influences the pathogenesis of obesity, diabetes and dyslipidemia [53, 137, 166], liver disease [33], inflammatory bowel disease (IBD) [132] and several cancers (see Sect. 5.2.1).

5.1.2.1 Structure of FXR

FXR is composed of an N-terminal AF1 domain, a DNA binding domain, a variable hinge region, followed by a C-terminal ligand binding domain (LBD) containing activation function 2 domain (AF2) [49]. There are two FXR genes, alpha and beta, but in humans only the alpha paralogue is protein coding, beta is a pseudogene [135]. Of the 49 mouse NRs, FXRβ is the only gene not expressed as one of the 48 human NRs. Alternative promoter usage and splice site slipping gives rise to four different transcript variants harbouring two different N-terminal AF1 domains, each of which can have inclusion or exclusion of a four amino acid addition to the hinge region. The organization of protein domains that are encoded by the FXRα gene and for the four splice variants experimentally validated in humans are shown in Fig. 5.1c.

5.1.2.2 FXR Ligands

BAs are cholesterol derived molecules synthesised in the liver, and activate multiple NRs including, but not limited to, farnesoid X receptor (FXR), pregnane X receptor (PXR), and vitamin D receptor (VDR). Here we focus on their role in FXR regulation, but further information regarding their role as PXR and VDR receptors can be found here [98, 123]. Around 90% of the primary BAs, cholic acid (CA) and chenodeoxycholic acid (CDCA), are synthesised via the classic pathway using microsomal cholesterol 7alpha-hydroxylase (CYP7A1). The remainder are produced via the alternative LXR and LRH1 mediated sterol-27-hydroxylase (CYP27A1) pathway. Liver BAs are conjugated with glycine or taurine and secreted into bile where they function as lipid emulsifiers that carry sterols and phospholipids. BAs are potent signalling molecules driving FXR dependent regulation of signalling pathways that converge to influence lipid, glucose and energy metabolism, drug detoxification, and liver regeneration. Once in the gallbladder, cholecystokinin (CCK) stimulates bile release in the duodenum to favour digestion of cholesterol, triglycerides and liposoluble vitamins. BAs can be reabsorbed in the ileum, and return into the liver via the portal vein (enterohepatic circulation), or proceed into the colon and be further metabolised by gut microbiota, which is responsible for their transformation into secondary BAs (for further information the reader is directed to [46]). BAs are natural ligands for FXR and CDCA is the most potent ligand, followed by deoxycholic acid (DCA), lithocholic acid (LCA), and finally CA [156]. Differences in BAs serum levels were observed among healthy subjects and on-alcoholic steatohepatitis (NASH) patients. Subjects with NASH showed total serum BAs levels 3 times higher than healthy controls. In particular, NASH patients have 4 times higher levels of serum DCA and lower levels of serum CDCA. These BAs differences are paired with upregulation of liver FXR gene expression [87].

5.1.2.3 Fine-Tuning of FXR Signaling

FXR is regulated by miRNAs and post-translational modifications. Reducing FXR activity is a common feature of liver disease and concequeently, several of these pathwyas have been explored in the context of steatosis, non-alcholic fatty liver disease (NAFLD), or hepatocellular carcinoma. MiR-192 [99], miR-194 [130], and niR-382-5p [131] all target and downregulate FXR expression – typically exacerbating liver pathologies such as NAFLD. Acetylation of FXR at K157 and K217 by p300 reduces dimerisation potential with RXRα and consequently its transcriptional activity [93]. SUMOylation of FXR occurs at several different amino acids and is also associated with suppression of FXR signaling, downregulation of FXR target genes [9] and progression of liver disease [201].

5.1.3 Peroxisome Proliferator-Activated Receptors (NR1C1, NR1C2, NR1C3)

PPAR target genes typically regulate carbohydrate and lipid metabolism and homeostasis, and give PPARs control and influence over cell fate decisions and tissue remodeling processes. The three genes show overlapping expression patterns in normal physiology, and all three have been implicated in a range of diseases. PPARα is expressed by metabolically active tissues that need high fatty acid oxidation to produce energy, like liver, brown adipose, and skeletal tissue. To facilitate the functions of these tissues, PPARα gene targets are involved in FA mobilization and oxidation, ketogenesis and plasma lipoprotein metabolism (comprehensively reviewed by [68, 139, 169]. PPARα suppresses inflammation by downregulating expression of pro-inflammatory genes and upregulating anti-inflammatory gene expression [150, 157, 203].

PPARδ is expressed more ubiquitously than the other PPARs with highest expression in gastrointestinal system, skeletal muscles, and kidneys. A major role of PPARδ is in coordinating reverse cholesterol transport and removal of triglycerides [134, 184]. PPARδ also promotes FA oxidation and energy uncoupling, reducing the risk to develop obesity [186]. PPARγ is mainly expressed by adipose tissue, and moderately expressed in intestine and spleen. PPARγ, especially the PPARγ2 variant (see Sect. 5.1.3.1), controls the balance of adiponectin and leptin secretion (adipokines that are typically out of balance in the adipose tissue of obese individuals), and helps to maintain insulin sensitivity. PPARγ also regulates the expression of genes involved in FA efflux, transport, and storage (e.g., LPL and FAT/CD36) and as such prevents lipotoxicity and lipid overload in liver, skeletal and other tissues (further details available [70].

5.1.3.1 Structure of PPARs

PPARs are ligand-activated NRs composed of an N-terminal DNA binding domain and C terminal ligand binding domain (LBD) containing activation function 2 domain (AF2) [58]. There are three paralogous PPAR genes in humans, which are further extended by expression of transcripts encoded from alternative promoters and splicing. PPARs are probably subject to significantly more alternative splicing than has been experimentally evaluated to date [6]. Annalora and colleagues annotated 28, 33, and 23 alteratively spliced transcripts for PPARα, PPARδ, and PPARγ respectively, which had been listed in Ensembl, AceView, or PubMed databases. Interestingly, and somewhat unusually, all three PPARs lack modular cassette exons. In the absence of modular cassette exons, alternative splicing is likely to result in a shift of the open reading frame creating truncated or non-functional variants. This perhaps explains why so few of these PPAR transcript variants have been experimentally observed at protein level to date. PPARγ has alternative promoters, resulting in the expression of four PPARγ transcripts that produce two distinct proteins (PPARγ1, 3, and 4 differ only in their 5'UTRs) (further details available [95]). PPARγ2 has an additional 30 amino acids at the N-terminal domain leading to an extended AF1 domain similar to the α4 variant of the LXR splice family. The organization of protein domains that are encoded by the PPAR genes and the major protein coding variants are shown in Fig. 5.1d.

5.1.3.2 PPAR Ligands

Essential fatty acids (EFAs) and eicosanoids are specific endogenous agonists of the Peroxisome Proliferator-Activated Receptor (PPAR) family. Fatty acids (FAs) are components of many lipids involved in energy storage and metabolism, cell structure and signaling. FAs are structurally composed by a terminal carboxyl group and a hydrocarbon chain of various lengths, made up by an even number of carbon atoms, with (unsaturated) or without (saturated) double bonds. FAs can be endogenously produced by fatty acid synthases (FAS). However, some FAs cannot be synthesized by animals and they must be introduced with diet. These EFAs are the two polyunsaturated fatty acids (PUFAs), alpha-linolenic acid (ALA) and linoleic acid (LA), belonging to the omega-3 (ω3) and omega-6 (ω6) families, respectively. Once ingested, ALA can be converted into eicosapentaenoic acid (EPA), and subsequently form docosahexaenoic acid (DHA). LA is converted to γ-linolenic acid (GLA) that can be elongated to dihomo-GLA (DGLA), which is the precursor of arachidonic acid (AA). AA, EPA and DGLA, can be further metabolized into eicosanoids, which have a physiological role in inflammation and immunity, circulatory and female reproductive system, kidneys and gastrointestinal functions (reviewed by [24]. PUFAs and especially, EPA and DHA, as well as eicosanoids (15-HETE, PGJ2, and 15-deoxy-Δ12,14-PGJ2) and oxidized metabolites of LA (9-HODE and 13-HODE) are ligands for PPAR. PPARα is activated by unsaturated FAs, especially omega-3, and the eicosanoids leukotriene B4 and (8S)-hydroxyeicosatetraenoic acid (8(S)-HETE) (further details available [70], and its activation plays a role in energy combustion and metabolism regulation. PPARδ is activated by unsaturated fatty acids, especially EPA, and eicosanoids.

5.1.3.3 Fine-Tuning of PPAR Signaling

The PPAR genes are predicted to produce relatively few alternative transcripts [6] and few have been experimentally validated to date. However, the PPARs are subject to extensive post-transcriptionally regulation by miRNAs [155], post-translational regulation by a variety of enzymes that modify the protein function through covalent modifications [22], and through co-factor expression levels. Expression of miR-9 in monocytes and macrophages downregulates PPARδ contributing to polarization of pro-inflammatory M1 macrophages [175]. M1 macrophages are further activated by increased miR-27b expression that targets PPARγ [83]. In adipocyte stem cells miR-138 [188], miR 130 [104], miR-548d-5p [165], and miR-27b [92] have all been shown to prevent differentiation via PPARγ. miR-27 is of particular interest in disease of lipid metabolism and has been implicated in the pathogenesis of NAFLD [197]. PPARγ can be phosphorylated by multiple kinases leading to pathway specific effects. For example, when phosphorylation occurs via the MAPK-JNK1/2-p38 pathway, transcriptional activity is reduced [3, 25, 75, 172], but when phosphorylation occurs via CDK7 or CDK9, transcriptional activity is enhanced [42, 80]. SUMOylation within the hinge region of PPARα enhances recruitment of NCOR leading to inhibition transcriptional activity [147].

5.1.4 Other Lipid-Sensing NR

Depending on how the term 'lipid' is defined, it is arguable that all NRs are lipid sensing. However, there are several examples of NRs that belong to the high affinity steroid hormone classes, such as glucocorticoid and estrogen receptor alpha, that can bind and be activated by the same lipids that also activate the lipid-sensing NRs described above.

The side-chain hydroxycholesterol 27-OHC is produced from cholesterol by a single hydroxylation on the final C atom by the CYP27A1 enzyme and is estrogenic. 27-OHC was originally identified as an antagonist for the ER when added in the presence of estradiol [178], but since emerged as a selective estrogen receptor modulator [50]. When bound to ERα, 27-OHC drives proliferation of ERα expressing cells in culture, and growth of ERα positive cancers *in vivo*. 27-OHC has an affinity for ERα orders of magnitude less that estradiol (27-OHC Ki = 1.32 μm, E2 Kd = 0.1 nm) in vascular endothelial cells [178], yet is present in circulation at concentrations orders of magnitude greater (ca. 30–990 ng/mL [163]) than estrodiol (30-400 pg/mL in premenopausal women). This balance between affinity and concentration likely means that endogenous 27-OHC can antagonise endogenous E2 activity.

The plasma membrane is a complex structure that contains a plethora of lipids such as sphingosines, phosphatidylinositols and phosphatidylcholines that act as ligands for the NR5A subgroup. The biophysical properties of phospholipids influence ordering of nanoregions in the membrane but are also released from the membrane, accumulating and acting within the nucleus, and even as part of chromatin and transcriptional complexes with LRH1 (NR5A2) [100, 105, 152]. This membrane-lipid NR links the plasma membrane composition to lipid homeostasis, lipid diversity [124], stem cell maintenance via pluripotency factor Oct4 [71], cellular stress responses [167], and estrogen synthesis [40]. Dysregulation of phospholipid and LRH1 function is therefore associated with several pathologies, including NAFLD [167], cancers of the liver [167] and breast [40, 161]. Steroidogenic Factor 1 (NR5A1/SF1) is also able to directly bind and respond to phospholipids [18]. When bound to SF1, sphingosine acts as an antagonist and suppresses expression of aromatase, the rate limiting enzyme in the synthesis of estrogen. Deeper discussions of how phospholipid influence nuclear receptor signalling and the plasma membrane structure and function are available in the following review articles [43, 128, 161].

5.2 Emerging Trends for Lipid-Sensing Nuclear Receptors

5.2.1 Cancer Theranostics

Cancers initiate from imbalances in proliferative and differentiation factors, driven by oncogenic transformation and loss of tumour suppressor function. However, metabolic imbalances are as readily measured as gene status by genomic profiling, and therefore the complexity of NR activity regulation means that simple measures of either NR expression or even genomic binding, may mean the centrality of lipid metabolites to many cancer processes is overlooked.

Expression of NR co-factors that strongly influence response to ligand are disrupted in cancers of the prostate [11, 48, 118], bladder [1], breast [78] and others. Accumulation or depletion of NR ligands can occur via changes in expression of the CYP family resulting in variation in NR ligand bioavailability. Such differences that impinge on NR activity can alter energy and cellular metabolism within the tumour mass. The actions of lipid-sensing, and indeed other NRs, in tumour cell energy regulation has been reviewed previously [174]. Assessing activity of the lipid-sensing NRs is perhaps more challenging than assessing activity of steroid hormone receptors levels (e.g. ERs and AR). Although all are subject modulation of activity by selective modulation, co-factors, and miRNAs, the mere presence or absence of hormone receptors such as ERα and AR is sufficient for clinical classifications. The greater complexity of measuring and understanding activity the NRs that sense a diverse range of ligands will most likely be more clearly resolved as technologies improve and allow them to be therapeutically and diagnostically exploited in the coming years.

5.2.1.1 LXR in Breast and Prostate Cancer

Breast tissue is rich with a heterogenous mixture of cell types that store and metabolise lipids. Epithelial and ductal cells, adipocytes, fibroblasts, tissue resident macrophages and others combine to regulate the synthesis, storage, metabolism, and movement of many lipid species. During tumouriogensis these non-cancer cells form the tumour microenvironment and may be co-opted to provide a range of selective advantages that enhance tumour growth, including utilization of lipids.

In fibroblasts CYP27A1 converts cholesterol into 27-OHC [7, 102], which owing to its estrogenic potential drives breast cancer (BCa) cell proliferation via activation of the estrogen receptor [190] and allows the tumour to evade anti-estrogen therapy. Contrary to this, when 27-OHC (or indeed several other oxysterols) activates LXR in BCa pro- and anti-tumourogenic effects occur. The LXR-oxysterol pathway slows proliferation [180] and activates apoptosis, yet exacerbates metastasis [8, 129] and drives chemotherapy resistance [79]. There appears to be a clear selective advantage for the tumour to maintaining these apparent idiosyncrasies, the OHC-LXR axis drives expression of both pro-tumour and anti-tumour pathways. Evaluating the cholesterol metabolic role of fibroblasts across the BCa subtypes has not yet been systematically evaluated, although their presence does drive activation of interferon-β1 signaling [20] which has recently been shown to be downstream of LXR in macrophages [101].

In macrophages, 25-hydroxylase (CH25H) and converts cholesterol into 25-OHC [17] which can be secreted [12]. This oxysterol is a potent activator of LXR in BCa cells [78], leading to acute chemotherapy resistance in triple negative BCa [79]. In the OXYTAM study 25-OHC was found to be a potential diagnostic indicator of disease relapse as it was significantly elevated in the serum of BCa patients who had disease recurrence compared to those with primary disease [44]. Interestingly, CH25H and LXR are involved in a feed-forward loop. LXR can bind the CH25H promoter and induce its transcription [116], leading to a rise in synthesis of 25-OHC from cholesterol. The CH25H enzyme may therefore be a useful clinical theranostic that indicates activity of the LXR pathway in a tumour, and could be targeted by existing therapeutics such as statins that lower circulating levels of several endogenous LXR agonists [173]. Other clinical studies, particularly centered around exploring the diagnostic, prognostic, and therapeutic potential of the oxysterol:LXR axis are required in the coming decade. Systematic reviews and meta-analyses support the proposal that the LXR pathway is important in breast cancer; pharmacological [115] interventions and dietary patterns [85] that lower circulating cholesterol, and by extension oxysterol levels [47], consistently indicate there is a reduced risk of developing and dying from breast and other cancers.

Contrary to a tumour promoting role for the side chain oxysterols in some BCa subtypes, the situation is reversed in prostate cancer. LXR is anti-proliferative in prostate cancer (PCa) cell lines [36–38] but in mouse models, LXR activity prevents features of benign prostate hyperplasia appearing [181] and can restrain hyper-proliferation induced by a high cholesterol diet [146]. In the absence of LXR, cholesterol esterification is drastically enhanced and expression of an array of genes involved in metabolism of cholesterol, fatty acids, and triglycerides are lost [146] and AR target genes involved in secretory cell-cell communication are induced [181]. Cross-talk between the androgen receptor (AR) and LXR is an established phenomenon. Treatment of PCa cells with AR ligands reduces expression of the canonical LXR target ABCA1 [57] and reciprocally, LXRα activation within the liver accentuates circulating testosterone levels, although perhaps not in the prostate directly [181]. These observations led to a series of human clinical studies that again indicate that PCa tumours are divergent to BCa tumours in the context of oxysterol and LXR signaling. In BCa, 27-OHC levels are significantly elevated in tumour tissue compared to adjacent normal breast tissue [190], whereas in PCa, results from the CHOMECAP study revealed the opposite, 27-OHC level were markedly lower in tumour tissue compared to adjacent normal tissues [28]. Low levels of both the 27-OHC metabolite and the gene that codes for its synthesizing enzyme *CYP27A1*, were predictive of higher grade PCa and relapse. The expression level of neither LXRα nor LXRβ were different between normal and tumour prostate tissue in the CHOMECAP study.

5.2.1.2 FXR in Cancer

In vitro models of BCa suggest that pharmacological activation of FXR reduces the tumour promoting effects of cancer associated fibroblasts [10, 65]. CDCA treatment and FXR consequent activation, increase BCa cells cytotoxicity [4], and in patients affected by invasive breast carcinoma, expression of FXR represents an independent prognostic factor of overall and disease patient's survival [62]. Conversely, other studies showed that CDCA increase BCa cells proliferation and metastasis [2, 90]. FXR activation has also pro tumorigenic potential, as well as pro proliferative and anti-apoptotic effects in gastric, esophageal, kidney and lung cancer [56, 72, 193, 200]. FXR expression is correlated with early colorectal cancer onset [194], proliferation and progression [55], and clinical outcome [103]. In FXR-null mouse models of hepatocellular carcinoma (HCC) however, 90% of mice developed liver tumours that was linked to significant increases in Myc oncogene expression [170]. FXR activation by CDCA increases chemosensitivity of biliary tract cancers [187], and in cholangiocarcinoma, CDCA correlates with tumour differentiation [52]. Processing of BA associated with FXR activation is therefore generally associated with reduced risk of oncogenic transformation.

5.2.1.3 PPAR in Cancer

An anti-tumorigenic effect of PPARγ activation has been proposed following the discoveries of multiple mechanisms where it can induce terminal differentiation, apoptotic signaling, cell cycle arrest promotion, and inhibition of pro-inflammatory signaling (reviewed in [142]). Despite these observations some synthetic PPARγ agonists may promote onset of colon and bladder cancer [142]. In murine models, long term administration of PPARα synthetic agonist induce liver cancer [73], however, this mechanism was not observed in humans [141, 142]. Currently, PPARα is under investigation in multiple cancer prevention studies owing to the ability to inhibit tumorigenesis [120, 127], cancer cell proliferation [32, 111, 127], angiogenesis [60] and interfere with the Warburg effect [30, 76]. PPARδ is consider pro-tumorigenic for several cancers, promoting cancer hallmarks, including angiogenesis, tumorigenesis, cell death resistance and metastasis (reviewed in [185]. The transcriptional coactivator peroxisome proliferator-activated receptor gamma coactivator 1-alpha (PPARGC1A) impairs PCa metastatic process in oestrogen-related receptor alpha (ERRα)-dependent mechanisms, [177, 179]. On the contrary, in BCa PPARGC1A promotes lung metastasis and helps adaptation to metabolic drugs [5].

5.2.2 Lipid-Sensing Nuclear Receptors and Immuno-Oncology

Immune evasion is considered as an emerging hallmark of cancer. Programmed death protein 1 (PD-1) and its ligands (PD-L1/2), as well as cytotoxic T-lymphocyte-associated protein 4 (CTLA-4) are the immune checkpoints that become activated and impair the immune system's response against cancer cells [192, 196]. During tumorigenesis, CD8+ cytotoxic T-cells, which play a significant role in cancer immune detection and elimination, lose the ability to produce effector molecules in a process termed exhaustion. These 'silenced' T-cells are marked by overexpression of PD-1, CTLA-4 and other inhibitory markers. There is now a clear association between cancer-initiated T-cell exhaustion with metabolism of cholesterol and fatty acid via the LXRs and PPARs [23, 39, 41, 66, 121].

Oxysterols can promote tumour growth by creating a pro-tumorigenic microenvironment. In fact, tumour cells derived oxysterols can recruit neutrophils in the tumour microenvironment (TME) promoting angiogenesis and immune suppression [149, 160]. Vice versa, oxysterols enzymatic inactivation by sulfotransferase 2B1b (SULT2B1b) reduce neutrophils presence in breast tumours and increase the ratio between CD8 + IFNγ+ cytotoxic T-cells and CD4 + IL-4+ T-cells, which are known to be pro tumorigenic, in favour of the effector cells [126]. Activated T-cells present elevated levels of SULT2B1 leading to suppression of LXR [15] and a consequential increase in proliferation, differentiation, and expansion of T-cells, and reduction of PD-1 expression [15, 27]. LXRα activation increases Treg populations and differentiation that maintain immune tolerance in cancer [27, 136, 189] and reduce dendritic cell migration to lymphoid organs and expression CC chemokine receptor-7 (CCR7) [182]. The oxysterol-LXR axis also mediates resistance to immune destruction via the myeloid population. LXR activation by 27-OHC in myeloid cells accelerates T-cell apoptosis thus enhancing immune suppression while blocking 27-OHC synthesis via CYP27A1 inhibition improves the efficacy of anti-PDL1 checkpoint inhibitors [122]. The expression of CYP27A1 in myeloid cells is needed to guarantee a pro-tumorigenic microenvironment in ovarian cancer, and mice treated with 27-OHC increase myeloid derived suppressor cells (MDSCs) production, which suppress cancer immune response, and alters T-cells population composition [74]. The pro-metastatic effect of 27-OHC also involves myeloid immune cell function. In fact, 27-OHC increases the activation/recruitment of polymorphonuclear-neutrophils and γδT-cells, with a reduction of CD8+ cytotoxic T-cells in distal metastatic sites [8]. Many studies therefore support the hypothesis that the oxysterol-LXR axis leads to immune resistance.

Evidence of a role for PPARs in cancer immune response is less clear than for LXR. PPARα activation supports anti-PD1 immunotherapy, enhances CD8+ T-cell activity, and increases survival time. This occurs via a change in their primary metabolism pattern from glycolysis to mitochondrial fatty acid oxidation and oxidative phosphorylation [35]. Activation of PPARγ impairs the inflammatory responses of MDSCs, mainly by production of reactive oxygen species and the RAGE pathway, leading to a reduction of tumour growth in melanoma cells [151, 198]. Moreover, PPARγ activation orients tumour associated macrophages (TAMs) toward an anti-inflammatory phenotype in BCa [64], reduces Treg response, and enhances GM-CSF–secreting tumour-cell vaccine (GVAX) power. Collectively this leads to an increase cancer rejection and improvement in tumour immunity when in combination with anti-CTLA-4 [67]. PPARγ activation also has synergistic activity with anti-PD1 in mouse colon cancer model [29]. Conversely, in lung cancer activation of PPARγ stimulates tumour progression and metastasis via promotion of Arginase 1 expression in TAMs, which is considered an effector and a marker of pro-inflammatory (M2) phenotype macrophages [106]. Also, the activation of PPARγ mediated by paracrine Wnt5a/β-catenin signalling in DCs increases IDO activity and Tregs production, leading to immunotolerance [199]. In bladder cancer, high PPARγ expression impairs CD8+ T-cell infiltration, and consequently sensitivity to immunotherapies [97]. In line with these findings, also PPARδ seems to promote cancer immune resistance. In fact, PPARδ macrophages recruitment and proliferation in colon cancer TME [84], and it is involved into TAMs pro-tumorigenic polarization in ovarian carcinoma [153].

5.2.3 Therapeutic ligands Targeting Lipid-Sensing Nuclear Receptors

Besides the classic natural and synthetic LXR ligands, several ligands have recently been developed or discovered during investigations to evaluate if ligands of lipid-sensing are effective therapeutic targets in metabolic and cardiovascular diseases. For example, N-(4-trifluoromethylphenyl) 3,4-dimethoxycinnamamide (TFCA) derived from cinnamide, and the plant triterpenoid ursolic acid, act as LXRα antagonists reducing lipogenic genes activation and drug induced cellular lipid content in hepatic cells, potentially decreasing the risk to develop fatty liver and drug-induced hepatic steatosis [113, 158]. Ouabagenin, an aglycone of the steroid hormone Ouabain isolated from Strophanthus gratus, was reported to be a selective agonist for LXRβ and downregulates the expression of the LXR target gene epithelial sodium channel (ENaC), making it a potential diuretic treatment for hypertension [171]. Activation of LXRs transcription by the dietary plant oxysterol 28-homobrassinolide (28-HB) has positive metabolic effects reducing glycaemia and cholesterol levels in diabetic rats [148].

Other plant derived lipids, phytosterols, directly activate the LXRs [143] and interfere with oxysterol mediated activation of LXR [77]. The array of LXR co-factors that can be recruited and/or exchanged in response to phytosterols remains largely unvalidated; only NCOA1/SRC1 has been experimentally confirmed as a phytosterol recruited co-activator and this was in a cell free assay [143]. In cardiovascular disease models several novel LXR agonists with potential anti-atherosclerotic effects have been developed. The LXRβ agonist E17110 and the LXRα agonist IMB-170 increase ABCA1 and ABCG1 gene expression, reduce lipid accumulation and enhance cholesterol efflux in macrophages [107, 108] and the dual LXRα/β agonist IMB-808 reduces macrophage lipid accumulation [109].

In cancer, LXR ligands may confer different effects depending on the tissue. In liver cancer, LXRs activation by plant-derived product bergapten inhibits hepatocarcinogenesis by regulating PI3K/Akt and IDOL/LDLR pathways [138], while the synthetic LXRs inverse agonist and degrader GAC0001E5 inhibits pancreatic cancer cells proliferation through the inhibition of oxidative stress and glutamine anaplerotic reactions [91, 162]. Similarly, two novel PPARγ ligands, lobeglitazone (LGZ) and CB11, were studied in papillary thyroid cancer (PTC) and non-small cell lung cancer (NSCLC). Through the inhibition of TGF-β1 and p38 MAPK phosphorylation, LGZ impairs epithelial to mesenchymal transition and reduced migration and invasion of PTC cells [89]. CB11 increased cell death, ROS production, cytotoxicity and cell cycle arrest in NSCLC cells via PPARγ activation [94].

In terms of FXR control of cholesterol and glucose metabolism, the anti-parasitic drug ivermectin, reduces glycaemia and cholesterol levels through the induction of FXR transcriptional activity [88]. At the hepatic level, activation of FXR by hedragonic acid protects form drug induced liver injury and reduce liver inflammation in mice [119], while the FXR steroidal agonist BAR704 protects liver from inflammation and fibrogenesis through the downregulation of genes involved in these pathways [26]. The plant-derived product isotschimgine activating FXR transcription has anti-steatotic and insulin-sensitizing properties in obese mice, suggesting it may be a potential NAFLD therapy [110]. Hepatic steatosis and inflammation in NAFLD can also be reduced by FXR ligand and immunomodulatory drug vidofludimus [202]. Obeticholic acid (OCA) was recently investigated as FXR agonist in liver diseases clinical trials, like NASH and biliary cirrhosis but its administration seems to increase cholesterol levels side-effects (further details available [61]. However, OCA in combination with nitazoxanide has a synergistic tumour suppressive effect in colon cancer [195].

PPAR agonists have been investigated and developed into therapies for dyslipidaemia and diabetes (e.g. fibrates). The wide range of PPAR functions in metabolically active tissues means their value as therapeutic targets should extend to multiple diseases. In disease of glucose and lipid imbalance, the dual PPARα/γ partial agonist LT175 is an attractive candidate as it impairs adipogenic activity and improves glucose and lipid metabolism [63]. Lipid levels are normalized with several dual PPARα/γ phytocannabinoid agonists derived from C. Sativa, which target PPAR gene transcription in adipocytes and hepatocytes [45]. Insulin-sensitizing and glucose-lowering properties have been attributed the PPARγ agonists UNIST HYUNDAI compound 1 (UHC1) and F12016 in obese and diabetic animal models respectively [34, 114].

5.3 Perspectives

There are both experimental challenges and gaps in understanding that need to be addressed in the coming decade. For example, experimentally it remains unclear of the best way to measure activity of lipid-sensing NRs; should this be done by measuring either recruitment of co-activators (in cell based or cell free assays), the concentration of ligand, expression of the NR itself with/without co-factors, expression of NR target genes, or a combination? Furthermore, distinct disciplines need to come together to address hypotheses that are intractable within a single discipline. For example, measuring lipid levels and type in cells, tissues or model lipid membranes, especially at the level of -omics or short angle x-ray scattering, is not in the domain of the typical clinical research setting or molecular biology wet lab. Computational modeling can give insight into lipid behaviour and protein function, but must be iteratively developed through pairing with 'wet-lab' experimentation to both validate findings and improve and revise hypotheses. Computational atomistic and coarse-grained modeling of intra and inter lipid-NR interactions, combined with methodologies such as x-ray scattering, atomic force microscopy, and cryo-EM will improve understanding at the molecular level of how lipid-NR moieties are formed and dissolved, and how they are regulated by other cellular processes and deregulated. Delicate bioinformatic approaches are required to combine different high dimensional data sets, so collaborative efforts between scientists in the fields of molecular biology, biophysics, and computational modeling and bioinformatics should be encouraged to address these challenges. More broadly, such combinatorial approaches most likely will have distinct statistical challenges including how data density is considered across sets, inevitable nomenclature issues (species-level, RNA, protein, metabolite relationships), and ultimately data visualization and availability.

There are also key gaps in understanding that remain. Newly identified nuclear receptor splice variants can be catalogued (in databases such as GTEX and TSVbd) but their experimental validation at peotein level is limited by detection of unique peptide sequences in proteomics, or antibodies that recognize alternative variants. Predictions based on genomic databases suggest extensive splicing in the NR family, with LXRα perhaps one of the most widely spliced genes, let alone members of the NR family. Only upon careful examination of primary tissues has the existence of some of these predicted splice variants been validated at the protein level. This suggests that either the spliceosome of cell lines are overly simplified, or that distinct cell populations within tissues contribute different splice variants. A major goal within the field therefore, is to define the exact range of protein variants that are expressed across different tissues and during development of different diseases. The methods to achieve such a goal for this are increasingly accessible. Specialized and targeted methods such as rapid immunoprecipitation mass spectrometry of endogenous proteins (RIME) [125], or integration of genomic and proteomic data (proteogenomics) [96] will yield more information than has been possible in previous efforts that had to rely on antibodies raised against peptides with a priori sequences. Indeed, unexpected bands in westerns may be a subtle clue that the NR of interest may be subject to unknown alternative splicing patterns and should not be immediately disregarded.

Another major gap in understanding is the role of pools of ligand. This is a particularly pertinent concept for the lipid-sensing NRs owing to their broad substrate repertoires, complexity in the type and quantity of different lipid species present in most tissues, and differing affinities or EC50 values for target NRs. Many studies to date have focused on the levels of individual NR ligands, but with improvements in lipid-omics and targeted mass-spectrometry methods, understanding how the pool of lipids is interacting the pool of lipid-sensing NRs is possible. The lipid-sensing NRs are broad affinity, and unlike the hormone or seco-steroid receptors, a cell may contain many variants of ligand-NR combinations leading to localized differences in transcription factor complexes. This again takes us back to the experimental challenge of how do we best measure genome wide activity of the lipid-sensing NRs.

The lipid-sensing NRs remain underexploited in the clinical setting. This is due to the complexity of their regulation rather than their functional involvement in disease processes. Indeed, given these complexities it is likely we may see the emergence of a wider range of lipid-based therapeutics and diagnostics as precision medicine becomes increasingly possible. Integrating disciplines and gaining a full understanding of ligand pools and the array of functional protein variants will be crucial in realizing the potential of these broad sensing mediators of cell and tissue physiology.

References

1. Abedin SA, Thorne JL, Battaglia S, Maguire O, Hornung LB, Doherty AP, Mills IG, Campbell MJ (2009) Elevated NCOR1 disrupts a network of dietary-sensing nuclear receptors in bladder cancer cells. Carcinogenesis 30:449–456
2. Absil L, Journé F, Larsimont D, Body J-J, Tafforeau L, Nonclercq D (2020) Farnesoid X receptor as marker of osteotropism of breast cancers through its role in the osteomimetism of tumor cells. BMC Cancer 20:1–15
3. Adams M, Reginato MJ, Shao D, Lazar MA, Chatterjee VK (1997) Transcriptional activation by peroxisome proliferator-activated receptor gamma is inhibited by phosphorylation at a consensus mitogen-activated protein kinase site. J Biol Chem 272:5128–5132
4. Alasmael N, Mohan R, Meira LB, Swales KE, Plant NJ (2016) Activation of the Farnesoid X-receptor in breast cancer cell lines results in cytotoxicity but not increased migration potential. Cancer Lett 370:250–259
5. Andrzejewski S, Klimcakova E, Johnson RM, Tabaries S, Annis MG, McGuirk S, Northey JJ, Chenard V, Sriram U, Papadopoli DJ et al (2017) PGC-1alpha promotes breast cancer metastasis and confers bioenergetic flexibility against metabolic drugs. Cell Metab 26:778–787 e775
6. Annalora AJ, Marcus CB, Iversen PL (2020) Alternative splicing in the nuclear receptor superfamily expands gene function to refine Endo-xenobiotic metabolism. Drug Metab Dispos 48:272–287
7. Axelson M, Larsson O (1995) Low density lipoprotein (LDL) cholesterol is converted to 27-hydroxycholesterol in human fibroblasts. Evidence that 27-hydroxycholesterol can be an important intracellular mediator between LDL and the suppression of cholesterol production. J Biol Chem 270:15102–15110
8. Baek AE, Yu YA, He S, Wardell SE, Chang CY, Kwon S, Pillai RV, McDowell HB, Thompson JW, Dubois LG et al (2017) The cholesterol metabolite 27 hydroxycholesterol facilitates breast cancer metastasis through its actions on immune cells. Nat Commun 8:864
9. Balasubramaniyan N, Luo Y, Sun A-Q, Suchy FJ (2013) SUMOylation of the farnesoid X receptor (FXR) regulates the expression of FXR target genes. J Biol Chem 288:13850–13862
10. Barone I, Vircillo V, Giordano C, Gelsomino L, Győrffy B, Tarallo R, Rinaldi A, Bruno G, Caruso A, Romeo F et al (2018) Activation of Farnesoid X receptor impairs the tumor-promoting function of breast cancer-associated fibroblasts. Cancer Lett 437:89–99
11. Battaglia S, Maguire O, Thorne JL, Hornung LB, Doig CL, Liu S, Sucheston LE, Bianchi A, Khanim FL, Gommersall LM et al (2010) Elevated NCOR1 disrupts PPARalpha/gamma signaling in prostate cancer and forms a targetable epigenetic lesion. Carcinogenesis 31:1650–1660
12. Bauman DR, Bitmansour AD, McDonald JG, Thompson BM, Liang G, Russell DW (2009) 25-Hydroxycholesterol secreted by macrophages in response to Toll-like receptor activation suppresses immunoglobulin A production. Proc Natl Acad Sci U S A 106:16764–16769
13. Becares N, Gage MC, Pineda-Torra I (2017) Posttranslational modifications of lipid-activated nuclear receptors: focus on metabolism. Endocrinology 158:213–225
14. Becares N, Gage MC, Voisin M, Shrestha E, Martin-Gutierrez L, Liang N, Louie R, Pourcet B, Pello OM, Luong TV et al (2019) Impaired LXRalpha phosphorylation attenuates progression of fatty liver disease. Cell Rep 26:984–995 e986
15. Bensinger SJ, Bradley MN, Joseph SB, Zelcer N, Janssen EM, Hausner MA, Shih R, Parks JS, Edwards PA, Jamieson BD, Tontonoz P (2008) LXR signaling couples sterol metabolism to proliferation in the acquired immune response. Cell 134:97–111
16. Bideyan L, Fan W, Kaczor-Urbanowicz KE, Priest C, Casero D, Tontonoz P (2022) Integrative analysis reveals multiple modes of LXR transcriptional regulation in liver. Proc Natl Acad Sci U S A 119:e2122683119

17. Blanc M, Hsieh WY, Robertson KA, Kropp KA, Forster T, Shui G, Lacaze P, Watterson S, Griffiths SJ, Spann NJ et al (2013) The transcription factor STAT-1 couples macrophage synthesis of 25-hydroxycholesterol to the interferon antiviral response. Immunity 38:106–118
18. Blind RD, Sablin EP, Kuchenbecker KM, Chiu HJ, Deacon AM, Das D, Fletterick RJ, Ingraham HA (2014) The signaling phospholipid PIP3 creates a new interaction surface on the nuclear receptor SF-1. Proc Natl Acad Sci U S A 111:15054–15059
19. Boergesen M, Pedersen TA, Gross B, van Heeringen SJ, Hagenbeek D, Bindesboll C, Caron S, Lalloyer F, Steffensen KR, Nebb HI et al (2012) Genome-wide profiling of liver X receptor, retinoid X receptor, and peroxisome proliferator-activated receptor alpha in mouse liver reveals extensive sharing of binding sites. Mol Cell Biol 32:852–867
20. Broad RV, Jones SJ, Teske MC, Wastall LM, Hanby AM, Thorne JL, Hughes TA (2021) Inhibition of interferon-signalling halts cancer-associated fibroblast-dependent protection of breast cancer cells from chemotherapy. Br J Cancer 124:1110–1120
21. Broekema MF, Hollman DAA, Koppen A, van den Ham HJ, Melchers D, Pijnenburg D, Ruijtenbeek R, van Mil SWC, Houtman R, Kalkhoven E (2018) Profiling of 3696 nuclear receptor-Coregulator interactions: a resource for biological and clinical discovery. Endocrinology 159:2397–2407
22. Brunmeir R, Xu F (2018) Functional regulation of PPARs through post-translational modifications. Int J Mol Sci 19:1738
23. Bu DX, Tarrio M, Maganto-Garcia E, Stavrakis G, Tajima G, Lederer J, Jarolim P, Freeman GJ, Sharpe AH, Lichtman AH (2011) Impairment of the programmed cell death-1 pathway increases atherosclerotic lesion development and inflammation. Arterioscler Thromb Vasc Biol 31:1100–1107
24. Calder PC (2020) Eicosanoids. Essays Biochem 64:423–441
25. Camp HS, Tafuri SR, Leff T (1999) c-Jun N-terminal kinase phosphorylates peroxisome proliferator-activated receptor-gamma1 and negatively regulates its transcriptional activity. Endocrinology 140:392–397
26. Carino A, Biagioli M, Marchiano S, Scarpelli P, Zampella A, Limongelli V, Fiorucci S (2018) Disruption of TFGbeta-SMAD3 pathway by the nuclear receptor SHP mediates the antifibrotic activities of BAR704, a novel highly selective FXR ligand. Pharmacol Res 131:17–31
27. Carpenter KJ, Valfort AC, Steinauer N, Chatterjee A, Abuirqeba S, Majidi S, Sengupta M, Di Paolo RJ, Shornick LP, Zhang J, Flaveny CA (2019) LXR-inverse agonism stimulates immune-mediated tumor destruction by enhancing CD8 T-cell activity in triple negative breast cancer. Sci Rep 9:19530
28. Celhay O, Bousset L, Guy L, Kemeny JL, Leoni V, Caccia C, Trousson A, Damon-Soubeyrant C, De Haze A, Sabourin L et al (2019) Individual comparison of cholesterol metabolism in Normal and tumour areas in radical prostatectomy specimens from patients with prostate cancer: results of the CHOMECAP study. Eur Urol Oncol 2:198–206
29. Chamoto K, Chowdhury PS, Kumar A, Sonomura K, Matsuda F, Fagarasan S, Honjo T (2017) Mitochondrial activation chemicals synergize with surface receptor PD-1 blockade for T cell-dependent antitumor activity. Proc Natl Acad Sci U S A 114:E761–E770
30. Chang NW, Huang YP (2019) The RNA degradation pathway is involved in PPARα-modulated anti-oral tumorigenesis. Biomedicine 9:27
31. Chen M, Beaven S, Tontonoz P (2005) Identification and characterization of two alternatively spliced transcript variants of human liver X receptor alpha. J Lipid Res 46:2570–2579
32. Chen L, Peng J, Wang Y, Jiang H, Wang W, Dai J, Tang M, Wei Y, Kuang H, Xu G et al (2020) Fenofibrate-induced mitochondrial dysfunction and metabolic reprogramming reversal: the anti-tumor effects in gastric carcinoma cells mediated by the PPAR pathway. Am J Transl Res 12:428–446
33. Chiang JYL, Ferrell JM (2020) Bile acid receptors FXR and TGR5 signaling in fatty liver diseases and therapy. Am J Physiol Gastrointest Liver Physiol 318:G554–g573
34. Choi SS, Kim ES, Koh M, Lee SJ, Lim D, Yang YR, Jang HJ, Seo KA, Min SH, Lee IH et al (2014) A novel non-agonist peroxisome proliferator-activated receptor gamma (PPARgamma) ligand UHC1 blocks PPARgamma phosphorylation by cyclin-dependent kinase 5 (CDK5) and improves insulin sensitivity. J Biol Chem 289:26618–26629
35. Chowdhury PS, Chamoto K, Kumar A, Honjo T (2018) PPAR-induced fatty acid oxidation in T cells increases the number of tumor-reactive CD8(+) T cells and facilitates anti-PD-1 therapy. Cancer Immunol Res 6:1375–1387
36. Chuu CP, Lin HP (2010) Antiproliferative effect of LXR agonists T0901317 and 22(R)-hydroxycholesterol on multiple human cancer cell lines. Anticancer Res 30:3643–3648
37. Chuu CP, Hiipakka RA, Kokontis JM, Fukuchi J, Chen RY, Liao S (2006) Inhibition of tumor growth and progression of LNCaP prostate cancer cells in athymic mice by androgen and liver X receptor agonist. Cancer Res 66:6482–6486
38. Chuu CP, Kokontis JM, Hiipakka RA, Liao SS (2007) Modulation of liver X receptor signaling as novel therapy for prostate cancer. J Biomed Sci 14:543–553
39. Cioccoloni G, Aquino A, Notarnicola M, Caruso MG, Bonmassar E, Zonfrillo M, Caporali S, Faraoni I, Villiva C, Fuggetta MP, Franzese O (2020) Fatty acid synthase inhibitor orlistat impairs cell growth and down-regulates PD-L1 expression of a human T-cell leukemia line. J Chemother 32:30–40

40. Clyne CD, Speed CJ, Zhou J, Simpson ER (2002) Liver receptor homologue-1 (LRH-1) regulates expression of aromatase in preadipocytes. J Biol Chem 277:20591–20597
41. Cochain C, Chaudhari SM, Koch M, Wiendl H, Eckstein HH, Zernecke A (2014) Programmed cell death-1 deficiency exacerbates T cell activation and atherogenesis despite expansion of regulatory T cells in atherosclerosis-prone mice. PLoS One 9:e93280
42. Compe E, Drané P, Laurent C, Diderich K, Braun C, Hoeijmakers JHJ, Egly J-M (2005) Dysregulation of the peroxisome proliferator-activated receptor target genes by XPD mutations. Mol Cell Biol 25:6065–6076
43. Crowder MK, Seacrist CD, Blind RD (2017) Phospholipid regulation of the nuclear receptor superfamily. Adv Biol Regul 63:6–14
44. Dalenc F, Iuliano L, Filleron T, Zerbinati C, Voisin M, Arellano C, Chatelut E, Marquet P, Samadi M, Roche H et al (2017) Circulating oxysterol metabolites as potential new surrogate markers in patients with hormone receptor-positive breast cancer: results of the OXYTAM study. J Steroid Biochem Mol Biol 169:210–218
45. D'Aniello E, Fellous T, Iannotti FA, Gentile A, Allara M, Balestrieri F, Gray R, Amodeo P, Vitale RM, Di Marzo V (2019) Identification and characterization of phytocannabinoids as novel dual PPARalpha/gamma agonists by a computational and in vitro experimental approach. Biochim Biophys Acta Gen Subj 1863:586–597
46. Di Ciaula A, Garruti G, Lunardi Baccetto R, Molina-Molina E, Bonfrate L, Wang DQH, Portincasa P (2017) Bile acid physiology. Ann Hepatol 16:S4–S14
47. Dias IHK, Milic I, Lip GYH, Devitt A, Polidori MC, Griffiths HR (2018) Simvastatin reduces circulating oxysterol levels in men with hypercholesterolaemia. Redox Biol 16:139–145
48. Doig CL, Singh PK, Dhiman VK, Thorne JL, Battaglia S, Sobolewski M, Maguire O, O'Neill LP, Turner BM, McCabe CJ et al (2012) Recruitment of NCOR1 to VDR target genes is enhanced in prostate cancer cells and associates with altered DNA methylation patterns. Carcinogenesis 34:248–256
49. Downes M, Verdecia MA, Roecker AJ, Hughes R, Hogenesch JB, Kast-Woelbern HR, Bowman ME, Ferrer J-L, Anisfeld AM, Edwards PA et al (2003) A chemical, genetic, and structural analysis of the nuclear bile acid receptor FXR. Mol Cell 11:1079–1092
50. DuSell CD, Umetani M, Shaul PW, Mangelsdorf DJ, McDonnell DP (2008) 27-hydroxycholesterol is an endogenous selective estrogen receptor modulator. Mol Endocrinol 22:65–77
51. Endo-Umeda K, Uno S, Fujimori K, Naito Y, Saito K, Yamagishi K, Jeong Y, Miyachi H, Tokiwa H, Yamada S, Makishima M (2012) Differential expression and function of alternative splicing variants of human liver X receptor alpha. Mol Pharmacol 81:800–810
52. Erice O, Labiano I, Arbelaiz A, Santos-Laso A, Munoz-Garrido P, Jimenez-Agüero R, Olaizola P, Caro-Maldonado A, Martín-Martín N, Carracedo A et al (2018) Differential effects of FXR or TGR5 activation in cholangiocarcinoma progression. Biochim Biophys Acta (BBA) – Mol Basis Dis 1864:1335–1344
53. Fang S, Suh JM, Reilly SM, Yu E, Osborn O, Lackey D, Yoshihara E, Perino A, Jacinto S, Lukasheva Y et al (2015) Intestinal FXR agonism promotes adipose tissue browning and reduces obesity and insulin resistance. Nat Med 21:159–165
54. Feldmann R, Fischer C, Kodelja V, Behrens S, Haas S, Vingron M, Timmermann B, Geikowski A, Sauer S (2013) Genome-wide analysis of LXRalpha activation reveals new transcriptional networks in human atherosclerotic foam cells. Nucleic Acids Res 41:3518–3531
55. Fu T, Coulter S, Yoshihara E, Oh TG, Fang S, Cayabyab F, Zhu Q, Zhang T, Leblanc M, Liu S et al (2019) FXR regulates intestinal cancer stem cell proliferation. Cell 176(1098–1112):e1018
56. Fujino T, Sakamaki R, Ito H, Furusato Y, Sakamoto N, Oshima T, Hayakawa M (2017) Farnesoid X receptor regulates the growth of renal adenocarcinoma cells without affecting that of a normal renal cell-derived cell line. J Toxicol Sci 42:259–265
57. Fukuchi J, Hiipakka RA, Kokontis JM, Hsu S, Ko AL, Fitzgerald ML, Liao S (2004) Androgenic suppression of ATP-binding cassette transporter A1 expression in LNCaP human prostate cancer cells. Cancer Res 64:7682–7685
58. Fyffe SA, Alphey MS, Buetow L, Smith TK, Ferguson MAJ, Sørensen MD, Björkling F, Hunter WN (2006) Recombinant human PPAR-β/δ ligand-binding domain is locked in an activated conformation by endogenous fatty acids. J Mol Biol 356:1005–1013
59. Gadaleta RM, Cariello M, Sabbà C, Moschetta A (2015) Tissue-specific actions of FXR in metabolism and cancer. Biochim Biophys Acta 1851:30–39
60. Garrido-Urbani S, Jemelin S, Deffert C, Carnesecchi S, Basset O, Szyndralewiez C, Heitz F, Page P, Montet X, Michalik L et al (2011) Targeting vascular NADPH oxidase 1 blocks tumor angiogenesis through a PPARα mediated mechanism. PLoS One 6:e14665–e14665
61. Gege C, Kinzel O, Steeneck C, Schulz A, Kremoser C (2014) Knocking on FXR's door: the "hammerhead" -structure series of FXR agonists – amphiphilic isoxazoles with potent in vitro and in vivo activities. Curr Top Med Chem 14:2143–2158
62. Giaginis C, Karandrea D, Alexandrou P, Giannopoulou I, Tsourouflis G, Troungos C, Danas E, Keramopoulos A, Patsouris E, Nakopoulou L, Theocharis S (2017) High Farnesoid X receptor (FXR) expression is a strong and independent prognosticator in invasive breast carcinoma. Neoplasma 64:633–639

63. Gilardi F, Giudici M, Mitro N, Maschi O, Guerrini U, Rando G, Maggi A, Cermenati G, Laghezza A, Loiodice F et al (2014) LT175 is a novel PPARalpha/gamma ligand with potent insulin-sensitizing effects and reduced adipogenic properties. J Biol Chem 289:6908–6920
64. Gionfriddo G, Plastina P, Augimeri G, Catalano S, Giordano C, Barone I, Morelli C, Giordano F, Gelsomino L, Sisci D et al (2020) Modulating tumor-associated macrophage polarization by synthetic and natural PPARγ ligands as a potential target in breast cancer. Cell 9:174
65. Giordano C, Barone I, Vircillo V, Panza S, Malivindi R, Gelsomino L, Pellegrino M, Rago V, Mauro L, Lanzino M et al (2016) Activated FXR inhibits leptin signaling and counteracts tumor-promoting activities of cancer-associated fibroblasts in breast malignancy. Sci Rep 6:21782
66. Gotsman I, Grabie N, Dacosta R, Sukhova G, Sharpe A, Lichtman AH (2007) Proatherogenic immune responses are regulated by the PD-1/PD-L pathway in mice. J Clin Invest 117:2974–2982
67. Goyal G, Wong K, Nirschl CJ, Souders N, Neuberg D, Anandasabapathy N, Dranoff G (2018) PPARgamma contributes to immunity induced by cancer cell vaccines that secrete GM-CSF. Cancer Immunol Res 6:723–732
68. Grabacka M, Pierzchalska M, Dean M, Reiss K (2016) Regulation of ketone Body metabolism and the role of PPARα. Int J Mol Sci 17:2093
69. Griffiths WJ, Wang Y (2021) Sterols, oxysterols, and accessible cholesterol: Signalling for homeostasis, in immunity and during development. Front Physiol 12:723224
70. Grygiel-Górniak B (2014) Peroxisome proliferator-activated receptors and their ligands: nutritional and clinical implications – a review. Nutr J 13:17
71. Gu P, Goodwin B, Chung AC, Xu X, Wheeler DA, Price RR, Galardi C, Peng L, Latour AM, Koller BH et al (2005) Orphan nuclear receptor LRH-1 is required to maintain Oct4 expression at the epiblast stage of embryonic development. Mol Cell Biol 25:3492–3505
72. Guan B, Li H, Yang Z, Hoque A, Xu X (2013) Inhibition of farnesoid X receptor controls esophageal cancer cell growth in vitro and in nude mouse xenografts. Cancer 119:1321–1329
73. Hays T, Rusyn I, Burns AM, Kennett MJ, Ward JM, Gonzalez FJ, Peters JM (2005) Role of peroxisome proliferator-activated receptor-α (PPARα) in bezafibrate-induced hepatocarcinogenesis and cholestasis. Carcinogenesis 26:219–227
74. He S, Ma L, Baek AE, Vardanyan A, Vembar V, Chen JJ, Nelson AT, Burdette JE, Nelson ER (2019) Host CYP27A1 expression is essential for ovarian cancer progression. Endocr Relat Cancer 26:659–675
75. Hu E, Kim JB, Sarraf P, Spiegelman BM (1996) Inhibition of adipogenesis through MAP kinase-mediated phosphorylation of PPARgamma. Science (New York, NY) 274:2100–2103
76. Huang YP, Chang NW (2016) PPARα modulates gene expression profiles of mitochondrial energy metabolism in oral tumorigenesis. Biomedicine 6:3
77. Hutchinson SA, Lianto P, Moore JB, Hughes TA, Thorne JL (2019a) Phytosterols inhibit side-chain oxysterol mediated activation of LXR in breast cancer cells. Int J Mol Sci 20:3241
78. Hutchinson SA, Lianto P, Roberg-Larsen H, Battaglia S, Hughes TA, Thorne JL (2019b) ER-negative breast cancer is highly responsive to cholesterol metabolite Signalling. Nutrients 11:2618
79. Hutchinson SA, Websdale A, Cioccoloni G, Roberg-Larsen H, Lianto P, Kim B, Rose A, Soteriou C, Pramanik A, Wastall LM et al (2021) Liver x receptor alpha drives chemoresistance in response to side-chain hydroxycholesterols in triple negative breast cancer. Oncogene 40:2872–2883
80. Iankova I, Petersen RK, Annicotte JS, Chavey C, Hansen JB, Kratchmarova I, Sarruf D, Benkirane M, Kristiansen K, Fajas L (2006) Peroxisome proliferator-activated receptor gamma recruits the positive transcription elongation factor b complex to activate transcription and promote adipogenesis. Mol Endocrinol (Baltimore, Md) 20:1494–1505
81. Janowski BA, Willy PJ, Devi TR, Falck JR, Mangelsdorf DJ (1996) An oxysterol signalling pathway mediated by the nuclear receptor LXR alpha. Nature 383:728–731
82. Janowski BA, Grogan MJ, Jones SA, Wisely GB, Kliewer SA, Corey EJ, Mangelsdorf DJ (1999) Structural requirements of ligands for the oxysterol liver X receptors LXRalpha and LXRbeta. Proc Natl Acad Sci U S A 96:266–271
83. Jennewein C, von Knethen A, Schmid T, Brüne B (2010) MicroRNA-27b contributes to lipopolysaccharide-mediated peroxisome proliferator-activated receptor γ (PPARγ) mRNA destabilization. J Biol Chem 285:11846–11853
84. Jeong E, Koo JE, Yeon SH, Kwak MK, Hwang DH, Lee JY (2014) PPARdelta deficiency disrupts hypoxia-mediated tumorigenic potential of colon cancer cells. Mol Carcinog 53:926–937
85. Jiang L, Zhao X, Xu J, Li C, Yu Y, Wang W, Zhu L (2019) The protective effect of dietary Phytosterols on cancer risk: a systematic meta-analysis. J Oncol 2019:7479518
86. Jiang L, Zhang H, Xiao D, Wei H, Chen Y (2021) Farnesoid X receptor (FXR): structures and ligands. Comput Struct Biotechnol J 19:2148–2159
87. Jiao N, Baker SS, Chapa-Rodriguez A, Liu W, Nugent CA, Tsompana M, Mastrandrea L, Buck MJ, Baker RD, Genco RJ et al (2018) Suppressed hepatic bile acid signalling despite elevated production of primary and secondary bile acids in NAFLD. Gut 67:1881–1891
88. Jin L, Feng X, Rong H, Pan Z, Inaba Y, Qiu L, Zheng W, Lin S, Wang R, Wang Z et al (2013) The anti-parasitic drug ivermectin is a novel FXR ligand that regulates metabolism. Nat Commun 4:1937

89. Jin JQ, Han JS, Ha J, Baek HS, Lim DJ (2021) Lobeglitazone, a peroxisome proliferator-activated receptor-gamma agonist, inhibits papillary thyroid cancer cell migration and invasion by suppressing p38 MAPK signaling pathway. Endocrinol Metab (Seoul) 36:1095–1110
90. Journe F, Durbecq V, Chaboteaux C, Rouas G, Laurent G, Nonclercq D, Sotiriou C, Body J-J, Larsimont D (2009) Association between farnesoid X receptor expression and cell proliferation in estrogen receptor-positive luminal-like breast cancer from postmenopausal patients. Breast Cancer Res Treat 115:523–535
91. Karaboga H, Huang W, Srivastava S, Widmann S, Addanki S, Gamage KT, Mazhar Z, Ebalunode JO, Briggs JM, Gustafsson JA et al (2020) Screening of focused compound library targeting liver X receptors in pancreatic cancer identified ligands with inverse agonist and degrader activity. ACS Chem Biol 15:2916–2928
92. Karbiener M, Fischer C, Nowitsch S, Opriessnig P, Papak C, Ailhaud G, Dani C, Amri EZ, Scheideler M (2009) microRNA miR-27b impairs human adipocyte differentiation and targets PPARgamma. Biochem Biophys Res Commun 390:247–251
93. Kemper JK, Xiao Z, Ponugoti B, Miao J, Fang S, Kanamaluru D, Tsang S, Wu S-Y, Chiang C-M, Veenstra TD (2009) FXR acetylation is normally dynamically regulated by p300 and SIRT1 but constitutively elevated in metabolic disease states. Cell Metab 10:392–404
94. Kim TW, Hong DW, Park JW, Hong SH (2020) CB11, a novel purine-based PPAR ligand, overcomes radio-resistance by regulating ATM signalling and EMT in human non-small-cell lung cancer cells. Br J Cancer 123:1737–1748
95. Knouff C, Auwerx J (2004) Peroxisome proliferator-activated receptor-γ calls for activation in moderation: lessons from genetics and pharmacology. Endocr Rev 25:899–918
96. Komor MA, Pham TV, Hiemstra AC, Piersma SR, Bolijn AS, Schelfhorst T, Delis-van Diemen PM, Tijssen M, Sebra RP, Ashby M et al (2017) Identification of differentially expressed splice variants by the Proteogenomic pipeline Splicify. Mol Cell Proteomics 16:1850–1863
97. Korpal M, Puyang X, Jeremy Wu Z, Seiler R, Furman C, Oo HZ, Seiler M, Irwin S, Subramanian V, Julie Joshi J et al (2017) Evasion of immunosurveillance by genomic alterations of PPARγ/RXRα in bladder cancer. Nat Commun 8:103
98. Koutsounas I, Theocharis S, Patsouris E, Giaginis C (2013) Pregnane X receptor (PXR) at the crossroads of human metabolism and disease. Curr Drug Metab 14:341–350
99. Krattinger R, Bostrom A, Schioth HB, Thasler WE, Mwinyi J, Kullak-Ublick GA (2016) microRNA-192 suppresses the expression of the farnesoid X receptor. Am J Physiol Gastrointest Liver Physiol 310:G1044–G1051
100. Krylova IN, Sablin EP, Moore J, Xu RX, Waitt GM, MacKay JA, Juzumiene D, Bynum JM, Madauss K, Montana V et al (2005) Structural analyses reveal phosphatidyl inositols as ligands for the NR5 orphan receptors SF-1 and LRH-1. Cell 120:343–355
101. la González A, Torresano M, Rosa J, Alonso B, Capa-Sardón E, Puig-Kroger A, Vega M, Castrillo A, Corbi A (2021) LXR nuclear receptors prompt a pro-inflammatory gene and functional profile in human macrophages. https://www.researchsquare.com/article/rs-169946/v1
102. Lange Y, Steck TL, Ye J, Lanier MH, Molugu V, Ory D (2009) Regulation of fibroblast mitochondrial 27-hydroxycholesterol production by active plasma membrane cholesterol. J Lipid Res 50:1881–1888
103. Lax S, Schauer G, Prein K, Kapitan M, Silbert D, Berghold A, Berger A, Trauner M (2012) Expression of the nuclear bile acid receptor/farnesoid X receptor is reduced in human colon carcinoma compared to nonneoplastic mucosa independent from site and may be associated with adverse prognosis. Int J Cancer 130:2232–2239
104. Lee EK, Lee MJ, Abdelmohsen K, Kim W, Kim MM, Srikantan S, Martindale JL, Hutchison ER, Kim HH, Marasa BS et al (2011a) miR-130 suppresses adipogenesis by inhibiting peroxisome proliferator-activated receptor gamma expression. Mol Cell Biol 31:626–638
105. Lee JM, Lee YK, Mamrosh JL, Busby SA, Griffin PR, Pathak MC, Ortlund EA, Moore DD (2011b) A nuclear-receptor-dependent phosphatidylcholine pathway with antidiabetic effects. Nature 474:506–510
106. Li H, Sorenson AL, Poczobutt J, Amin J, Joyal T, Sullivan T, Crossno JT Jr, Weiser-Evans MC, Nemenoff RA (2011) Activation of PPARgamma in myeloid cells promotes lung cancer progression and metastasis. PLoS One 6:e28133
107. Li N, Wang X, Zhang J, Liu C, Li Y, Feng T, Xu Y, Si S (2014) Identification of a novel partial agonist of liver X receptor alpha (LXRalpha) via screening. Biochem Pharmacol 92:438–447
108. Li N, Wang X, Liu P, Lu D, Jiang W, Xu Y, Si S (2016) E17110 promotes reverse cholesterol transport with liver X receptor beta agonist activity in vitro. Acta Pharm Sin B 6:198–204
109. Li N, Wang X, Xu Y, Lin Y, Zhu N, Liu P, Lu D, Si S (2017) Identification of a novel liver X receptor agonist that regulates the expression of key cholesterol homeostasis genes with distinct pharmacological characteristics. Mol Pharmacol 91:264–276
110. Li Y, Chen H, Ke Z, Huang J, Huang L, Yang B, Fan S, Huang C (2020) Identification of isotschimgine as a novel farnesoid X receptor agonist with potency for the treatment of obesity in mice. Biochem Biophys Res Commun 521:639–645
111. Liang H, Kowalczyk P, Junco JJ, Klug-De Santiago HL, Malik G, Wei SJ, Slaga TJ (2014) Differential effects on lung cancer cell proliferation by ago-

nists of glucocorticoid and PPARα receptors. Mol Carcinog 53:753–763
112. Lianto P, Hutchinson SA, Moore JB, Hughes TA, Thorne JL (2021) Characterization and prognostic value of LXR splice variants in triple-negative breast cancer. iScience 24:103212
113. Lin YN, Wang CCN, Chang HY, Chu FY, Hsu YA, Cheng WK, Ma WC, Chen CJ, Wan L, Lim YP (2018) Ursolic acid, a novel liver X receptor alpha (LXRalpha) antagonist inhibiting ligand-induced nonalcoholic fatty liver and drug-induced lipogenesis. J Agric Food Chem 66:11647–11662
114. Liu C, Feng T, Zhu N, Liu P, Han X, Chen M, Wang X, Li N, Li Y, Xu Y, Si S (2015) Identification of a novel selective agonist of PPARgamma with no promotion of adipogenesis and less inhibition of osteoblastogenesis. Sci Rep 5:9530
115. Liu B, Yi Z, Guan X, Zeng YX, Ma F (2017) The relationship between statins and breast cancer prognosis varies by statin type and exposure time: a meta-analysis. Breast Cancer Res Treat 164:1–11
116. Liu Y, Wei Z, Ma X, Yang X, Chen Y, Sun L, Ma C, Miao QR, Hajjar DP, Han J, Duan Y (2018) 25-hydroxycholesterol activates the expression of cholesterol 25-hydroxylase in an LXR-dependent mechanism. J Lipid Res 59:439–451
117. Lizard G, Poirot M, Iuliano L (2021) European network for oxysterol research (ENOR): 10th anniversary. J Steroid Biochem Mol Biol 214:105996
118. Long MD, Thorne JL, Russell J, Battaglia S, Singh PK, Sucheston-Campbell LE, Campbell MJ (2014) Cooperative behavior of the nuclear receptor superfamily and its deregulation in prostate cancer. Carcinogenesis 35:262–271
119. Lu Y, Zheng W, Lin S, Guo F, Zhu Y, Wei Y, Liu X, Jin S, Jin L, Li Y (2018) Identification of an Oleanane-type triterpene Hedragonic acid as a novel Farnesoid X receptor ligand with liver protective effects and anti-inflammatory activity. Mol Pharmacol 93:63–72
120. Luo Y, Xie C, Brocker CN, Fan J, Wu X, Feng L, Wang Q, Zhao J, Lu D, Tandon M et al (2019) Intestinal PPARα protects against colon carcinogenesis via regulation of methyltransferases DNMT1 and PRMT6. Gastroenterology 157:744–759.e744
121. Ma X, Bi E, Lu Y, Su P, Huang C, Liu L, Wang Q, Yang M, Kalady MF, Qian J et al (2019) Cholesterol induces CD8(+) T cell exhaustion in the tumor microenvironment. Cell Metab 30:143–156 e145
122. Ma L, Wang L, Nelson AT, Han C, He S, Henn MA, Menon K, Chen JJ, Baek AE, Vardanyan A et al (2020) 27-Hydroxycholesterol acts on myeloid immune cells to induce T cell dysfunction, promoting breast cancer progression. Cancer Lett 493:266–283
123. Makishima M, Lu TT, Xie W, Whitfield GK, Domoto H, Evans RM, Haussler MR, Mangelsdorf DJ (2002) Vitamin D receptor as an intestinal bile acid sensor. Science 296:1313–1316
124. Miranda DA, Krause WC, Cazenave-Gassiot A, Suzawa M, Escusa H, Foo JC, Shihadih DS, Stahl A, Fitch M, Nyangau E et al (2018) LRH-1 regulates hepatic lipid homeostasis and maintains arachidonoyl phospholipid pools critical for phospholipid diversity. JCI Insight 3:e96151
125. Mohammed H, Taylor C, Brown GD, Papachristou EK, Carroll JS, D'Santos CS (2016) Rapid immunoprecipitation mass spectrometry of endogenous proteins (RIME) for analysis of chromatin complexes. Nat Protoc 11:316–326
126. Moresco MA, Raccosta L, Corna G, Maggioni D, Soncini M, Bicciato S, Doglioni C, Russo V (2018) Enzymatic inactivation of oxysterols in breast tumor cells constraints metastasis formation by reprogramming the metastatic lung microenvironment. Front Immunol 9:2251
127. Morinishi T, Tokuhara Y, Ohsaki H, Ibuki E, Kadota K, Hirakawa E (2019) Activation and expression of peroxisome proliferator-activated receptor alpha are associated with tumorigenesis in colorectal carcinoma. PPAR Res 2019:7486727
128. Musille PM, Pathak M, Lauer JL, Griffin PR, Ortlund EA (2013) Divergent sequence tunes ligand sensitivity in phospholipid-regulated hormone receptors. J Biol Chem 288:20702–20712
129. Nelson ER, Wardell SE, Jasper JS, Park S, Suchindran S, Howe MK, Carver NJ, Pillai RV, Sullivan PM, Sondhi V et al (2013) 27-hydroxycholesterol links hypercholesterolemia and breast cancer pathophysiology. Science 342:1094–1098
130. Nie H, Song C, Wang D, Cui S, Ren T, Cao Z, Liu Q, Chen Z, Chen X, Zhou Y (2017) MicroRNA-194 inhibition improves dietary-induced non-alcoholic fatty liver disease in mice through targeting on FXR. Biochim Biophys Acta Mol basis Dis 1863:3087–3094
131. Nie X, Liu H, Wei X, Li L, Lan L, Fan L, Ma H, Liu L, Zhou Y, Hou R, Chen WD (2021) miRNA 382 5p suppresses the expression of Farnesoid X receptor to promote progression of liver cancer. Cancer Manag Res 13:8025–8035
132. Nijmeijer RM, Gadaleta RM, van Mil SW, van Bodegraven AA, Crusius JB, Dijkstra G, Hommes DW, de Jong DJ, Stokkers PC, Verspaget HW et al (2011) Farnesoid X receptor (FXR) activation and FXR genetic variation in inflammatory bowel disease. PLoS One 6:e23745
133. Oishi Y, Spann NJ, Link VM, Muse ED, Strid T, Edillor C, Kolar MJ, Matsuzaka T, Hayakawa S, Tao J et al (2017) SREBP1 contributes to resolution of pro-inflammatory TLR4 signaling by reprogramming fatty acid metabolism. Cell Metab 25:412–427
134. Ooi EM, Watts GF, Sprecher DL, Chan DC, Barrett PH (2011) Mechanism of action of a peroxisome proliferator-activated receptor (PPAR)-delta agonist on lipoprotein metabolism in dyslipidemic subjects with central obesity. J Clin Endocrinol Metab 96:E1568–E1576

135. Otte K, Kranz H, Kober I, Thompson P, Hoefer M, Haubold B, Remmel B, Voss H, Kaiser C, Albers M et al (2003) Identification of farnesoid X receptor beta as a novel mammalian nuclear receptor sensing lanosterol. Mol Cell Biol 23:864–872
136. Pardoll DM (2012) The blockade of immune checkpoints in cancer immunotherapy. Nat Rev Cancer 12:252–264
137. Pathak P, Xie C, Nichols RG, Ferrell JM, Boehme S, Krausz KW, Patterson AD, Gonzalez FJ, Chiang JYL (2018) Intestine farnesoid X receptor agonist and the gut microbiota activate G-protein bile acid receptor-1 signaling to improve metabolism. Hepatology (Baltimore, Md) 68:1574–1588
138. Pattanayak SP, Bose P, Sunita P, Siddique MUM, Lapenna A (2018) Bergapten inhibits liver carcinogenesis by modulating LXR/PI3K/Akt and IDOL/LDLR pathways. Biomed Pharmacother 108:297–308
139. Pawlak M, Lefebvre P, Staels B (2015) Molecular mechanism of PPARα action and its impact on lipid metabolism, inflammation and fibrosis in non-alcoholic fatty liver disease. J Hepatol 62:720–733
140. Pehkonen P, Welter-Stahl L, Diwo J, Ryynanen J, Wienecke-Baldacchino A, Heikkinen S, Treuter E, Steffensen KR, Carlberg C (2012) Genome-wide landscape of liver X receptor chromatin binding and gene regulation in human macrophages. BMC Genomics 13:50
141. Peters JM, Cheung C, Gonzalez FJ (2005) Peroxisome proliferator-activated receptor-α and liver cancer: where do we stand? J Mol Med 83:774–785
142. Peters JM, Shah YM, Gonzalez FJ (2012) The role of peroxisome proliferator-activated receptors in carcinogenesis and chemoprevention. Nat Rev Cancer 12:181–195
143. Plat J, Nichols JA, Mensink RP (2005) Plant sterols and stanols: effects on mixed micellar composition and LXR (target gene) activation. J Lipid Res 46:2468–2476
144. Poirot M, Silvente-Poirot S (2013) Cholesterol-5,6-epoxides: chemistry, biochemistry, metabolic fate and cancer. Biochimie 95:622–631
145. Poirot M, Silvente-Poirot S (2018) The tumor-suppressor cholesterol metabolite, dendrogenin A, is a new class of LXR modulator activating lethal autophagy in cancers. Biochem Pharmacol 153:75–81
146. Pommier AJ, Dufour J, Alves G, Viennois E, De Boussac H, Trousson A, Volle DH, Caira F, Val P, Arnaud P et al (2013) Liver x receptors protect from development of prostatic intra-epithelial neoplasia in mice. PLoS Genet 9:e1003483
147. Pourcet B, Pineda-Torra I, Derudas B, Staels B, Glineur C (2010) SUMOylation of human peroxisome proliferator-activated receptor alpha inhibits its trans-activity through the recruitment of the nuclear corepressor NCoR. J Biol Chem 285:5983–5992
148. Premalatha R, Srikumar K, Vijayalaksmi D, Kumar GN, Mathur PP (2014) 28-Homobrassinolide: a novel oxysterol transactivating LXR gene expression. Mol Biol Rep 41:7447–7461
149. Raccosta L, Fontana R, Maggioni D, Lanterna C, Villablanca EJ, Paniccia A, Musumeci A, Chiricozzi E, Trincavelli ML, Daniele S et al (2013) The oxysterol-CXCR2 axis plays a key role in the recruitment of tumor-promoting neutrophils. J Exp Med 210:1711–1728
150. Ramanan S, Kooshki M, Zhao W, Hsu F-C, Robbins ME (2008) PPARα ligands inhibit radiation-induced microglial inflammatory responses by negatively regulating NF-κB and AP-1 pathways. Free Radical Biol Med 45:1695–1704
151. Riehl A, Nemeth J, Angel P, Hess J (2009) The receptor RAGE: bridging inflammation and cancer. Cell Commun Signal 7:12
152. Sablin EP, Blind RD, Uthayaruban R, Chiu HJ, Deacon AM, Das D, Ingraham HA, Fletterick RJ (2015) Structure of Liver Receptor Homolog-1 (NR5A2) with PIP3 hormone bound in the ligand binding pocket. J Struct Biol 192:342–348
153. Schumann T, Adhikary T, Wortmann A, Finkernagel F, Lieber S, Schnitzer E, Legrand N, Schober Y, Nockher WA, Toth PM et al (2015) Deregulation of PPARbeta/delta target genes in tumor-associated macrophages by fatty acid ligands in the ovarian cancer microenvironment. Oncotarget 6:13416–13433
154. Segala G, David M, de Medina P, Poirot MC, Serhan N, Vergez F, Mougel A, Saland E, Carayon K, Leignadier J et al (2017) Dendrogenin A drives LXR to trigger lethal autophagy in cancers. Nat Commun 8:1903
155. Seiri P, Abi A, Soukhtanloo M (2019) PPAR-gamma: its ligand and its regulation by microRNAs. J Cell Biochem. https://doi.org/10.1002/jcb.28419
156. Sepe V, Festa C, Renga B, Carino A, Cipriani S, Finamore C, Masullo D, del Gaudio F, Monti MC, Fiorucci S, Zampella A (2016) Insights on FXR selective modulation. Speculation on bile acid chemical space in the discovery of potent and selective agonists. Sci Rep 6:19008
157. Shin MH, Lee S-R, Kim M-K, Shin C-Y, Lee DH, Chung JH (2016) Activation of peroxisome proliferator-activated receptor alpha improves aged and UV-irradiated skin by catalase induction. PloS One 11:e0162628
158. Sim WC, Kim DG, Lee KJ, Choi YJ, Choi YJ, Shin KJ, Jun DW, Park SJ, Park HJ, Kim J et al (2015) Cinnamamides, novel liver X receptor antagonists that inhibit ligand-induced lipogenesis and fatty liver. J Pharmacol Exp Ther 355:362–369
159. Solheim S, Hutchinson SA, Lundanes E, Wilson SR, Thorne JL, Roberg-Larsen H (2019) Fast liquid chromatography-mass spectrometry reveals side chain oxysterol heterogeneity in breast cancer tumour samples. J Steroid Biochem Mol Biol 192:105309

160. Soncini M, Corna G, Moresco M, Coltella N, Restuccia U, Maggioni D, Raccosta L, Lin CY, Invernizzi F, Crocchiolo R et al (2016) 24-hydroxycholesterol participates in pancreatic neuroendocrine tumor development. Proc Natl Acad Sci U S A 113:E6219–E6227
161. Soteriou C, Kalli AC, Connell SD, Tyler AII, Thorne JL (2021) Advances in understanding and in multi-disciplinary methodology used to assess lipid regulation of signalling cascades from the cancer cell plasma membrane. Prog Lipid Res 81:101080
162. Srivastava S, Widmann S, Ho C, Nguyen D, Nguyen A, Premaratne A, Gustafsson J, Lin CY (2020) Novel liver X receptor ligand GAC0001E5 disrupts glutamine metabolism and induces oxidative stress in pancreatic cancer cells. Int J Mol Sci 21:9622
163. Stiles AR, Kozlitina J, Thompson BM, McDonald JG, King KS, Russell DW (2014) Genetic, anatomic, and clinical determinants of human serum sterol and vitamin D levels. Proc Natl Acad Sci U S A 111:E4006–E4014
164. Sun D, Zhang J, Xie J, Wei W, Chen M, Zhao X (2012) MiR-26 controls LXR-dependent cholesterol efflux by targeting ABCA1 and ARL7. FEBS Lett 586:1472–1479
165. Sun J, Wang Y, Li Y, Zhao G (2014) Downregulation of PPARγ by miR-548d-5p suppresses the adipogenic differentiation of human bone marrow mesenchymal stem cells and enhances their osteogenic potential. J Transl Med 12:168
166. Sun L, Xie C, Wang G, Wu Y, Wu Q, Wang X, Liu J, Deng Y, Xia J, Chen B et al (2018) Gut microbiota and intestinal FXR mediate the clinical benefits of metformin. Nat Med 24:1919–1929
167. Sun Y, Demagny H, Schoonjans K (2021) Emerging functions of the nuclear receptor LRH-1 in liver physiology and pathology. Biochim Biophys Acta Mol basis Dis 1867:166145
168. Svensson S, Ostberg T, Jacobsson M, Norström C, Stefansson K, Hallén D, Johansson IC, Zachrisson K, Ogg D, Jendeberg L (2003) Crystal structure of the heterodimeric complex of LXRalpha and RXRbeta ligand-binding domains in a fully agonistic conformation. EMBO J 22:4625–4633
169. Tahri-Joutey M, Andreoletti P, Surapureddi S, Nasser B, Cherkaoui-Malki M, Latruffe N (2021) Mechanisms mediating the regulation of Peroxisomal fatty acid Beta-oxidation by PPARα. Int J Mol Sci 22:8969
170. Takahashi S, Tanaka N, Fukami T, Xie C, Yagai T, Kim D, Velenosi TJ, Yan T, Krausz KW, Levi M (2018) Role of farnesoid X receptor and bile acids in hepatic tumor development. Hepatol Commun 2:1567–1582
171. Tamura S, Okada M, Kato S, Shinoda Y, Shioda N, Fukunaga K, Ui-Tei K, Ueda M (2018) Ouabagenin is a naturally occurring LXR ligand without causing hepatic steatosis as a side effect. Sci Rep 8:2305
172. Tang X, Guilherme A, Chakladar A, Powelka AM, Konda S, Virbasius JV, Nicoloro SM, Straubhaar J, Czech MP (2006) An RNA interference-based screen identifies MAP 4K4/NIK as a negative regulator of PPARgamma, adipogenesis, and insulin-responsive hexose transport. Proc Natl Acad Sci U S A 103:2087–2092
173. Thelen KM, Lutjohann D, Vesalainen R, Janatuinen T, Knuuti J, von Bergmann K, Lehtimaki T, Laaksonen R (2006) Effect of pravastatin on plasma sterols and oxysterols in men. Eur J Clin Pharmacol 62:9–14
174. Thorne JL, Campbell MJ (2015) Nuclear receptors and the Warburg effect in cancer. Int J Cancer 137:1519–1527
175. Thulin P, Wei T, Werngren O, Cheung L, Fisher RM, Grander D, Corcoran M, Ehrenborg E (2013) MicroRNA-9 regulates the expression of peroxisome proliferator-activated receptor delta in human monocytes during the inflammatory response. Int J Mol Med 31:1003–1010
176. Torra IP, Ismaili N, Feig JE, Xu CF, Cavasotto C, Pancratov R, Rogatsky I, Neubert TA, Fisher EA, Garabedian MJ (2008) Phosphorylation of liver X receptor alpha selectively regulates target gene expression in macrophages. Mol Cell Biol 28:2626–2636
177. Torrano V, Valcarcel-Jimenez L, Cortazar AR, Liu X, Urosevic J, Castillo-Martin M, Fernández-Ruiz S, Morciano G, Caro-Maldonado A, Guiu M et al (2016) The metabolic co-regulator PGC1α suppresses prostate cancer metastasis. Nat Cell Biol 18:645–656
178. Umetani M, Domoto H, Gormley AK, Yuhanna IS, Cummins CL, Javitt NB, Korach KS, Shaul PW, Mangelsdorf DJ (2007) 27-Hydroxycholesterol is an endogenous SERM that inhibits the cardiovascular effects of estrogen. Nat Med 13:1185–1192
179. Valcarcel-Jimenez L, Macchia A, Crosas-Molist E, Schaub-Clerigué A, Camacho L, Martín-Martín N, Cicogna P, Viera-Bardón C, Fernández-Ruiz S, Rodriguez-Hernandez I et al (2019) PGC1α suppresses prostate cancer cell invasion through ERRα transcriptional control. Cancer Res 79:6153–6165
180. Vedin LL, Lewandowski SA, Parini P, Gustafsson JA, Steffensen KR (2009) The oxysterol receptor LXR inhibits proliferation of human breast cancer cells. Carcinogenesis 30:575–579
181. Viennois E, Esposito T, Dufour J, Pommier A, Fabre S, Kemeny JL, Guy L, Morel L, Lobaccaro JM, Baron S (2012) Lxralpha regulates the androgen response in prostate epithelium. Endocrinology 153:3211–3223
182. Villablanca EJ, Raccosta L, Zhou D, Fontana R, Maggioni D, Negro A, Sanvito F, Ponzoni M, Valentinis B, Bregni M et al (2010) Tumor-mediated liver X receptor-α activation inhibits CC chemokine receptor-7 expression on dendritic cells and dampens antitumor responses. Nat Med 16:98–105

183. Voisin M, de Medina P, Mallinger A, Dalenc F, Huc-Claustre E, Leignadier J, Serhan N, Soules R, Segala G, Mougel A et al (2017) Identification of a tumor-promoter cholesterol metabolite in human breast cancers acting through the glucocorticoid receptor. Proc Natl Acad Sci U S A 114:E9346–E9355
184. Vrins CL, van der Velde AE, van den Oever K, Levels JH, Huet S, Oude Elferink RP, Kuipers F, Groen AK (2009) Peroxisome proliferator-activated receptor delta activation leads to increased transintestinal cholesterol efflux. J Lipid Res 50:2046–2054
185. Wagner N, Wagner K-D (2020) PPAR Beta/Delta and the Hallmarks of cancer. Cells 9:1133
186. Wang Y-X, Lee C-H, Tiep S, Yu RT, Ham J, Kang H, Evans RM (2003) Peroxisome-proliferator-activated receptor δ activates fat metabolism to prevent obesity. Cell 113:159–170
187. Wang W, Zhan M, Li Q, Chen W, Chu H, Huang Q, Hou Z, Man M, Wang J (2016) FXR agonists enhance the sensitivity of biliary tract cancer cells to cisplatin via SHP dependent inhibition of Bcl-xL expression. Oncotarget 7:34617
188. Wang Y, Lin L, Huang Y, Sun J, Wang X, Wang P (2019) MicroRNA-138 suppresses Adipogenic differentiation in human adipose tissue-derived mesenchymal stem cells by targeting lipoprotein lipase. Yonsei Med J 60:1187–1194
189. Wood KJ, Sawitzki B (2006) Interferon gamma: a crucial role in the function of induced regulatory T cells in vivo. Trends Immunol 27:183–187
190. Wu Q, Ishikawa T, Sirianni R, Tang H, McDonald JG, Yuhanna IS, Thompson B, Girard L, Mineo C, Brekken RA et al (2013a) 27-hydroxycholesterol promotes cell-autonomous, ER-positive breast cancer growth. Cell Rep 5:637–645
191. Wu C, Hussein MA, Shrestha E, Leone S, Aiyegbo MS, Lambert WM, Pourcet B, Cardozo T, Gustafson JA, Fisher EA et al (2015) Modulation of macrophage gene expression via liver X receptor alpha serine 198 phosphorylation. Mol Cell Biol 35:2024–2034
192. Yearley JH, Gibson C, Yu N, Moon C, Murphy E, Juco J, Lunceford J, Cheng J, Chow LQM, Seiwert TY et al (2017) PD-L2 expression in human tumors: relevance to anti-PD-1 therapy in cancer. Clin Cancer Res 23:3158–3167
193. You W, Chen B, Liu X, Xue S, Qin H, Jiang H (2017) Farnesoid X receptor, a novel proto-oncogene in non-small cell lung cancer, promotes tumor growth via directly transactivating CCND1. Sci Rep 7:1–13
194. Yu J, Li S, Guo J, Xu Z, Zheng J, Sun X (2020) Farnesoid X receptor antagonizes Wnt/β-catenin signaling in colorectal tumorigenesis. Cell Death Dis 11:640
195. Yu J, Yang K, Zheng J, Zhao W, Sun X (2021) Synergistic tumor inhibition of colon cancer cells by nitazoxanide and obeticholic acid, a farnesoid X receptor ligand. Cancer Gene Ther 28:590–601
196. Zerdes I, Matikas A, Bergh J, Rassidakis GZ, Foukakis T (2018) Genetic, transcriptional and post-translational regulation of the programmed death protein ligand 1 in cancer: biology and clinical correlations. Oncogene 37:4639–4661
197. Zhang Z, Moon R, Thorne JL, Moore JB (2021) NAFLD and vitamin D: evidence for intersection of microRNA-regulated pathways. Nutr Res Rev:1–20. https://doi.org/10.1017/S095442242100038X
198. Zhao T, Du H, Blum JS, Yan C (2016) Critical role of PPARgamma in myeloid-derived suppressor cell-stimulated cancer cell proliferation and metastasis. Oncotarget 7:1529–1543
199. Zhao F, Xiao C, Evans KS, Theivanthiran T, DeVito N, Holtzhausen A, Liu J, Liu X, Boczkowski D, Nair S et al (2018) Paracrine Wnt5a-β-catenin signaling triggers a metabolic program that drives dendritic cell Tolerization. Immunity 48:147–160.e147
200. Zhou H, Ni Z, Li T, Su L, Zhang L, Liu N, Shi Y (2018) Activation of FXR promotes intestinal metaplasia of gastric cells via SHP-dependent upregulation of the expression of CDX2. Oncol Lett 15:7617–7624
201. Zhou J, Cui S, He Q, Guo Y, Pan X, Zhang P, Huang N, Ge C, Wang G, Gonzalez FJ et al (2020) SUMOylation inhibitors synergize with FXR agonists in combating liver fibrosis. Nat Commun 11:240
202. Zhu Y, Xu S, Lu Y, Wei Y, Yao B, Guo F, Zheng X, Wang Y, He Y, Jin L, Li Y (2020) Repositioning an immunomodulatory drug Vidofludimus as a Farnesoid X receptor modulator with therapeutic effects on NAFLD. Front Pharmacol 11:590
203. Zúñiga J, Cancino M, Medina F, Varela P, Vargas R, Tapia G, Videla LA, Fernández V (2011) N-3 PUFA supplementation triggers PPAR-α activation and PPAR-α/NF-κB interaction: anti-inflammatory implications in liver ischemia-reperfusion injury. PLoS ONE 6:e28502
204. Bodin K et al (2001) Antiepileptic drugs increase plasma levels of 4beta-hydroxycholesterol in humans: evidence for involvement of cytochrome p450 3A4. J Biol Chem 276(42):38685–38689
205. Norlin M et al (2000) 24-Hydroxycholesterol is a substrate for hepatic cholesterol 7α-hydroxylase (CYP7A). J Lipid Res 41(10):1629–1639
206. Strushkevich N et al (2011) Structural basis for pregnenolone biosynthesis by the mitochondrial monooxygenase system. Proc Natl Acad Sci 108(25):10139–10143
207. Lund EG, Guileyardo JM, Russell DW (1999) cDNA cloning of cholesterol 24-hydroxylase, a mediator of cholesterol homeostasis in the brain. Proc Natl Acad Sci 96(13):7238–7243
208. Andersson S et al (1989) Cloning, structure, and expression of the mitochondrial cytochrome P-450 sterol 26-hydroxylase, a bile acid biosynthetic enzyme. J Biol Chem 264(14):8222–8229
209. Lund EG et al (1998) cDNA cloning of mouse and human cholesterol 25-hydroxylases, polytopic membrane proteins that synthesize a potent oxy-

sterol regulator of lipid metabolism. J Biol Chem 273(51):34316–34327
210. Bodin K et al (2002) Metabolism of 4 beta-hydroxycholesterol in humans. J Biol Chem 277(35):31534–31540
211. Norlin M et al (2003) On the substrate specificity of human CYP27A1: implications for bile acid and cholestanol formation. J Lipid Res 44(8):1515–1522
212. Pikuleva IA et al (1998) Activities of recombinant human cytochrome P450c27 (CYP27) which produce intermediates of alternative bile acid biosynthetic pathways. J Biol Chem 273(29):18153–18160
213. Li-Hawkins J et al (2000) Expression cloning of an oxysterol 7α-hydroxylase selective for 24-hydroxycholesterol. J Biol Chem 275(22):16543–16549
214. Yantsevich AV et al (2014) Human steroid and oxysterol 7α-hydroxylase CYP 7B1: substrate specificity, azole binding and misfolding of clinically relevant mutants. FEBS J 281(6):1700–1713
215. Griffiths WJ et al (2019) Additional pathways of sterol metabolism: Evidence from analysis of Cyp27a1-/- mouse brain and plasma. Biochim Biophys Acta Mol Cell Biol Lipids 1864(2):191–211
216. Cases S et al (1998) ACAT-2, a second mammalian acyl-CoA: cholesterol acyltransferase its cloning, expression, and characterization. J Biol Chem 273(41):26755–26764
217. Szedlacsek SE et al (1995) Esterification of oxysterols by human plasma lecithin-cholesterol acyltransferase. J Biol Chem 270(20):11812–11819
218. La Marca V et al (2016) Lecithin-cholesterol acyltransferase in brain: Does oxidative stress influence the 24-hydroxycholesterol esterification? Neurosci Res 105:19–27
219. Wang Z et al (2017) Upregulation of hydroxysteroid sulfotransferase 2B1b promotes hepatic oval cell proliferation by modulating oxysterol-induced LXR activation in a mouse model of liver injury. Arch Toxicol 91(1):271–287
220. Bai Q et al (2011) Sulfation of 25-hydroxycholesterol by SULT2B1b decreases cellular lipids via the LXR/SREBP-1c signaling pathway in human aortic endothelial cells. Atherosclerosis 214(2):350–356
221. Javitt NB et al (2001) Cholesterol and hydroxycholesterol sulfotransferases: identification, distinction from dehydroepiandrosterone sulfotransferase, and differential tissue expression. Endocrinology 142(7):2978–2984

Part III

Central Systems

6 Corticosteroid Receptors in Cardiac Health and Disease

Jessica R. Ivy, Gillian A. Gray, Megan C. Holmes, Martin A. Denvir, and Karen E. Chapman

Abstract

Nuclear receptors play a central role in both energy metabolism and cardiomyocyte death and survival in the heart. Recent evidence suggests they may also influence cardiomyocyte endowment. Although several members of the nuclear receptor family play key roles in heart maturation (including thyroid hormone receptors) and cardiac metabolism, here, the focus will be on the corticosteroid receptors, the glucocorticoid receptor (GR) and mineralocorticoid receptor (MR). The heart is an important target for the actions of corticosteroids, yet the homeostatic role of GR and MR in the healthy heart has been elusive. However, MR antagonists are important in the treatment of heart failure, a condition associated with mitochondrial dysfunction and energy failure in cardiomyocytes leading to mitochondria-initiated cardiomyocyte death (Ingwall and Weiss, Circ Res 95:135–145, 2014; Ingwall , Cardiovasc Res 81:412–419, 2009; Zhou and Tian , J Clin Invest 128:3716–3726, 2018). In contrast, animal studies suggest GR activation in cardiomyocytes has a cardioprotective role, including in heart failure.

J. R. Ivy · G. A. Gray · M. C. Holmes · M. A. Denvir
K. E. Chapman (✉)
University/BHF Centre for Cardiovascular Science, The Queen's Medical Research Institute, The University of Edinburgh, Edinburgh, UK
e-mail: Karen.Chapman@ed.ac.uk

Keywords

Heart failure · Glucocorticoid · Corticosteroid · Cardiomyocyte · DOHAD (developmental origins of health and disease) · Fetal programming · Preterm birth · Antenatal corticosteroids

6.1 Introduction

The importance of corticosteroid action in the heart has long been recognised. In his 1855 monograph "On the constitutional and local effects of disease of the supra-renal capsules", Thomas Addison described the "remarkable feebleness of the heart's action" when adrenal gland function is impaired [1]. This suggests weak contractile function in adrenal insufficiency and may be related to the general fatigue that is often amongst the first symptoms of the condition. Indeed, synthetic glucocorticoids like dexamethasone have been assessed in Duchenne Muscular Dystrophy, where they may slow cardiac disease and progression to heart failure in patients [54], probably through pro-ergogenic metabolic programming [59]. Glucocorticoids are well known effectors of adaptive responses [57] and may mediate, for example, some of the effects of exercise upon cardiac remodelling as well as contributing to the adverse effects of chronic stress [21, 47]. Understanding the actions of glucocorticoids and their receptors in cardiomyocytes – including on energy metabolism and calcium handling - are

M. J. Campbell, C. L. Bevan (eds.), *Nuclear Receptors in Human Health and Disease*, Advances in Experimental Medicine and Biology 1390, https://doi.org/10.1007/978-3-031-11836-4_6

likely key to elucidating their role in cardiac homeostasis. Here, we focus on recent advances that are providing insights into the homeostatic role of corticosteroid receptors in the heart; the reader is also referred to other recent reviews [4, 51, 62, 76, 98].

Cardiac contraction is governed by calcium flux [25]. Contraction is initiated by the entry of Ca^{2+} into the cardiomyocyte via L-type Ca^{2+} channels located in t tubules, invaginations of the plasma membrane that reach deep into the myocyte and are closely juxtaposed to the sarcoplasmic reticulum. The increase in Ca^{2+} in the dyadic space between the t tubule and the sarcoplasmic reticulum induces the opening of ryanodine receptors (RyR) causing a large release of Ca^{2+} from the sarcoplasmic reticulum, termed Ca^{2+}-induced-Ca^{2+}-release. This triggers the contractile process, with myosin sliding relative to actin. To release myosin from actin and relax the muscle, Ca^{2+} must be returned to the sarcoplasmic reticulum via the sarcoplasmic/endoplasmic reticulum Ca^{2+} adenosine triphosphatase-2 (SERCA2). The sodium/calcium exchanger-1 (NCX1) provides an export pathway for Ca^{2+} to exit the cell. A sustained failure to maintain the sarcoplasmic reticulum-cytoplasmic difference in Ca^{2+} during diastole can lead to arrhythmia and/or heart failure. This Ca^{2+} movement requires energy: ATP is required to release myosin from actin and is also required for SERCA2 to return Ca^{2+} to the sarcoplasmic reticulum. The requirement for ATP to maintain a sufficiently low diastolic cytoplasmic Ca^{2+} concentration to enable contraction and relaxation creates an interdependency between Ca^{2+} handling, metabolism and mitochondrial function, also regulated by Ca^{2+} [25]. Indeed, mitochondrial Ca^{2+} uptake through the mitochondrial Ca^{2+} uniporter (MCU) acutely matches cardiac workload with mitochondrial metabolism and ATP generation in a "fight-or-flight response" [48]. Whether mitochondrial uptake of Ca^{2+} impacts cytoplasmic concentrations is currently unknown [26]. However, elevations in mitochondrial Ca^{2+} concentrations, especially in combination with reactive oxygen species, can lead to cell death via opening of the mitochondrial permeability transition pore (PTP) [8]. It is currently not known if glucocorticoids impact mitochondrial Ca^{2+} concentrations, either through MR or GR or the balance between them.

The heart produces and consumes large amounts of ATP, daily consuming in the order of 10 times its weight in ATP [39]. If mitochondria are dysfunctional or unable to meet the demand for ATP production to fuel cardiomyocyte contraction and relaxation - for example following an ischemic insult or other injury - then cardiomyocyte death can be triggered [100]. This, in turn, increases the workload on the remaining cardiomyocytes, potentially leading to heart failure as there is very limited regeneration of cardiomyocytes [35]. Indeed, there is very little cardiomyocyte proliferation beyond the neonatal period so the lifelong number of cardiomyocytes – the cardiomyocyte endowment – is determined shortly after birth. Any deficit in cardiomyocyte number, due either to low endowment or cardiomyocyte death, increases workload and stress for the remaining cardiomyocytes, creating a vulnerability to further stress or injury. Recent evidence is providing mechanistic insight into how glucocorticoids may influence heart disease, through the balance of cardiomyocyte death and survival in the adult heart, discussed below. Whether related mechanisms underly the early life glucocorticoid programming of cardiovascular risk in adulthood is an intriguing, but as yet unanswered, question.

There are 2 receptors for glucocorticoids. The mineralocorticoid receptor (MR) has higher affinity for cortisol and corticosterone than does the glucocorticoid receptor (GR). However, in mineralocorticoid-target tissues like the distal nephron, MR is co-expressed with 11β-hydroxysteroid dehydrogenase type 2 (11β-HSD2), the enzyme that inactivates cortisol and corticosterone to intrinsically inert cortisone and 11-dehydrocorticosterone, respectively (reviewed, [15]). By inactivating competing glucocorticoids in MR target cells, 11β-HSD2 allows aldosterone, which circulates at lower levels than glucocorticoids, to bind to MR, and thus confers mineralocorticoid specificity upon MR. Cardiomyocytes express both GR and MR but show negligible expression of 11β-HSD2 (reviewed, [76]). Thus, in the heart, MR is normally occupied by glucocorticoids, rather than mineralocorticoids [40]. Receptor density is an important contributor to the relative balance between MR and GR activity in cardiomyocytes, as is, crucially, the diurnal and stress variation in glucocorticoid levels. Whereas the higher affinity MR is likely to be fully occupied even at nadir levels of cortisol/corticosterone, GR is only likely to be fully occupied at the diurnal peak or following stress or exercise; potent stimuli to the hypothalamic-pituitary-adrenal (HPA) axis. Accordingly, the relative balance between active GR and active MR in cardiomyocytes will vary in a circadian manner or with stress. Moreover, whereas dexamethasone, a potent synthetic glucocorticoid frequently used in experimental paradigms, is 'MR-sparing', preferentially activating GR, it suppresses the HPA axis and thus endogenous glucocorticoid production, depriving MR of its glucocorticoid ligand in cardiomyocytes and other cells in which it is primarily a glucocorticoid receptor [23]. This is an important consideration when interpreting experimental evidence. In other cell types, such as brain and macrophages that express MR without 11β-HSD2, the outcomes of MR and GR activation are distinct and frequently opposing, despite the strong conservation of the DNA binding domain ([7, 22, 58] and reviewed [76, 78]). The same appears true of the heart. Mice with cardiomyocyte-specific knockout of GR or MR - "CardioGRKO" and "CardioMRKO" mice, respectively, show distinct cardiac phenotypes (CardioGRKO mice develop heart disease whereas CardioMRKO mice do not, see below) and distinct differential expression of gene sets in their hearts, prior to the development of any overt pathology [64]. Similarly, although transgenic over-expression of either GR or MR in cardiomyocytes alters the surface electrocardiogram (ECG) and induces ion channel remodelling, each receptor does so in a distinct way and only MR over-expression results in lethal arrhythmias [67, 85]. Thus, receptor density and glucocorticoid levels dynamically determine the balance between MR and GR activation and cardiac outcomes, including in early life "programming" of cardiovascular risk in adulthood. The consequences of the dynamic 'fine tuning' of this balance in heart are a long way from being understood. Whilst considerable interest has focussed on the role of MR activation in the heart, especially its pathophysiological role in heart failure, until recently the role of GR in the heart has been less explored.

6.2 MR and Heart Failure

The role of MR in heart disease - and the mechanistic insight forthcoming from investigations in genetically altered mice - has been extensively reviewed (e.g. see [9, 11, 13, 43, 72, 76, 98]) and will not be reviewed in detail here. Briefly, elevated cardiac expression of MR is associated with heart failure and arrhythmias in humans and in animal models [65, 91]. MR antagonists clearly reduce morbidity and mortality in patients with chronic heart failure with reduced ejection fraction, a measure of left ventricular contraction [70, 99] and in patients with left ventricular dysfunction after myocardial infarction [71]. These benefits appear to be independent of their effect on blood pressure [45, 72, 87]. MR antagonists may also be of benefit in obese patients who have heart failure with preserved ejection fraction [27]. Exactly why MR antagonists are effective in heart failure is still not completely clear, but in mice, knock-out of MR in cardiomyocytes (leaving GR unopposed) improves myocardial infarct healing and prevents contractile dysfunction and adverse cardiac remodelling in ischemic heart failure as well as in a mouse model of pressure overload [31, 52] without affecting the associated cardiac fibrosis [52]. MR is, nevertheless, implicated in cardiac fibrosis in a mouse model in which renin-angiotension-aldosterone activity is inappropriately high for salt status, with MR in cardiomyocytes sustaining inflammatory cell recruitment to the heart [80]. More significantly perhaps, MR knock-out in macrophages (which also express MR without 11β-HSD2) is protective against cardiac fibrosis in two different mouse models of altered renin-angiotension-aldosterone activity relative to salt status [79, 92]. This effect has been attributed to altered macrophage polarisation in the absence of MR rather than an effect on blood pressure, which was unaffected (or even exacerbated) by macrophage MR knock-out in one study [92] but reduced in the other [79]. The alteration in macrophage polarisation could, conceivably, result in part from unopposed GR activation, promoting pro-reparative polarisation [17]. ~~Endothelial~~ Within the vasculature, endothelial MR also contributes to lifestyle- or age-related heart disease, probably also independently of effects on blood pressure [42, 53, 81]. The effects of MR antagonists in human heart failure are thus likely to be complex and involve actions within the heart, vasculature and immune system, as well as the kidney. The role played by MR in cardiomyocytes in the healthy heart is less clear, with cardiomyocyte knock-out of MR having only minor effects on heart function and size in mice [31]. However, transgenic over-expression of MR in cardiomyocytes causes ion-channel remodelling, increases action potential duration and amplitude of Ca^{2+} transients and leads to cardiac arrhythmias and sudden death [67]. This suggests that an elevation in MR activity alone (or relative to GR) is sufficient to trigger dysrrhythmic cardiac events. Interestingly, these mice do not go on to develop cardiac fibrosis [67], supporting a non-cardiomyocyte origin for MR-mediated cardiac fibrosis. Although over-expression of GR in cardiomyocytes causes ion-channel remodelling associated with conduction defects and atrio-ventricular block, this mainly involves Na^+ and K^+ currents [85] and is distinct from the ion channel remodelling induced by MR, which mainly alters Ca^{2+} currents [67]. GR over-expression does not cause arrhythmia or early death, nor does it cause cardiac hypertrophy or fibrosis [85]. Thus, although GR and MR alter many of the same processes in heart, the outcomes are distinct, reflecting the regulation of differing gene sets.

6.3 GR Limits Cardiac Injury and Subsequent Pathophysiology

Whilst much attention has focussed on cardiomyocyte MR in cardiac pathophysiology, more recently a protective role for GR in countering heart disease has been uncovered [55, 63, 64, 77]. Glucocorticoids are important for maintenance of cardiac health in animals. Over several months, adrenalectomised mice develop cardiac hypertrophy and dysfunction, associated with ECG abnormalities and gene expression changes indicative of abnormal Ca^{2+} handling [19]. The ECG abnormalities are rescued by aldosterone replacement, whereas corticosterone replacement or betamethasone treatment restores cardiac function and gene expression without affecting the ECG [19]. However, these effects on the heart may be indirect. CardioGRKO mice, with cardiomyocyte-specific deletion of GR generated using an αMHC-Cre transgene, develop spontaneous heart failure at around 6 months of age [63, 64] (Table 6.1). This raises the question of whether this simply reflects a lack of GR, or whether it is due to unopposed (and detrimental) MR activation. Remarkably, concurrent deletion of MR in cardiomyocytes in "CardioGRMRdKO" mice alleviates the heart disease seen in CardioGRKO mice lacking just GR, with the onset of contractile dysfunction and overt heart failure delayed by several months [64]. This is associated with an increase in expression of cell survival and a decrease in cell death pathways. Moreover, it is independent of the inflammation and cardiomyocyte hypertrophy seen in CardioGRKO mice, which are unchanged or even worsened in hearts of CardioGRMRdKO mice [64]. These findings indicate: firstly, that GR restrains cardiomyocyte hypertrophy and inflammation independently of MR and secondly, that GR signalling in cardiomyocytes is cardioprotective by countering the adverse effects of MR activation to promote death and reduce survival of cardiomyocytes, the latter through a mechanism that is entirely or at least substantially distinct from cardiomyocyte hypertrophy and cardiac inflammation. As well as protection against age-related heart disease,

Table 6.1 Comparison of the main phenotypes of independent lines of mice with GR knockout in cardiomyocytes

GR KO in cardiomyocytes	MHCα-Cre [20, 64, 64]	MHCα-Cre [55]	SM22α-Cre [3, 75, 77]
	1–6 months	*10 weeks*	*6, 10–12 weeks, 1 year*
Survival at birth	Normal	Normal	Reduced
Mortality after birth	Die from 3-6mo; LV systolic dysfunction and heart failure	Normal	Normal
Heart size (wet weight)	Increased	Not different	Increased
Cardiomyocyte hypertrophy?	Increased MHCβ, BNP	Fibre area not different (Trend for increased ANP, BNP)	Yes, in adult males. Not in females or juvenile males
Response to pressure overload	Not determined	Worse remodelling/ fibrosis in cardioGRKO In WT mice, GR down-regulated folllowing TAC	Not determined
Fibrosis	No	No	Yes. Partially attenuated by MR antagonism from birth
ECG	Normal/modest increase in QRS complex duration	Not determined	Not determined
Ryr2 mRNA	Reduced	Reduced	Reduced in adult females but not males
Caffeine response (in vitro in primary adult cardiomyocytes)	No difference with 10 mM At 0.5 mM: Failure to sustain increase in Ca2+, increased frequency of oscillations	Not determined	Preliminary data suggest reduced SR Ca2+ content; slower rate of decay of Ca2+ transient[1]
MR expression	Unchanged	Elevated	Unchanged at P2, then elevated
Effect of CardioMRKO?	Versus GRKO, in DKO, HF is delayed and apoptosis/ necrosis reduced without affecting inflammatory pathways or *Ryr2*	Versus GRKO, DKO has no effect on pressure overload phenotype	Not determined
Gene expression	Pathways - cell death/survival, inflammation, mitochondrial dysfunction, oxidative stress Reduced *Dmd* Reduced *Klf15*		Reduced SERCA in adult females Reduced *Agt* Increased *Bcl2*

cardiomyocyte GR is also protective against pressure overload cardiac injury in younger mice. In "GRcKO" mice, a different line of mice with GR knockout in cardiomyocytes (also generated using αMHC-Cre), adverse cardiac remodelling is considerably worse in a model of pressure overload, with GRcKO mice showing a greater increase in heart weight and cardiomyocyte hypertrophy, reduced contractile function and greater reactivation of a fetal gene expression profile compared to control mice [55] (Table 6.1). There is no impact of concurrent knock-out of MR in this model, with "GRMRdcKO" mice showing similar cardiac remodelling to GRcKO mice after pressure overload [55]. Thus, in this case the greater adverse remodelling in GRcKO mice appears due to a lack of GR rather than unopposed MR activation, though it is consistent with the similar (or worse) cardiomyocyte hypertrophy in CardioGRMRdKO compared to CardioGRKO in the Oakley study [64]. GRcKO mice were examined at a relatively young age (10–12 weeks) and whether they are at increased risk of heart failure following pressure overload or go on to develop spontaneous heart failure as CardioGRKO mice do once they are older than 3 months of age is currently unclear [55].

6.4 Glucocorticoid Action in Heart Is Sexually Dimorphic

There is sexual dimorphism in the risk and presentation of cardiovascular disease in humans [50]. It is therefore to be expected, given sexual dimorphism in corticosteroid physiology, that there may be sexual dimorphism in corticosteroid action in the heart. Corticosteroid actions on the cardiovascular system are complex and reflect the many contributions from heart, vasculature, autonomic nervous system, immune system, kidney etc. as well as the effects of developmental programming (this latter is discussed further below). Moreover, many animal studies report outcomes on males only, but there are important insights from those that have investigated outcomes in females as well as males. Sexually dimorphic effects of corticosteroid action within cardiomyocytes have been investigated in mice with cardiomyocyte knockout of GR. Female CardioGRKO mice (with αMHC-Cre mediated deletion of GR) die of heart failure at a later age than do males. Cardiac function is preserved for longer and there is less dysregulation of cardiac Ca^{2+} handling in females compared to males [20]. Importantly, expression of βMHC, a marker of cardiomyocyte hypertrophy, is increased in hearts of male CardioGRKO mice, but not in females; this is reversed by gonadectomy of males suggesting an interaction with sex hormones [20]. Similarly, male SMGRKO mice (with SM22-Cre mediated GR knockout in cardiomyocytes and vascular smooth muscle cells but without Cre expression in adult cardiomyocytes) develop cardiomyocyte hypertrophy in association with an increase in βMHC expression though only after 6 weeks of age; females show neither cardiomyocyte hypertrophy nor elevated βMHC expression [77] (Table 6.1). This points to a role for GR in restraining the cardiomyocyte hypertrophy in males that is driven by the increase in testosterone following puberty [50]. Moreover, this phenotype is independent of the Cre used to delete GR and whether it is expressed in adult cardiomyocytes. Elucidation of the underlying mechanisms, and the consequences, could be helpful in the understanding of how heart disease manifests and progresses in humans.

6.5 Cardiac Injury and Repair: The Balance of GR and MR Action as a Determinant of Cell Death or Survival

Although chronic glucocorticoid excess is associated with adverse cardiovascular outcomes [69, 94], acutely, glucocorticoids are protective against ischemia/reperfusion injury and reduce myocardial infarct size in animals [88, 96]. However, if administered after myocardial infarction in pigs, glucocorticoids worsen cardiac remodelling [46]. Unpicking the mechanisms underlying the outcomes of the balance between GR and MR in mediating cardioprotective (or otherwise) effects of glucocorticoids in the heart has proved tricky, particularly given the complex actions of both receptors within the cardiovascular system as a whole. However, pathway analysis on hearts of mice with cardiomyocyte knockout of GR, MR, or both, is providing valuable information on the importance of GR/MR balance in determining cardiomyocyte death or survival. Pathways uniquely dysregulated in hearts of young pre-symptomatic CardioGRKO mice that are not dysregulated in CardioGRMRdKO mice, point to increased cell death, reduced cell survival and mitochondrial dysfunction/oxidative stress [64]. Impaired Ca^{2+} handling in CardioGRKO mice is also largely mitigated by concurrent knockout of MR [64].

Glucocorticoids are protective against apoptosis in cardiomyocytes *in vivo* and *in vitro* [63, 74, 96]. Dexamethasone treatment of cultured primary neonatal cardiomyocytes prevents apoptosis triggered by serum deprivation or an inflammatory milieu [74]. Various mechanisms have been implicated, including transcriptional activation of anti-apoptotic Bcl-xL [74, 96], down-regulation of pro-apototic Gas2 [74] and non-transcriptional activation of eNOS and/or prostaglandin signalling [36, 90].

The use of αMHC-Cre to delete GR in cardiomyocytes is highly relevant to the spontaneous heart failure that develops in CardioGRKO mice. The high expression of Cre recombinase in adult cardiomyocytes directed by the αMHC promoter causes accumulation of DNA damage through cleavage at cryptic LoxP sites (including in the *Dmd* gene, encoding dystrophin) leading to cardiomyocyte hypertrophy, dilated cardiomyopathy and, ultimately, heart failure [12, 33, 93]. Indeed, contractile dysfunction is evident in the 6 month old αMHC-Cre control mice in the Oakley study [64]. αMHC-Cre mice, but not mice expressing Cre recombinase under the control of the *Tnnt2* promoter, show alterations in L-type Ca^{2+} current, possibly as a consequence of reduced expression of dystrophin [33]. *Dmd* mRNA is reduced in hearts of CardioGRKO, CardioMRKO and CardioGRMRdKO mice [64], consistent with this being due to Cre-mediated DNA damage. A different line of mice, "SMGRKO" mice, with GR knock-out in cardiomyocytes and vascular smooth muscle cells, was generated using SM22-Cre, which directs only transient expression of Cre recombinase during cardiomyocyte development and is not expressed in the adult heart. These mice do not show the spontaneous heart disease up to 1 year of age [77], indicating that spontaneous heart failure is a consequence of GR deficiency in the context of lifelong high Cre recombinase expression in cardiomyocytes. The mitigation of the phenotype in CardioGRMRKO mice additionally deficient in MR suggests the balance between GR and MR activation is important in cardiomyocyte survival following cardiac injury, certainly in the context of older age and/or DNA damage. GR induction of *Klf13* is critical to protect cardiomyocytes from DNA damage and cell death, including following an ischemic insult or doxorubicin, an anthracycline drug used as an anti-cancer agent [18], suggesting a plausible contributory mechanism. Furthermore, consistent with a role in cell survival after DNA damage, following a mild genotoxic challenge there are fewer double strand DNA breaks in the livers and colon of GR haploinsufficient mice than in controls, an effect attributed to increased cell death [56]. DNA damage is a common mechanism in cardiac injury: it is DNA damage that mediates the cardiomyocyte hypertrophy that occurs in pressure overload [60]. Oxidative stress associated with heart failure also causes DNA damage. MR may act in an opposing manner to GR, promoting cell death after DNA damage. Consistent with this, knockout of MR alleviates the systolic dysfunction in 6 month old αMHC-Cre mice with intact GR signalling [64]. MR antagonism may promote cardiomyocyte survival in other situations associated with cardiac injury or DNA damage, including anthracycline induced cardiotoxicity [5, 97]; whether MR antagonism is cardioprotective in drug-induced cardiotoxicity will be an important question to answer in the future. It is intriguing to speculate that the contribution of GR and MR signalling to cell death and survival following DNA damage may be related to cellular energy (ATP) levels. Poly(ADP-ribose)polymerase-1 (PARP1) is an enzyme critical in DNA repair and chromatin remodelling pathways with capacity to promote pro-inflammatory gene expression [89]. PARP1 utilises NAD^+ as a substrate for ADP-ribosylation, thus extensive activation of PARP depletes cellular NAD^+ levels and ultimately ATP, leading to cell death. PARP inhibitors preserve myocardial contractility and relaxation and limit ventricular remodelling in models of heart failure [37]. Perhaps the GR/MR balance is a determinant of ATP levels under cardiomyocyte stress. Whether GR and MR action impact DNA repair through maintenance of cellular energy stores and redox status or by other mechanisms is also worthy of future investigation.

Genetic evidence is consistent with the balance between GR and MR density being an important determinant of cardiac health. A common haplotype of the glucocorticoid receptor gene (GR) that leads to relative glucocorticoid resistance is associated with increased risk of heart failure [66]. Whilst this increase in risk has been attributed to an association with low-grade inflammation, it is also plausible that it reflects vulnerability to cardiac insult due to reduced GR activity. Intriguingly, people who are heterozygous for a loss-of-function allele in *NR3C2* (encoding MR) have favourable diastolic left ventricular function despite lifelong higher circulating aldosterone levels [28], consistent with better preservation of cardiomyocyte function with an increase in the relative ratio of GR to MR. Thus, even in the absence of obvious cardiac insult, GR/MR balance is likely to play an important homeostatic role in the heart.

6.6 Glucocorticoids Regulate calcium Handling and Metabolism in Cardiomyocytes

It has been known for many years that adrenal insufficiency is associated with hypercalcaemia [49]. This, and many other lines of evidence, supports a role for glucocorticoids in Ca^{2+} homeostasis, yet the mechanisms remain poorly characterised and are likely to differ depending on tissue. Corticosteroids affect all aspects of Ca^{2+} handling in cardiomyocytes - the influx of Ca^{2+} into the cell via L-type Ca^{2+} channels, sarcoplasmic calcium release via RyRs as well as affecting SERCA2, and Ca^{2+} extrusion from cardiomyocytes via NCX1 and PMCA. Similarly, glucocorticoids are well known for their effects on metabolism and energy partitioning. Amongst their actions, they stimulate lipolysis to release fatty acids into the circulation. In the heart, glucocorticoids regulate energy metabolism, promoting an ergogenic metabolic programme requiring induction of *Klf15*. They affect mitochondrial capacity and function in a variety of ways, thus impacting ATP generation, which in turn is integrated with Ca^{2+} homeostasis. Untangling what is directly regulated by GR and/or MR and what is indirect is challenging.

Over-expression of GR in cardiomyocytes alters their electrical properties, distinct to alterations induced by MR over-expression [67, 85]. It alters Ca^{2+} homeostasis, including an increase in L-type Ca^{2+} current, in Ca^{2+} transients and in sarcoplasmic reticulum Ca^{2+} load, associated with increased cardiomyocyte contractility [85]. Hearts of aged male rats treated with dexamethasone show improved myocardial contractile performance, associated with higher rates of SERCA2 mediated Ca^{2+} uptake into the sarcoplasmic reticulum [61]. These changes in Ca^{2+} handling suggest glucocorticoids increase capacity for Ca^{2+}-induced-Ca^{2+}-release and thus cardiomyocyte contractility. Conversely, although young male CardioGRKO mice have a normal surface ECG prior to the appearance of symptoms, their hearts show impaired contractility with reduced expression of *Ryr2* and *Atp2a2* (SERCA2) (though both are normally expressed in females at the same age), also reduced t tubule density and L-type Ca^{2+} channel-RyR coupling [20, 63, 86]. Furthermore, in adult human and rat cardiomyocytes cultured *in vitro* in serum free medium to induce t tubule loss, glucocorticoids - acting via GR rather than MR - prevent loss of t tubules and preserve excitation-contraction coupling and contractile function [86]. This is consistent with developmental effects of glucocorticoids; dexamethasone, in combination with thyroid hormone, promotes t tubule development in human induced pluripotent stem cells differentiated into cardiomyocytes, though interestingly neither are effective alone [68]. Enhanced autophagy has been suggested as the mechanism for the preservative effects of glucocorticoids on t-tubules in adult cardiomyocytes [86]. It is interesting that dexamethasone abolishes the apoptosis induced in cultured cardiomyocytes by serum-deprivation and TNFα treatment *in vitro* [74]. Whether these cardiomyocyte-preserving effects with serum deprivation involve effects of glucocorticoids on t tubules, mitochondria and/or energy metabolism will be interesting questions to explore.

In vitro experiments in cardiomyocytes have provided some insight into mechanisms by which corticosteroids affect energy metabolism, mitochondrial function and induce ion channel remodelling. Treatment of mouse fetal cardiomyocytes with glucocorticoids promotes maturation of Ca^{2+} handling, contractile function and myofibril structure, consistent with their antenatal maturational effects [84]. These effects are blocked by knock-down of GR expression or by

GR antagonism, but unaffected by antagonism of MR [84]. In fetal cardiomyocytes, glucocorticoids increase capacity for basal mitochondrial respiration as well as capacity for mitochondria fatty acid oxidation, without affecting mitochondrial number or morphology [41], effects that presumably contribute to the maturation of energy metabolism in the developing heart. Amongst the pro-maturational effects of glucocorticoids on mouse fetal cardiomyocytes, the induction of *Ppargc1a* (encoding PGC-1α, a master regulator of mitochondrial capacity) is essential for at least some of the maturational effects of glucocorticoids in fetal cardiomyocytes *in vitro* [84] and possibly also *in vivo* [41]. As well as PGC-1α, GR activation is directly responsible for inducing a host of additional metabolic regulators including PPARα, Lipin-1, Klf15 and CD36 [59, 84]. Given that genes encoding Ca^{2+} handling proteins do not appear to be direct GR targets in cardiomyocytes, it will be interesting to establish if glucocorticoid effects on Ca^{2+} homeostasis in cardiomyocytes is secondary to effects on mitochondrial metabolism and/or changes in ion channels, a number of which are implicated as direct targets of GR in cardiomyocytes [84].

6.7 Early Life Programming of Cardiovascular Disease: Creating Vulnerability?

It has been established for several decades that excessive exposure to glucocorticoids in early life is associated with increased susceptibility to cardiovascular disease in adulthood [4, 83]. Whilst many associations have been suggested as causal, whether these contribute to the underlying mechanisms remains unconfirmed. In late gestation, endogenous fetal glucocorticoid levels rise markedly: this is essential for maturation of fetal tissues and organs, including the heart [82], for survival after birth. Indeed, this is the rationale for administration of antenatal corticosteroids (typically the MR-sparing synthetic glucocorticoids, betamethasone or dexamethasone) to women considered at risk of preterm delivery, with the aim of maturing the fetal lungs in particular [4]. If GR activation is excessive, it is easy to see how this normal glucocorticoid-promoted maturational process could be accelerated or exaggerated, resulting in a greater or an earlier switch from tissue accretion to differentiation [30, 76]. It should be noted that thyroid hormones affect many of the same maturational pathways regulated by glucocorticoids [16]; given that *Dio2*, encoding a thyroid hormone activating enzyme, is a glucocorticoid target gene in fetal cardiomyoyctes [84], some actions of glucocorticoids may be mediated by (or synergise with) thyroid hormone. Because there is little proliferation of cardiomyocytes beyond the neonatal period in mammals, any effect on cardiomyocyte proliferation in the perinatal period directly impacts the number of cardiomyocytes for the remainder of the life-time (the cardiomyocyte endowment). Having a reduced cardiomyocyte endowment may lower functional reserve, thereby increasing vulnerability to cardiac stress or injury, as suggested in adults born preterm [10]. There are reports of early life glucocorticoid treatment promoting perinatal cardiomyocyte proliferation [34], reducing cardiomyocyte proliferation [24, 44] or having no effect. Even in zebrafish (circumventing any maternal effects), findings are confusing: one study reported that dexamethasone treatment during the embryonic period increases proliferation of cardiomyocytes and accelerates maturation of the heart, with these features carried forward into adulthood as larger hearts [95]. On the other hand, another found stress or dexamethasone treatment interferes with the normal trajectory of heart development, reducing cardiomyocyte numbers [6]. The differences may reflect the choice of steroid (synthetic MR sparing vs cortisol/corticosterone) and dose, the timing of administration or indeed, whether GR itself is autoregulated [41]. In the sheep fetal heart, a low dose of cortisol stimulates cardiomyocyte proliferation, but this is via MR activation rather than GR [29]. Whether antenatal administration of MR-sparing GR agonists affects cardiomyocyte proliferation is currently unclear (though this could reduce corticosterone activation of MR, via HPA axis suppression).

However, in neonatal rats, dexamethasone treatment reduces cardiomyocyte proliferation and subsequent cardiomyocyte endowment in adults [32]. Conversely, neonatal SMGRKO mice, with cardiomyocyte and vascular smooth muscle deletion of GR, show increased Ki67 staining of cardiomyocytes [77], suggesting neonatal cardiomyocyte proliferation is increased in the absence of GR signalling. Dexamethasone causes oxidative stress in neonatal rat hearts [2]; given the role of mitochondria and oxidative stress (and the DNA damage response) in the cessation of cardiomyocyte proliferation [73], this suggests a plausible mechanism worth exploring further. The neonatal increase in mitochondrial fatty acid oxidation in cardiomyocytes elevates production of reactive oxygen species, potentially contributing to cardiomyocyte cell cycle arrest mediated through the DNA damage response [14]. It is tempting to speculate that the increase in cardiac capacity for mitochondrial fatty acid oxidation promoted by glucocorticoid treatment in neonatal mice [41] leads to an increase in production of reactive oxygen species and thus DNA damage that may hasten the cessation of proliferation, thus reducing cardiomyocyte endowment. This is predicted to increase vulnerability to future cardiac insult, potentially contributing to the well-known increase in cardiovascular disease risk.

6.8 Concluding Remarks and Future Perspectives

Although much of the attention has been focussed on how MR activation contributes to heart disease, increasingly it is becoming apparent that GR plays an important role in cardiac health. The extent to which GR and MR oppose the actions of each other in the heart - and how - is still extremely unclear and is a key question to address. Whilst activation of GR is cardioprotective, improving survival of adult cardiomyocytes after cytotoxic or other stressful insult, how this is mediated will be important to establish. To some extent, GR activation may be cardioprotective by opposing MR actions (as in apoptosis of cardiomyocytes, for example) and to some extent cardiac protection may be independent of MR and mediated via other mechanisms (as might happen in cardiomyocyte hypertrophy). Other critical questions concern ion channel remodelling induced by corticosteroids - the different outcomes resulting from MR and GR activation raise questions about the extent to which this is influenced by the balance between the receptors *versus* the intracellular number of each receptor. There are similar questions about the extent to which MR and GR induce metabolic remodelling – and how. Unpicking what is due to GR, MR or the balance between the receptors is likely to be vital to tailor better therapies for the future prevention and treatment of heart failure. Just as importantly, given the widespread use of antenatal corticosteroids in women at risk of preterm delivery, is establishing the extent to which exogenous glucocorticoids interfere with the normal trajectory of heart development and harnessing beneficial maturation effects of corticosteroids on the fetal heart, to optimise benefits and minimise harms of this important life-saving treatment.

Acknowledgements Research related to this topic in the Authors' labs has been funded by the Medical Research Council (MR/P002811/1), the British Heart Foundation (TG/18/1/33408) and Heart Research UK (RG2694).

References

1. Addison T (1855) On the constitutional and local effects of disease of the supra-renal capsules. Samuel Highley, London
2. Adler A, Camm EJ, Hansell JA, Richter HG, Giussani DA (2010) Investigation of the use of antioxidants to diminish the adverse effects of postnatal glucocorticoid treatment on mortality and cardiac development. Neonatology 98:73–83
3. Agnew EJ (2017) The effect of antenatal glucocorticoid treatment on fetal heart maturation in mice. PhD thesis, the University of Edinburgh
4. Agnew EJ, Ivy JR, Stock SJ, Chapman KE (2018) Glucocorticoids, antenatal corticosteroid therapy and fetal heart maturation. J Mol Endocrinol 61:R61–R73
5. Akpek M, Ozdogru I, Sahin O, Inanc M, Dogan A, Yazici C, Berk V, Karaca H, Kalay N, Oguzhan A, Ergin A (2015) Protective effects of spironolactone against anthracycline-induced cardiomyopathy. Eur J Heart Fail 17:81–89

6. Apaydin DC, Jaramillo PAM, Corradi L, Cosco F, Rathjen FG, Kammertoens T, Filosa A, Sawamiphak S (2020) Early-life stress regulates cardiac development through an IL-4-glucocorticoid signaling balance. Cell Rep 33:108404
7. Avital A, Segal M, Richter-Levin G (2006) Contrasting roles of corticosteroid receptors in hippocampal plasticity. J Neurosci 26:9130–9134
8. Bauer TM, Murphy E (2020) Role of mitochondrial calcium and the permeability transition pore in regulating cell death. Circ Res 126:280–293
9. Bauersachs J, Jaisser F, Toto R (2015) Mineralocorticoid receptor activation and mineralocorticoid receptor antagonist treatment in cardiac and renal diseases. Hypertension 65:257–263
10. Bensley JG, Moore L, De Matteo R, Harding R, Black MJ (2018) Impact of preterm birth on the developing myocardium of the neonate. Pediatr Res 83:880–888
11. Bienvenu LA, Reichelt ME, Delbridge LM, Young MJ (2013) Mineralocorticoid receptors and the heart, multiple cell types and multiple mechanisms: a focus on the cardiomyocyte. Clin Sci (Lond) 125:409–421
12. Buerger A, Rozhitskaya O, Sherwood MC, Dorfman AL, Bisping E, Abel ED, Pu WT, Izumo S, Jay PY (2006) Dilated cardiomyopathy resulting from high-level myocardial expression of Cre-recombinase. J Card Fail 12:392–398
13. Buonafine M, Bonnard B, Jaisser F (2018) Mineralocorticoid receptor and cardiovascular disease. Am J Hypertens 31:1165–1174
14. Cardoso AC, Lam NT, Savla JJ, Nakada Y, Pereira AHM, Elnwasany A, Menendez-Montes I, Ensley EL, Bezan Petric U, Sharma G, Sherry AD, Malloy CR, Khemtong C, Kinter MT, Tan WLW, Anene-Nzelu CG, Foo RS-Y, Nguyen NUN, Li S, Ahmed MS, Elhelaly WM, Abdisalaam S, Asaithamby A, Xing C, Kanchwala M, Vale G, Eckert KM, Mitsche MA, McDonald JG, Hill JA, Huang L, Shaul PW, Szweda LI, Sadek HA (2020) Mitochondrial substrate utilization regulates cardiomyocyte cell-cycle progression. Nat Metab 2:167–178
15. Chapman KE, Holmes MC, Seckl JR (2013) 11β-hydroxysteroid dehydrogenases: intracellular gate-keepers of tissue glucocorticoid action. Physiol Rev 93:1139–1206
16. Chattergoon NN, Giraud GD, Louey S, Stork P, Fowden AL, Thornburg KL (2012) Thyroid hormone drives fetal cardiomyocyte maturation. FASEB J 26:397–408
17. Coutinho AE, Chapman KE (2011) The anti-inflammatory and immunosuppressive effects of glucocorticoids, recent developments and mechanistic insights. Mol Cell Endocrinol 335:2–13
18. Cruz-Topete D, He B, Xu X, Cidlowski JA (2016a) Kruppel-like factor 13 is a major mediator of glucocorticoid receptor-signaling in cardiomyocytes and protects these cells from DNA damage and death. J Biol Chem 291:19374–19386
19. Cruz-Topete D, Myers PH, Foley JF, Willis MS, Cidlowski JA (2016b) Corticosteroids are essential for maintaining cardiovascular function in male mice. Endocrinology 157:2759–2771
20. Cruz-Topete D, Oakley RH, Carroll NG, He B, Myers PH, Xu X et al (2019) Deletion of the cardiomyocyte glucocorticoid receptor leads to sexually dimorphic changes in cardiac gene expression and progression to heart failure. J Am Heart Assoc 8:e011012. https://doi.org/10.1161/JAHA.118.011012
21. da Rocha AL, Teixeira GR, Pinto AP, de Morais GP, Oliveira LDC, de Vicente LG, da Silva L, Pauli JR, Cintra DE, Ropelle ER, de Moura LP, Mekary RA, de Freitas EC, da Silva ASR (2018) Excessive training induces molecular signs of pathologic cardiac hypertrophy. J Cell Physiol 233:8850–8861
22. Datson NA, van der Perk J, de Kloet ER, Vreugdenhil E (2001) Identification of corticosteroid-responsive genes in rat hippocampus using serial analysis of gene expression. Eur J Neurosci 14:675–689
23. de Kloet ER (2014) From receptor balance to rational glucocorticoid therapy. Endocrinology 155:en20141048
24. de Vries WB, van der Leij FR, Bakker JM, Kamphuis PJ, van Oosterhout MF, Schipper ME, Smid GB, Bartelds B, van Bel F (2002) Alterations in adult rat heart after neonatal dexamethasone therapy. Pediatr Res 52:900–906
25. Eisner DA, Caldwell JL, Kistamas K, Trafford AW (2017) Calcium and excitation-contraction coupling in the heart. Circ Res 121:181–195
26. Eisner DA, Caldwell JL, Trafford AW, Hutchings DC (2020) The control of diastolic calcium in the heart. Circ Res 126:395–412
27. Elkholey K, Papadimitriou L, Butler J, Thadani U, Stavrakis S (2021) Effect of obesity on response to spironolactone in patients with heart failure with preserved ejection fraction. Am J Cardiol 146:36–47
28. Escoubet B, Couffignal C, Laisy JP, Mangin L, Chillon S, Laouenan C, Serfaty JM, Jeunemaitre X, Mentre F, Zennaro MC (2013) Cardiovascular effects of aldosterone: insight from adult carriers of mineralocorticoid receptor mutations. Circ Cardiovasc Genet 6:381–390
29. Feng X, Reini S, Richards E, Wood CE, Keller-Wood M (2013) Cortisol stimulates proliferation and apoptosis in the late gestation fetal heart: differential effects of mineralocorticoid and glucocorticoid receptors. Am J Physiol Regul Integr Comp Physiol 305:343–350
30. Fowden AL, Valenzuela OA, Vaughan OR, Jellyman JK, Forhead AJ (2016) Glucocorticoid programming of intrauterine development. Domest Anim Endocrinol 56(Suppl):S121–S132
31. Fraccarollo D, Berger S, Galuppo P, Kneitz S, Hein L, Schutz G, Frantz S, Ertl G, Bauersachs J (2011) Deletion of cardiomyocyte mineralocorticoid receptor ameliorates adverse remodeling after myocardial infarction. Circulation 123:400–408

32. Gay MS, Li Y, Xiong F, Lin T, Zhang L (2015) Dexamethasone treatment of newborn rats decreases cardiomyocyte endowment in the developing heart through epigenetic modifications. PLoS One 10:e0125033
33. Gillet L, Guichard S, Essers MC, Rougier JS, Abriel H (2019) Dystrophin and calcium current are decreased in cardiomyocytes expressing Cre enzyme driven by alphaMHC but not TNT promoter. Sci Rep 9:19422
34. Giraud GD, Louey S, Jonker S, Schultz J, Thornburg KL (2006) Cortisol stimulates cell cycle activity in the cardiomyocyte of the sheep fetus. Endocrinology 147:3643–3649
35. Gunthel M, Barnett P, Christoffels VM (2018) Development, proliferation, and growth of the mammalian heart. Mol Ther 26:1599–1609
36. Hafezi-Moghadam A, Simoncini T, Yang Z, Limbourg FP, Plumier JC, Rebsamen MC, Hsieh CM, Chui DS, Thomas KL, Prorock AJ, Laubach VE, Moskowitz MA, French BA, Ley K, Liao JK (2002) Acute cardiovascular protective effects of corticosteroids are mediated by non-transcriptional activation of endothelial nitric oxide synthase. Nat Med 8:473–479
37. Henning RJ, Bourgeois M, Harbison RD (2018) Poly(ADP-ribose) polymerase (PARP) and PARP inhibitors: mechanisms of action and role in cardiovascular disorders. Cardiovasc Toxicol 18:493–506
38. Ingwall JS (2009) Energy metabolism in heart failure and remodelling. Cardiovasc Res 81:412–419
39. Ingwall JS, Weiss RG (2004) Is the failing heart energy starved? On using chemical energy to support cardiac function. Circ Res 95:135–145
40. Iqbal J, Andrew R, Cruden NL, Kenyon CJ, Hughes KA, Newby DE, Hadoke PW, Walker BR (2014) Displacement of cortisol from human heart by acute administration of a mineralocorticoid receptor antagonist. J Clin Endocrinol Metab 99:915–922
41. Ivy JR, Carter RN, Zhao JF, Buckley C, Urquijo H, Rog-Zielinska EA, Panting E, Hrabalkova L, Nicholson C, Agnew EJ, Kemp MW, Morton NN, Stock SJ, Wyrwoll C, Ganley IG, Chapman KE (2021) Glucocorticoids regulate mitochondrial fatty acid oxidation in fetal cardiomyocytes. J Physiol. https://doi.org/10.1113/JP281860. Online ahead of print
42. Jia G, Habibi J, DeMarco VG, Martinez-Lemus LA, Ma L, Whaley-Connell AT, Aroor AR, Domeier TL, Zhu Y, Meininger GA, Mueller KB, Jaffe IZ, Sowers JR (2015) Endothelial mineralocorticoid receptor deletion prevents diet-induced cardiac diastolic dysfunction in females. Hypertension 66:1159–1167
43. Jia G, Jia Y, Sowers JR (2017) Role of mineralocorticoid receptor activation in cardiac diastolic dysfunction. Biochim Biophys Acta Mol basis Dis 1863:2012–2018
44. Kim MY, Eiby YA, Lumbers ER, Wright LL, Gibson KJ, Barnett AC, Lingwood BE (2014) Effects of glucocorticoid exposure on growth and structural maturation of the heart of the preterm piglet. PLoS One 9:e93407
45. Kolkhof P, Jaisser F, Kim SY, Filippatos G, Nowack C, Pitt B (2017) Steroidal and novel non-steroidal mineralocorticoid receptor antagonists in heart failure and Cardiorenal diseases: comparison at bench and bedside. Handb Exp Pharmacol 243:271–305
46. Kuster D, Merkus D, Kremer A, van Ijcken W, de Beer V, Verhoeven A, Duncker D (2011) Left ventricular remodeling in swine after myocardial infarction: a transcriptional genomics approach. Basic Res Cardiol 106:1269–1281
47. Kuster DW, Merkus D, Blonden LA, Kremer A, van IWF, Verhoeven AJ, Duncker DJ (2014) Gene reprogramming in exercise-induced cardiac hypertrophy in swine: a transcriptional genomics approach. J Mol Cell Cardiol 77:168–174
48. Kwong JQ, Lu X, Correll RN, Schwanekamp JA, Vagnozzi RJ, Sargent MA, York AJ, Zhang J, Bers DM, Molkentin JD (2015) The mitochondrial calcium uniporter selectively matches metabolic output to acute contractile stress in the heart. Cell Rep 12:15–22
49. Leeksma CHW, De Graeff J, De Cock J (1957) Hypercalcaemia in adrenal insufficiency. Acta Med Scand 156:455–458
50. Leinwand LA (2003) Sex is a potent modifier of the cardiovascular system. J Clin Invest 112:302–307
51. Liu B, Zhang TN, Knight JK, Goodwin JE (2019) The glucocorticoid receptor in cardiovascular health and disease. Cell 8:1227
52. Lother A, Berger S, Gilsbach R, Rosner S, Ecke A, Barreto F, Bauersachs J, Schutz G, Hein L (2011) Ablation of mineralocorticoid receptors in myocytes but not in fibroblasts preserves cardiac function. Hypertension 57:746–754
53. Lother A, Furst D, Bergemann S, Gilsbach R, Grahammer F, Huber TB, Hilgendorf I, Bode C, Moser M, Hein L (2016) Deoxycorticosterone acetate/salt-induced cardiac but not renal injury is mediated by endothelial mineralocorticoid receptors independently from blood pressure. Hypertension 67:130–138
54. Markham LW, Kinnett K, Wong BL, Woodrow Benson D, Cripe LH (2008) Corticosteroid treatment retards development of ventricular dysfunction in Duchenne muscular dystrophy. Neuromuscul Disord 18:365–370
55. Matsuhashi T, Endo J, Katsumata Y, Yamamoto T, Shimizu N, Yoshikawa N et al (2019) Pressure overload inhibits glucocorticoid receptor transcriptional activity in cardiomyocytes and promotes pathological cardiac hypertrophy. J Mol Cell Cardiol 130:122–130. https://doi.org/10.1016/j.yjmcc.2019.03.019
56. Matthews LC, Berry AA, Morgan DJ, Poolman TM, Bauer K, Kramer F, Spiller DG, Richardson RV, Chapman KE, Farrow SN, Norman MR, Williamson AJ, Whetton AD, Taylor SS, Tuckermann JP, White MR, Ray DW (2015) Glucocorticoid receptor regulates accurate chromosome segregation and is asso-

ciated with malignancy. Proc Natl Acad Sci U S A 112:5479–5484
57. McEwen BS (1998) Stress, adaptation, and disease. Allostasis and allostatic load. Ann N Y Acad Sci 840:33–44
58. Mifsud KR, Kennedy CLM, Salatino S, Sharma E, Price EM, Haque SN, Gialeli A, Goss HM, Panchenko PE, Broxholme J, Engledow S, Lockstone H, Cordero Llana O, Reul J (2021) Distinct regulation of hippocampal neuroplasticity and ciliary genes by corticosteroid receptors. Nat Commun 12:4737
59. Morrison-Nozik A, Anand P, Zhu H, Duan Q, Sabeh M, Prosdocimo DA, Lemieux ME, Nordsborg N, Russell AP, MacRae CA, Gerber AN, Jain MK, Haldar SM (2015) Glucocorticoids enhance muscle endurance and ameliorate Duchenne muscular dystrophy through a defined metabolic program. Proc Natl Acad Sci U S A 112:E6780–E6789
60. Nakada Y, Nhi Nguyen NU, Xiao F, Savla JJ, Lam NT, Abdisalaam S, Bhattacharya S, Mukherjee S, Asaithamby A, Gillette TG, Hill JA, Sadek HA (2019) DNA damage response mediates pressure overload-induced cardiomyocyte hypertrophy. Circulation 139:1237–1239
61. Narayanan N, Yang C, Xu A (2004) Dexamethasone treatment improves sarcoplasmic reticulum function and contractile performance in aged myocardium. Mol Cell Biochem 266:31–36
62. Oakley RH, Cidlowski JA (2015) Glucocorticoid signaling in the heart: a cardiomyocyte perspective. J Steroid Biochem Mol Biol 153:27–34
63. Oakley RH, Ren R, Cruz-Topete D, Bird GS, Myers PH, Boyle MC et al (2013) Essential role of stress hormone signaling in cardiomyocytes for the prevention of heart disease. Proc Natl Acad Sci U S A 110:17035–17040. https://doi.org/10.1073/pnas.1302546110
64. Oakley RH, Cruz-Topete D, He B, Foley JF, Myers PH, Xu X et al (2019) Cardiomyocyte glucocorticoid and mineralocorticoid receptors directly and antagonistically regulate heart disease in mice. Sci Signal 12:eaau9685. https://doi.org/10.1126/scisignal.aau9685
65. Ohtani T, Ohta M, Yamamoto K, Mano T, Sakata Y, Nishio M, Takeda Y, Yoshida J, Miwa T, Okamoto M, Masuyama T, Nonaka Y, Hori M (2007) Elevated cardiac tissue level of aldosterone and mineralocorticoid receptor in diastolic heart failure: beneficial effects of mineralocorticoid receptor blocker. Am J Physiol Regul Integr Comp Physiol 292:R946–R954
66. Otte C, Wust S, Zhao S, Pawlikowska L, Kwok P-Y, Whooley MA (2010) Glucocorticoid receptor gene, low-grade inflammation, and heart failure: the heart and soul study. J Clin Endocrinol Metab 95:2885–2891
67. Ouvrard-Pascaud A, Sainte-Marie Y, Benitah JP, Perrier R, Soukaseum C, Nguyen Dinh Cat A, Royer A, Le Quang K, Charpentier F, Demolombe S, Mechta-Grigoriou F, Beggah AT, Maison-Blanche P, Oblin ME, Delcayre C, Fishman GI, Farman N, Escoubet B, Jaisser F (2005) Conditional mineralocorticoid receptor expression in the heart leads to life-threatening arrhythmias. Circulation 111:3025–3033
68. Parikh SS, Blackwell DJ, Gomez-Hurtado N, Frisk M, Wang L, Kim K, Dahl CP, Fiane AE, Tonnessen T, Kryshtal DO, Louch WE, Knollmann BC (2017) Thyroid and glucocorticoid hormones promote functional T-tubule development in human-induced pluripotent stem cell derived cardiomyocytes. Circ Res 121:1323–1330
69. Pimenta E, Wolley M, Stowasser M (2012) Adverse cardiovascular outcomes of corticosteroid excess. Endocrinology 153:5137–5142
70. Pitt B, Zannad F, Remme WJ, Cody R, Castaigne A, Perez A, Palensky J, Wittes J (1999) The effect of spironolactone on morbidity and mortality in patients with severe heart failure. Randomized Aldactone evaluation study investigators. N Engl J Med 341:709–717
71. Pitt B, Remme W, Zannad F, Neaton J, Martinez F, Roniker B, Bittman R, Hurley S, Kleiman J, Gatlin M (2003) Eplerenone, a selective aldosterone blocker, in patients with left ventricular dysfunction after myocardial infarction. N Engl J Med 348:1309–1321
72. Pitt B, Pedro Ferreira J, Zannad F (2017) Mineralocorticoid receptor antagonists in patients with heart failure: current experience and future perspectives. Eur Heart J Cardiovasc Pharmacother 3:48–57
73. Puente BN, Kimura W, Muralidhar SA, Moon J, Amatruda JF, Phelps KL, Grinsfelder D, Rothermel BA, Chen R, Garcia JA, Santos CX, Thet S, Mori E, Kinter MT, Rindler PM, Zacchigna S, Mukherjee S, Chen DJ, Mahmoud AI, Giacca M, Rabinovitch PS, Aroumougame A, Shah AM, Szweda LI, Sadek HA (2014) The oxygen-rich postnatal environment induces cardiomyocyte cell-cycle arrest through DNA damage response. Cell 157:565–579
74. Ren R, Oakley RH, Cruz-Topete D, Cidlowski JA (2012) Dual role for glucocorticoids in cardiomyocyte hypertrophy and apoptosis. Endocrinology 153:5346–5360
75. Richardson RV (2014) The role of the glucocorticoid receptor in cardiac growth and remodelling. PhD thesis, University of Edinburgh
76. Richardson RV, Batchen EJ, Denvir MA, Gray GA, Chapman KE (2016) Cardiac GR and MR: from development to pathology. Trends Endocrinol Metab 27:35–43
77. Richardson R, Batchen E, Thomson A, Darroch R, Pan X, Rog-Zielinska E et al (2017) Glucocorticoid receptor alters isovolumetric contraction and restrains cardiac fibrosis. J Endocrinol 232:437–450. https://doi.org/10.1530/JOE-16-0458
78. Rickard AJ, Young MJ (2009) Corticosteroid receptors, macrophages and cardiovascular disease. J Mol Endocrinol 42:449–459

79. Rickard AJ, Morgan J, Tesch G, Funder JW, Fuller PJ, Young MJ (2009) Deletion of mineralocorticoid receptors from macrophages protects against deoxycorticosterone/salt-induced cardiac fibrosis and increased blood pressure. Hypertension 54:537–543
80. Rickard AJ, Morgan J, Bienvenu LA, Fletcher EK, Cranston GA, Shen JZ, Reichelt ME, Delbridge LM, Young MJ (2012) Cardiomyocyte mineralocorticoid receptors are essential for deoxycorticosterone/salt-mediated inflammation and cardiac fibrosis. Hypertension 60:1443–1450
81. Rickard AJ, Morgan J, Chrissobolis S, Miller AA, Sobey CG, Young MJ (2014) Endothelial cell mineralocorticoid receptors regulate deoxycorticosterone/salt-mediated cardiac remodeling and vascular reactivity but not blood pressure. Hypertension 63:1033–1040
82. Rog-Zielinska EA, Thomson A, Kenyon CJ, Brownstein DG, Moran CM, Szumska D, Michailidou Z, Richardson J, Owen E, Watt A, Morrison H, Forrester LM, Bhattacharya S, Holmes MC, Chapman KE (2013) Glucocorticoid receptor is required for fetal heart maturation. Hum Mol Genet 22:3269–3282
83. Rog-Zielinska EA, Richardson RV, Denvir MA, Chapman KE (2014) Glucocorticoids and foetal heart maturation; implications for prematurity and foetal programming. J Mol Endocrinol 52:R125–R135
84. Rog-Zielinska EA, Craig MA, Manning JR, Richardson RV, Gowans GJ, Dunbar DR, Gharbi K, Kenyon CJ, Holmes MC, Hardie DG, Smith GL, Chapman KE (2015) Glucocorticoids promote structural and functional maturation of foetal cardiomyocytes: a role for PGC-1alpha. Cell Death Differ 22:1106–1116
85. Sainte-Marie Y, Nguyen Dinh Cat A, Perrier R, Mangin L, Soukaseum C, Peuchmaur M, Tronche F, Farman N, Escoubet B, Benitah JP, Jaisser F (2007) Conditional glucocorticoid receptor expression in the heart induces atrio-ventricular block. FASEB J 21:3133–3141
86. Seidel T, Fiegle DJ, Baur TJ, Ritzer A, Nay S, Heim C, Weyand M, Milting H, Oakley RH, Cidlowski JA, Volk T (2019) Glucocorticoids preserve the t-tubular system in ventricular cardiomyocytes by upregulation of autophagic flux. Basic Res Cardiol 114:47
87. Serenelli M, Jackson A, Dewan P, Jhund PS, Petrie MC, Rossignol P, Campo G, Pitt B, Zannad F, Ferreira JP, McMurray JJV (2020) Mineralocorticoid receptor antagonists, blood pressure, and outcomes in heart failure with reduced ejection fraction. JACC Heart Failure 8:188–198
88. Spath JA, Lefer AM (1975) Effects of dexamethasone on myocardial cells in the early phase of acute myocardial infarction. Am Heart J 90:50–55
89. Szántó M, Gupte R, Kraus WL, Pacher P, Bai P (2021) PARPs in lipid metabolism and related diseases. Prog Lipid Res 84:101117
90. Tokudome S, Sano M, Shinmura K, Matsuhashi T, Morizane S, Moriyama H, Tamaki K, Hayashida K, Nakanishi H, Yoshikawa N, Shimizu N, Endo J, Katayama T, Murata M, Yuasa S, Kaneda R, Tomita K, Eguchi N, Urade Y, Asano K, Utsunomiya Y, Suzuki T, Taguchi R, Tanaka H, Fukuda K (2009) Glucocorticoid protects rodent hearts from ischemia/reperfusion injury by activating lipocalin-type prostaglandin D synthase-derived PGD2 biosynthesis. J Clin Invest 119:1477–1488
91. Tsai CT, Chiang FT, Tseng CD, Hwang JJ, Kuo KT, Wu CK, Yu CC, Wang YC, Lai LP, Lin JL (2010) Increased expression of mineralocorticoid receptor in human atrial fibrillation and a cellular model of atrial fibrillation. J Am Coll Cardiol 55:758–770
92. Usher MG, Duan SZ, Ivaschenko CY, Frieler RA, Berger S, Schutz G, Lumeng CN, Mortensen RM (2010) Myeloid mineralocorticoid receptor controls macrophage polarization and cardiovascular hypertrophy and remodeling in mice. J Clin Invest 120:3350–3364
93. Wang X, Lauth A, Wan TC, Lough JW, Auchampach JA (2020) Myh6-driven Cre recombinase activates the DNA damage response and the cell cycle in the myocardium in the absence of loxP sites. Dis Model Mech 13:dmm046375
94. Wei L, MacDonald TM, Walker BR (2004) Taking glucocorticoids by prescription is associated with subsequent cardiovascular disease. Ann Intern Med 141:764–770
95. Wilson KS, Baily J, Tucker CS, Matrone G, Vass S, Moran C, Chapman KE, Mullins JJ, Kenyon C, Hadoke PW, Denvir MA (2015) Early-life perturbations in glucocorticoid activity impacts on the structure, function and molecular composition of the adult zebrafish (Danio rerio) heart. Mol Cell Endocrinol 414:120–131
96. Xu B, Strom J, Chen QM (2011) Dexamethasone induces transcriptional activation of Bcl-xL gene and inhibits cardiac injury by myocardial ischemia. Eur J Pharmacol 668:194–200
97. Yavas G, Celik E, Yavas C, Elsurer C, Afsar RE (2017) Spironolactone ameliorates the cardiovascular toxicity induced by concomitant trastuzumab and thoracic radiotherapy. Rep Pract Oncol Radiother 22:295–302
98. Young MJ, Rickard AJ (2015) Mineralocorticoid receptors in the heart: lessons from cell-selective transgenic animals. J Endocrinol 224:R1–R13
99. Zannad F, McMurray JJV, Krum H, van Veldhuisen DJ, Swedberg K, Shi H, Vincent J, Pocock SJ, Pitt B (2011) Eplerenone in patients with systolic heart failure and mild symptoms. N Engl J Med 364:11–21
100. Zhou B, Tian R (2018) Mitochondrial dysfunction in pathophysiology of heart failure. J Clin Invest 128:3716–3726

7 Physiological Convergence and Antagonism Between GR and PPARγ in Inflammation and Metabolism

Marija Dacic, Gayathri Shibu, and Inez Rogatsky

Abstract

Nuclear receptors (NRs) are transcription factors that modulate gene expression in a ligand-dependent manner. The ubiquitously expressed glucocorticoid receptor (GR) and peroxisome proliferator-activated receptor gamma (PPARγ) represent steroid (type I) and non-steroid (type II) classes of NRs, respectively. The diverse transcriptional and physiological outcomes of their activation are highly tissue-specific. For example, in subsets of immune cells, such as macrophages, the signaling of GR and PPARγ converges to elicit an anti-inflammatory phenotype; in contrast, in the adipose tissue, their signaling can lead to reciprocal metabolic outcomes. This review explores the cooperative and divergent outcomes of GR and PPARγ functions in different cell types and tissues, including immune cells, adipose tissue and the liver. Understanding the coordinated control of these NR pathways should advance studies in the field and potentially pave the way for developing new therapeutic approaches to exploit the GR:PPARγ crosstalk.

Keywords

Glucocorticoid receptor (GR) · Peroxisome proliferator-activated receptor gamma (PPARγ) · Inflammation · Transcription · Immune cells · Metabolic tissues

7.1 Introduction

Nuclear receptors (NRs), such as the glucocorticoid receptor (GR) and peroxisome proliferator activated receptor-γ (PPARγ) are a versatile superfamily of structurally conserved transcription factors (TFs) that regulate numerous homeostatic physiological processes, largely in a ligand-modulated manner, thereby adapting gene expression programs to environmental changes.

GR, or NR3C1, named for its role in regulating glucose metabolism, is an archetypal steroid hormone receptor (type I) involved in numerous signaling circuits that maintain metabolic homeostasis. GR is activated by its endogenous gluco-

M. Dacic
Hospital for Special Surgery Research Institute, The David Rosenzweig Genomics Center, New York, NY, USA

Graduate Program in Physiology, Biophysics and Systems Biology, Weill Cornell Graduate School of Medical Sciences, New York, NY, USA

G. Shibu · I. Rogatsky (✉)
Hospital for Special Surgery Research Institute, The David Rosenzweig Genomics Center, New York, NY, USA

Graduate Program in Immunology and Microbial Pathogenesis, Weill Cornell Graduate School of Medical Sciences, New York, NY, USA
e-mail: rogatskyi@hss.edu

M. J. Campbell, C. L. Bevan (eds.), *Nuclear Receptors in Human Health and Disease*, Advances in Experimental Medicine and Biology 1390, https://doi.org/10.1007/978-3-031-11836-4_7

corticoid (GC) ligands, whose levels are controlled by the hypothalamic-pituitary-adrenal (HPA) axis. Upon ligand binding, the cytoplasmic GR multiprotein complex, also containing immunophilins and chaperones, undergoes conformational changes and translocates into the nucleus (Reviewed in [1]), where GR binds to specific palindromic DNA sequences called GC-response elements (GRE) or tethers to other DNA-bound TFs, recruits cofactors (coactivators and corepressors) and regulates transcription of associated genes [2, 3].

Peroxisome proliferator activated receptor-γ (PPARγ, also known as NR1C3) is a non-steroid (type II) NR that senses oxidized fatty acids (FA). It is mainly implicated in homeostatic maintenance of lipid metabolism and insulin sensitivity [4, 5]. Similar to the related PPARs and other type II NRs, PPARγ exerts its biological functions by forming heterodimeric complexes with another member of the NR family, retinoic acid receptor α (RXRα). In the absence of a ligand, the PPARγ/RXRα complex binds to specific DNA sequences known as PPAR response elements (PPRE) or direct repeat (DR)1 sequences together with a corepressor complex (reviewed in [6]). Upon ligand binding, the corepressor complex is released, and a coactivator complex is recruited [7].

NRs have been linked to the regulation and maintenance of metabolic homeostasis for decades. Both GR and PPARγ were initially described as regulators of metabolic functions in the liver and adipose tissue, respectively. Since then, a myriad of non-metabolic roles have been described for each receptor, with one of the most renowned functions being the regulation of immune responses and inflammation. Interestingly, despite representing two different families of NRs, GR and PPARγ exhibit a striking functional overlap in the immune system while having disparate roles in healthy liver and divergent ones in lipid metabolism. These overlapping yet distinct outcomes of GR and PPARγ activation stem from differences at multiple levels of regulation, ranging from the ligand-binding events to the engagement of other TFs, co-regulators and components of basal transcriptional machinery and chromatin. In this Chapter, we will discuss the tissue-specific convergence of GR and PPARγ signaling in the immune system and briefly contrast it with some of their antagonistic roles in metabolic tissues. It should be noted that many of these functions have been deduced using NR knock-out (KO) mouse strains and *in vitro* studies with endogenous or synthetic ligands, often at super-physiological concentrations, which remains a limitation to our understanding of NR biology.

7.2 GR and PPARγ in Monocytes and Macrophages

During inflammation, both GR and PPARγ play crucial roles in regulating macrophage responses. Indeed, GCs have long been known to exert potent immunosuppressive effects on monocytes and macrophages. Mice lacking GR in macrophages produce more inflammatory cytokines, including IL-1β, IL-6, TNF, and IL-12, and display higher mortality rates during bacterial lipopolysaccharide (LPS)-induced sepsis relative to their wild-type (WT) counterparts [8–11]. Although the role of PPARγ in this context is less understood, it negatively regulates macrophage activation by down-regulating synthesis of TNF, IL-6 and other pro-inflammatory cytokines [12] and decreasing macrophage migration *in vitro* [13]. Myeloid-specific deletion of PPARγ exacerbates inflammation in mouse models of inflammatory bowel disease (IBD) [14]. Consistently, treating mice with pioglitazone, a synthetic PPARγ agonist, reduced systemic inflammatory response during cecal ligation and puncture-induced sepsis [15]. Thus, both receptors down-regulate pro-inflammatory mediators at the nexus of pro-inflammatory responses and effectively curb inflammation *in vivo*.

GR acts on macrophages to dampen inflammation in a variety of ways. One broadly established mechanism of action is direct tethering of liganded GR to effector TFs downstream of Toll-like receptor (TLR) signaling, including NF-κB, AP-1 and interferon regulatory factor 3 (IRF3), and repression of their activity (Fig. 7.1a;

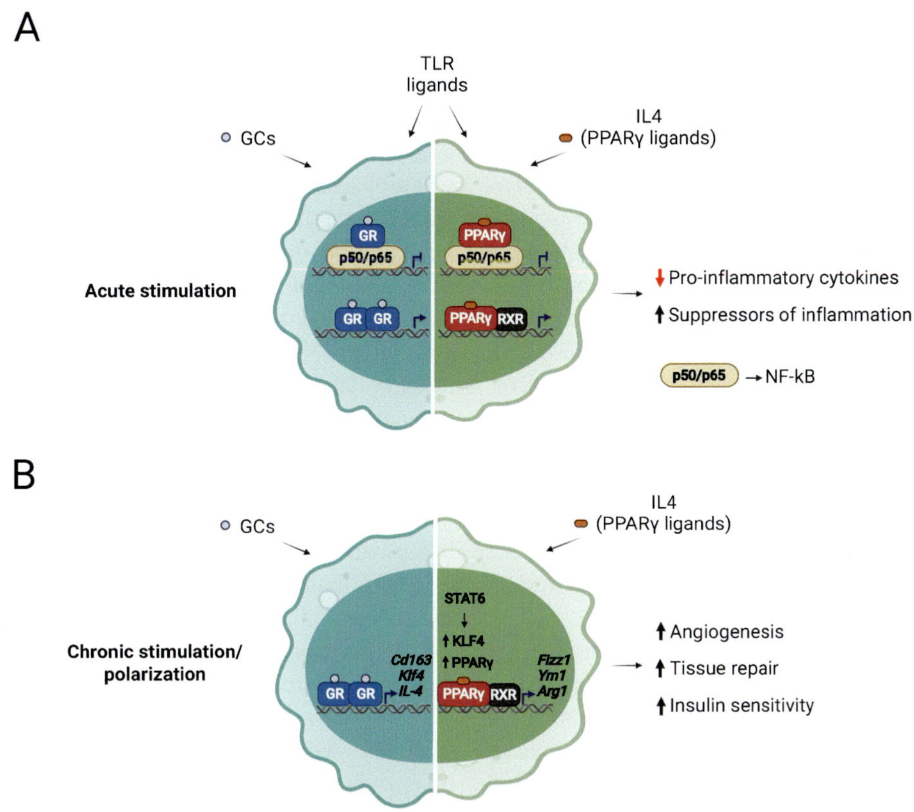

Fig. 7.1 GR and PPARγ mediate both short-term and long-term anti-inflammatory responses in macrophages. (**a**) Upon short-term treatment with GCs or PPARγ ligands, and in the presence of inflammatory toll-like receptor (TLR) ligands, GR and PPARγ are recruited to their genomic binding sites and inhibit pro-inflammatory gene transcription (often by binding to the p50/p65 NF-kB heterodimers) and up-regulate suppressors of inflammation. (**b**) Chronic stimulation with GCs or PPARγ ligands up-regulates GR and STAT6 signaling, respectively, and STAT6 in turn increases KLF4 and PPARγ expression. GR and PPARγ promote expression of M2 genes and help establish a stable macrophage sub-type that promotes angiogenesis, tissue repair and increases sensitivity to insulin

reviewed in [16]). Conversely, many genes encoding inhibitors of TLR signaling are activated by GR, such as IL-1 receptor-associated kinase 3 (IRAK3), which negatively regulate mitogen-activated protein kinase 1 (MAPK1) and IL-1 receptor signaling [17]. GILZ is another well-known GR-inducible target that can bind c-Jun and c-Fos components of the AP-1 complex [18] as well as NF-κB [19] and antagonize their actions. GR-activated anti-inflammatory genes also encode proteins that can function at steps further removed from transcriptional modulation. For instance, GR-upregulated ZFP36 facilitates mRNA degradation of several pro-inflammatory genes, most notably *TNF* [20]. Suppressor of cytokine signaling 1 (SOCS1) – encoded by another GC-inducible gene – is an inhibitor of Janus kinase (JAK)–STAT cascade downstream of cytokines binding to their cell surface receptors [21]. GR can also act to suppress inflammation by altering the epigenetic state of chromatin at target promoters through mitogen- and stress-activated protein (MSK1) kinase and GR-interacting protein (GRIP)1 (nuclear receptor coactivator 2, Ncoa2) recruitment, which affects components of basal transcriptional machinery and the rate-limiting steps in RNA polymerase II transcription cycle such as promoter-proximal pausing [22–24]. The opposite arm of regulation includes chromatin modulators such as BRD9, which attenuates GR-mediated repression of inflammatory genes [25].

Similar to GR, PPARγ represses transcription of pro-inflammatory genes by directly binding NF-kB and AP-1 and interfering with their activities (Fig. 7.1a; [26]). PPARγ directly binds the p65 subunit of NF-kB under basal conditions in human colonic HT29 cells and mouse embryonic fibroblasts (MEFs), and the binding in MEFs increases after stimulation with LPS and TNF [27]. Additionally, PPARγ-deficient macrophages that are unstimulated *in vitro* [28] or sorted from tissues during perinatal development [29] are pro-inflammatory. Contrary to these findings, however, mice lacking PPARγ in the myeloid lineage express less IL-1 than WT after NLRP3 activation *in vivo* and in primary macrophages [30].

In addition to acute actions of each receptor that lead to rapid and dramatic, yet reversible changes in the inflammatory transcriptome, a sustained exposure to pro- or anti-inflammatory signals, including NR ligands, results in a stable change of epigenomic landscape and associated macrophage phenotype, which alters responses to subsequent acute stimuli. Historically, macrophages were thought to have the capacity to be 'polarized' to two distinct phenotypic states. Bacterial products such as LPS and the T helper-1 (Th1) cytokine interferon-γ (IFNγ) bias macrophages toward the inflammatory state termed 'M1'. Conversely, a tissue repair/wound healing phenotype of an 'M2' macrophage was originally described as a polarization state conferred by the Th2 cytokine IL-4 [31]. These macrophages are implicated in the Th2-driven response to parasitic infection or allergies, as well as in homeostatic functions such as wound healing, angiogenesis and insulin-sensitizing metabolic functions (Fig. 7.1b). Signaling downstream of IL-4 involves activation of the TFs STAT6 and KLF4 that cooperatively facilitate the gradual acquisition of the M2 transcriptional state [32]. Depending on the stimuli used *in vitro*, the populations of M2-like macrophages were further classified as M2a (after exposure to IL-4 or IL-13), M2b (immune complexes in combination with IL-1β or LPS) and M2c (IL-10, TGFβ or, importantly, GCs) [33]. This binary M1/M2 view of polarization was later challenged by extensive expression profiling studies that arrived at a spectral model of macrophage activation states whereby every signal or a combination of signals yields a distinct transcriptional make-up [34]. Nonetheless, transcriptomes resulting from stimulation with LPS or IFNγ vs. those produced by IL-4, IL-10 or GCs did cluster at the opposite ends of the spectrum, supporting the idea that M1-like and M2-like phenotypes represent the two extremes of macrophage transcriptional states.

Thus, the anti-inflammatory effects of GC signaling in macrophages range from the acute upregulation of anti-inflammatory and repression of pro-inflammatory genes to more sustained phenotypic changes upon prolonged (beyond 24 h) GC exposure. The latter involves upregulated phagocytosis of apoptotic cells and debris while the production of inflammatory mediators subsides, which together drive the resolution phase of inflammation [35, 36]. GC-polarized macrophages are characterized by high expression of scavenger receptors such as CD163 and type 2 and anti-inflammatory cytokines IL-4 and IL-10 [37].

Similarly, PPARγ is reportedly essential for transitioning to an anti-inflammatory macrophage [38]. Indeed, pharmacological activation of PPARγ increases the expression of *Fizz1, Ym1* and *Arg1*, typical 'M2 genes' in macrophage-like RAW264.7 cells and human peripheral blood mononuclear cells [39]. Conversely, mice with PPARγ-deficient macrophages display impaired wound healing *in vivo* [40]. Thus, PPARγ and GR both drive the M2-like macrophage phenotype with resolving properties, even though the direct gene targets are not fully shared.

Genomic studies revealed that sustained IL-4 signaling leads to the binding of transcription factors: STAT6, and subsequently RXR and PU.1, and to the recruitment of cofactors P300 and RAD21 to a subset of new RXR sites; 60% of them need PPARγ binding to open, and the majority of new RXR sites are PPARγ-dependent irrespective of STAT6 binding (Fig. 7.1b) [41].

IL-4 itself induces the expression of the PPARγ-encoding gene *Pparg,* highlighting the importance of PPARγ for the M2-like phenotype [42]. Notably, these changes are driven by IL-4, not a specific PPARγ ligand, which contrasts with the strict dependence of GR on GCs to drive the M2-like phenotype.

Cofactors provide an additional level of convergence between NR-driven and IL4-induced macrophage polarization. GRIP1/NCoA2 is a member of the p160 family of NR coregulators shared by GR and PPARγ [43]. GRIP1 has further been shown to serve as a coactivator for KLF4, thereby directly contributing to the IL-4:STAT6:KLF4 pathway [44]. Indeed, macrophage-specific GRIP1 deletion in mice shifted their macrophage balance toward the more inflammatory M1-like phenotype *in vitro* and in an obesity-induced model of metabolic inflammation *in vivo* [44]. The role of GRIP1 in facilitating both GR-mediated activation and repression is well established [45, 46]. It is tempting to speculate that GRIP1 may serve as a platform for integrating pathways involved in M2-like macrophage polarization in response to distinct physiological stimuli.

In the context of the human *in vitro* model of atherosclerosis, PPARγ reduces inflammatory cytokine secretion in human umbilical vein endothelial cells exposed to oxidized-low-density lipoproteins (oxLDLs) [47]. In line with these findings, peritoneal macrophages from conditional PPARγ KO mice had more foam cell formation after treatment with oxLDLs *in vitro* [48] suggesting that PPARγ reduces inflammation and pathogenesis of atherosclerosis. The function of PPARγ in atherosclerosis is consistent with *in vitro* effect of PPARγ in macrophages, as well as with GR actions in macrophages *in vivo* in inflammatory settings.

The predominantly immunosuppressive effects of GCs on the immune system contrast observations that, at low doses, GCs can enhance pro-inflammatory signaling [49], in part by upregulating TLR2, TLR4, components of the inflammasome and certain cytokines [50]. On the basis of these studies, it was proposed that low-level GR signaling may sensitize cells to harmful stimuli by promoting the expression of pattern-recognition and cytokine receptors, thus enabling a prompt response to pathogens [9]. These pro-inflammatory effects of GCs mirror the up-regulation of IL-1 expression by PPARγ after inflammasome activation – the pro-inflammatory functions of these TFs are also convergent.

7.3 GR and PPARγ in Non-Macrophage Immune Cell Subsets

7.3.1 T Cells

It is well established that GCs inhibit CD4$^+$ T cell activity, however, it remains unclear if GCs predominantly affect CD4$^+$ helper T cells, CD4$^+$Foxp3$^+$ regulatory (T_{reg}) cells or both. GCs inhibit T cell activation directly by inhibiting the TFs downstream of TCR signaling: an extensive body of literature has documented a direct repression of NF-kB, AP-1 and nuclear factor of activated T cells (NF-AT) activity by GR via tethering in numerous cell types [51]. GCs were also proposed to inhibit T cell activation through non-genomic effects, by disrupting the TCR-associated GR protein complexes which include the lymphocyte-specific protein tyrosine kinase (LCK) and FYN kinase, ultimately leading to impaired TCR signaling [52]. GCs also affect T cell activation in an indirect manner, by interfering with the function of dendritic cells (DCs; discussed in detail later in the Chapter) in a GILZ-dependent manner and promoting their tolerogenic phenotype, marked by decreased levels of co-stimulatory CD86, CD83 and CD80, decreased secretion of chemokines CCL3, CCL5 and CXCL8 in activated DCs and a subsequent reduction of CD4$^+$ T cell proliferation [53]. Indeed, IFNγ production by CD4$^+$ T lymphocytes was no longer inhibited when DCs were transfected with GILZ siRNA [53]. Thus, GCs reduce the responsiveness of T cells to antigens and regulate the balance between activating and tolerogenic DCs, thereby suppressing effector T (T_{eff})

cell activity through both direct cell-intrinsic and indirect mechanisms.

Unexpectedly, a recent study suggested that the CD4+ T_{eff} subset might not be the primary target of therapeutic actions of GCs in T cells. Absence of GR specifically in Foxp3+ T_{reg} cells abrogated therapeutic effects of the GC dexamethasone (Dex) in murine experimental autoimmune encephalomyelitis (EAE) and allergic airway inflammation (AAI) models, suggesting that T_{regs} were necessary for GCs to exert their anti-inflammatory effects [54]. Mechanistically, GR was shown to induce microRNA miR-342-3p expression, leading to inhibition of Rictor, an adaptor protein of the glycolysis-favoring mTORC2 complex; this led to metabolic reprogramming of T_{regs} and induction of oxidative phosphorylation, which ultimately reinforces their suppressive functions [54]. In support of this study, GR-deficient T_{reg} cells were impaired in their ability to suppress T cell-dependent colitis in mice and acquired features typical of Th1 cells [55]. In the house dust mite-induced AAI model, treatment with synthetic GCs reduced T_{reg} recruitment to the lungs [56]. Mice with a T cell-specific GILZ KO had decreased absolute numbers of peripheral T_{reg} cells, an effect reversed by GILZ overexpression [57]. Effects of GR on T_{reg} cells are thus multifaceted, stimulating their activity, metabolism, proliferation and recruitment to inflammatory sites.

Among the CD4+ T_{eff} cell subsets, GCs inhibit Th1 as well as Th17, but up-regulate Th2 cell differentiation [58]. Similarly, in mice overexpressing GILZ in the T cell lineage, CD4+ T cells stimulated with CD3/CD28 antibodies secreted more Th2 and less Th1 cytokines compared to WT, an effect mirrored by up-regulation of Th2-specific TFs GATA-3 and STAT6 and down-regulation of the Th1-specific T-bet [59]. Finally, GILZ in Th17 cells localized to genomic sites in the proximity of Irf4, Batf, Stat3, and RORγt binding sites – TFs that drive Th17 activation and differentiation – suggesting that GC-induced GILZ may act as a transcriptional repressor of Th17-activating TFs [60] and that by upregulating GILZ, GCs shift the balance toward Th2-mediated humoral immunity (Fig. 7.2).

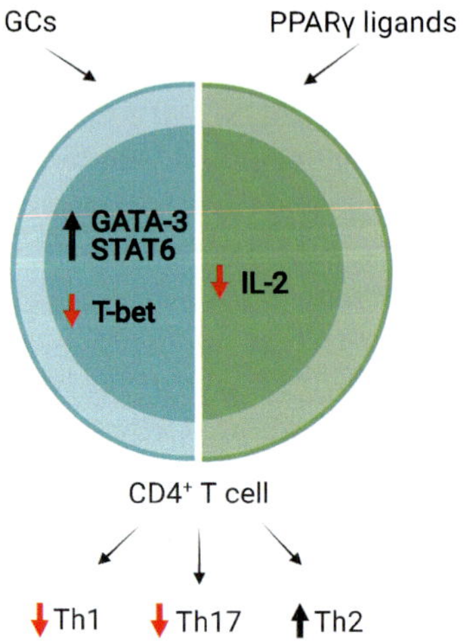

Fig. 7.2 GCs and PPARγ ligands promote Th2 and inhibit Th1 and Th17 immunity. Stimulation of CD4+ T cells with GCs increases transcription of TFs GATA-3 and STAT6, and down-regulates T-bet expression, which biases CD4+ T cells toward Th2 immunity and away from Th1 and Th17 responses. Similarly, treatment with PPARγ ligands decreases transcription of IL-2, which favors Th2 responses

Notably, although the predominant view is that GCs primarily affect the CD4+ T cell subset, in some disease contexts, GC-mediated suppression of CD8+ T cells is essential. In a mouse model of acute graft-versus-host disease (aGVHD), for example, lethally irradiated mice receiving a bone marrow transplant with GR-deficient T cells displayed much greater CD8+ T cell infiltration into the jejunum and their CD8+ T cells had augmented cytolytic activity compared to mice with WT T-cell transfer [61]. Thus, GR activity in CD8+ T cells in the context of aGVHD attenuates their inflammatory phenotype, mirroring the effects in CD4+ T_{eff} cells.

During development, pharmacological GCs induce caspase-dependent apoptosis of thymocytes [62–65] with GR deletion rendering GR-KO thymocytes GC-resistant. The mechanism of GC-induced apoptosis was shown to involve the activation of caspase-9 [66–68]. The physiological role of GC-induced thymocyte apoptosis continues to be debated. Although CD4+CD8+ double-positive thymocytes are particularly sensitive to GC-induced apoptosis, GCs at physiological levels do not appear to regulate death-by-neglect of these cells [69]. Rather, GCs are needed for optimal TCR repertoire and T cell responses to foreign antigens, thus contributing to negative selection [69]. In other studies, however, absence of GR had no effect on adult thymocyte development, as mice on a mixed background (129sv/C57BL/6) with a whole-body GR deletion had normal numbers of mature $CD4^{+}CD8^{-}$ and $CD4^{-}CD8^{+}$ cells, suggesting that positive selection was occurring normally [70]. It is yet to be determined if GC-induced thymocyte apoptosis indeed broadly affects T-cell development, or if it is limited to specific mouse models.

In contrast to GR, the overall contribution of PPARγ to the survival of T cells awaits further investigation. Both synthetic and endogenous PPARγ agonists stimulate apoptosis of murine T cells when administered in high doses [71]. Similarly, T cells stimulated with the proliferative agent, lectin phytohaemagglutinin P, undergo apoptosis after treatment with synthetic PPARγ agonists [72]. However, PPARγ-deficient, but not WT $CD4^{+}$ T cells, showed increased apoptosis after transfer into RAG1 KO mice, suggesting that PPARγ promotes $CD4^{+}$ T cell survival under conditions of low lymphocyte numbers [73]. Thus, the role of PPARγ in T cell survival remains controversial with net effect relatively poorly defined [74].

With respect to the balance of effector T-cell subsets, the PPARγ function appears similar to that of GR. At pharmacological concentrations, PPARγ ligands inhibit T cell, especially Th1, proliferation and decrease their viability [75], in part, by decreasing the transcription [76, 77] or protein expression [78] of IL-2. In addition, PPARγ ligands downregulate Th1 pro-inflammatory cytokines and augment the production of Th2 cytokines thereby shifting immune responses toward type-2 (Fig. 7.2). *In vivo*, PPARγ was shown to contribute to type-2 responses in T cells and DCs in an AAI model [79]. Specifically, in lung-resident $CD11b^{+}$ DCs, IL-4 and IL-33 signaling upregulated PPARγ levels, correlating with enhanced DC migration to draining lymph nodes and Th2 priming capacity. *In vitro*, production of IL-12 by DCs after stimulation with CD40 ligand, which normally induces Th1 responses, was inhibited by both endogenous and synthetic PPARγ ligands [80]. Thus, PPARγ mediates DC-T cell interactions in type-2 immunity in the context of *in vivo* Th2 responses, as well as promoting DC phenotypes associated with Th2-immunity *in vitro*.

Interestingly, PPARγ has been recently reported to facilitate group 2 innate lymphoid cell (ILC2)-induced AAI [81]. Loss of PPARγ in hematopoietic cells in mice diminished the function of ILC2 in the lungs, reducing the airway inflammation upon challenge with IL-33 or Papain. The transcriptional target of PPARγ in ILC2s was shown to be the IL-33 receptor ST2, such that overexpressing ST2 rescued the functional defects of PPARγ deficiency. Given that ILC2s and Th2 cells have been shown to collaborate in multiple AAI models [82–84], it appears that PPARγ can enhance both innate and adaptive arms of Th2 immunity.

In non-allergic models of inflammation, PPARγ has been generally shown to exert protective effects. Indeed, in a dextran sodium sulfate (DSS) colitis model, mice lacking PPARγ specifically in T cells exhibited reduced recruitment of T_{reg} cells to mesenteric lymph nodes, decrease in IL-10-producing $CD4^{+}$ T cells and increase in $CD8^{+}$ T cells, which together augmented colitis severity [85]. Similarly, in the EAE model of neuroinflammation, T-cell-specific PPARγ KO mice had higher clinical scores and enhanced infiltration of Th17 cells into the CNS [86]. The latter was consistent with *in vitro* data whereby naïve PPARγ KO $CD4^{+}$ T cells showed enhanced Th17 differentiation, suggesting that PPARγ constrains the Th17 cell lineage commitment [86]. Thus, endogenous PPARγ serves as an important brake on the inflammatory response *in vivo* in different organ systems.

In addition to the transcriptional effects on immune cell-specific genes, as discussed below, PPARγ is a key regulator of lipid metabolism across cell types and, therefore, impacts T cell biology by altering their bioenergetics and metabolic state. For example, the mechanistic target of rapamycin complex 1 (mTORC1)-PPARγ pathway is crucial for the FA uptake program in activated CD4+ T cells in mice [87]. PPARγ directly binds to promoters of genes associated with FA uptake in CD4+ T cells, leading to their metabolic reprogramming and rapid antigen-induced proliferation *in vivo*. Unlike its effect on genes specific to immune cell functions, the effect of PPARγ on metabolism of CD4+ T cells does not favor their differentiation toward a specific subset, but merely activates them.

7.3.2 Dendritic Cells (DCs)

DCs are often viewed as a bridge between the innate and adaptive immune system. Their role is to present pathogen-derived antigens on the cell surface, which get recognized by and activate T cells. Thus, aside from the direct effects of GR or PPARγ on T cells, the two NRs can affect DC activity, thereby potentially producing a less specific effect on T cell immunity.

Mice with a DC-specific KO of GR (GRCD11c-cre) were shown to be highly susceptible to septic shock induced by LPS, as evidenced by augmented production of inflammatory cytokines, a greater susceptibility to hypothermia and higher mortality [11]. Endogenous GCs inhibit LPS-induced inflammation and enhance tolerance by reducing IL-12 production by CD8+ DCs, and consequently, decreasing IFNγ secretion by natural killer cells [11]. The molecular mechanisms underlying GC actions specifically in CD8+ DCs have not been elucidated. However, GCs up-regulate the transcription of GILZ [58] and inhibit NF-kB and AP-1 activities and the MAPK pathway, thereby reducing production of IL-6, IL-12, and TNF [88, 89] in DCs similar to that seen in other cell types (Fig. 7.3). As discussed previously [53], GCs down-regulate co-stimulatory molecules on DCs and decrease their secretion of chemokines in a GILZ-dependent manner, in this way reducing the inflammatory phenotype of DCs (Fig. 7.3).

The prominent functions of PPARγ in DCs have been studied extensively. Over 1000 transcripts, including those of key lipid regulators FABP4 and ABCG2, were modulated by the PPARγ agonist rosiglitazone during GM-CSF- and IL-4-induced DC differentiation from monocytes *in vitro* [90], and PPARγ itself was

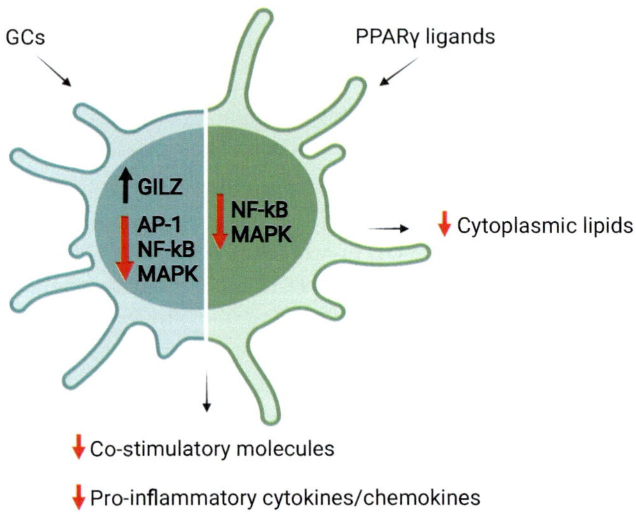

Fig. 7.3 Effect of GCs and PPARγ on DC activity. Activation of GR and PPARγ with their respective ligands leads to DC inactivation, manifested as decreased production of co-stimulatory molecules and pro-inflammatory cytokines. Both GR and PPARγ inactivate DCs by down-regulating MAPK and NF-kB pathways, with GR additionally decreasing AP-1 activity. The effects of GR are mediated by GILZ. Unlike GR, PPARγ also affects the lipid metabolism of DCs, decreasing their cytoplasmic lipid content

markedly up-regulated at both the mRNA and protein level [91]. Interestingly, FABP4 expression was elevated when human monocytes were differentiated to DCs in the presence of human serum, rather than specific ligand, suggesting that the endogenous PPARγ ligands were sufficient to drive PPARy-dependent gene transcription. In human monocyte derived DCs, PPARγ activation inhibited NF-kB and MAPK pathways, down-regulating co-stimulatory molecules and dampening TLR-induced secretion of pro-inflammatory cytokines ([92], Fig. 7.3). Genes linked to lipid metabolism were also up-regulated such that PPARγ-activated DCs had increased capacity to metabolize and re-distribute lipids, resulting in decreased cytoplasmic lipid content (Fig. 7.3). PPARγ hence connects lipid processing in DCs with their immune function. In a mouse model of asthma, knocking out PPARγ in DCs attenuated recruitment of eosinophils to the airways, IL-4 secretion by $CD4^+$ cells and histopathological changes, demonstrating that PPARγ in DCs orchestrates Th2 immunity in the lungs [79]. Given the previously described role of PPARγ agonists in reducing inflammation in asthma, this study demonstrated that endogenous PPARγ in DCs may have the opposite role [93]. The PPARγ-dependent skewing of DCs toward Th2 immunity is concordant with the preference of PPARγ for type-2 responses in both innate and adaptive arms.

The examples above illustrate that both GR and PPARγ exert primarily anti-inflammatory actions in macrophages, T cells and DCs and bias the immune system toward type 2 responses. Likewise, both NRs can induce thymocyte apoptosis, although PPARγ can favor $CD4^+$ cell survival. Some of the effects of these two receptors on immune cells are conferred via metabolic reprogramming. A well-known GR transcriptional target GILZ is an important effector of downstream responses in DCs, T_{regs} and Th2 subsets. The specific targets of PPARγ in immune cells appear more diverse and cell type-specific.

7.4 GR and PPARγ in Adipocytes

The most well-studied cell type-specific role of PPARγ is in adipocytes, where it serves as the master regulator that is necessary and sufficient to induce adipogenic gene expression and lipid accumulation [94]. Adipocyte-specific KO of PPARγ using the aP2-Cre, a target of PPARγ, and resulting in unhindered adipocyte differentiation, allows for assessing the role of PPARγ in the mature cells [95]. PPARγ deletion led to enlargement of white and brown adipocytes and reduction in their numbers. At the systemic level, adipocyte-specific loss of PPARγ resulted in elevated free FA and triglyceride (TG) plasma levels, fatty liver with increased gluconeogenesis, as well as reduced levels of leptin and adipocyte complement-related protein of 30 kDa (ACRP30), known to be secreted exclusively by differentiated adipocytes. Ablation of PPARγ in fat with a more specific Adipoq-Cre, which uses a regulatory region of adiponectin, resulted in severe adipose tissue loss, insulin resistance and other metabolic abnormalities [96].

Multiple TFs and coregulators – coactivators and corepressors – modulate the function of PPARγ in adipocytes [97]. PPARγ and the TF C/EBPα bind to the majority of the genes upregulated during adipogenesis, and both TFs, as well as C/EBPβ, were required for the expression of adipogenesis-inducing genes, suggesting that cooperativity between PPARγ and C/EBP (Fig. 7.4a) is needed for adipogenesis [98]. Coactivators affect PPARγ function in adipocytes by directly binding PPARγ and facilitating the recruitment of additional components of transcriptional machinery or chromatin modifiers, such as thyroid hormone receptor interacting protein 3 (TRIP3) and members of the NCoA/p160 family, e.g., NCoA2/TIF2/GRIP1, NCoA1/SRC-1 and the PPARγ coactivator 1-alpha (PGC-1a). Knock-down of TRIP3 leads to diminished differentiation of adipocytes, so TRIP3 acts as a positive regulator of PPARγ-mediated adipocyte differentiation [99]. NCoA2 promotes

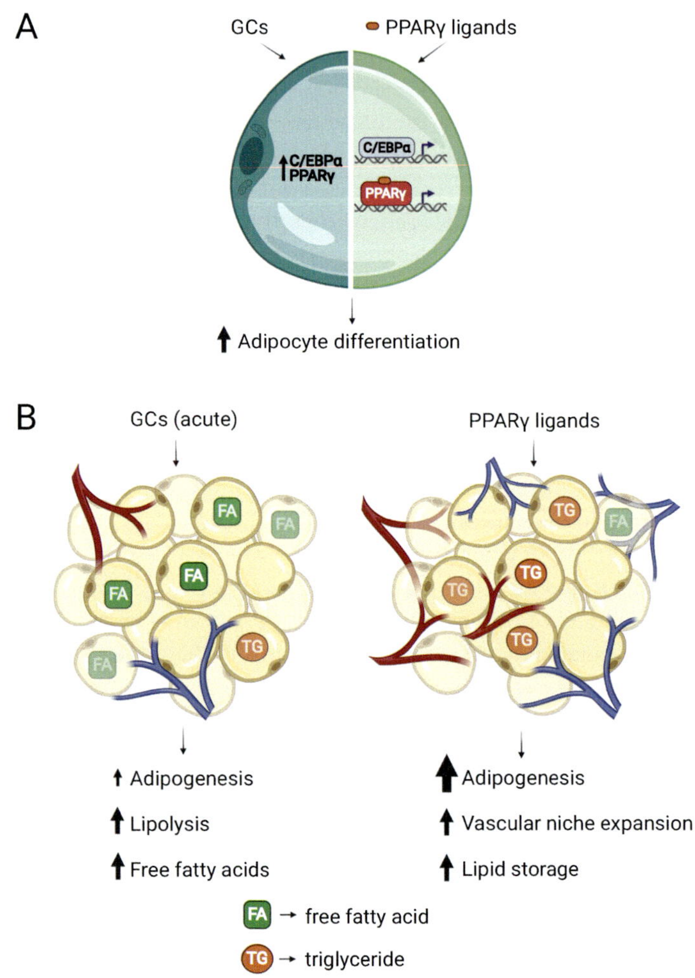

Fig. 7.4 GCs and PPARγ ligands affect adipose tissue in distinct ways. (**a**) In adipocytes, GR increases the expression of PPARγ and C/EBPα; PPARγ is the major driver of adipogenesis in cooperation with C/EBPα. (**b**) At the adipose tissue level, GR and PPARγ have disparate functions, with GR up-regulating lipolysis and the levels of free FA in addition to adipogenesis upon acute exposure. PPARγ affects the adipose tissue on multiple levels, by promoting adipogenesis, angiogenesis and lipid storage

PPARγ activity and fat accumulation in white adipose tissue (WAT), whereas NCoA1 enhances energy expenditure and protects from obesity [100]. Mediator complex subunit 14 (MED14) is another direct interactor of PPARγ, which tethers the Mediator complex to PPARγ to activate PPARγ-specific lipogenic genes [101]. In mature 3 T3-L1 adipocytes, the histone acetyltransferase coactivator Tip60 is recruited to PPARγ target genes, and reduction of Tip60 protein levels impedes 3T3-L1 preadipocyte differentiation [102]. These studies indicate that coactivators affect multiple and diverse aspects of the PPARγ function in adipocytes.

Transcription activation by PPARγ is negatively regulated by corepressors such as NCoR/SMRT; these are recruited by PPARγ in the absence of ligand, and dissociate upon ligand binding when they are replaced by coactivators due to a change in PPARγ conformation [103, 104]. In 3T3-L1 cells, knocking down NCoR and SMRT leads to increased expression of adipocyte-specific genes [103]. NCoR deletion in adipocytes was shown to enhance adipogenesis, reduce inflammation and improve insulin sensitivity at the organismal level [105]. Mechanistically, NCoR and SMRT recruit HDAC3 to induce histone deacetylation of PPARγ-bound regulatory regions [106]. However, in adipocytes, NCoR facilitates the recruitment of cyclin dependent kinase (CDK)5, which binds to and phosphorylates PPARγ at S-273 (inhibitory site that reduces recruitment of PGC-1 and GRIP1/NCoA2 and increases interactions with SMRT and NCoR), leading to impaired regulation of metabolic genes, such as insulin-sensitizing adiponectin [107]. Conversely, ring finger protein 20 (RNF20), which was shown to target NCoR for proteasomal degradation, acts as a positive regulator of PPARγ activity during adipogenesis [108]. A transcriptional cofactor with PDZ-binding motif (TAZ) was shown to act as a PPARγ corepressor [109]. TAZ deletion in adipocytes led to constitutive activity of PPARγ, and improved glucose tolerance and sensitivity to insulin in obese mice [110]. The functions of PPARγ in adipocytes are, thus, modulated by direct repression, which itself may be modulated by secondary cofactors.

In addition to direct gene regulation in adipocytes, PPARγ affects adipose tissue physiology by acting in its resident immune cells. For example, PPARγ modulates T_{reg} accumulation, phenotype and function in the visceral adipose tissue (VAT) [111]. PPARγ cooperates with Foxp3 to upregulate a large number of T_{reg}-specific genes in the VAT, as shown by analyzing gene expression of naïve $CD4^+$ T cells retrovirally transduced with *Pparg* and *Foxp3*. Additionally, VAT T_{reg} cells were found to uptake lipids upon stimulation with PPARγ ligand pioglitazone [111]. PPARγ is therefore necessary for the maintenance and accumulation of T_{reg} cells in the VAT, and mediates the insulin-sensitizing activity of pioglitazone. PPARγ activation also promotes anti-inflammatory VAT phenotype by inhibiting resident conventional DC maturation and T_{eff} cell recruitment in both lean and obese mice [112]. In addition, PPARγ directs the establishment and maintenance of the adipose vascular niche. *In vivo*, PPARγ overexpression in the adipose lineage upregulates PDGFRβ and VEGF in adipose progenitor cells, and both of these genes contribute to endothelial cell proliferation and adipose niche expansion [113]. Another important function of PPARγ is promoting the conversion of subcutaneous WAT to brown adipose tissue (BAT) [114]. Nuclear factor I-A (NFIA) assists PPARγ in WAT browning by facilitating the binding of PPARγ to BAT-specific enhancers, as shown in mouse C2C12 myoblasts treated with adipocyte differentiation cocktail that included rosiglitazone [115]. Thus, PPARγ acts as a broad regulator of adipose tissue physiology and metabolism.

GR performs several key functions in adipose tissue, many of which are opposite to those of PPARγ, but there is an overlap with respect to adipogenesis. GCs were shown to promote adipogenesis *in vitro*. Specifically, GR facilitated the up-regulation of C/EBPα and PPARγ mRNA and protein levels in 3T3-L1 cells upon stimulation with Dex and other compounds that promote adipogenesis (Fig. 7.4a, [116]). Consistently, GR KO MEFs failed to up-regulate CEBPα and PPARγ after treatment with a Dex-containing differentiation cocktail [117]. Mechanistically, in response to stimulation of pre-adipocytes with a Dex-containing cocktail, GR binds to transiently acetylated regions to establish a new gene expression program, including upregulation of PPARγ [118]. *In vivo*, however, GCs may facilitate adipogenesis without being absolutely required for it. Indeed, mice with a GR deletion in the BAT (using Myf5-Cre) had normal BAT size and morphology as well as normal expression of adipogenesis marker genes including *Cebpa* and, notably, *Pparg* [119]. Additionally, white and brown GR KO pre-adipocytes undergoing differentiation *in vitro* had reduced levels of adipogenesis markers early on, but eventually reached the levels of the WT [119]. Furthermore, in adrenalectomized (ADX) mice, largely lacking endogenous GCs, injection of MEFs into subcutaneous tissue did result in fat pad formation, although reduced in size compared to those in intact mice [117]. In the same study, injection of both WT MEFs into ADX mice, and GR KO or WT MEFs into WT mice, led to fat pad formation with comparable expression of adipocyte-specific genes, not significantly different from that in inguinal WAT of WT mice. During adipogenesis, therefore, GR and PPARγ may cooperate, thereby accelerating the PPARγ-dependent processes (Fig. 7.4a).

A broadly lipolytic effect of GC exposure in the adipose tissue, opposite to that of PPARγ activation, was reported over 40 years ago (Fig. 7.4b) and confirmed in multiple studies thereafter [120]. Typically, GC-induced lipolysis in the

WAT is associated with an acute hormone exposure due to stress response or fasting [121]. Prolonged or chronic exposure in rats, however, resulted in visceral fat accumulation, adipocyte hyperplasia and reduction in adipocyte size [122]. GR ligands can also enhance lipid storage, but only under specific, often, pathological conditions. For instance, hypercortisolemia during Cushing's syndrome is known to cause an expansion of visceral fat depots due to the synergistic effects of GCs with insulin, whereby GCs upregulate genes involved in lipid deposition [123].

Finally, GR activity in the liver, discussed below, exerts secondary effects on the adipose tissue. Crossing adult STAT5a/b KO mice with Alfp-Cre GR KO generated mice with a combined deletion of GR and STAT5 in hepatocytes [124]. These double KO mice had smaller adipocytes and fat depots, displayed hypercortisolism and aggravated steatosis compared to WT or STAT5 single KO mice.

Thus, outcomes of GC action upon the adipose tissue are complex, dependent on ligand concentration and duration of exposure, and further modulated by the systemic effects of GCs in other tissues, ultimately leading to adipogenesis and lipid storage, or lipolysis (Fig. 7.4b).

7.5 GR and PPARγ in the Liver

GCs were originally named for their ability to promote gluconeogenesis in the liver (Fig. 7.5a). Indeed, liver is a major target organ for GC action and plays a central role in glucose metabolism. In mice, a conditional liver-specific deletion of GR led to hypoglycemic lethality within days of birth [125]. In the clinical setting, excess GC levels during Cushing's syndrome or as a result of GC therapy have been associated with hyperglycemia and central obesity [126].

Two critical rate-limiting enzymes involved in gluconeogenesis, glucose-6-phosphatase and phosphoenolpyruvate carboxykinase, are encoded by the *G6pc* and *Pck1* genes, respectively, both of which are known to be direct GR targets (Fig. 7.5a, [16, 127]). The *Pck1* gene has been studied extensively and has a GRE site upstream of the TSS [128]. Interestingly, later studies pointed to roles of NRs other than GR in *G6pc* and *Pck1* regulation. Specifically, the dominant PPAR in the liver – PPARα, rather than PPARγ – is recruited to both genes and contributes to their transcriptional regulation in addition to GR [129, 130].

Apart from gluconeogenesis, GCs have also been linked to the regulation of FA metabolism in the liver. Patients with Cushing's syndrome often develop dyslipidemia that manifests as high TG and systemic cholesterol levels [131]. Intriguingly, liver-specific KO of GR in mice with hepatic steatosis led to a notable reduction in hepatic TGs and elevated ketone levels in circulation, along with upregulation of genes involved in FA oxidation and TG hydrolysis [132]. Genes mediating lipid storage and transport (e.g., FA transporter *Cd36*) were also significantly downregulated. Thus, liver-specific GR KO ameliorated hepatic steatosis by increasing hydrolysis of TG stores, indicating that under conditions of fatty liver, GR promotes TG storage (Fig. 7.5b).

Analyses of the tissue-specific distribution of PPARs position PPARα as the primary PPAR expressed in the liver; in contrast, PPARγ levels are relatively low. Thus, numerous studies suggested that metabolic effects of PPARγ stem primarily from its action in adipose tissue, with indirect secondary effects on the liver. However, a common phenotype of the adipocyte-specific PPARγ KO, in addition to lipodystrophy, is a substantial increase in hepatic PPARγ along with accumulation of TG in the liver [96]. Interestingly, hepatocyte-specific deletion of *Pparg* alleviated steatosis phenotypes in various animal models [133–135], further indicating that hepatocyte-expressed rather than adipocyte PPARγ was responsible for the fat accrual. Alb-Cre-mediated deletion of PPARγ in the liver markedly diminished the expression of the *Pparg2*, but not *Pparg1* isoform, so PPARγ2 appears to be the major isoform in hepatocytes contributing to fat accumulation [133]. Thus, in the context of liver steatosis, PPARγ can promote TG accumulation similar to GR (Fig. 7.5b).

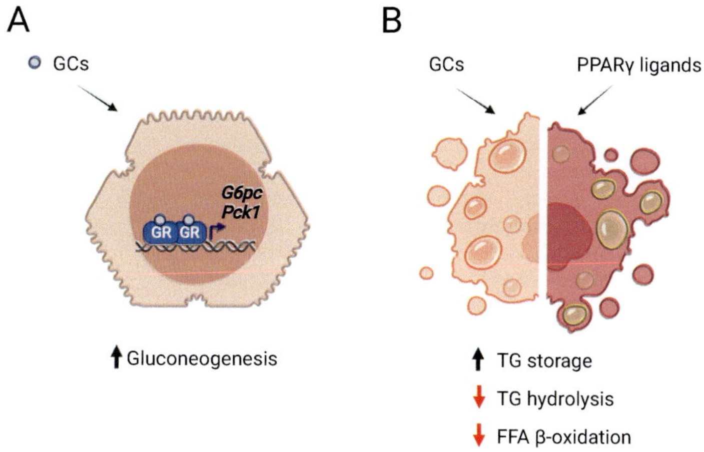

Fig. 7.5 Effects of GCs and PPARγ ligands in hepatocytes. (**a**) In healthy hepatocytes GR is the main driver of gluconeogenesis. (**b**) Under conditions of hepatic steatosis, both GR and PPARγ increase TG storage by decreasing TG hydrolysis and FA oxidation

Thus, in healthy liver, GR is a dominant regulator of glucose metabolism which up-regulates *de novo* glucose production, with little to no contribution from PPARγ. Under conditions of liver steatosis, both GR and PPARγ inhibit lipid hydrolysis and FA oxidation, thereby augmenting an increase in liver mass.

7.6 Concluding Remarks

GR and PPARγ are highly divergent NRs from steroid and non-steroid families, respectively, both viewed as critical therapeutic targets with a range of actions in the immune system and in metabolic homeostasis. Interestingly, the two NRs share many functions in immune cells at homeostasis and under pathogenic conditions. These TFs are anti-inflammatory during acute and chronic inflammation, and act as drivers of the Th2 response by promoting the M2-like macrophage subtype, biasing T cells towards Th2 and DCs towards tolerogenic state. Apart from a more pronounced role of GR in thymocyte selection, and that of PPARγ in DC development, the functional overlap of GR and PPARγ in immune cells eclipses isolated examples of their distinct roles. In the adipose tissue, however, the differences are striking: PPARγ is essential for adipogenesis and enhances lipid storage in adipocytes, whereas GR is mostly lipolytic upon acute hormone exposure. Finally, in the liver, GR is the uniquely critical regulator of normal glucose metabolism, while the two NRs have overlapping roles in TG metabolism during liver steatosis. Given that these TFs are invaluable therapeutic targets for, among others, autoimmune diseases and type 2 diabetes, novel insights on the consequences of activating both NRs, and understanding the effects their ligands may have at super-physiological doses *in vivo,* could potentially inform the use of combined treatments in clinical settings.

Acknowledgments We thank Dr. Y. Chinenov (HSS Genomics Center) for critical feedback on the manuscript. The figures in this chapter were created with BioRender.com. This work was supported by the NIH R01DK099087, NIH R21NS110520, NIH R01AI148129 and The Hospital for Special Surgery David Rosensweig Genomics Center.

References

1. Oakley RH, Cidlowski JA (2013) The biology of the glucocorticoid receptor: new signaling mechanisms in health and disease. J Allergy Clin Immunol 132(5):1033–1044. https://doi.org/10.1016/j.jaci.2013.09.007
2. Meijsing SH, Pufall MA, So AY, Bates DL, Chen L, Yamamoto KR (2009) DNA binding site sequence directs glucocorticoid receptor structure and activity. Science 324(5925):407–410. https://doi.org/10.1126/science.1164265
3. Ratman D, Vanden Berghe W, Dejager L, Libert C, Tavernier J, Beck IM et al (2013) How glucocorticoid receptors modulate the activity of other transcription factors: a scope beyond tethering. Mol Cell Endocrinol 380(1–2):41–54. https://doi.org/10.1016/j.mce.2012.12.014

4. Itoh T, Fairall L, Amin K, Inaba Y, Szanto A, Balint BL et al (2008) Structural basis for the activation of PPARgamma by oxidized fatty acids. Nat Struct Mol Biol 15(9):924–931. https://doi.org/10.1038/nsmb.1474
5. Varga T, Czimmerer Z, Nagy L (2011) PPARs are a unique set of fatty acid regulated transcription factors controlling both lipid metabolism and inflammation. Biochim Biophys Acta 1812(8):1007–1022. https://doi.org/10.1016/j.bbadis.2011.02.014
6. Brunmeir R, Xu F (2018) Functional regulation of PPARs through post-translational modifications. Int J Mol Sci 19(6). https://doi.org/10.3390/ijms19061738
7. Nagy L, Schwabe JW (2004) Mechanism of the nuclear receptor molecular switch. Trends Biochem Sci 29(6):317–324. https://doi.org/10.1016/j.tibs.2004.04.006
8. Bhattacharyya S, Brown DE, Brewer JA, Vogt SK, Muglia LJ (2007) Macrophage glucocorticoid receptors regulate Toll-like receptor 4-mediated inflammatory responses by selective inhibition of p38 MAP kinase. Blood 109(10):4313–4319. https://doi.org/10.1182/blood-2006-10-048215
9. Cain DW, Cidlowski JA (2017) Immune regulation by glucocorticoids. Nat Rev Immunol 17(4):233–247. https://doi.org/10.1038/nri.2017.1
10. Kleiman A, Hubner S, Rodriguez Parkitna JM, Neumann A, Hofer S, Weigand MA et al (2012) Glucocorticoid receptor dimerization is required for survival in septic shock via suppression of interleukin-1 in macrophages. FASEB J 26(2):722–729. https://doi.org/10.1096/fj.11-192112
11. Li CC, Munitic I, Mittelstadt PR, Castro E, Ashwell JD (2015) Suppression of dendritic cell-derived IL-12 by endogenous glucocorticoids is protective in LPS-induced sepsis. PLoS Biol 13(10):e1002269. https://doi.org/10.1371/journal.pbio.1002269
12. Nagy L, Szanto A, Szatmari I, Szeles L (2012) Nuclear hormone receptors enable macrophages and dendritic cells to sense their lipid environment and shape their immune response. Physiol Rev 92(2):739–789. https://doi.org/10.1152/physrev.00004.2011
13. Babaev VR, Yancey PG, Ryzhov SV, Kon V, Breyer MD, Magnuson MA et al (2005) Conditional knockout of macrophage PPARγIncreases atherosclerosis in C57BL/6 and low-density lipoprotein receptor–deficient mice. Arterioscler Thromb Vasc Biol 25(8):1647–1653
14. Shah YM, Morimura K, Gonzalez FJ (2007) Expression of peroxisome proliferator-activated receptor-gamma in macrophage suppresses experimentally induced colitis. Am J Physiol Gastrointest Liver Physiol 292(2):G657–G666. https://doi.org/10.1152/ajpgi.00381.2006
15. Ferreira AE, Sisti F, Sonego F, Wang S, Filgueiras LR, Brandt S et al (2014) PPAR-gamma/IL-10 axis inhibits MyD88 expression and ameliorates murine polymicrobial sepsis. J Immunol 192(5):2357–2365. https://doi.org/10.4049/jimmunol.1302375
16. Sacta MA, Chinenov Y, Rogatsky I (2016) Glucocorticoid signaling: an update from a genomic perspective. Annu Rev Physiol 78:155–180. https://doi.org/10.1146/annurev-physiol-021115-105323
17. Miyata M, Lee JY, Susuki-Miyata S, Wang WY, Xu H, Kai H et al (2015) Glucocorticoids suppress inflammation via the upregulation of negative regulator IRAK-M. Nat Commun 6:6062. https://doi.org/10.1038/ncomms7062
18. Mittelstadt PR, Ashwell JD (2001) Inhibition of AP-1 by the glucocorticoid-inducible protein GILZ. J Biol Chem 276(31):29603–29610
19. Ayroldi E, Migliorati G, Bruscoli S, Marchetti C, Zollo O, Cannarile L et al (2001) Modulation of T-cell activation by the glucocorticoid-induced leucine zipper factor via inhibition of nuclear factor kappaB. Blood 98(3):743–753. https://doi.org/10.1182/blood.v98.3.743
20. Smoak K, Cidlowski JA (2006) Glucocorticoids regulate tristetraprolin synthesis and posttranscriptionally regulate tumor necrosis factor alpha inflammatory signaling. Mol Cell Biol 26(23):9126–9135. https://doi.org/10.1128/MCB.00679-06
21. Bhattacharyya S, Zhao Y, Kay TW, Muglia LJ (2011) Glucocorticoids target suppressor of cytokine signaling 1 (SOCS1) and type 1 interferons to regulate Toll-like receptor-induced STAT1 activation. Proc Natl Acad Sci U S A 108(23):9554–9559. https://doi.org/10.1073/pnas.1017296108
22. Beck IM, Vanden Berghe W, Vermeulen L, Bougarne N, Vander Cruyssen B, Haegeman G et al (2008) Altered subcellular distribution of MSK1 induced by glucocorticoids contributes to NF-kappaB inhibition. EMBO J 27(12):1682–1693. https://doi.org/10.1038/emboj.2008.95
23. Gupte R, Muse GW, Chinenov Y, Adelman K, Rogatsky I (2013) Glucocorticoid receptor represses proinflammatory genes at distinct steps of the transcription cycle. Proc Natl Acad Sci U S A 110(36):14616–14621. https://doi.org/10.1073/pnas.1309898110
24. Sacta MA, Tharmalingam B, Coppo M, Rollins DA, Deochand DK, Benjamin B et al (2018) Gene-specific mechanisms direct glucocorticoid-receptor-driven repression of inflammatory response genes in macrophages. elife 7. https://doi.org/10.7554/eLife.34864
25. Wang L, Oh TG, Magida J, Estepa G, Obayomi SB, Chong L-W et al (2021) Bromodomain containing 9 (BRD9) regulates macrophage inflammatory responses by potentiating glucocorticoid receptor activity. Proc Natl Acad Sci 118(35):e2109517118
26. Pascual G, Glass CK (2006) Nuclear receptors versus inflammation: mechanisms of transrepression. Trends Endocrinol Metab 17(8):321–327. https://doi.org/10.1016/j.tem.2006.08.005

27. Hou Y, Moreau F, Chadee K (2012) PPARγ is an E3 ligase that induces the degradation of NFκB/p65. Nat Commun 3(1):1–11
28. Nelson VL, Nguyen HC, Garcìa-Cañaveras JC, Briggs ER, Ho WY, DiSpirito JR et al (2018) PPARγ is a nexus controlling alternative activation of macrophages via glutamine metabolism. Genes Dev 32(15–16):1035–1044
29. Okreglicka K, Iten I, Pohlmeier L, Onder L, Feng Q, Kurrer M et al (2021) PPARγ is essential for the development of bone marrow erythroblastic island macrophages and splenic red pulp macrophages. J Exp Med 218(5):e20191314
30. Weber KJ, Sauer M, He L, Tycksen E, Kalugotla G, Razani B et al (2018) PPARγ deficiency suppresses the release of IL-1β and IL-1α in macrophages via a type 1 IFN–dependent mechanism. J Immunol 201(7):2054–2069
31. Gordon S, Taylor PR (2005) Monocyte and macrophage heterogeneity. Nat Rev Immunol 5(12):953–964. https://doi.org/10.1038/nri1733
32. Liao X, Sharma N, Kapadia F, Zhou G, Lu Y, Hong H et al (2011) Kruppel-like factor 4 regulates macrophage polarization. J Clin Invest 121(7):2736–2749. https://doi.org/10.1172/JCI45444
33. Martinez FO, Sica A, Mantovani A, Locati M (2008) Macrophage activation and polarization. Front Biosci 13:453–461. https://doi.org/10.2741/2692
34. Xue J, Schmidt SV, Sander J, Draffehn A, Krebs W, Quester I et al (2014) Transcriptome-based network analysis reveals a spectrum model of human macrophage activation. Immunity 40(2):274–288. https://doi.org/10.1016/j.immuni.2014.01.006
35. Giles KM, Ross K, Rossi AG, Hotchin NA, Haslett C, Dransfield I (2001) Glucocorticoid augmentation of macrophage capacity for phagocytosis of apoptotic cells is associated with reduced p130Cas expression, loss of paxillin/pyk2 phosphorylation, and high levels of active Rac. J Immunol 167(2):976–986. https://doi.org/10.4049/jimmunol.167.2.976
36. Liu Y, Cousin JM, Hughes J, Van Damme J, Seckl JR, Haslett C et al (1999) Glucocorticoids promote nonphlogistic phagocytosis of apoptotic leukocytes. J Immunol 162(6):3639–3646
37. Tugal D, Liao X, Jain MK (2013) Transcriptional control of macrophage polarization. Arterioscler Thromb Vasc Biol 33(6):1135–1144
38. Abdalla HB, Napimoga MH, Lopes AH, de Macedo Maganin AG, Cunha TM, Van Dyke TE et al (2020) Activation of PPAR-γ induces macrophage polarization and reduces neutrophil migration mediated by heme oxygenase 1. Int Immunopharmacol 84:106565
39. Yao Q, Liu J, Zhang Z, Li F, Zhang C, Lai B et al (2018) Peroxisome proliferator-activated receptor (PPAR) induces the gene expression of integrin (V5) to promote macrophage M2 polarization. J Biol Chem 293(43):16572–16582
40. Chen H, Shi R, Luo B, Yang X, Qiu L, Xiong J et al (2015) Macrophage peroxisome proliferator-activated receptor gamma deficiency delays skin wound healing through impairing apoptotic cell clearance in mice. Cell Death Dis 6:e1597. https://doi.org/10.1038/cddis.2014.544
41. Daniel B, Nagy G, Horvath A, Czimmerer Z, Cuaranta-Monroy I, Poliska S et al (2018) The IL-4/STAT6/PPARgamma signaling axis is driving the expansion of the RXR heterodimer cistrome, provid ing complex ligand responsiveness in macrophages. Nucleic Acids Res 46(9):4425–4439. https://doi.org/10.1093/nar/gky157
42. Lawrence T, Natoli G (2011) Transcriptional regulation of macrophage polarization: enabling diversity with identity. Nat Rev Immunol 11(11):750–761. https://doi.org/10.1038/nri3088
43. Viswakarma N, Jia Y, Bai L, Vluggens A, Borensztajn J, Xu J et al (2010) Coactivators in PPAR-regulated gene expression. PPAR Res 2010:250126
44. Coppo M, Chinenov Y, Sacta MA, Rogatsky I (2016) The transcriptional coregulator GRIP1 controls macrophage polarization and metabolic homeostasis. Nat Commun 7:12254. https://doi.org/10.1038/ncomms12254
45. Rollins DA, Kharlyngdoh JB, Coppo M, Tharmalingam B, Mimouna S, Guo Z et al (2017) Glucocorticoid-induced phosphorylation by CDK9 modulates the coactivator functions of transcriptional cofactor GRIP1 in macrophages. Nat Commun 8(1):1739. https://doi.org/10.1038/s41467-017-01569-2
46. Chinenov Y, Gupte R, Dobrovolna J, Flammer JR, Liu B, Michelassi FE et al (2012) Role of transcriptional coregulator GRIP1 in the anti-inflammatory actions of glucocorticoids. Proc Natl Acad Sci 109(29):11776–11781
47. Wang J, Xu X, Li P, Zhang B, Zhang J (2021) HDAC3 protects against atherosclerosis through inhibition of inflammation via the microRNA-19b/PPARγ/NF-κB axis. Atherosclerosis 323:1–12
48. Gao Q, Wei A, Chen F, Chen X, Ding W, Ding Z et al (2020) Enhancing PPARγ by HDAC inhibition reduces foam cell formation and atherosclerosis in ApoE deficient mice. Pharmacol Res 160:105059
49. Lim HY, Muller N, Herold MJ, van den Brandt J, Reichardt HM (2007) Glucocorticoids exert opposing effects on macrophage function dependent on their concentration. Immunology 122(1):47–53. https://doi.org/10.1111/j.1365-2567.2007.02611.x
50. Busillo JM, Cidlowski JA (2013) The five Rs of glucocorticoid action during inflammation: ready, reinforce, repress, resolve, and restore. Trends Endocrinol Metab 24(3):109–119. https://doi.org/10.1016/j.tem.2012.11.005
51. De Bosscher K, Haegeman G (2009) Minireview: latest perspectives on antiinflammatory actions of glucocorticoids. Mol Endocrinol 23(3):281–291

52. Löwenberg M, Verhaar AP, Bilderbeek J, van Marle J, Buttgereit F, Peppelenbosch MP et al (2006) Glucocorticoids cause rapid dissociation of a T-cell-receptor-associated protein complex containing LCK and FYN. EMBO Rep 7(10):1023–1029
53. Cohen N, Mouly E, Hamdi H, Maillot M-C, Pallardy M, Vr G et al (2006) GILZ expression in human dendritic cells redirects their maturation and prevents antigen-specific T lymphocyte response. Blood 107(5):2037–2044
54. Kim D, Nguyen QT, Lee J, Lee SH, Janocha A, Kim S et al (2020) Anti-inflammatory roles of glucocorticoids are mediated by Foxp3+ regulatory T cells via a miR-342-dependent mechanism. Immunity 53(3):581–96. e5
55. Rocamora-Reverte L, Tuzlak S, von Raffay L, Tisch M, Fiegl H, Drach M et al (2019) Glucocorticoid receptor-deficient Foxp3+ regulatory T cells fail to control experimental inflammatory bowel disease. Front Immunol 10:472
56. Olsen P, Kitoko J, Ferreira T, De-Azevedo C, Arantes A, Martins M (2015) Glucocorticoids decrease Treg cell numbers in lungs of allergic mice. Eur J Pharmacol 747:52–58
57. Bereshchenko O, Coppo M, Bruscoli S, Biagioli M, Cimino M, Frammartino T et al (2014) GILZ promotes production of peripherally induced Treg cells and mediates the crosstalk between glucocorticoids and TGF-β signaling. Cell Rep 7(2):464–475
58. Liberman AC, Budziñski ML, Sokn C, Gobbini RP, Steininger A, Arzt E (2018) Regulatory and mechanistic actions of glucocorticoids on T and inflammatory cells. Front Endocrinol 9:235
59. Cannarile L, Fallarino F, Agostini M, Cuzzocrea S, Mazzon E, Vacca C et al (2006) Increased GILZ expression in transgenic mice up-regulates Th-2 lymphokines. Blood 107(3):1039–1047
60. Yosef N, Shalek AK, Gaublomme JT, Jin H, Lee Y, Awasthi A et al (2013) Dynamic regulatory network controlling TH 17 cell differentiation. Nature 496(7446):461–468
61. Theiss-Suennemann J, Jorss K, Messmann JJ, Reichardt SD, Montes-Cobos E, Luhder F et al (2015) Glucocorticoids attenuate acute graft-versus-host disease by suppressing the cytotoxic capacity of CD8(+) T cells. J Pathol 235(4):646–655. https://doi.org/10.1002/path.4475
62. Cifone MG, Migliorati G, Parroni R, Marchetti C, Millimaggi D, Santoni A et al (1999) Dexamethasone-induced thymocyte apoptosis: apoptotic signal involves the sequential activation of phosphoinositide-specific phospholipase C, acidic sphingomyelinase, and caspases. Blood 93(7):2282–2296
63. Hakem R, Hakem A, Duncan GS, Henderson JT, Woo M, Soengas MS et al (1998) Differential requirement for caspase 9 in apoptotic pathways in vivo. Cell 94(3):339–352
64. Kuida K, Haydar TF, Kuan C-Y, Gu Y, Taya C, Karasuyama H et al (1998) Reduced apoptosis and cytochrome c–mediated caspase activation in mice lacking caspase 9. Cell 94(3):325–337
65. McColl KS, He H, Zhong H, Whitacre CM, Berger NA, Distelhorst CW (1998) Apoptosis induction by the glucocorticoid hormone dexamethasone and the calcium-ATPase inhibitor thapsigargin involves Bc1-2 regulated caspase activation. Mol Cell Endocrinol 139(1–2):229–238
66. Jamieson CA, Yamamoto KR (2000) Crosstalk pathway for inhibition of glucocorticoid-induced apoptosis by T cell receptor signaling. Proc Natl Acad Sci 97(13):7319–7324
67. Prenek L, Boldizsár F, Kugyelka R, Ugor E, Berta G, Németh P et al (2017) The regulation of the mitochondrial apoptotic pathway by glucocorticoid receptor in collaboration with Bcl-2 family proteins in developing T cells. Apoptosis 22(2):239–253
68. Dong L, Vaux DL (2020) Glucocorticoids can induce BIM to trigger apoptosis in the absence of BAX and BAK1. Cell Death Dis 11(6):1–15
69. Taves MD, Ashwell JD (2021) Glucocorticoids in T cell development, differentiation and function. Nat Rev Immunol 21(4):233–243. https://doi.org/10.1038/s41577-020-00464-0
70. Purton JF, Zhan Y, Liddicoat DR, Hardy CL, Lew AM, Cole TJ et al (2002) Glucocorticoid receptor deficient thymic and peripheral T cells develop normally in adult mice. Eur J Immunol 32(12):3546–3555
71. Wang YL, Frauwirth KA, Rangwala SM, Lazar MA, Thompson CB (2002) Thiazolidinedione activation of peroxisome proliferator-activated receptor γ can enhance mitochondrial potential and promote cell survival. J Biol Chem 277(35):31781–31788
72. Schmidt S, Moric E, Schmidt M, Sastre M, Feinstein DL, Heneka MT (2004) Anti-inflammatory and antiproliferative actions of PPAR-γ agonists on T lymphocytes derived from MS patients. J Leukoc Biol 75(3):478–485
73. Housley WJ, Adams CO, Vang AG, Brocke S, Nichols FC, LaCombe M et al (2011) Peroxisome proliferator-activated receptor γ is required for CD4+ T cell-mediated lymphopenia-associated autoimmunity. J Immunol 187(8):4161–4169
74. Choi J-M, Bothwell AL (2012) The nuclear receptor PPARs as important regulators of T-cell functions and autoimmune diseases. Mol Cells 33(3):217–222
75. da Rocha Junior LF, Dantas AT, Duarte AL, de Melo Rego MJ, Pitta Ida R, Pitta MG (2013) PPARgamma agonists in adaptive immunity: what do immune disorders and their models have to tell us? PPAR Res 2013:519724. https://doi.org/10.1155/2013/519724
76. Rockwell CE, Snider NT, Thompson JT, Heuvel JPV, Kaminski NE (2006) Interleukin-2 suppression by 2-arachidonyl glycerol is mediated through peroxisome proliferator-activated receptor γ independently of cannabinoid receptors 1 and 2. Mol Pharmacol 70(1):101–111

77. Yang XY, Wang LH, Chen T, Hodge DR, Resau JH, DaSilva L et al (2000) Activation of human T lymphocytes is inhibited by peroxisome proliferator-activated receptor γ (PPARγ) agonists: PPARγ co-association with transcription factor NFAT. J Biol Chem 275(7):4541–4544
78. Marx N, Kehrle B, Kohlhammer K, Grub M, Koenig W, Hombach V et al (2002) PPAR activators as anti-inflammatory mediators in human T lymphocytes: implications for atherosclerosis and transplantation-associated arteriosclerosis. Circ Res 90(6):703–710
79. Nobs SP, Natali S, Pohlmeier L, Okreglicka K, Schneider C, Kurrer M et al (2017) PPARγ in dendritic cells and T cells drives pathogenic type-2 effector responses in lung inflammation. J Exp Med 214(10):3015–3035
80. Faveeuw C, Fougeray S, Angeli V, Fontaine J, Chinetti G, Gosset P et al (2000) Peroxisome proliferator-activated receptor γ activators inhibit interleukin-12 production in murine dendritic cells. FEBS Lett 486(3):261–266
81. Xiao Q, He J, Lei A, Xu H, Zhang L, Zhou P et al (2021) PPARgamma enhances ILC2 function during allergic airway inflammation via transcription regulation of ST2. Mucosal Immunol 14(2):468–478. https://doi.org/10.1038/s41385-020-00339-6
82. Halim TY, Steer CA, Mathä L, Gold MJ, Martinez-Gonzalez I, McNagny KM et al (2014) Group 2 innate lymphoid cells are critical for the initiation of adaptive T helper 2 cell-mediated allergic lung inflammation. Immunity 40(3):425–435
83. Li BW, de Bruijn MJ, Tindemans I, Lukkes M, KleinJan A, Hoogsteden HC et al (2016) T cells are necessary for ILC2 activation in house dust mite-induced allergic airway inflammation in mice. Eur J Immunol 46(6):1392–1403
84. Liu B, Lee J-B, Chen C-Y, Hershey GKK, Wang Y-H (2015) Collaborative interactions between type 2 innate lymphoid cells and antigen-specific CD4+ Th2 cells exacerbate murine allergic airway diseases with prominent eosinophilia. J Immunol 194(8):3583–3593
85. Guri AJ, Mohapatra SK, Horne WT, Hontecillas R, Bassaganya-Riera J (2010) The role of T cell PPAR γ in mice with experimental inflammatory bowel disease. BMC Gastroenterol 10(1):1–13
86. Klotz L, Burgdorf S, Dani I, Saijo K, Flossdorf J, Hucke S et al (2009) The nuclear receptor PPARγ selectively inhibits Th17 differentiation in a T cell–intrinsic fashion and suppresses CNS autoimmunity. J Exp Med 206(10):2079–2089
87. Angela M, Endo Y, Asou HK, Yamamoto T, Tumes DJ, Tokuyama H et al (2016) Fatty acid metabolic reprogramming via mTOR-mediated inductions of PPARgamma directs early activation of T cells. Nat Commun 7:13683. https://doi.org/10.1038/ncomms13683
88. Ronchetti S, Migliorati G, Riccardi C (2015) GILZ as a mediator of the anti-inflammatory effects of glucocorticoids. Front Endocrinol 6:170
89. Ricci E, Ronchetti S, Gabrielli E, Pericolini E, Gentili M, Roselletti E et al (2019) GILZ restrains neutrophil activation by inhibiting the MAPK pathway. J Leukoc Biol 105(1):187–194
90. Szatmari I, Töröcsik D, Agostini M, Nagy T, Gurnell M, Barta E et al (2007) PPARγ regulates the function of human dendritic cells primarily by altering lipid metabolism. Blood 110(9):3271–3280
91. Gosset P, Charbonnier AS, Delerive P, Fontaine J, Staels B, Pestel J et al (2001) Peroxisome proliferator-activated receptor γ activators affect the maturation of human monocyte-derived dendritic cells. Eur J Immunol 31(10):2857–2865
92. Appel S, Mirakaj V, Bringmann A, Weck MM, Grünebach F, Brossart P (2005) PPAR-γ agonists inhibit toll-like receptor-mediated activation of dendritic cells via the MAP kinase and NF-κB pathways. Blood 106(12):3888–3894
93. Nobs SP, Kopf M (2018) PPAR-γ in innate and adaptive lung immunity. J Leukoc Biol 104(4):737–741
94. Lefterova MI, Haakonsson AK, Lazar MA, Mandrup S (2014) PPARγ and the global map of adipogenesis and beyond. Trends Endocrinol Metab 25(6):293–302
95. He W, Barak Y, Hevener A, Olson P, Liao D, Le J et al (2003) Adipose-specific peroxisome proliferator-activated receptor γ knockout causes insulin resistance in fat and liver but not in muscle. Proc Natl Acad Sci 100(26):15712–15717
96. Wang F, Mullican SE, DiSpirito JR, Peed LC, Lazar MA (2013) Lipoatrophy and severe metabolic disturbance in mice with fat-specific deletion of PPARγ. Proc Natl Acad Sci 110(46):18656–18661
97. Siersbæk R, Nielsen R, Mandrup S (2012) Transcriptional networks and chromatin remodeling controlling adipogenesis. Trends Endocrinol Metab 23(2):56–64
98. Lefterova MI, Zhang Y, Steger DJ, Schupp M, Schug J, Cristancho A et al (2008) PPARγ and C/EBP factors orchestrate adipocyte biology via adjacent binding on a genome-wide scale. Genes Dev 22(21):2941–2952
99. Koppen A, Houtman R, Pijnenburg D, Jeninga EH, Ruijtenbeek R, Kalkhoven E (2009) Nuclear receptor-coregulator interaction profiling identifies TRIP3 as a novel peroxisome proliferator-activated receptor γ cofactor. Mol Cell Proteomics 8(10):2212–2226
100. Picard F, Géhin M, Annicotte J-S, Rocchi S, Champy M-F, O'Malley BW et al (2002) SRC-1 and TIF2 control energy balance between white and brown adipose tissues. Cell 111(7):931–941
101. Grøntved L, Madsen MS, Boergesen M, Roeder RG, Mandrup S (2010) MED14 tethers mediator to the N-terminal domain of peroxisome proliferator-activated receptor γ and is required for full transcriptional activity and adipogenesis. Mol Cell Biol 30(9):2155–2169
102. van Beekum O, Brenkman AB, Grøntved L, Hamers N, van den Broek NJ, Berger R et al (2008) The

adipogenic acetyltransferase Tip60 targets activation function 1 of peroxisome proliferator-activated receptor γ. Endocrinology 149(4):1840–1849
103. Yu C, Markan K, Temple KA, Deplewski D, Brady MJ, Cohen RN (2005) The nuclear receptor corepressors NCoR and SMRT decrease peroxisome proliferator-activated receptor γ transcriptional activity and repress 3T3-L1 adipogenesis. J Biol Chem 280(14):13600–13605
104. Shang J, Mosure SA, Zheng J, Brust R, Bass J, Nichols A et al (2020) A molecular switch regulating transcriptional repression and activation of PPARγ. Nat Commun 11(1):1–14
105. Li P, Fan W, Xu J, Lu M, Yamamoto H, Auwerx J et al (2011) Adipocyte NCoR knockout decreases PPARγ phosphorylation and enhances PPARγ activity and insulin sensitivity. Cell 147(4):815–826
106. Perissi V, Jepsen K, Glass CK, Rosenfeld MG (2010) Deconstructing repression: evolving models of co-repressor action. Nat Rev Genet 11(2):109–123
107. Choi JH, Banks AS, Estall JL, Kajimura S, Boström P, Laznik D et al (2010) Anti-diabetic drugs inhibit obesity-linked phosphorylation of PPARγ by Cdk5. Nature 466(7305):451–456
108. Jeon YG, Lee JH, Ji Y, Sohn JH, Lee D, Kim DW et al (2020) RNF20 functions as a transcriptional coactivator for PPARγ by promoting NCoR1 degradation in adipocytes. Diabetes 69(1):20–34
109. Hong J-H, Hwang ES, McManus MT, Amsterdam A, Tian Y, Kalmukova R et al (2005) TAZ, a transcriptional modulator of mesenchymal stem cell differentiation. Science 309(5737):1074–1078
110. El Ouarrat D, Isaac R, Lee YS, Wollam J, Lackey D, Riopel M et al (2020) TAZ is a negative regulator of PPARγ activity in adipocytes and TAZ deletion improves insulin sensitivity and glucose tolerance. Cell Metab 31(1):162–73. e5
111. Cipolletta D, Feuerer M, Li A, Kamei N, Lee J, Shoelson SE et al (2012) PPAR-γ is a major driver of the accumulation and phenotype of adipose tissue T reg cells. Nature 486(7404):549–553
112. Macdougall CE, Wood EG, Loschko J, Scagliotti V, Cassidy FC, Robinson ME et al (2018) Visceral adipose tissue immune homeostasis is regulated by the crosstalk between adipocytes and dendritic cell subsets. Cell Metab 27(3):588–601. e4
113. Jiang Y, Berry DC, Jo A, Tang W, Arpke RW, Kyba M et al (2017) A PPARγ transcriptional cascade directs adipose progenitor cell-niche interaction and niche expansion. Nat Commun 8(1):1–16
114. Ohno H, Shinoda K, Spiegelman BM, Kajimura S (2012) PPARγ agonists induce a white-to-brown fat conversion through stabilization of PRDM16 protein. Cell Metab 15(3):395–404
115. Hiraike Y, Waki H, Yu J, Nakamura M, Miyake K, Nagano G et al (2017) NFIA co-localizes with PPARγ and transcriptionally controls the brown fat gene program. Nat Cell Biol 19(9):1081–1092
116. Pantoja C, Huff JT, Yamamoto KR (2008) Glucocorticoid signaling defines a novel commitment state during adipogenesis in vitro. Mol Biol Cell 19(10):4032–4041
117. Bauerle KT, Hutson I, Scheller EL, Harris CA (2018) Glucocorticoid receptor signaling is not required for in vivo adipogenesis. Endocrinology 159(5):2050–2061
118. Steger DJ, Grant GR, Schupp M, Tomaru T, Lefterova MI, Schug J et al (2010) Propagation of adipogenic signals through an epigenomic transition state. Genes Dev 24(10):1035–1044
119. Park Y-K, Ge K (2017) Glucocorticoid receptor accelerates, but is dispensable for, adipogenesis. Mol Cell Biol 37(2):e00260–e00216
120. Swarbrick M, Zhou H, Seibel M (2021) MECHANISMS IN ENDOCRINOLOGY: local and systemic effects of glucocorticoids on metabolism: new lessons from animal models. Eur J Endocrinol 185(5):R113–RR29
121. Beaupere C, Liboz A, Fève B, Blondeau B, Guillemain G (2021) Molecular mechanisms of glucocorticoid-induced insulin resistance. Int J Mol Sci 22(2):623
122. Campbell JE, Peckett AJ, D'souza AM, Hawke TJ, Riddell MC (2011) Adipogenic and lipolytic effects of chronic glucocorticoid exposure. Am J Phys Cell Phys 300(1):C198–C209
123. Lee M-J, Pramyothin P, Karastergiou K, Fried SK (2014) Deconstructing the roles of glucocorticoids in adipose tissue biology and the development of central obesity. Biochimica et Biophysica Acta (BBA)-Molecular Basis of Disease 1842(3):473–481
124. Mueller KM, Kornfeld JW, Friedbichler K, Blaas L, Egger G, Esterbauer H et al (2011) Impairment of hepatic growth hormone and glucocorticoid receptor signaling causes steatosis and hepatocellular carcinoma in mice. Hepatology 54(4):1398–1409
125. Opherk C, Tronche F, Kellendonk C, Kohlmuller D, Schulze A, Schmid W et al (2004) Inactivation of the glucocorticoid receptor in hepatocytes leads to fasting hypoglycemia and ameliorates hyperglycemia in streptozotocin-induced diabetes mellitus. Mol Endocrinol 18(6):1346–1353. https://doi.org/10.1210/me.2003-0283
126. Patel R, Williams-Dautovich J, Cummins CL (2014) Minireview: new molecular mediators of glucocorticoid receptor activity in metabolic tissues. Mol Endocrinol 28(7):999–1011. https://doi.org/10.1210/me.2014-1062
127. Jitrapakdee S (2012) Transcription factors and coactivators controlling nutrient and hormonal regulation of hepatic gluconeogenesis. Int J Biochem Cell Biol 44(1):33–45. https://doi.org/10.1016/j.biocel.2011.10.001
128. Yang J, Reshef L, Cassuto H, Aleman G, Hanson RW (2009) Aspects of the control of phosphoenolpyruvate carboxykinase gene transcription. J Biol Chem 284(40):27031–27035

129. Boergesen M, Pedersen TA, Gross B, van Heeringen SJ, Hagenbeek D, Bindesboll C et al (2012) Genome-wide profiling of liver X receptor, retinoid X receptor, and peroxisome proliferator-activated receptor alpha in mouse liver reveals extensive sharing of binding sites. Mol Cell Biol 32(4):852–867. https://doi.org/10.1128/MCB.06175-11
130. Lee JM, Wagner M, Xiao R, Kim KH, Feng D, Lazar MA et al (2014) Nutrient sensing nuclear receptors coordinate autophagy. Nature 516(7529):112–115. https://doi.org/10.1038/nature13961
131. Arnaldi G, Scandali VM, Trementino L, Cardinaletti M, Appolloni G, Boscaro M (2010) Pathophysiology of dyslipidemia in Cushing's syndrome. Neuroendocrinology 92(Suppl 1):86–90. https://doi.org/10.1159/000314213
132. Lemke U, Krones-Herzig A, Berriel Diaz M, Narvekar P, Ziegler A, Vegiopoulos A et al (2008) The glucocorticoid receptor controls hepatic dyslipidemia through Hes1. Cell Metab 8(3):212–223. https://doi.org/10.1016/j.cmet.2008.08.001
133. Gavrilova O, Haluzik M, Matsusue K, Cutson JJ, Johnson L, Dietz KR et al (2003) Liver peroxisome proliferator-activated receptor gamma contributes to hepatic steatosis, triglyceride clearance, and regulation of body fat mass. J Biol Chem 278(36):34268–34276. https://doi.org/10.1074/jbc.M300043200
134. Matsusue K, Haluzik M, Lambert G, Yim SH, Gavrilova O, Ward JM et al (2003) Liver-specific disruption of PPARgamma in leptin deficient mice improves fatty liver but aggravates diabetic phenotypes. J Clin Invest 111(5):737–747. https://doi.org/10.1172/JCI17223
135. Moran-Salvador E, Lopez-Parra M, Garcia-Alonso V, Titos E, Martinez-Clemente M, Gonzalez-Periz A et al (2011) Role for PPARgamma in obesity-induced hepatic steatosis as determined by hepatocyte- and macrophage-specific conditional knockouts. FASEB J 25(8):2538–2550. https://doi.org/10.1096/fj.10-173716

Circadian Rhythm and Nuclear Receptors

8

David W. Ray

Abstract

All life of Earth has evolved mechanisms to track time. This permits anticipation of predictable changes in light/dark, and in most cases also directs fed/fasted cycles, and sleep/wake. The nuclear receptors enjoy a close relationship with the molecular machinery of the clock. Some play a core role within the circadian machinery, other respond to ligands which oscillate in concentration, and physical cross-talk between clock transcription factors, eg cryptochromes, and multiple nuclear receptors also enable coupling of nuclear receptor function to time of day. Essential processes including inflammation, and energy metabolism are strongly regulated by both the circadian machinery, and rhythmic behaviour, and also by multiple members of the nuclear receptor family. An emerging theme is reciprocal regulation of key processes by different members of the nuclear receptor family, for example NR1D1/2, and NR1F1, in regulation of the core circadian clock transcription factor BMAL1.

D. W. Ray (✉)
University of Oxford, Oxford, UK
e-mail: david.ray@ocdem.ox.ac.uk

Keywords

Circadian · Nuclear receptor · Energy metabolism · Inflammation · Suprachiasmatic nucleus · Sleep

8.1 Ligands and NR Expression Through Time

The evolutionary origins of the nuclear receptors likely lie as sensors of nutrients, and products of intermediary metabolism. As life on Earth is subject to predictable changes in the environment from light to dark there have been major adaptations to anticipate such changes. In simple organisms this may have conferred protection from the actions of reactive oxygen species by allowing temporal segregation of DNA synthesis away from peak ATP generation, for example. In more advanced organisms cycles of rest, and activity, and feeding/fasting tend to be directed by light cycle [1]. The emergence of a circadian clock, which allows anticipation of the fed/fasted state offers a clear survival advantage, and there may be further advantages in terms of tuning immune responses, cancer susceptibility, and brain function. The convergence between the direct actions of the circadian machinery, and various members of the nuclear receptor superfamily of important

M. J. Campbell, C. L. Bevan (eds.), *Nuclear Receptors in Human Health and Disease*, Advances in Experimental Medicine and Biology 1390, https://doi.org/10.1007/978-3-031-11836-4_8

regulatory pathways suggested the existence of crosstalk between the two super-systems [2, 3].

Ligands: The nuclear receptors are ligand activated transcription factors. A number of the ligands for nuclear receptors show a circadian oscillation in abundance, predicting a periodic change in activity. In particular, the production of steroid hormones from the adrenal cortex varies strongly by time of day, and for other ligands, such as saturated fatty acids, or bile salts the abundance is indirectly regulated by time of day by means of feeding behaviour [2].

Nuclear Hormone Receptors (NHR) expression: An obvious, and direct mechanism for the circadian clock to capture nuclear receptor activity is by regulating gene expression. Indeed, the majority of nuclear receptors show circadian expression changes, but with differences between tissues [3] (Fig. 8.1). As at least 10% of the genome in any given tissue is under circadian control this suggests that there is particular enrichment for nuclear receptors amongst the targets for the circadian clock machinery [4, 5].

Tissue sensitivity: There is evidence that the spectrum, and amplitude of response to nuclear receptor function varies by time of day, in addition to the changes seen by cell type. The cell type differences are well-described, with an important role for cell-type specific transcription factors in preparing chromatin for nuclear receptor recruitment. The mechanisms required to capture time of day variation in response are less well defined, but at least two mechanisms have been proposed. The direct crosstalk between members of the nuclear receptor super family and core circadian clock proteins affords a direct means for reciprocal regulation [6–8]. In addition, the actions of the clock components on enhancers and chromatin accessibility may also be regulating how easily nuclear receptors can navigate to their binding sites, in a time-of-day dependent manner [5]. This is further complicated as a tripartite mechanism including both core clock components, and cell-type determining factors may be required.

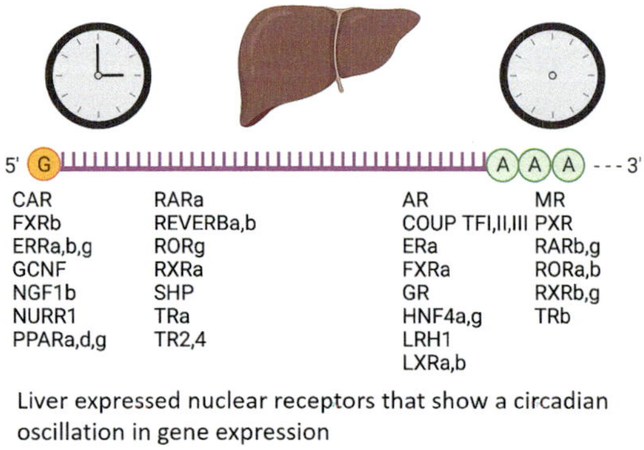

Fig. 8.1 Liver expressed nuclear receptors that show a circadian oscillation in gene expression

8.2 Circadian Clocks

Organisation of behaviour, and physiology through time is a consistent feature through all kingdoms of life. This is dependent in an internal time keeping mechanism which allows the organism to anticipate predictable changes in the light-dark environment. There are clear survival advantages in partitioning particular functions to specific phases of the solar day. In mammals the central clock is located in the hypothalamus in the suprachiasmatic nucleus (SCN) [1, 9]. This small, bilateral structure receives light information from the retina, through the retinohypothalamic tract, an independent pathway, and not involved in image formation [10–12]. The suprachiasmatic nucleus then relays time of day information through neural, and humoral signals. The core mechanism explaining time keeping in the suprachiasmatic nucleus is a transcription translation negative feedback look (TTFL), in which transactivating components BMAL1 and CLOCK drive expression of repressors, the PERIOD and CRYPTOCHROME protein-encoding genes [1, 13, 14] (Fig. 8.2). As PERIOD and CRYPTOCHROME proteins accumulate in the cytoplasm they are subject to post translational modification, and eventually reach threshold concentrations to inhibit the transactivation function of the BMAL1/CLOCK heterodimer.

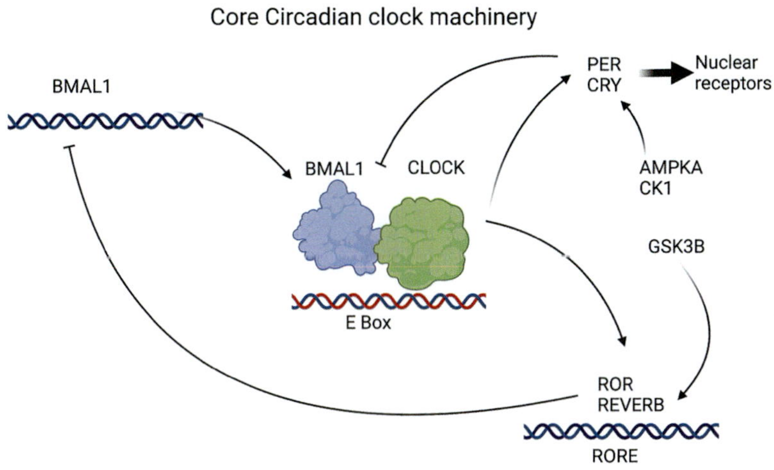

Fig. 8.2 Core circadian clock machinery

In addition to this loop an additional negative feedback loop exists. This is driven by transactivation of the orphan nuclear receptors NR1F1/RORa, and NR1D1/REVERBa. These proteins bind to a common recognition DNA sequence, the RORE, and either transactivate (RORa), or transrepress (REVERBa) [15]. A dynamic competition is established between these two transcription factors. RORE elements are found on the proximal *Bmal1* gene promoter, and so *Bmal1* transcription is captured by the second negative feedback loop.

Taken together the existence of two, complementary, negative feedback loops contribute to the robust nature of the core molecular oscillator. All the components of the clock exist as multiple functional paralogues, with the exception of BMAL1, so that deletion of single genes has a minimal impact of the function of the oscillator. Although there is a BMAL2 its function remains elusive. For example, the REVERBa null mouse has a minor shortening of its circadian period length when kept under constant conditions (no light-dark cycles) [16]. However, deletion of both REVERBA, and NR1D2/REVERBB results in an arrhythmic mouse, thereby confirming the role of REVERBs as core circadian clock transcription factors [15].

All peripheral tissues also possess intrinsic circadian clock activity, and so what emerges is that the central clock in the SCN is an entraining centre, which keeps the phase of clocks operating in peripheral tissues aligned [14]. It is though that such internal circadian alignment is important for healthy function, allowing optimal tuning of activity between organs involved in specialised activities eg gut, liver, adipose and muscle retain circadian coherence. The core circadian machinery in peripheral tissues is the same as in the SCN, but in most cases the oscillators retain less robust internal time keeping compared to the SCN. This suggests a hierarchy of clocks, with the SCN being dominant, and the only one with light input [17–20].

8.2.1 Clocks, Entrainment and Misalignment

It is thought that the purpose of internal time-keeping is to enhance engagement with the environment; to optimise coincidence of active periods with food availability, and optimise sleep periods for safety. In natural systems the major entraining influence is sunlight, with retinal illumination engaging the central clock in the suprachiasmatic nucleus. However, in conditions of modern life people spend less time in natural daylight, and furthermore extend the illuminated day late into the evening. The abundance of artificial light has increased massively over the last century and is associated with many of the scourges of modern living, including rates of obesity, and type 2 diabetes [21–25]. A potential mediator of the modern society risk of metabolic disease is artificial light, and its regulation of the core circadian clock. This raises important questions for nuclear receptor researchers working on metabolic diseases, when many of the molecules and circuits under investigation are changing with circadian phase [26].

Although the SCN drives light-entrained physiological rhythms there are other important timing cues or zeitgebers (time givers). Of these feeding times are particularly important in mammals, and especially in humans who now extend the fed period late into the evening, with consequences for circadian control of energy metabolism. There are a number of mediators of feeding entrainment, including gut microbiome derived metabolites, nutrients, and hormones, including insulin, glucagon [27, 28]. Important nuclear receptors engaged in food entrainment include the fasting-activated PPARs, with both PPARα, and PPARγ being important [29]. The fed state is signalled by activation of FXR by bile salts after absorption in the distal small bowel, and re-circulation to the liver in the blood stream [30]. PPARs can act on the *REVERBa* gene to drive expression, providing a functional circuit to the core of the circadian clock. In addition, the rise in glucocorticoids that accompanies the fasting state activates the glucocorticoid receptor (NR3C1/GR), which drives a programme of gluconeogenesis and hepatic insulin resistance to promote mobilisation of glucose to feed tissues including most prominently the brain, during the fasted state [31]. Importantly, the activated GR is a powerful transactivator of the *PERIOD2* gene, again offering a route into the clock machinery. Here we see that there is productive synergy between the fasting activated nuclear receptors, and the signalling pathways activated by catecholaminergic, growth hormone, and glucagon signalling.

A prevalent change in modern living is extension of the fed period into the late evening, and also mis-timed feeding, as for example seen during night shiftwork. These changes in feeding behaviour are associated with metabolic diseases including obesity, and type 2 diabetes [24, 25]. There is a complexity here, with mis-timed feeding also being associated with sleep deprivation, shiftworkers typically sleep for at least an hour less per day, as well as activation of stress responses [21, 22]. However, even experimental mistimed feeding results in altered energy metabolism, with changes to insulin sensitivity and glucose tolerance by time of day, and so the role of the circadian machinery in explaining the risks of mistimed feeding is emerging [32–34]. This has further implications for circulating biomarkers, where we have to consider not only the time of day the sample was taken, but also the sleep-wake, and feeding-fasted cycle of the individual [35, 36]. It is estimated that at least 10% of circulating metabolites show a circadian variation, with sleep, and feeding also playing important roles in affecting amplitude, and phase of the oscillation [37].

As the two major zeitgebers for the circadian clock are light and food, there is potential for conflict if these signals become misaligned. The two signals are typically aligned, and reinforce one another, with feeding time being determined by the SCN, which is itself entrained to the light cycles. However, if food is time-restricted to an atypical circadian phase a conflict is established. Some peripheral organs, such as liver, will follow feeding time, with a progressive change in circadian acrophase of about an hour a day. Thus a 12-h shift in feeding time will take about 12 days to result in complete liver re-setting. After that point the metabolic clock oscillates in anti-phase to the central clock in the SCN, which follows light cycles. Surprisingly, some non-metabolic tissues, including lung will also follow feeding time [38]. But, for some organs, such as the adrenal, a biphasic response results, with food driving one oscillation, and sympathetic innervation conveying SCN time cues. It has been proposed that the misalignment between phase between organs, or between the internal phase, and the external activity cycle is particularly damaging for energy metabolism. Such misalignment is far more prevalent with modern living than at any time during human evolution and may provide an attractive explanation for many of the ills seen in modern societies [39].

A further driver of misalignment is the move to fixed time zones, which cover a broad range of longitude. In this way people living at the western edge of time zones live in a state of partial jet-lag, or shiftwork, and indeed have been shown to have an excess risk of many of the same metabolic disturbances as seen in shiftwork. Even within a given location there is a range of chronotype within a population, that is preference for morning, or evening activity [40, 41]. Chronotype varies by sex, and age, and is affected by genetic variation, with many loci associated with chronotype in GWAS studies [41]. The functional importance of chronotype is demonstrated by the altered risk of disease associated with chronotype, including metabolic disease, neuropsychiatric disease, and parameters including educational attainment. Within human populations chronotype variation covers about a 4 h interval, but the very late chronotypes seen in adolescence, especially in boys, can lead to major disruption of education, and social organisation. This has led to campaigns for innovations such as late school start times in secondary education.

8.3 Nuclear Receptors Within the Clock

A. **Core clock**: The orphan nuclear receptors RORa and REVERBa play essential roles within the core circadian oscillator. The RORs are considered to be essential regulators of Th17 lymphocytes, a subset of CD4+ T helper cells with a pathogenic role in autoimmunity. The RORs include three receptors RORa, NR1F2/RORb, and NR1F3/RORg. The RORs transactivate as monomers through ROR response elements (AGGTCA), with a 5′ AT-rich extension. The same DNA element is a target for the REVERBs, which are obligate repressors. As RORs, and REVERBs tend to be co-expressed in the same cells this sets up a dynamic competition between them [42].

The RORs exist in multiple splice variants, with variation in the N terminal portion of the encoded protein resulting. RORa is widely expressed in liver, muscle, skin, adipose, and brain. RORb is very restricted to the brain. RORgt plays a major role in T helper 17 (Th17) cells, and has been extensively studied in this context. RORg, and especially RORgt (or RORg2), is highly expressed in immune tissues including thymus, but RORg is also found in liver, muscle, adipose and kidney [42].

The "staggerer" mouse results from a naturally occurring intragenic insertion within the RORa gene, which results in a premature stop codon. These mice have a severe cerebellar development problem, and under constant conditions show a shortened circadian period length. The RORb null mice in contrast have a lengthened circadian period. No circadian abnormalities result from loss of RORg. As RORa affects circadian parameters and as the circadian clock is linked to development of energy metabolic abnormalities, targeting RORa is an attractive idea [42].

Experimental evidence for the benefits of targeting RORa was provided from a drug screen for circadian rhythm amplitude enhancing compounds. Here, a compound screen was conducted against circadian PER2-luc oscillations in a CLOCK haploinsufficient mouse model, in which the amplitude of circadian oscillations was found to be attenuated. The lead compound was found to be a ligand of RORa, and was a natural component of citrus fruit skin, nobiletin. In further work nobiletin treatment affected circadian amplitude in-vivo, and was able to preserve glucose homeostasis in response to high fat diet feeding. Further work since then has identified a role of nobiletin activated RORa in opposing glucagon action in the liver, possibly part of the mechanism explaining its impact on metabolic disease [43].

A number of purely synthetic ROR agonists have been described, and their actions studied in both circadian, and also metabolic contexts. Issues with the compounds remain, in that the affinity remains relatively low, and the off-target effects may be hard to control [44].

REVERBa was considered an orphan receptor until it was noted that the drosophila homologue,

D75 was found to bind heme when crystallised. Indeed, mammalian REVERBs were also found to bind heme, and moreover the function of REVERB was affected by the ligand concentration, supporting the idea that heme was a true ligand, and not just an inert binding partner [45]. It is interesting that REVERB offers negative feedback to the synthesis of heme, and with heme serving as a REVERB ligand this completes a classical negative feedback loop. A number of synthetic compound ligands for the REVERBs have been identified, with actions which include phase shifting the circadian clock, and regulation of processes considered to be REVERB targets, including inflammation, and energy metabolism [46–48]. A metabolic circuit affecting cancer cell function has also been identified as potentially regulated by REVERB ligands, but the presence of off-target effects, and low efficacy has resulted in some confusion in the literature about which events are primary REVERB regulated, and which result from pleiotropic actions of the compounds, which are often tested at very high concentrations/doses [47, 49].

The competition between RORs, and REVERBs for shared DNA response elements may also be further complicated by the idea that the binding of one affects the subsequent binding of the other, in an assisted loading type model [50].

B. **Clock acting NHRs:** The discovery that circadian clocks were operating in peripheral cells, even in cell lines, as well as in the SCN was a seminal moment for the circadian biology field [14]. It was recognised that a number of extracellular stimuli were capable of entraining the cellular clock, even in a cell growing in a dish. Amongst these serum shock, and glucocorticoids were powerful interventions, with the activated GR transactivating the *Period* genes 1 and 2. In this way glucocorticoids can reset the clock, or in a population of cells cycling asynchronously a glucocorticoid challenge results in synchronisation, and entrainment.

In vivo the role of glucocorticoids as a time-giver is less clear. For some rhythmic processes the presence of glucocorticoids is important, but not the rhythm of the glucocorticoid concentration. This has been shown in zebrafish, and also in mammalian systems [51]. Analysis of the role of the hypothalamic-pituitary-adrenal axis suggests that the long negative feedback loop of CRH-ACTH-corticosteroid results in stabilisation of the circadian rhythm in physiology. Loss of the adrenal glands results in more rapid phase shifting responses to altered light cycle phasing, or relative protection from jet lag [52]. The intersection of the circadian machinery, and glucocorticoid action on inflammation raises the possibility of productive cross talk between the systems. Indeed, circadian variation in tissue inflammation could be abolished by adrenalectomy [53]. However, even tonic replacement of corticosteroid, by using slow-release pellets, was sufficient for restoration of circadian rhythmic inflammatory responses. Therefore, ligand-activation of the GR is required, but the timing sensitivity of tissue response is conferred by the target cells, and not the adrenal production of ligand. This result is supported by work in zebrafish which showed a strong gain in circadian amplitude in glucocorticoid deficient fish, resulting from a genetic defect in steroidogenesis, even with tonic, long-acting glucocorticoid Dexamethasone applied to the water [54].

Many lines of evidence point to differential engagement of transcriptional programmes by the clock in response to altered physiology. For example, in high fat diet induced obesity a number of liver genes lose, and some gain a circadian rhythm [29]. Nuclear receptors have emerged as key regulators of this switch, including receptors responding to nutritional cues such as the PPARs. There are two important considerations here. One is the extent to which the re-wiring is tissue autonomous, and results from altered circadian clock function, or downstream coupling, and the second is the role of systemic, or circulating cues, including metabolites and hormones reflecting nutritional state. A major problem affecting the field is the comparison of rhythmicity between states. Perhaps most transcripts, proteins, and metabolites vary through time, in response to delays in feedback, and feedforward loops, and changes in physiological state, but for a molecular species to be assigned as oscillating with circadian rhythm additional criteria need to be met. There are reviews which address this in more detail, but identifying circadian oscillations in noisy data, and extraction of key parameters including amplitude, phase, and period are all important [55]. The concern is that false positive and false negative rates can be high, especially if highly dimensional data, such as RNA-Seq are analysed. This leads to some paradoxical results where circadian rhythms are apparently found to persist in the absence of a core circadian clock. Computational solutions to these problems include JTK-cycle, which is freely available, widely used and understood, and does address some of the false discovery concerns, as well as detecting rhythmic species with low amplitude. In addition to the analysis of data, care needs to be taken in the experimental set up, with attempts made to define minimal standards for the elucidation of circadian signals [55]. Until recently it was not possible to directly compare circadian rhythmicity between states, rather rhythmicity was defined for each, and differences were inferred from differentially rhythmic lists. More recently new computational approaches have been developed which allow direct comparison, including compareRhythms, and DryR [55]. These new approaches will make circadian science more robust in the future.

The response to high fat diet obesity has been studied extensively. Here, a number of changes in circadian rhythmicity in the liver are reported, and analysis suggests that a gain of rhythmicity results from a gain of PPARα function. The numerous links between the core cellular circadian clock and energy sensing also suggest the importance of the clock in coping with fluctuating energy supply in complex organisms. Here, AMPKA activity, a cellular energy-deprivation sensor, modifies components of the core clock negative feedback loop, including CRYPTOCHROMES, and thereby reduces their stability [56, 57]. This leads to a gain in circadian amplitude, and the proposal that fasting or starvation may lead to a gain in circadian amplitude, and that perhaps this offers a survival advantage under conditions of low food supply. Therefore, a loss of circadian amplitude was proposed to result from obesity, and loss of AMPKA activity. However, a number of newly rhythmic genes were found in liver to be PPARα targets, suggesting acquisition of a new circuit, linking free fatty acid mobilisation and circadian rhythmic gene expression in the liver. In other systems the emergence of new transcriptional master regulators of state with major impacts on circadian rhythmicity have emerged, including KLF4 in monocytes as a driver of age-related changes in rhythmicity, and STAT3 in liver in response to distal cancer [58, 59].

8.4 Circadian Impacts of NHR Function

As indicated above the function of NHRs may be affected by circadian controls on ligand abundance, or receptor expression, but in addition cross-talk between the clock machinery and NHRs is extensive. It appears that components of the negative feedback arm of the clock are capable of binding to and regulating the function of many NHRs. Here, there is evidence for PER, CRY, and REVERB proteins [2, 6–8, 60] (Fig. 8.3). The profile of target NHRs is surprisingly broad, to the extent that it is not possible to consider NHR action in the absence of considering circadian factors.

The GR has received close attention with regard to timing, in part because its actions on metabolism are so dependent on time of day, and the cycles between fed and fasted state. In seminal work the CRYPTOCHROME proteins were found to bind to the GR, and serve as transcriptional co-repressors. Essentially the CRYPTOCHROMES were conferring the same repressive influence of the GR that they do on the BMAL1/CLOCK heterodimer. It was further shown that loss of the CRYPTOCHROMES resulted in a gain in glucocorticoid transactivation, but not transrepression. This result is supported by earlier ChIP-SEQ analysis for the

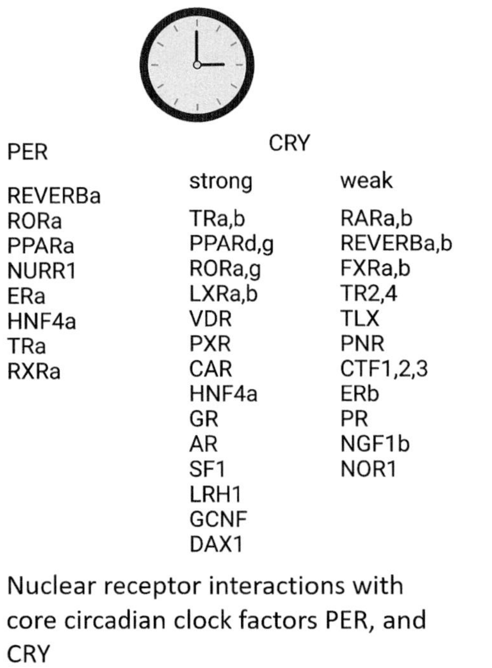

Fig. 8.3 Nuclear receptor interactions with core circadian clock factors PER, and CRY

CRYPTOCHROME cistrome in liver, which was found to be very extensive, and significantly enriched by nuclear receptor binding sequences. Further crosstalk between the clock and the GR was found to involve REVERBa. Here, GR transactivation in the liver was found to vary by time of day, but in a REVERBa dependent manner, with GR and REVERB binding sites found to be closely aligned [7]. A number of the metabolic actions of GR in the liver were found to be time of day dependent, and importantly to be co-regulated by GR, and REVERBa. In further analysis of the time-dependent GR action loss of REVERB caused a phase inversion of some of the glucocorticoid gene regulatory effects, effectively turned night regulated gene to daytime regulated. The GR is a post-translationally modified protein, and it was detected to cycle between nucleus and cytoplasm, and to under circadian phase dependent phosphorylation [61]. Further hints to the complex cross-regulation by clock on GR is the data identifying GR acetylation by the CLOCK core clock transcription factor [62].

8.5 Nuclear Hormone Receptor Chronotherapeutics

The idea that timing of drug administration impacts on efficacy, and side effect profile has been around for decades. Good examples include timing statin use at night-time to target the strongly circadian HMG CoA reductase, and the administration of glucocorticoids in the morning to avoid sleep disruption. However, with new insights into circadian control of NHR expression, and action, it is timely to re-visit this concept.

Regulation of liver lipid metabolism is strongly circadian, and also under very strong fed/fasted control. The emergence of PPARα as a dominant regulator of circadian liver gene expression in obesity in animal models suggests that therapeutic targeting of the PPARs should consider time of day. The anti-inflammatory actions of GR are a major drug target, with data emerging from new analysis of human trials suggesting that time of day affects anti-inflammatory efficacy. This has been further tested by using re-formulated prednisolone, to alter the pharmacokinetics of the steroid [63]. In this way enhanced efficacy against symptoms of rheumatoid arthritis was detected. In addition, analysis of old trials using inhaled glucocorticoids for asthma revealed that time of day affected therapeutic response, with immediate implications for steroid sparing therapeutic strategies.

However, there are practical difficulties in embedding clock logic into NHR therapeutics. It is hard to persuade people to take their prescribed medication at all, let alone at specific times of day. In addition, an error in time-of-day dosing that results in increased toxicity is a major risk to patient safety. Nonetheless it is likely that some promising drug candidates fail in clinic due to testing at the wrong time of day.

This was recently identified in the context of new therapies for stroke, which has seen many promising approaches with pre-clinical model support fail in clinic.

8.6 Conclusions

There is a tight co-regulatory link between the core circadian clock machinery and the family of NHRs. Analysis and interpretation of NHR experiments should take into account the clock time, or circadian phase. Reporting of in-vivo model data, and human clinical trials will need a clear description of timing, to allow maximal benefit to flow. The pervasive influence of the circadian clock on physiology and health are exemplified by the excess risks of metabolic disease and cancer seen in shiftworkers, who can be considered to be living against their circadian clock, and are misaligned with the external light-dark cycle. In addition, testing for biomarkers requires a record of clock time, and ideally in the future some measure of the underlying circadian phase of the donor. In this way we have identified how clinically used biomarkers can vary by time of day so that treatment decisions become captured by the time of clinical assessment. This field is mature in terms of the underlying biology, and mechanisms, but the translation to the clinic and to clinical trials remains undeveloped. The challenge now is to bring the promise recognised by the Nobel prize committee in 2017 to the benefit of human populations.

References

1. Allada R, Bass J (2021) Circadian mechanisms in medicine. N Engl J Med 384:550–561. https://doi.org/10.1056/NEJMra1802337
2. Yang X, Lamia KA, Evans RM (2007) Nuclear receptors, metabolism, and the circadian clock. Cold Spring Harb Symp Quant Biol 72:387–394. https://doi.org/10.1101/sqb.2007.72.058
3. Yang X et al (2006) Nuclear receptor expression links the circadian clock to metabolism. Cell 126:801–810
4. Zhang R, Lahens NF, Ballance HI, Hughes ME, Hogenesch JB (2014) A circadian gene expression atlas in mammals: implications for biology and medicine. Proc Natl Acad Sci U S A 111:16219–16224. https://doi.org/10.1073/pnas.1408886111
5. Koike N et al (2012) Transcriptional architecture and chromatin landscape of the core circadian clock in mammals. Science 338:349–354. https://doi.org/10.1126/science.1226339
6. Kriebs A et al (2017) Circadian repressors CRY1 and CRY2 broadly interact with nuclear receptors and modulate transcriptional activity. Proc Natl Acad Sci U S A 114:8776–8781. https://doi.org/10.1073/pnas.1704955114
7. Caratti G et al (2018) REVERBa couples the circadian clock to hepatic glucocorticoid action. J Clin Invest 128:4454–4471. https://doi.org/10.1172/JCI96138
8. Schmutz I, Ripperger JA, Baeriswyl-Aebischer S, Albrecht U (2010) The mammalian clock component PERIOD2 coordinates circadian output by interaction with nuclear receptors. Genes Dev 24:345–357. https://doi.org/10.1101/gad.564110
9. Carter SJ et al (2016) A matter of time: study of circadian clocks and their role in inflammation. J Leukoc Biol 99:549–560. https://doi.org/10.1189/jlb.3RU1015-451R
10. Jagannath A et al (2021) Adenosine integrates light and sleep signalling for the regulation of circadian timing in mice. Nat Commun 12:2113. https://doi.org/10.1038/s41467-021-22179-z
11. Foster RG, Hankins MW, Peirson SN (2007) Light, photoreceptors, and circadian clocks. Methods Mol Biol 362:3–28. https://doi.org/10.1007/978-1-59745-257-1_1
12. Foster RG, Helfrich-Forster C (2001) The regulation of circadian clocks by light in fruitflies and mice. Philos Trans R Soc Lond Ser B Biol Sci 356:1779–1789. https://doi.org/10.1098/rstb.2001.0962
13. Takahashi JS (2017) Transcriptional architecture of the mammalian circadian clock. Nat Rev Genet 18:164–179. https://doi.org/10.1038/nrg.2016.150
14. Balsalobre A et al (2000) Resetting of circadian time in peripheral tissues by glucocorticoid signaling. Science 289:2344–2347
15. Cho H et al (2012) Regulation of circadian behaviour and metabolism by REV-ERB-alpha and REV-ERB-beta. Nature 485:123–127
16. Preitner N et al (2002) The orphan nuclear receptor REV-ERBalpha controls circadian transcription within the positive limb of the mammalian circadian oscillator. Cell 110:251–260. https://doi.org/10.1016/s0092-8674(02)00825-5
17. Sinturel F et al (2021) Circadian hepatocyte clocks keep synchrony in the absence of a master pacemaker in the suprachiasmatic nucleus or other extrahepatic clocks. Genes Dev 35:329–334. https://doi.org/10.1101/gad.346460.120
18. Schibler U et al (2015) Clock-talk: interactions between central and peripheral circadian oscillators in mammals. Cold Spring Harb Symp Quant Biol 80:223–232. https://doi.org/10.1101/sqb.2015.80.027490

19. Gerber A et al (2015) The systemic control of circadian gene expression. Diabetes Obes Metab 17(Suppl 1):23–32. https://doi.org/10.1111/dom.12512
20. Gerber A et al (2013) Blood-borne circadian signal stimulates daily oscillations in actin dynamics and SRF activity. Cell 152:492–503. https://doi.org/10.1016/j.cell.2012.12.027
21. Maidstone R et al (2021) Shift work is associated with positive COVID-19 status in hospitalised patients. Thorax 76:601–606. https://doi.org/10.1136/thoraxjnl-2020-216651
22. Maidstone RJ et al (2021) Night shift work is associated with an increased risk of asthma. Thorax 76:53–60. https://doi.org/10.1136/thoraxjnl-2020-215218
23. Daghlas I et al (2019) Sleep duration and myocardial infarction. J Am Coll Cardiol 74:1304–1314. https://doi.org/10.1016/j.jacc.2019.07.022
24. Tam SKE et al (2021) Dim light in the evening causes coordinated realignment of circadian rhythms, sleep, and short-term memory. Proc Natl Acad Sci U S A 118. https://doi.org/10.1073/pnas.2101591118
25. Vetter C et al (2018) Night shift work, genetic risk, and type 2 diabetes in the UK biobank. Diabetes Care 41:762–769. https://doi.org/10.2337/dc17-1933
26. Pariollaud M, Lamia KA (2020) Cancer in the fourth dimension: what is the impact of circadian disruption? Cancer Discov 10:1455–1464. https://doi.org/10.1158/2159-8290.CD-20-0413
27. Hunter AL et al (2020) Nuclear receptor REVERBα is a state-dependent regulator of liver energy metabolism. Proc Natl Acad Sci U S A 117:25869–25879. https://doi.org/10.1073/pnas.2005330117
28. Crosby P et al (2019) Insulin/IGF-1 drives PERIOD Synthesis to entrain circadian rhythms with feeding time. Cell 177:896–909 e820. https://doi.org/10.1016/j.cell.2019.02.017
29. Eckel-Mahan KL et al (2013) Reprogramming of the circadian clock by nutritional challenge. Cell 155:1464–1478. https://doi.org/10.1016/j.cell.2013.11.034
30. Preidis GA, Kim KH, Moore DD (2017) Nutrient-sensing nuclear receptors PPARalpha and FXR control liver energy balance. J Clin Invest 127:1193–1201. https://doi.org/10.1172/JCI88893
31. Goldstein I et al (2017) Transcription factor assisted loading and enhancer dynamics dictate the hepatic fasting response. Genome Res 27:427–439. https://doi.org/10.1101/gr.212175.116
32. Arble DM et al (2015) Impact of sleep and circadian disruption on energy balance and diabetes: a summary of workshop discussions. Sleep 38:1849–1860. https://doi.org/10.5665/sleep.5226
33. Dashti HS et al (2015) Habitual sleep duration is associated with BMI and macronutrient intake and may be modified by CLOCK genetic variants. Am J Clin Nutr 101:135–143. https://doi.org/10.3945/ajcn.114.095026
34. Mattson MP et al (2014) Meal frequency and timing in health and disease. Proc Natl Acad Sci U S A 111:16647–16653. https://doi.org/10.1073/pnas.1413965111
35. Wilkinson M et al (2019) Circadian rhythm of exhaled biomarkers in health and asthma. Eur Respir J 54. https://doi.org/10.1183/13993003.01068-2019
36. Durrington HJ et al (2018) Time of day affects eosinophil biomarkers in asthma: implications for diagnosis and treatment. Am J Respir Crit Care Med 198:1578–1581. https://doi.org/10.1164/rccm.201807-1289LE
37. Skene DJ et al (2018) Separation of circadian- and behavior-driven metabolite rhythms in humans provides a window on peripheral oscillators and metabolism. Proc Natl Acad Sci U S A 115:7825–7830. https://doi.org/10.1073/pnas.1801183115
38. Zhang Z et al (2019) Genome-wide effect of pulmonary airway epithelial cell-specific Bmal1 deletion. FASEB J 33:6226–6238. https://doi.org/10.1096/fj.201801682R
39. West AC et al (2017) Misalignment with the external light environment drives metabolic and cardiac dysfunction. Nat Commun 8:417. https://doi.org/10.1038/s41467-017-00462-2
40. Chang AM et al (2019) Chronotype genetic variant in PER2 is associated with intrinsic circadian period in humans. Sci Rep 9:5350. https://doi.org/10.1038/s41598-019-41712-1
41. Jones SE et al (2019) Genome-wide association analyses of chronotype in 697,828 individuals provides insights into circadian rhythms. Nat Commun 10:343. https://doi.org/10.1038/s41467-018-08259-7
42. Duez H, Staels B (2008) The nuclear receptors Rev-erbs and RORs integrate circadian rhythms and metabolism. Diab Vasc Dis Res 5:82–88
43. He B et al (2016) The small molecule Nobiletin targets the molecular oscillator to enhance circadian rhythms and protect against metabolic syndrome. Cell Metab 23:610–621. https://doi.org/10.1016/j.cmet.2016.03.007
44. Chen Z, Yoo SH, Takahashi JS (2018) Development and therapeutic potential of small-molecule modulators of circadian systems. Annu Rev Pharmacol Toxicol 58:231–252. https://doi.org/10.1146/annurev-pharmtox-010617-052645
45. Yin L, Wu N, Lazar MA (2010) Nuclear receptor Rev-erbalpha: a heme receptor that coordinates circadian rhythm and metabolism. Nucl Recept Signal 8:e001
46. Meng QJ et al (2008) Ligand modulation of REV-ERBalpha function resets the peripheral circadian clock in a phasic manner. J Cell Sci 121:3629–3635. https://doi.org/10.1242/jcs.035048
47. Trump RP et al (2013) Optimized chemical probes for REV-ERBα. J Med Chem 56:4729–4737. https://doi.org/10.1021/jm400458q
48. Solt LA et al (2012) Regulation of circadian behaviour and metabolism by synthetic REV-ERB agonists. Nature 485:62–68
49. Dierickx P et al (2019) SR9009 has REV-ERB-independent effects on cell proliferation and metabolism. Proc Natl Acad Sci U S A 116:12147–12152. https://doi.org/10.1073/pnas.1904226116
50. Zhu B et al (2015) Coactivator-dependent oscillation of chromatin accessibility dictates circadian gene

amplitude via REV-ERB loading. Mol Cell 60:769–783. https://doi.org/10.1016/j.molcel.2015.10.024
51. Ince LM et al (2019) Circadian variation in pulmonary inflammatory responses is independent of rhythmic glucocorticoid signaling in airway epithelial cells. FASEB J 33:126–139. https://doi.org/10.1096/fj.201800026RR
52. Le Minh N, Damiola F, Tronche F, Schutz G, Schibler U (2001) Glucocorticoid hormones inhibit food-induced phase-shifting of peripheral circadian oscillators. EMBO J 20:7128–7136. https://doi.org/10.1093/emboj/20.24.7128
53. Gibbs J et al (2014) An epithelial circadian clock controls pulmonary inflammation and glucocorticoid action. Nat Med 20:919–926. https://doi.org/10.1038/nm.3599
54. Dickmeis T et al (2007) Glucocorticoids play a key role in circadian cell cycle rhythms. PLoS Biol 5:e78. https://doi.org/10.1371/journal.pbio.0050078
55. Weger BD et al (2021) Systematic analysis of differential rhythmic liver gene expression mediated by the circadian clock and feeding rhythms. Proc Natl Acad Sci U S A 118. https://doi.org/10.1073/pnas.2015803118
56. Jordan SD, Lamia KA (2013) AMPK at the crossroads of circadian clocks and metabolism. Mol Cell Endocrinol 366:163–169. https://doi.org/10.1016/j.mce.2012.06.017
57. Lamia KA et al (2009) AMPK regulates the circadian clock by cryptochrome phosphorylation and degradation. Science 326:437–440. https://doi.org/10.1126/science.1172156
58. Masri S et al (2016) Lung adenocarcinoma distally rewires hepatic circadian homeostasis. Cell 165:896–909. https://doi.org/10.1016/j.cell.2016.04.039
59. Blacher E et al (2022) Aging disrupts circadian gene regulation and function in macrophages. Nat Immunol 23:229–236. https://doi.org/10.1038/s41590-021-01083-0
60. Lamia KA et al (2011) Cryptochromes mediate rhythmic repression of the glucocorticoid receptor. Nature 480:552–556. https://doi.org/10.1038/nature10700
61. Robles MS, Humphrey SJ, Mann M (2017) Phosphorylation is a central mechanism for circadian control of metabolism and physiology. Cell Metab 25:118–127. https://doi.org/10.1016/j.cmet.2016.10.004
62. Nader N, Chrousos GP, Kino T (2009) Circadian rhythm transcription factor CLOCK regulates the transcriptional activity of the glucocorticoid receptor by acetylating its hinge region lysine cluster: potential physiological implications. FASEB J 23:1572–1583. https://doi.org/10.1096/fj.08-117697
63. Buttgereit F et al (2008) Efficacy of modified-release versus standard prednisone to reduce duration of morning stiffness of the joints in rheumatoid arthritis (CAPRA-1): a double-blind, randomised controlled trial. Lancet 371:205–214

Vitamin D and Gut Health

9

James C. Fleet

Abstract

Vitamin D is a conditionally required nutrient that can either be obtained from skin synthesis following UVB exposure from the diet. Once in the body, it is metabolized to produce the endocrine hormone, 1,25 dihydroxyvitamin D (1,25(OH)2D), that regulates gene expression in target tissues by interacting with a ligand-activated transcription factor, the vitamin D receptor (VDR). The first, and most responsive, vitamin D target tissue is the intestine. The classical intestinal role for vitamin D is the control of calcium metabolism through the regulation of intestinal calcium absorption. However, studies clearly show that other functions of the intestine are regulated by the molecular actions of 1,25$(OH)_2$ D that are mediated through the VDR. This includes enhancing gut barrier function, regulation of intestinal stem cells, suppression of colon carcinogenesis, and inhibiting intestinal inflammation. While research demonstrates that there are both classical, calcium-regulating and non-calcium regulating roles for vitamin D in the intestine, the challenge facing biomedical researchers is how to translate these findings in ways that optimize human intestinal health.

Keywords

Vitamin D · Calcium · Absorption · Inflammation · Stem cell · Tight junction · Cancer · VDR

9.1 Introduction

In 1922, E.V. McCollum first coined the term “vitamin D” to describe the fat-soluble vitamin with a critical role in bone health. However, by 1937, work by Nicolaysen made it clear that intestinal calcium absorption is dependent on vitamin D [1] and others showed that intestinal calcium absorption efficiency is reduced by more than 75% during vitamin D deficiency [2, 3]. The molecular era of vitamin D research began when the active metabolite of vitamin D, 1,25 dihydroxyvitamin D (1,25$(OH)_2D_3$) [4, 5], and its nuclear receptor, the vitamin D receptor (VDR) [6], were isolated from the intestinal mucosa. Since then, research on the molecular actions of vitamin D has revealed how 1,25$(OH)_2D_3$ acts through the VDR to regulate gene transcription (see [7, 8] for a detailed discussion of this topic). While the highest expression of VDR is seen in the intestinal epithelium [9, 10], VDR protein and VDR-mediated gene expression has been

J. C. Fleet (✉)
Department of Nutritional Sciences, Dell Pediatric Research Institute, University of Texas, Austin, TX, USA
e-mail: james.fleet@austin.utexas.edu

M. J. Campbell, C. L. Bevan (eds.), *Nuclear Receptors in Human Health and Disease*, Advances in Experimental Medicine and Biology 1390, https://doi.org/10.1007/978-3-031-11836-4_9

identified is many different tissues [11]. In the intestine, VDR gene expression is regulated by glucocorticoids [12] and estrogens [13], increases in the late post-natal period [12, 14], and declines with aging [15, 16]. In this chapter, I will build upon the critical role that vitamin D signaling has on specific intestinal target cells. This information is critical to understand the biological role that vitamin D has on bone health, colon cancer, and inflammatory bowel disease.

9.2 Classical Role of Vitamin D as a Regulator of Intestinal Ca Absorption

A number of studies show that vitamin D-mediated intestinal calcium absorption is the single most important role for vitamin D and VDR during growth. Global VDR gene knockout reduces calcium absorption efficiency by 70% in growing mice [17, 18] and this causes reduced serum calcium, high serum levels of both $1,25(OH)_2D_3$ and PTH, and osteomalacia. Mice with intestine-specific VDR deletion also have the same phenotype as global VDR knockout mice [19], thereby demonstrating the critical importance of intestine for whole body calcium metabolism. As proof of this concept, my research group found that intestine-specific transgenic expression of VDR could normalize calcium absorption efficiency in VDR knockout mice and this was enough to prevent the changes in serum PTH, serum calcium, and bone mineral density that is normally seen in these animals [20].

Careful examination of basal and vitamin D-regulated calcium absorption in rodents and in Caco-2 cells shows that calcium movement across the intestinal barrier occurs through both saturable (transcelluar) and non-saturable (paracelluar) pathways [21–24]. $1,25(OH)_2D_3$ regulates the saturable component of calcium absorption [2, 25–27] and this pathway is energy dependent [28], highest in the proximal small intestine (i.e. the duodenum and proximal jejunum) but also occurs in the large intestine [29–33]. Several groups have shown that VDR expression in the colon is also necessary for normal calcium homeostasis [34, 35] . A comprehensive review of vitamin D mediated Ca absorption is available elsewhere [36].

(a) *Models of vitamin D regulated intestinal Ca absorption:* The best studied model to describe vitamin D-induced Ca absorption is the facilitated diffusion model [37]. In this model, the transient receptor potential cation channel vanilloid family member 6 (TRPV6) mediates basal and vitamin D-induced apical membrane calcium uptake [38–41]. Although $1,25(OH)_2D_3$-induced intestinal calcium absorption was not reduced in TRPV6 knockout mice [42, 43], the increase in calcium absorption induced by a low Ca diet was reduced in mice with a non-functional D541A variant TRPV6 [44]. Also, my group has shown that intestine-specific transgenic expression of TRPV6 increased Ca absorption and recovered the abnormal bone phenotype VDR knockout mice [45], thus proving that TRPV6 is a bona fide mediator of intestinal Ca uptake. The proposed mediator of intracellular diffusion of calcium during absorption is the cytoplasmic calcium binding protein calbindin D_{9k} [37]. However calbindin D_{9k} is not essential for basal or vitamin D regulated Ca absorption [43, 46] and data from a number of studies suggest that calbindins are more likely intracellular calcium buffers than intracellular calcium ferries [47] [41] [45]. The final step in the facilitated diffusion model is the extrusion of calcium from the cell, a process that is mediated by the plasma membrane calcium ATPase 1b (PMCA1b) [28, 48, 49]. Deletion of PMCA1b (Atp2b1) or 4.1R, a protein that stabilizes PMCA1b in the basolateral membrane, reduces both basal and $1,25(OH)_2D_3$-induced intestinal calcium absorption [50, 51].

Several other models for vitamin D regulated intestinal Ca absorption exist and have interesting features, but are less well supported by data than the facilitated diffusions model. In the vesicular transport model, Ca is sequestered into vesicles within the cell as an alternative to the ferry/buffer role proposed for calbindin D. Consistent with a role for vesicles in Ca absorption, $1,25(OH)_2D_3$ treatment increases the number of lysosomes in chick intestine [52], the release of lysosomal enzymes from isolated rat enterocytes [53], the cycling of lysosomes [54], and the level of lysosomal calcium [55]. Although these data support a role for vesicular movement during intestinal Ca absorption, it isn't clear what makes it specific for calcium. Transcaltachia has been described as the rapid absorption of calcium that occurs after exposing chick enterocytes to $1,25(OH)_2D_3$ [56]. Transcaltachia occurs only in response to serosal $1,25(OH)_2D_3$ exposure which suggests either that VDR has a novel membrane signaling role [57] or that transcaltachia is mediated by a multi-functional protein, the membrane associated rapid response steroid binding protein (MARRS) [58]. Intestine-specific deletion of MARRS in mice reduced cellular $1,25(OH)_2D_3$ binding, disrupted $1,25(OH)_2D_3$ regulated calcium and phosphate uptake into isolated enterocytes [59, 60], and reduced basal calcium absorption in by 30% [61]. However, there have been no reported adverse effects of MARRS deletion on bone, despite the critical importance of vitamin D mediated intestinal calcium absorption for normal bone growth [20]. In addition to the transcellular calcium absorption models, some studies show that vitamin D signaling increases paracellular Ca transport in the jejunum and ileum [24, 62, 63] due to $1,25(OH)_2D_3$-mediated induction of two tight junction proteins, claudin 2 and claudin 12 [64]. This may be why non-saturable ileal calcium absorption is reduced in chronic renal disease patients with low serum $1,25(OH)_2D_3$ levels [24].

Finally, we have conducted research that suggests additional mechanisms may control vitamin D-regulated Ca absorption. By using a forward-genetics approach in recombinant inbred lines from a cross of C57BL/6 J and DBA/2 J (BXD) mice, we mapped multiple loci where genetic variation controls intestinal Ca absorption [65]. None of these loci contained genes that encode the proteins that are central to the Ca absorption models described above. As such, our genetic mapping study suggests that novel mechanisms for Ca absorption exist that have not yet been described.

9.3 Gut Absorption and Excretion of Vitamin D

While the intestine is a target organ for $1,25(OH)_2D_3$ action, it is also important for the management of vitamin D status by mediating absorption of dietary/supplemental vitamin D and by mediating the excretion of vitamin D metabolites.

Gastrointestinal and hepatobiliary diseases that cause fat malabsorption also cause vitamin D deficiency in humans [66]. This suggests that vitamin D "follows the fat" during its intestinal absorption, i.e. it is incorporated into micelles, repackaged into chylomicrons, and absorbed into the lymphatic system. Consistent with the "follow the fat" model, when rats were given radiolabeled vitamin D, the label appeared within chylomicrons in the lymph [67–69] and this required the presence of bile acids [70, 71]. In contrast, recent studies show that intestinal absorption of vitamin D may also be an active process that requires the cholesterol transporters SR-BI and NPC1L1 [72]. Regardless of the route of absorption, there doesn't appear to be any regulation of vitamin D absorption. Rats with experimental nephrotic syndrome lose large amounts of vitamin D metabolites in urine and have reduced serum 25OHD levels [73] but even under these conditions, intestinal vitamin D absorption is not elevated. This suggests there is no homeostatic mechanism to upregulate vitamin D absorption in times of need.

There is no evidence to suggest vitamin D_2 and vitamin D_3 are absorbed by different mechanisms. However, 25 hydroxyvitamin D (25OHD) and $1,25(OH)_2D_3$ are absorbed more efficiently than vitamin D [69, 70] and 25OHD is better absorbed than vitamin D in subjects with stearorrhea [74]. This suggests that hydroxylated vitamin D metabolites don't use the fat absorption pathway. Instead, the higher absorption efficiency for 25OHD is due to chylomicro-independent absorption [70] into the lymph where it is associated with an alpha globulin like the Vitamin D Binding Protein (DBP) [69].

While 1,25(OH)$_2$ D can be metabolized to the terminal compound calcitroic acid [75], 1,25(OH)$_2$ D can also be sulfonated and glucuronidated in the liver [76, 77]. These metabolites are then excreted through the bile [78], which is the primary route of excretion for vitamin D metabolites. Although the modified 1,25(OH)$_2$D$_3$ forms are not biologically active, the glucuronide residue can be removed by colonic micro-organisms and act locally [79]. This releases the active 1,25(OH)$_2$D$_3$ in the colonic lumen which can then either be reabsorbed (i.e. making an entero-hepatic cycle [80, 81]) or act locally on colonocytes. In fact, while duodenal gene expression is strongly upregulated by increases in circulating 1,25(OH)$_2$D$_3$, colonic gene expression is more strongly upregulated by apical delivery of the hormone [82]. As such, the release of glucuronidated 1,25(OH)$_2$D$_3$ into the bile may be an important mechanism for activating vitamin D mediated gene expression in the colon.

9.4 Cellular Targets of Vitamin D Action in the Intestine

Although the bulk of cells in the intestine are epithelial, it is important to recognize that there is significant diversity in the cell populations that exist within the intestine. At the base of the crypt are multipotent stem cells that give rise to proliferating daughter cells that then receive signals to differentiate into either absorptive epithelial cells or to secretory lineage cells (i.e. goblet cells and enteroendocrine cells) [83]. In addition, underneath the epithelial layer are stromal cells, vascular endothelial cells, and gut associated immune cells. Nonetheless, the bulk of research conducted on vitamin D action in the intestine has focused on the epithelial cells. In this section I will discuss the molecular actions of vitamin D that affect the intestinal epithelium, the stem cells, and, briefly, the gut associated immune cells.

(a) <u>Molecular Targets in Intestinal Epithelial Cells.</u> Before the genomics era, only a few genes involved in intestinal Ca absorption had been examined for transcriptional regulation following 1,25(OH)$_2$D$_3$ -dependent activation of the VDR (e.g. [84, 85]). However, genomic profiling permits a more comprehensive view of vitamin D action on the intestine.

The earliest genomic report was a microarray experiment using 1,25(OH)$_2$D$_3$ -treated (24 h, 100 nM) Caco-2 cells that had been differentiated in culture to resemble the cells of the small intestinal villus [86]. Using an Affymetrix array (12,635 probesets), this report identified 234 probesets that were expressed in all samples, significant at $p < 0.05$, and differentially expressed by vitamin D treatment; only 13 of these probesets (representing 12 distinct genes) changed by more than two-fold. This analysis identified several known vitamin D regulated genes (i.e. CYP24A1) but also potential new vitamin D target genes like amphiregulin, ceruloplasmin, sorcin, and Jun b. As such, this was important "proof of principle" that vitamin D has broader effects on intestinal biology than simply to regulate intestinal Ca absorption.

The largest reported genomic analysis of vitamin D action reported to date is one by Lee et al. [87] who examined the impact of 1,25(OH)$_2$D$_3$ treatment (10 ng/g BW, 6 h) on small intestinal gene expression in CYP27B1 knockout mice using RNA-seq. In mice fed a normal diet, 1,25(OH)$_2$D$_3$ regulated 599 genes while in mice fed a rescue diet to prevent hypocalcemia, it regulated 119 genes. 45 genes were in common across the two diet groups (86% up-regulated), including Cyp24a1, Trpv6, S100G, and Atp2b1. Ion binding was the most enriched GO term for the full vitamin D-regulated gene list, reflecting the large number of mineral transporters that were upregulated. They subsequently looked for enrichment of VDR binding to DNA following 1,25(OH)$_2$D$_3$ treatment (10 ng/g BW, 1 h) of wild-type mice using ChIP-seq. This revealed more than 4000 basal and 17,000 vitamin D-induced VDR binding sites. The genes for the bulk of the vitamin D regulated transcripts identified in Cyp27b1 KO mice had VDR binding sites associated with them (75% for normal diet, 84% for rescue diet; 87% for the overlapping genes). This included binding sites in all of the traditional intestinal vitamin D target genes (Trpv6, S100g, Apt2b1, Cyp24a1, Cldn2) and for a number of new target genes that includes transporters like Slc30a10 (Mn export), Slc30a1 (the Zn exporter ZnT1), Slc30a5 (the Zn transporter ZnT5) and Slc37a2 (a glucose-6-phosphate transporter), ion channels/sensors like Lrrc26 and Mctp2, and transcription factors like Pdx1, Bach1, and Ppard. The breadth of the functions of these new gene targets suggests vitamin D signaling may control some aspects of lipid metabolism (e.g. Pdx1 and Ppard), mineral toxicity (Slc30a10, Slc30a1, Slc30a5), and the biological response to oxidative stress (Bach1).

Building from the observations of Lee et al. [87], we generated RNA-seq data that reveals vitamin D has distinct gene targets depending upon the state of cell differentiation or the intestinal segment examined. For this, we cultured human duodenal stem cells under conditions that either promote differentiation (to make villus-like enteroids) or to maintain the proliferating stem-cell properties of the culture [88]. When treated with $1,25(OH)_2D_3$ (10 nM, 24 h), the villus-like enteroids had 387 differentially regulated transcripts while the undifferentiated cultures had 130 differentially regulated transcripts; 86 transcripts were in both groups and this overlap group included the classic intestinal vitamin D gene targets. Critically, this experiment demonstrated that intestinal differentiation alters the vitamin D target gene profile and suggests that *in vivo* studies that use mucosal scrapings underestimate the complexity of intestinal vitamin D action. Consistent with this, we have unpublished data that show distinct differences in the vitamin D regulated transcriptome among the small intestine crypts, small intestine villi, and colon epithelium. While this work confirms some of the earlier target genes from Lee et al. [87], it provides more clarity that some gene regulatory events are specific to different functional compartments of the intestine.

(b) <u>Intestinal stem cells</u>

In the base of the crypts of all intestinal segments, 4–8 multipotent Lgr5+ stem cells exist that are the precursor for all of the epithelial cell types in intestine [89]. In the colon, these cells are also where cancer originates [90]. Lgr5+ stem cells express the VDR and are thus vitamin D target cells [91, 92]. Several groups have recently examined how vitamin D signaling impacts the biology of intestinal stem cells. When Peregrina et al. [91] examined the impact of low vitamin D diets (i.e. the New Western Diet 1 or NWD1) or stem cell specific deletion of VDR on Lgr5+ stem cells they found that the these interventions reduced the percentage of Lgr5+ cells (e.g. by 30% after 3 months on the NWD1) and that there were fewer progeny from Lgr5 cells in the villus of these mice. Others have shown that Bmi1+ cells are a reserve stem cell population in the intestine that expands when the Lgr5+ stem cells are damaged (e.g. following radiation) [93]. Consistent with this, Li et al. [94] found that as the NWD1 reduced Lrg5+ cells, it expanded the population of Bmi1+ cells. Feeding the NWD1 also significantly altered the transcript profile of both Lgr5+ and Bmi1+ stem cells. Collectively, these data suggest that vitamin D signaling is required for the maintenance and balance of healthy intestinal stem cell population.

Consistent with a role for vitamin D in the regulation of stem cell biology, Sittipo et al. [95] found that treatment of small intestinal stem cell cultures with $1,25(OH)_2D_3$ for 3 days increased markers of lineage differentiation for goblet cells (Muc2), Paneth cells (Lyz), enteroendocrine cells (Chga), and epithelial cells (Villin). In addition, $1,25(OH)_2D_3$ reduced the number of budding organoids and that this was associated with fewer Ki-67 and Lgr5-labeled cells as well as reduced proliferation and increased apoptosis. Similarly, Fernandez-Barral et al. [92] reported results from an RNA-seq study that shows $1,25(OH)_2D_3$ treatment (10 d, 100 nM) suppressed cell proliferation in human colonic organoids and promoted a differentiated phenotype in colon tumor organoids (3 d, 100 nM). $1,25(OH)_2D_3$-treatment also regulated a wide variety of genes involved in pathways that control suppression of proliferation and tumorigenesis, promotion of differentiation, and maintenance of stemness. These data fit the traditional model of $1,25(OH)_2D_3$ as an antiproliferative, pro-differentiating agent.

(c) Vitamin D regulates tight junctions and barrier function

In addition to its role in regulating nutrient, drug, and fluid movement into and out of the body, the intestine has a primary role in protecting the body by forming tight junctions between cells [96] and acting as a barrier to foreign invaders [97]. A number of studies have clearly shown that vitamin D signaling is crucial to the maintenance of barrier function under normal and inflammatory conditions. The earliest study to make this connection was by Kong et al. [98] who showed that $1,25(OH)_2D_3$ treatment enhanced transepithelial electrical resistance (TEER) in Caco-2 cell monolayers through a VDR-dependent mechanism that induced expression of the tight junction proteins ZO-1 and claudins, 1, 2, and 5. Consistent with a physiological role for this VDR-dependent strengthening of tight junctions, Kong et al. also found that TEER was reduced sooner and more severely in VDR knockout mice following treatment with dextran sulfate sodium (DSS), an agent that damages the epithelium and induces colitis. Several other studies are consistent with a tight junction protective effect of $1,25(OH)_2D$ in intestine. Zhao et al. [99] found that in Caco-2 cells, $1,25(OH)_2D_3$ increased TEER, protein and mRNA levels for tight junction proteins, and decreased monolayer permeability following DSS treatment. Chen et al. [100] later reported that $1,25(OH)_2D_3$ treatment prevented lipopolysaccharide (LPS)-induced Caco-2 cell monolayer damage and prevented LPS-induced redistribution of tight junction proteins.

Several groups have reported that VDR knockout mice are more susceptible to DSS-induced mucosal injury [98, 101, 102]. However, studies in global VDR knockout mice are confounded by the disruption of calcium metabolism and hair loss-related thermoregulation that are central phenotypes to this model. To overcome this problem, we conducted research in two unique mouse models, a mouse with colon-epithelial cell specific deletion of VDR and a VDR knockout mouse with transgenic expression of VDR in the intestinal epithelium [103]. Using these models, we found that intestinal epithelial cell deletion of VDR made the intestinal epithelium more susceptible to DSS induced damage, but that recovery from the damage was normal. In contrast, loss of VDR in the cells outside of the epithelium prevented recovery from DSS-induced barrier damage. Other data from our group [103] and others [101] suggest that intestinal epithelial healing is regulated by innate immune cells like M2 macrophages, and that healing is enhanced by activating $1,25(OH)_2D_3$ signaling in these cells.

(d) Regulation of gut associated immune cells

While the focus of vitamin D action in the intestine has been on its role in modulating epithelial cell function, the intestine also contains a robust mucosal immune system [104]. However, the impact of vitamin D on this system has not been extensively studied. In contrast, there is a body of literature on the role of vitamin D in the regulation of the systemic immune cells (see [105, 106] for overviews) that is driven by the observation that vitamin D deficiency is associated with increased autoimmunity and an increased susceptibility to infection [107].

Although the findings related to the systemic immune system may not apply to the mucosal immune cells, a brief evaluation of vitamin D's impact on systemic immunity is warranted. The vitamin D receptor is expressed in immune cells (e.g. T and B cells, and antigen presenting cells) and when T cells and monocytes/macrophages are activated they can synthesize $1,25(OH)_2D_3$ and use it as an autocrine signal [108–110]. Activation of vitamin D signaling can impact both the innate and adaptive immune responses.

In the innate immune system $1,25(OH)_2D_3$ stimulates differentiation of monocytes to macrophages [111] and regulates genes crucial for autophagy and anti-microbial actions [112–115]. In addition, it reprograms dendritic cells (DC) to become tolerogenic in an inflammatory setting [116] by altering DC differentiation as well as the function of tolerogenic DC [117]. In adaptive immunity, vitamin D may create a more tolerogenic T helper cell profile. However, neither the number nor the type of T cells are grossly abnormal in mice lacking VDR [118]. Similarly, the function of mature T-cells is not strongly influenced by VDR deletion [119, 120]. This suggests that VDR does not have a primary role for normal T-cell development but that $1,25(OH)_2D_3$ may be a modulator of T-cell mediated immune responses. Consistent with this idea, *in vitro* $1,25(OH)_2D$ suppresses NFkB signaling necessary for T_1 helper cell activation [121] and blocks development of Th_{17} and Th_9 cells implicated in the pathogenesis of different types of autoimmunity and inflammatory diseases [122].

The bulk of research on vitamin D and immunity relevant to the intestine has been on vitamin D's role in reducing the severity or duration of colitis and inflammatory bowel disease. This has been reviewed recently elsewhere (see [123]). However, several research groups have examined the role of vitamin D signaling in the biology of type 3 innate lymphoid cells (ILC3), a gut resident immune cell population that participates in innate defense of the intestinal mucosa by producing IL-17 and IL-22 to regulate the production of antimicrobial agents like beta defensin. An early study by Chen et al. [124] showed that global VDR deletion increased ILC3 cell number in small intestine, increased production of antimicrobial peptides, and caused resistance to C. Rodentium infection. However, several later studies have reported opposite findings. Konya et al. [125] found that the pro-inflammatory cytokines IL-23 and IL-6 increased ILC3 VDR expression and that $1,25(OH)_2D$ treatment suppressed IL-22 and IL-17F production by ILC3 cells. He et al. [126] then found that global or ILC3-specific VDR deletion or vitamin D deficiency reduced colonic ILC3 cell number and proliferation while increasing susceptibility to C. Rodentium infection. These findings were confirmed by Lin et al [127] None of these studies characterized the subtype of ILC3. Thus, while it is clear that vitamin D signaling regulates these gut-resident innate immune cells, it is also clear that additional research is necessary to clarify the molecular mechanisms of action in these cells and the physiologic relevance of this regulation.

9.5 Conclusions

In this chapter I reviewed the cellular and molecular actions of vitamin D in the intestine. It has been clear from the beginning of vitamin D research, that the intestine is an important target tissue. The earliest studies on vitamin D and intestine revealed its critical role as a regulator of intestinal calcium absorption, and thus indirectly in the development and maintenance of bone mass. However, genomics studies clearly show that vitamin D has broader intestinal actions that the regulation of calcium absorption. Vitamin D signaling has distinct actions on intestinal stem cells, one undifferentiated crypt cells, on differentiated villus cells, and on gut associated intestinal cells. Thus, in addition to its important role in calcium metabolism, vitamin D also regulates the cell biology of the intestine in ways that protect the stem cell from cancer, limit gut leakiness, and both suppresses epithelial injury, promotes epithelial recovery from injury, and reduces intestinal inflammation (Summarized in Fig. 9.1). Yet while research suggests a variety of important biological roles for vitamin D in the intestine, the challenge facing biomedical researchers is how to translate these findings in ways that optimize human intestinal health.

Conflicts of Interest Dr. Fleet has no conflicts to report.

Funding The work on this review was supported by NIH grants DK118036 and DK112365 to JCF.

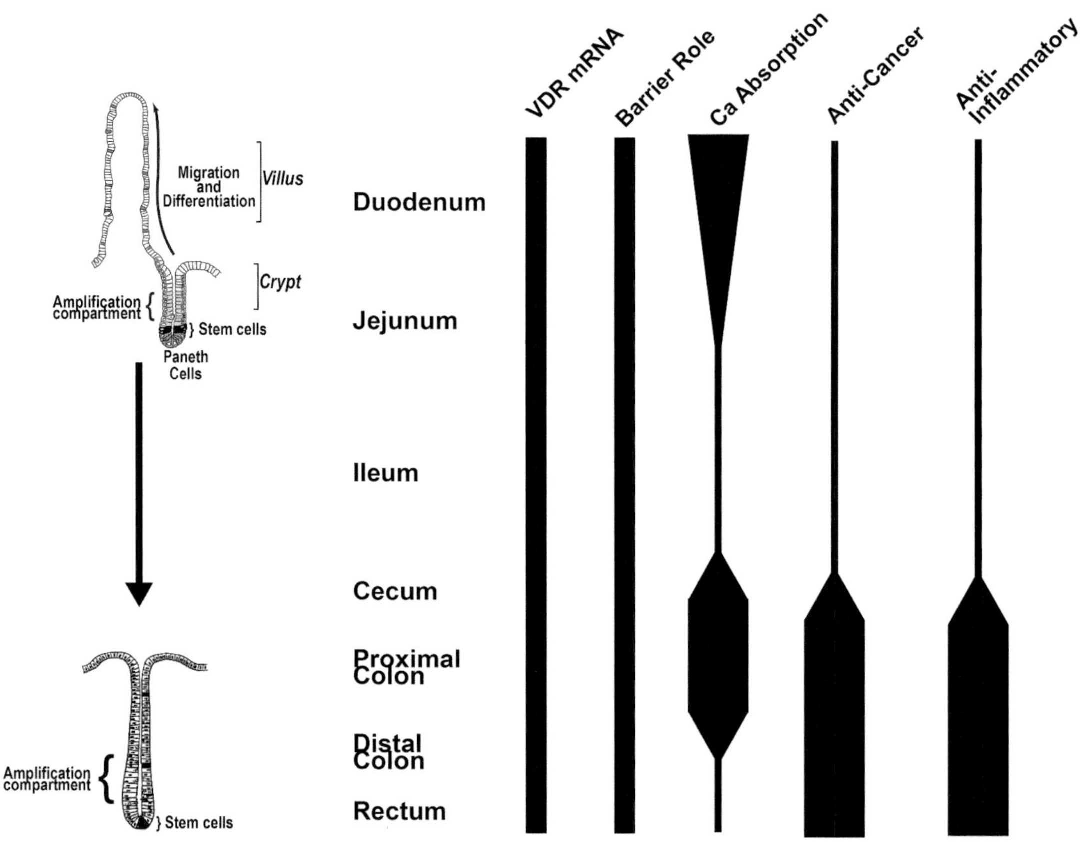

Fig. 9.1 A summary of Vitamin D action across the intestinal tract

References

1. Nicolaysen R (1937) Studies upon the mode of action of Vitamin D. The influence of vitamin D on the absorption of calcium and phosphorus in the rat. Biochem J 37:122–129
2. Pansu D, Bellaton C, Roche C, Bronner F (1983) Duodenal and ileal calcium absorption in the rat and effects of vitamin D. Am J Phys 244(6):G695–G700
3. Sheikh MS, Ramirez A, Emmett M, Santa AC, Schiller LR, Fordtran JS (1988) Role of vitamin D-dependent and vitamin D-independent mechanisms in absorption of food calcium. J Clin Invest 81(1):126–132
4. Holick MF, Schnoes HK, DeLuca HF, Suda T, Cousins RJ (1971) Isolation and identification of 1,25-dihydroxycholecalciferol. A metabolite of vitamin D active in intestine. Biochemistry 10(14):2799–2804
5. Norman AW, Myrtle JF, Midgett RJ, Nowicki HG, Williams V, Popjak G (1971) 1,25-dihydroxycholecalciferol: identification of the proposed active form of vitamin D3 in the intestine. Science 173(3991):51–54
6. Brumbaugh PF, Haussler MR (1973) Nuclear and cytoplasmic receptors for 1,25-dihydroxycholecalciferol in intestinal mucosa. Biochem Biophys Res Commun 51(1):74–80
7. Carlberg C (2017) Molecular endocrinology of vitamin D on the epigenome level. Mol Cell Endocrinol 453:14–21
8. Pike JW, Meyer MB (2014) Fundamentals of vitamin D hormone-regulated gene expression. J Steroid Biochem Mol Biol 144(Pt A):5–11
9. Lee SM, Bishop KA, Goellner JJ, O'Brien CA, Pike JW (2014) Mouse and human BAC transgenes recapitulate tissue-specific expression of the vitamin D receptor in mice and rescue the VDR-null phenotype. Endocrinology 155(6):2064–2076
10. Cartwright JA, Gow AG, Milne E et al (2018) Vitamin D receptor expression in dogs. J Vet Intern Med 32(2):764–774
11. Walters MR (1992) Newly identified actions of the vitamin D endocrine system. Endocr Rev 13(4):719–764
12. Massaro E, Simpson R, DeLuca H (1983) Quantification of endogenously occupied and unoccupied binding sites for 1,25 dihydroxyvita-

min D3 in rat intestine. Proc Natl Acad Sci U S A 80:2549–2553
13. Liel Y, Shany S, Smirnoff P, Schwartz B (1999) Estrogen increases 1,25-dihydroxyvitamin D receptors expression and bioresponse in the rat duodenal mucosa. Endocrinology 140(1):280–285
14. Pierce EA, DeLuca HF (1988) Regulation of the intestinal 1,25-dihydroxyvitamin D3 receptor during neonatal development in the rat. Arch Biochem Biophys 261:241–249
15. Takamoto S, Seino Y, Sacktor B, Liang CT (1990) Effect of age on duodenal 1,25-dihydroxyvitamin D-3 receptors in Wistar rats. Biochim Biophys Acta 1034:22–28
16. Horst RL, Goff JP, Reinhardt TA (1990) Advancing age results in reduction of intestinal and bone 1,25 dihydroxyvitamin D receptor. Endocrinology 126:1053–1057
17. Van Cromphaut SJ, Dewerchin M, Hoenderop JG et al (2001) Duodenal calcium absorption in vitamin D receptor-knockout mice: functional and molecular aspects. Proc Natl Acad Sci U S A 98(23):13324–13329
18. Song Y, Kato S, Fleet JC (2003) Vitamin D receptor (VDR) knockout mice reveal VDR-independent regulation of intestinal calcium absorption and ECaC2 and calbindin D9k mRNA. J Nutr 133(2):374–380
19. Lieben L, Masuyama R, Torrekens S et al (2012) Normocalcemia is maintained in mice under conditions of calcium malabsorption by vitamin D-induced inhibition of bone mineralization. J Clin Invest 122(5):1803–1815
20. Xue YB, Fleet JC (2009) Intestinal Vitamin D receptor is required for Normal calcium and bone metabolism in mice. Gastroenterology 136(4):1317–1327
21. Wasserman RH, Taylor AN (1969) Some aspects of the intestinal absorption of calcium, with special reference to vitamin D. In: Comar CL, Bronner F (eds) Mineral metabolism, an advanced treatise. 3. Academic, New York, pp 321–403
22. Pansu D, Bellaton C, Bronner F (1981) Effect of Ca intake on saturable and nonsaturable components of duodenal Ca transport. Am J Phys 240(1):32–37
23. Heaney RP, Saville PD, Recker RR (1975) Calcium absorption as a function of calcium intake. J Lab Clin Med 85(6):881–890
24. Sheikh MS, Schiller LR, Fordtran JS (1990) In vivo intestinal absorption of calcium in humans. Miner Electrolyte Metab 16(2–3):130–146
25. Chandra S, Fullmer CS, Smith CA, Wasserman RH, Morrison GH (1990) Ion microscopic imaging of calcium transport in the intestinal tissue of vitamin D-deficient and vitamin D-replete chickens: a 44Ca stable isotope study. Proc Natl Acad Sci U S A 87(15):5715–5719
26. Fullmer CS, Chandra S, Smith CA, Morrison GH, Wasserman RH (1996) Ion microscopic imaging of calcium during 1,25-dihydroxyvitamin D-mediated intestinal absorption. Histochem Cell Biol 106(2):215–222
27. Giuliano AR, Wood RJ (1991) Vitamin D-regulated calcium transport in Caco-2 cells: unique in vitro model. Am J Phys 260(2 Pt 1):G207–GG12
28. Favus MJ, Angeid-Backman E, Breyer MD, Coe FL (1983) Effects of trifluoperazine,ouabain, and ethacrynic acid on intestinal calcium. Am J Phys 244:G111–G1G5
29. Favus MJ, Kathpalia SC, Coe FL (1981) Kinetic characteristics of calcium absorption and secretion by rat colon. Am J Phys 240(5):G350–G3G4
30. Favus MJ, Langman CB (1984) Effects of 1,25 dihydroxyvitamin D3 on colonic calcium transport in vitamin D-deficient and normal rats. Am J Phys 246:G268–GG73
31. Karbach U, Rummel W (1987) Calcium transport across the colon ascendens and the influence of 1,25-dihydroxyvitamin D3 and dexamethasone. Eur J Clin Invest 17(4):368–374
32. Karbach U, Feldmeier H (1993) The cecum is the site with the highest calcium absorption in rat intestine. Dig Dis Sci 38(10):1815–1824
33. Barger-Lux MJ, Heaney RP, Recker RR (1989) Time course of calcium absorption in humans: evidence for a colonic component. Calcif Tissue Int 44(5):308–311
34. Christakos S, Dhawan P, Verstuyf A, Verlinden L, Carmeliet G (2016) Vitamin D: metabolism, molecular mechanism of action, and pleiotropic effects. Physiol Rev 96(1):365–408
35. Reyes-Fernandez PC, Fleet JC (2016) Compensatory changes in calcium metabolism accompany the loss of Vitamin D receptor (VDR) from the distal intestine and kidney of mice. J Bone Miner Res 31(1):143–151
36. Fleet JC (2018) Regulation of intestinal calcium and phosphate absorption. In: JWP DF, Bouillon R, Giovannucci E, Goltzman D, Hewison M (eds) Vitamin D. 1, 4th edn. Academic, pp 329–342
37. Bronner F, Pansu D, Stein WD (1986) An analysis of intestinal calcium transport across the rat intestine. Am J Phys 250(5 Pt 1):G561–G5G9
38. Peng JB, Chen XZ, Berger UV et al (1999) Molecular cloning and characterization of a channel-like transporter mediated intestinal calcium absorption. J Biol Chem 274:22739–22746
39. Meyer MB, Zella LA, Nerenz RD, Pike JW (2007) Characterizing early events associated with the activation of target genes by 1,25-dihydroxyvitamin D3 in mouse kidney and intestine in vivo. J Biol Chem 282:22344–22352
40. Fleet JC, Eksir F, Hance KW, Wood RJ (2002) Vitamin D-inducible calcium transport and gene expression in three Caco-2 cell lines. Am J Phys 283(3):G618–GG25
41. Song Y, Peng X, Porta A et al (2003) Calcium transporter 1 and epithelial calcium channel messenger ribonucleic acid are differentially regulated by 1,25 dihydroxyvitamin D3 in the intestine and kidney of mice. Endocrinology 144(9):3885–3894

42. Kutuzova GD, Sundersingh F, Vaughan J et al (2008) TRPV6 is not required for 1alpha,25-dihydroxyvitamin D3-induced intestinal calcium absorption in vivo. Proc Natl Acad Sci U S A 105(50):19655–19659
43. Benn BS, Ajibade D, Porta A et al (2008) Active intestinal calcium transport in the absence of transient receptor potential vanilloid type 6 and calbindin-D9k. Endocrinology 149(6):3196–3205
44. Woudenberg-Vrenken TE, Lameris AL, Weissgerber P et al (2012) Functional TRPV6 channels are crucial for transepithelial Ca2+ absorption. Am J Physiol Gastrointest Liver Physiol 303(7):G879–G885
45. Cui M, Li Q, Johnson R, Fleet JC (2012) Villin promoter-mediated transgenic expression of transient receptor potential cation channel, subfamily V, member 6 (TRPV6) increases intestinal calcium absorption in wild-type and vitamin D receptor knockout mice. J Bone Miner Res 27(10):2097–2107
46. Akhter S, Kutuzova GD, Christakos S, DeLuca HF (2007) Calbindin D9k is not required for 1,25-dihydroxyvitamin D3-mediated Ca2+ absorption in small intestine. Arch Biochem Biophys 460(2):227–232
47. Spencer R, Charman M, Wilson PW, Lawson DEM (1978) The relationship between vitamin D-stimulated calcium transport and intestinal calcium-binding protein in the chicken. Biochem J 170:93–101
48. Wasserman RH, Smith CA, Brindak ME et al (1992) Vitamin-D and mineral deficiencies increase the plasma membrane calcium pump of chicken intestine. Gastroenterology 102(3):886–894
49. Cai Q, Chandler JS, Wasserman RH, Kumar R, Penniston JT (1993) Vitamin D and adaptation to dietary calcium and phosphate deficiencies increase intestinal plasma membrane calcium pump gene expression. Proc Natl Acad Sci U S A 90(4):1345–1349
50. Liu C, Weng H, Chen L et al (2013) Impaired intestinal calcium absorption in protein 4.1R-deficient mice due to altered expression of plasma membrane calcium ATPase 1b (PMCA1b). J Biol Chem 288(16):11407–11415
51. Ryan ZC, Craig TA, Filoteo AG et al (2015) Deletion of the intestinal plasma membrane calcium pump, isoform 1, Atp2b1, in mice is associated with decreased bone mineral density and impaired responsiveness to 1, 25-dihydroxyvitamin D3. Biochem Biophys Res Commun 467(1):152–156
52. Davis WL, Jones RG (1982) Lysosomal proliferation in rachitic avian intestinal absorptive cells following 1,25-dihydroxycholecalciferol. Tissue Cell 14:585–595
53. Nemere I, Szego CM (1981) Early actions of parathyroid hormone and 1,25-dihydroxycholecalciferol on isolated epithelial cells from rat intestine: 1. Limited lysosomal enzyme release and calcium uptake. Endocrinology 108:1450–1462
54. Warner RR, Coleman JR (1975) Electron probe analysis of calcium transport by small intestine. J Cell Biol 64(1):54–74
55. Nemere I, Leathers V, Norman AW (1986) 1, 25 dihydroxyvitamin D3-mediated intestinal calcium transport. Biochemical identification of lysozomes containing calcium and calcium-binding protein (calbindin-D 28k). J Biol Chem 261:16106–16114
56. Nemere I, Yoshimoto Y, Norman AW (1984) Calcium transport in perfused duodena from normal chicks: enhancement within fourteen minutes of exposure to 1,25 dihydroxyvitamin D3. Endocrinology 115:1476–1483
57. Huhtakangas JA, Olivera CJ, Bishop JE, Zanello LP, Norman AW (2004) The vitamin D receptor is present in caveolae-enriched plasma membranes and binds 1 alpha,25(OH)(2)-vitamin D-3 in vivo and in vitro. Mol Endocrinol 18(11):2660–2671
58. Nemere I, Safford SE, Rohe B, DeSouza MM, Farach-Carson MC (2004) Identification and characterization of 1,25D(3)-membrane-associated rapid response, steroid (1,25D(3)-MARRS) binding protein. J Steroid Biochem Mol Biol 89–90:281–285
59. Nemere I, Garbi N, Hammerling GJ, Khanal RC (2010) Intestinal cell calcium uptake and the targeted knockout of the 1,25D3-MARRS (membrane-associated, rapid response steroid-binding) receptor/PDIA3/Erp57. J Biol Chem 285(41):31859–31866
60. Nemere I, Garcia-Garbi N, Hammerling GJ, Winger Q (2012) Intestinal cell phosphate uptake and the targeted knockout of the 1,25D(3)-MARRS receptor/PDIA3/ERp57. Endocrinology 153(4):1609–1615
61. Nemere I, Garbi N, Hammerling G, Hintze KJ (2012) Role of the 1,25D(3)-MARRS receptor in the 1,25(OH)(2)D(3)-stimulated uptake of calcium and phosphate in intestinal cells. Steroids 77(10):897–902
62. Karbach U (1992) Paracellular calcium transport across the small intestine. J Nutr 122(3):672–677
63. Tudpor K, Teerapornpuntakit J, Jantarajit W, Krishnamra N, Charoenphandhu N (2008) 1,25-dihydroxyvitamin d(3) rapidly stimulates the solvent drag-induced paracellular calcium transport in the duodenum of female rats. J Physiol Sci 58(5):297–307
64. Fujita H, Sugimoto K, Inatomi S et al (2008) Tight junction proteins claudin-2 and -12 are critical for vitamin D-dependent Ca2+ absorption between enterocytes. Mol Biol Cell 19(5):1912–1921
65. Reyes Fernandez PC, Replogle RA, Wang L, Zhang M, Fleet JC (2016) Novel genetic loci control calcium absorption and femur bone mass as well as their response to low calcium intake in male BXD recombinant inbred mice. J Bone Miner Res 31(5):994–1002
66. Sitrin M, Meredith S, Rosenberg IH (1978) Vitamin D deficiency and bone disease in gastrointestinal disorders. Arch Intern Med 138(Suppl_5):886–888
67. Schachter D, Finkelstein JD, Kowarski S (1964) Metabolism of Vitamin D. I. Preparation of radioac-

tive Vitamin D and its intestinal absorption in the rat. J Clin Invest 43:787–796
68. Hollander D (1981) Intestinal absorption of vitamins A, E, D, and K. J Lab Clin Med 97(4):449–462
69. Dueland S, Pedersen JI, Helgerud P, Drevon CA (1983) Absorption, distribution, and transport of vitamin D3 and 25-hydroxyvitamin D3 in the rat. Am J Phys 245(5 Pt 1):E463–E467
70. Sitrin MD, Pollack KL, Bolt MJ, Rosenberg IH (1982) Comparison of vitamin D and 25-hydroxyvitamin D absorption in the rat. Am J Phys 242(4):G326–G332
71. Watkins DW, Khalafi R, Cassidy MM, Vahouny GV (1985) Alterations in calcium, magnesium, iron, and zinc metabolism by dietary cholestyramine. Dig Dis Sci 30(5):477–482
72. Reboul E, Goncalves A, Comera C et al (2011) Vitamin D intestinal absorption is not a simple passive diffusion: evidences for involvement of cholesterol transporters. Mol Nutr Food Res 55(5):691–702
73. Khamiseh G, Vaziri ND, Oveisi F, Ahmadnia MR, Ahmadnia L (1991) Vitamin D absorption, plasma concentration and urinary excretion of 25-hydroxyvitamin D in nephrotic syndrome. Proc Soc Exp Biol Med 196(2):210–213
74. Krawitt EL, Chastenay BF (1980) 25-hydroxy vitamin D absorption test in patients with gastrointestinal disorders. Calcif Tissue Int 32(3):183–187
75. Bikle D (2000) Vitamin D: Production, Metabolism, and Mechanisms of Action. In: Feingold KR, Anawalt B, Boyce A et al (eds) Endotext. South Dartmouth (MA)
76. Kurogi K, Sakakibara Y, Suiko M, Liu MC (2017) Sulfation of vitamin D3 -related compounds-identification and characterization of the responsible human cytosolic sulfotransferases. FEBS Lett 591(16):2417–2425
77. Hashizume T, Xu Y, Mohutsky MA et al (2008) Identification of human UDP-glucuronosyltransferases catalyzing hepatic 1alpha,25-dihydroxyvitamin D3 conjugation. Biochem Pharmacol 75(5):1240–1250
78. Larsson SE, Lorentzon R (1977) Excretion of active metabolites of vitamin D in urine and bile of the adult rat. Clin Sci Mol Med 53(4):373–377
79. Zimmerman DR, Koszewski NJ, Hoy DA, Goff JP, Horst RL (2015) Targeted delivery of 1,25-dihydroxyvitamin D3 to colon tissue and identification of a major 1,25-dihydroxyvitamin D3 glycoside from Solanumglaucophyllum plant leaves. J Steroid Biochem Mol Biol 148:318–325
80. Wiesner RH, Kumar R, Seeman E, Go VL (1980) Enterohepatic physiology of 1,25-dihydroxyvitamin D3 metabolites in normal man. J Lab Clin Med 96(6):1094–1100
81. Kumar R (1984) Metabolism of 1,25-dihydroxyvitamin D3. Physiol Rev 64(2):478–504
82. Koszewski NJ, Horst RL, Goff JP (2012) Importance of apical membrane delivery of 1,25-dihydroxyvitamin D3 to vitamin D-responsive gene expression in the colon. Am J Physiol Gastrointest Liver Physiol 303(7):G870–G878
83. Crosnier C, Stamataki D, Lewis J (2006) Organizing cell renewal in the intestine: stem cells, signals and combinatorial control. Nat Rev Genet 7(5):349–359
84. Wang L, Klopot A, Freund JN, Dowling LN, Krasinski SD, Fleet JC (2004) Control of Differentiation-Induced Calbindin-D 9k Gene Expression in Caco-2 Cells by Cdx-2 adn HNF-1à. Am J Phys 287:G943–G953
85. Meyer MB, Watanuki M, Kim S, Shevde NK, Pike JW (2006) The human transient receptor potential vanilloid type 6 distal promoter contains multiple vitamin D receptor binding sites that mediate activation by 1,25-dihydroxyvitamin D3 in intestinal cells. Mol Endocrinol 20(6):1447–1461
86. Wood RJ, Tchack L, Angelo G, Pratt RE, Sonna LA (2004) DNA microarray analysis of vitamin D-induced gene expression in a human colon carcinoma cell line. Physiol Genomics 17(2):122–129
87. Lee SM, Riley EM, Meyer MB et al (2015) 1,25-Dihydroxyvitamin D3 controls a cohort of Vitamin D receptor target genes in the proximal intestine that is enriched for calcium-regulating components. J Biol Chem 290(29):18199–18215
88. Li S, De La Cruz J, Hutchens S et al (2020) Analysis of 1,25-dihydroxyvitamin D3 genomic action reveals calcium-regulating and calcium-independent effects in mouse intestine and human enteroids. Mol Cell Biol 41(1):e00372-20
89. Barker N, van Es JH, Kuipers J et al (2007) Identification of stem cells in small intestine and colon by marker gene Lgr5. Nature 449(7165):1003–1007
90. Barker N, Ridgway RA, van Es JH et al (2009) Crypt stem cells as the cells-of-origin of intestinal cancer. Nature 457(7229):608–611
91. Peregrina K, Houston M, Daroqui C, Dhima E, Sellers RS, Augenlicht LH (2015) Vitamin D is a determinant of mouse intestinal Lgr5 stem cell functions. Carcinogenesis 36(1):25–31
92. Costales-Carrera A, Fernandez-Barral A, Bustamante-Madrid P et al (2020) Comparative study of organoids from patient-derived normal and tumor colon and rectal tissue. Cancers (Basel) 12(8):2302
93. Yan KS, Chia LA, Li X et al (2012) The intestinal stem cell markers Bmi1 and Lgr5 identify two functionally distinct populations. Proc Natl Acad Sci U S A 109(2):466–471
94. Li W, Zimmerman SE, Peregrina K et al (2019) The nutritional environment determines which and how intestinal stem cells contribute to homeostasis and tumorigenesis. Carcinogenesis 40(8):937–946
95. Sittipo P, Kim HK, Han J, Lee MR, Lee YK (2021) Vitamin D3 suppresses intestinal epithelial stemness via ER stress induction in intestinal organoids. Stem Cell Res Ther 12(1):285
96. Laukoetter MG, Bruewer M, Nusrat A (2006) Regulation of the intestinal epithelial barrier by the

apical junctional complex. Curr Opin Gastroenterol 22(2):85–89
97. Watson AJ, Chu S, Sieck L et al (2005) Epithelial barrier function in vivo is sustained despite gaps in epithelial layers. Gastroenterology 129(3):902–912
98. Kong J, Zhang Z, Musch MW et al (2008) Novel role of the vitamin D receptor in maintaining the integrity of the intestinal mucosal barrier. AJP – Gastrointest Liver Physiol 294(1):G208–GG16
99. Zhao H, Zhang H, Wu H et al (2012) Protective role of 1,25(OH)2 vitamin D3 in the mucosal injury and epithelial barrier disruption in DSS-induced acute colitis in mice. BMC Gastroenterol 12:57
100. Chen SW, Wang PY, Zhu J et al (2015) Protective effect of 1,25-dihydroxyvitamin d3 on lipopolysaccharide-induced intestinal epithelial tight junction injury in caco-2 cell monolayers. Inflammation 38(1):375–383
101. Froicu M, Cantorna MT (2007) Vitamin D and the vitamin D receptor are critical for control of the innate immune response to colonic injury. BMC Immunol 8:5
102. Reich KM, Fedorak RN, Madsen K, Kroeker KI (2014) Vitamin D improves inflammatory bowel disease outcomes: basic science and clinical review. World J Gastroenterol 20(17):4934–4947
103. Wang F, Johnson RL, DeSmet ML, Snyder PW, Fairfax KC, Fleet JC (2017) Vitamin D receptor-dependent signaling protects mice from dextran sulfate sodium-induced colitis. Endocrinology 158(6):1951–1963
104. Brown H, Esterhazy D (2021) Intestinal immune compartmentalization: implications of tissue specific determinants in health and disease. Mucosal Immunol 14(6):1259–1270
105. Charoenngam N, Holick MF (2020) Immunologic effects of Vitamin D on human health and disease. Nutrients 12(7):2097
106. Mailhot G, White JH (2020) Vitamin D and immunity in infants and children. Nutrients 12(5):1233
107. Prietl B, Treiber G, Pieber TR, Amrein K (2013) Vitamin D and immune function. Nutrients 5(7):2502–2521
108. Ooi JH, McDaniel KL, Weaver V, Cantorna MT (2014) Murine CD8+ T cells but not macrophages express the vitamin D 1alpha-hydroxylase. J Nutr Biochem 25(1):58–65
109. Overbergh L, Decallonne B, Valckx D et al (2000) Identification and immune regulation of 25-hydroxyvitamin D-1-alpha-hydroxylase in murine macrophages. Clin Exp Immunol 120(1):139–146
110. Stoffels K, Overbergh L, Giulietti A, Verlinden L, Bouillon R, Mathieu C (2006) Immune regulation of 25-hydroxyvitamin-D3-1alpha-hydroxylase in human monocytes. J Bone Miner Res 21(1):37–47
111. Hewison M (2010) Vitamin D and the intracrinology of innate immunity. Mol Cell Endocrinol 321(2):103–111
112. Wang TT, Nestel FP, Bourdeau V et al (2004) Cutting edge: 1,25-dihydroxyvitamin D3 is a direct inducer of antimicrobial peptide gene expression. J Immunol 173(5):2909–2912
113. Gombart AF, Borregaard N, Koeffler HP (2005) Human cathelicidin antimicrobial peptide (CAMP) gene is a direct target of the vitamin D receptor and is strongly up-regulated in myeloid cells by 1,25-dihydroxyvitamin D3. FASEB J 19(9):1067–1077
114. Wang TT, Dabbas B, Laperriere D et al (2010) Direct and indirect induction by 1,25-dihydroxyvitamin D3 of the NOD2/CARD15-defensin beta2 innate immune pathway defective in Crohn disease. J Biol Chem 285(4):2227–2231
115. Lagishetty V, Misharin AV, Liu NQ et al (2010) Vitamin D deficiency in mice impairs colonic antibacterial activity and predisposes to colitis. Endocrinology 151(6):2423–2432
116. Szeles L, Keresztes G, Torocsik D et al (2009) 1,25-dihydroxyvitamin D3 is an autonomous regulator of the transcriptional changes leading to a tolerogenic dendritic cell phenotype. J Immunol 182(4):2074–2083
117. Adams JS, Liu PT, Chun R, Modlin RL, Hewison M (2007) Vitamin D in defense of the human immune response. Ann N Y Acad Sci 1117:94–105
118. Mathieu C, van Etten E, Gysemans C et al (2001) In vitro and in vivo analysis of the immune system of vitamin D receptor knockout mice. J Bone Miner Res 16(11):2057–2065
119. Yu S, Bruce D, Froicu M, Weaver V, Cantorna MT (2008) Failure of T cell homing, reduced CD4/CD8alphaalpha intraepithelial lymphocytes, and inflammation in the gut of vitamin D receptor KO mice. Proc Natl Acad Sci U S A 105(52):20834–20839
120. Froicu M, Weaver V, Wynn TA, McDowell MA, Welsh JE, Cantorna MT (2003) A crucial role for the vitamin D receptor in experimental inflammatory bowel diseases. Mol Endocrinol 17(12):2386–2392
121. Griffin MD, Dong X, Kumar R (2007) Vitamin D receptor-mediated suppression of RelB in antigen presenting cells: a paradigm for ligand-augmented negative transcriptional regulation. Arch Biochem Biophys 460(2):218–226
122. Palmer MT, Lee YK, Maynard CL et al (2011) Lineage-specific effects of 1,25-dihydroxyvitamin D(3) on the development of effector CD4 T cells. J Biol Chem 286(2):997–1004
123. Fletcher J, Cooper SC, Ghosh S, Hewison M (2019) The role of Vitamin D in inflammatory bowel disease: mechanism to management. Nutrients 11(5):1019
124. Chen J, Waddell A, Lin YD, Cantorna MT (2015) Dysbiosis caused by vitamin D receptor deficiency confers colonization resistance to Citrobacter rodentium through modulation of innate lymphoid cells. Mucosal Immunol 8(3):618–626

125. Konya V, Czarnewski P, Forkel M et al (2018) Vitamin D downregulates the IL-23 receptor pathway in human mucosal group 3 innate lymphoid cells. J Allergy Clin Immunol 141(1):279–292
126. He L, Zhou M, Li YC (2019) Vitamin D/Vitamin D receptor signaling is required for normal development and function of group 3 innate lymphoid cells in the Gut. iScience 17:119–131
127. Lin YD, Arora J, Diehl K, Bora SA, Cantorna MT (2019) Vitamin D is required for ILC3 derived IL-22 and protection from Citrobacter rodentium infection. Front Immunol 10:1

Part IV

Cancer

Estrogen Receptor Alpha and *ESR1* Mutations in Breast Cancer

10

Jaymin M. Patel and Rinath M. Jeselsohn

Abstract

The estrogen receptor alpha (ERα) is a nuclear transcription factor that is expressed in more than 70% of all breast cancers. Key genes involved in proliferation and tumor progression are transcriptionally regulated by ERα making it an important therapeutic target. Indeed, the first class of targeted treatments in cancer are endocrine treatments that target ERα either by competitive inhibition, reduced ligand production or receptor degradation. Despite the efficacy of these drugs, resistance to endocrine treatment remains a key clinical challenge. Only about 50% of patients treated with endocrine treatment in early-stage disease will benefit from adjuvant endocrine treatment and nearly all patients treated in the metastatic setting will develop disease progression while on endocrine treatment. Multiple mechanisms of resistance to endocrine treatment have been identified in pre-clinical models and clinical samples. These include both intrinsic (*de novo*) mechanisms and adaptive, acquired mechanisms. Over the past few years, gain-of-function missense mutations of *ESR1*, the gene encoding ERα, have been unveiled and identified as the most common genomic mechanism of acquired resistance to endocrine treatments. These mutations are clustered in a "hot spot" region within the ligand binding domain and engender constitutive, ligand-independent activity. Clinical studies evaluating these *ESR1* mutations in metastatic ERα positive breast cancer demonstrate decreased overall survival which also highlights their prognostic role. In this chapter, we will provide a detailed review of structural and biophysical characteristics, functional consequences and clinical implications of the *ESR1* mutations. We will also discuss potential therapeutic strategies to overcome treatment resistance in the context of *ESR1* mutations and implications for future treatment selection.

Keywords

ESR1 · Breast cancer · Estrogen receptor alpha · Hormone positive · Endocrine resistance

J. M. Patel
Beth Israel Deaconess Medical Center, Harvard Medical School, Boston, MA, USA

R. M. Jeselsohn (✉)
Dana-Farber Cancer Institute, Harvard Medical School, Boston, MA, USA
e-mail: Rinath_Jeselsohn@dfci.harvard.edu

M. J. Campbell, C. L. Bevan (eds.), *Nuclear Receptors in Human Health and Disease*, Advances in Experimental Medicine and Biology 1390, https://doi.org/10.1007/978-3-031-11836-4_10

10.1 Introduction to Estrogen Receptor

10.1.1 Structure

The estrogen receptor alpha, also known as ERα was first discovered in 1950s by Jenson and Jacobsen [1, 2] as a mechanism through which the biologic effects of 17β-estradiol could be mediated. In 1995, a second estrogen receptor, ERβ was discovered [3, 4]. These estrogen receptors are among a family of ligand-activated nuclear receptors which share similarities in primary structure, receptor activation and downstream regulation [5–8]. Wild-type ERα is a 595 amino acids long protein with a molecular weight of 66.2 kDa that is encoded by the *ESR1* gene located on chromosome 6 [9, 10]. Wild-type ERβ is comprised of 530 amino acids with a molecular weight of 59.2 kDa that is encoded by the *ESR2* gene located on chromosome 14 [11]. Each gene has 8 coding regions interrupted by introns that encodes for the 5 primary domains seen in all members of the nuclear receptor family: A/B, C, D, E and F [12, 13]. In this chapter, we will focus on ERα that is encoded by *ESR1*, generally considered the most relevant gene in breast cancer (Fig. 10.1a).

10.1.2 Functional Domains

ESR1 encodes 5 functional components or domains [14]. The N-terminus encoded by **A/B domain** is responsible for interactions between other proteins and estrogen receptor (ER) domains to either activate or repress transcription [15, 16]. These interactions may occur in the presence or absence of ligand [17–19]. Notable hormone-independent interactions include growth-factor induced, kinase-based phosphorylation for transactivation via activation factor-1 (AF-1) and inhibitory function of distal N-terminus end with hydrophobic surface of C-terminal ligand binding domain [20–23] (Fig. 10.1b).

The **C domain** encodes for a 70–80 amino acid DNA-binding domain. This region, as with other members of nuclear receptor family, is

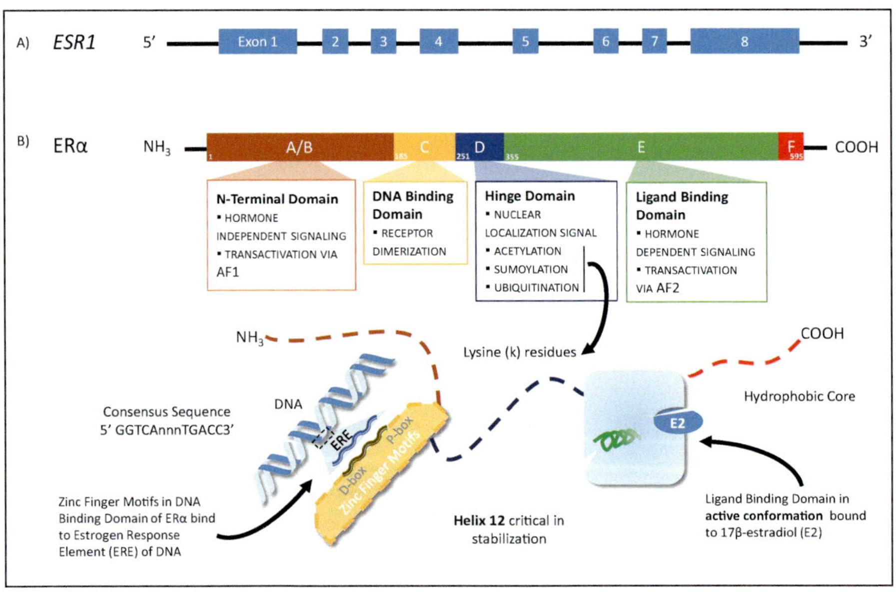

Fig. 10.1 **Schematic representation of structure and function.** (**a**) *ESR1* gene schematic of 8 ERα protein coding exons (**b**) Unfolded ERα protein from N-terminal end to C-terminal with primary domains A-F, corresponding functional roles and key features in visual schematic containing DNA with estrogen response element (ERE) and 17β-estradiol (E2) bound to ligand binding domain

responsible for DNA binding and receptor dimerization through two zinc-binding motifs. The first contains a sequence-specific DNA binding region called the P-box [13, 15, 24]. The second contains, distal or D-box which provides surface for ERs to dimerize on DNA and contributes to binding of two ER monomers to the estrogen response element. DNA (Fig. 10.1b) binding sites in ERα have a higher density of TA-rich motifs and Forkhead (FoxA1) transcription factor binding sites while ERβ had a predominance of classical estrogen response elements and GC-rich motifs. Inhibition of FoxA1 blocks ERα from associating with chromatin and subsequent modulation of target gene transcription [24–28].

The hinge or **D domain** encodes a part of the nuclear translocation signal and lysine (K) residues that serve as a target site for acetylation, sumoylation and ubiquitination (Fig. 10.1b) [29, 30]. Sumolyation also enhances transcriptional activity and ubiquitination targets receptor for proteasome degradation [31, 32].

The ligand binding domain (LBD) or **E domain** is a multifunctional protein folded into a three-layered, antiparallel α-helical sandwich composed of 12 α-helices (H1–12) and a small two-stranded anti-parallel β-sheet [33]. The ligand binding cavity consists of a large internal hydrophobic core protected from the external environment. Ligand binding is supported through a combination of complementary hydrophobic interactions and hydrogen bonds of a wide variety of non-polar ligands that are significantly smaller in size relative to binding cavity [33–35] (Fig. 10.1b). This highly conserved region between ERα and ERβ differs by only two amino acids, Leu384 and Met421 versus Met336 and Ile373 respectively [36]. This domain is responsible for hormone-dependent transactivation (AF-2) through helix 12, which acts as a dynamic molecular switch capable of binding coactivators (ie SRC-1) and corepressors that promote "agonist" or "antagonist" conformations, respectively (Fig. 10.2). Other key components of transcriptional activation include H3, H4, and H5 [34, 35, 37, 38]. It is also responsible for dimerization which is essential to their function as transcription factors. Hydrophobic N-terminal end of H11 from each monomer intertwine to form a strong backbone and then H8 from one monomer and parts of H9 and H10 from the other help to provide additional support

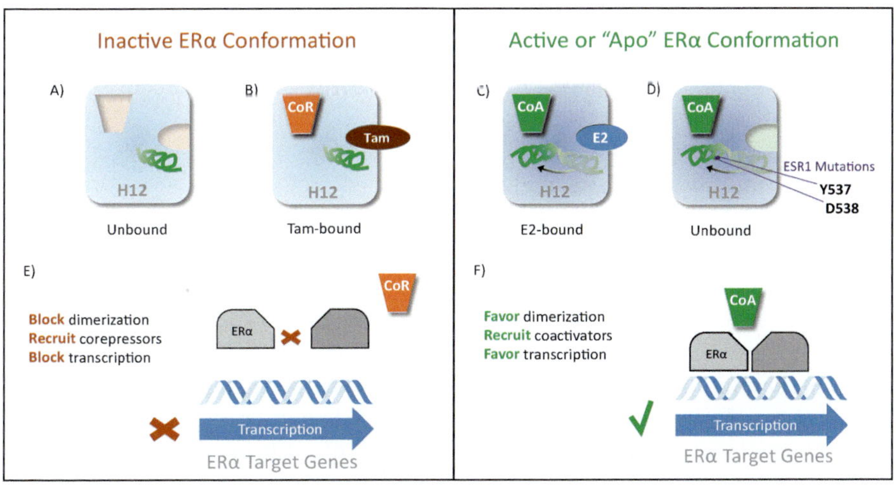

Fig. 10.2 Schematic of ERα LBD receptor conformations. Left Column: Inactive ERα conformation (**a**) Unbound ERα LBD or (**b**) Tamoxifen-bound ERα LBD with helix 12 (H12) in unfavorable orientation promoting inactive conformation that (**e**) blocks dimerization, promotes corepressor recruitment and bocks ligand-dependent transcription of ERα target genes; Right Column: Active or "Apo" ERα conformation (**b**) E2-bound ERα LBD or (**c**) ERα-mutant LBD with H12 in favorable orientation promoting active conformation that (**f**) promotes dimerization, cofactor recruitment and activation of ERα transcriptional program. (Adapted from Fig. 10.1b. In Jeselsohn et al. [149])

[34, 39]. Although possible in the absence of ligand and co-activators, dimerization is most stable when present [40]. Additional factors such as coactivator NR-box act to lock ligand binding in cavity, thereby reducing rate of dissociation and improving potency in transcription activation [41].

Lastly, the **F-domain** at the C-terminal end plays an array of modulatory roles and is least understood among the 5 domains. It can affect the agonist/antagonist activity of antiestrogens and E2-dependent transcriptional activation [42, 43].

10.1.3 Genomic Signaling

The primary mechanism of signaling for Type I nuclear receptors such as ER is through ligand-induced receptor activation followed by direct genomic interaction via response element on target gene. However, there are alternative pathways such as ligand-independent receptor activation and possibly non-genomic signaling [44]. The latter is beyond the scope of this chapter.

Ligand-Dependent Activation
The AF-2 domain of ER is key to ligand-induced activation [45]. It is able to switch between "agonist" and "antagonist" conformations based on ligand binding [46]. Agonist ligand binding leads to a confirmation change that orients H12 to seal LBD and stabilizes receptor confirmation for efficient interaction with α-helical nuclear receptor (NR)-box motifs (LXXLL), which in turn facilitates transcription activation [47, 48] (Fig. 10.2c). In contrast, antagonist binding results in a conformational change that orients H12 along hydrophobic coactivator groove which favors corepressor binding to antagonize ligand-dependent transcriptional activation [34, 38, 46] (Fig. 10.2b). Crystallographic studies have revealed two mechanisms by which ligands can induce "antagonist" conformation. In one scenario the ligand has bulky-side chain extensions which sterically inhibit "agonist" confirmation. Alternatively, a less bulky ligand that easily fits in ligand binding cavity but fails to connect with key amino acids lining its surface will destabilize "agonist" orientation of H12 [34, 35, 38, 49]. In addition to ligand structure, additional studies reveal the importance of NR-box sequence [50], flanking sequences [51] and additional surfaces outside of AF-2 core that are critical in coactivator interaction with ER [52]. SRC-1 and GRIP-1 interact with AF-1 and AF-2 coactivators separately, but also can promote cooperative association of AF-1 and AF-2 to combine N-terminal and C-terminal function in the presence of an agonist ligand [53–56]. The variability in ER ligand binding and receptor confirmations impact ER binding kinetics to various ERE sequences which ultimately impacts gene selection and transcription [57, 58] (Fig. 10.2e, f).

Ligand-Independent Activation
ER is a phosphoprotein like most nuclear receptors. Therefore, its activity can be altered through phosphorylation at specific serine and tyrosine residues, even in the absence of a ligand. The site of phosphorylation is dependent on the protein kinase and its activator [59, 60]. Receptor tyrosine kinases and mitogen-induced kinases have been shown to transduce a variety of extracellular signals through membrane receptors [44], such as dopamine [61], epidermal growth factor, transforming growth factor α [62], insulin or insulin-like growth factor-1 [63], heregulin [64] and cyclin AMP [64]. See Sect. 10.2.4 for further detail.

10.1.4 Estrogen Response Elements and Gene Regulation

Ligand activated ERα regulates gene expression as a transcription factor in a dimeric form through its interaction with specific sequences of DNA known as estrogen response elements (EREs) [65]. ER binds to ERE in sequence specific manner using the first zinc finger while the second zinc finger is responsible for dimerization and ERE half-site spacing recognition [66, 67]. This interaction is dependent on multiple factors including, ER subtypes (ERα and ERβ), and ligand, dimerization partner, anatomy of

ERE and location of ERE which in turn impacts co-regulatory recruitment [67–69]. The consensus ERE sequence, derived from the African clawed frog *Xenopus laevis* gene and chicken apo-VLDL II gene is 5′GGTCAnnnTGACC3′ (Fig. 10.1b). It is a 13 base pair perfect palindromic inverted repeat with a 3 base pair spacing of variable bases [70, 71] However, this highly estrogen-responsive sequence is only found in a small number of estrogen-inducible genes whereas most target genes contain a variant with reduced affinity [72]. Furthermore, some EREs are separated by many base pairs and act in combination to form estrogen-response units [65].

Although there is no clear consensus due to variability in assay, analysis pipeline and biologic sample used, CHIP-seq analysis performed by Palaniappan et al. of mouse mammary cells revealed 6237 ERα binding sites and 3686 unique genes that recruit ER in response to estradiol [73]. Similar analysis by Grober et al. in MCF7 cells revealed 9702 ERβ and 6024 ERα binding sites in estrogen-stimulated cells with 921 genes differentially regulated by ERβ [74]. In general, most studies show that ER binding sites are in ranges of thousands and genes regulated by ER are in the range of hundreds. Thus, many of the sites detected to be bound by ER most likely do not impact transcription.

10.1.5 Normal Physiologic Functions and Tissue Expression

ER is a key regulatory factor in organ development and numerous biologic activities that include reproduction, bone modeling, cardiovascular function, metabolism and behavior [75]. Signaling activity is likely a function of variable ERα and ERβ expression densities and ratios across tissues. In particular, ER is critical to normal mammary development and differentiation, but scientists are still unraveling how ER signaling is altered to drive tumorigenesis [76].

10.2 Wild-Type ERα in Breast Cancer

Clinical laboratories at the time of this writing evaluate ERα expression using an antibody targeting the ERα receptor [77]. There is some data that suggests ERβ plays a role in response and resistance, however staining is not currently being used in routine clinical care or clinical studies due to the unclear role of ERβ in treatment decisions and prognosis.

10.2.1 Prognostic and Predictive Implications

American Society of Clinical Oncology and the College of American Pathologists for ER assessment in breast cancer require immunohistochemical (IHC) staining of ≥1 percent using an internally validated protocol against a clinically validated assay. Recommended antibody clones such as SP-1, 6F11 and ID5 are all specific to ERα only [77]. The primary purpose of ER status is to predict if the patient will benefit from anti-estrogen or hormone therapy. Numerous studies have demonstrated benefit from hormone therapy in both early stage and metastatic breast cancer with a qualitatively positive ER status by IHC [78–82]. Conversely, ER negative (ER-) status predicts no benefit from hormone therapy [83]. In addition, the prognosis of ER positive (ER+) patients is in general better than those who are ER-. This is further impacted by progesterone receptor (PR) status, histologic features and tumor stage [84].

10.2.2 Molecular Subtypes or Heterogeneity in ER Positive

Heterogeneity among ER+ patients can be described in several ways. The most clinically accessible and utilized is based on ER/PR percent and strength of staining. Higher staining percent and strength of staining suggests a more favorable prognosis and greater likelihood of response to endocrine therapy [85, 86]. Reduced or absence of staining in either receptor suggests poorer prognosis with potential for reduced benefit from endocrine therapy [87].

The advent of molecular sequencing and microarray assays had led to the distinction of two molecular subtypes for hormone receptor-positive (HR+) or ERα and/or PR positive breast cancer [88, 89]. Luminal A – lower grade, best prognosis, least metastatic potential and most endocrine sensitive – versus Luminal B – higher grade, good but poor prognosis, some metastatic potential and less endocrine sensitive [90, 91]. Similarly, a number of clinically validated molecular assays, the most prominent being Oncotype Dx assay [92] and MammaPrint [93], offer both prognostic and predictive information on the benefit or lack thereof for adjuvant chemotherapy in the curative setting [94]. Oncotype Dx is a 21 gene recurrence assay which primarily assesses and weighs factors related to estrogen signaling and proliferation. Based on the score and other clinical factors such as stage, age and menopausal status, patients are deemed likely or unlikely to benefit from adjuvant chemotherapy [95, 96].

10.2.3 Evidence for Tumorigenic Signaling

ER signaling in breast cancer, even in the absence of a mutation, is shown to be distinct from ER signaling in normal mammary cells. In 2019, Chi et al. reported transcriptional profiling of normal ER+ mature luminal mammary epithelial cells and ER+ breast tumors. Gene expression profiles in response to estrogen stimulation were distinct enough to segregate normal tissue from cancer tissue. Specifically, their group discovered increased DNA binding motif grainyhead-like transcription factor 2 (GRHL2) enrichment in response to estradiol in tumor cells compared with normal resulting in distinct transcriptional programs. In addition Rho GTPase-activating protein, deleted in liver cancer 1(DLC1), which exhibited tumor suppressor function in normal estrogen stimulated tissue was found to have decreased expression in breast cancer. Clinical datasets demonstrate DLC1 loss and GRHL2 gain predicts for lower survival rate [76]. Furthermore, our group has demonstrated *ESR1* allele specific differences in transcriptional programs as well as acquisition of neomorphic properties that favor a metastatic phenotype [97].

10.2.4 Ligand-Independent Activation Through Phosphorylation

Canonical ligand-dependent activation of ERα requires binding of estrogen to LBD to facilitate structural changes and co-factor recruitment that promotes binding of dimerized ERs to target DNA sequence for modulation of transcriptional activity [98, 99]. However, in ligand-independent activation, this canonical pathway is circumvented through phosphorylation via growth factors and cytokines [100, 101]. In particular the N-terminal is highly susceptible to phosphorylation by multiple different kinases [19, 102]. One of the most characterized events is phosphorylation of Ser118(S118) by CDK7 upon estrogen stimulation [103, 104]. However, growth factors such as epidermal growth factor (EGF) can also phosphorylate Ser118 in the absence of estrogen through mitogen-activated protein kinase (MAPK) [18, 19]. The site of phosphorylation has been shown to influence ERα chromatin-binding profile through cooperation with different transcription factor complexes such AP-1 and PBX-1 [105, 106]. In addition, S167 is another important phosphorylation site in N-terminus of ER. This is usually induced by phosphatidylinositol 3-kinase (PI3K)/AKT/mammalian target of rapamycin (mTOR) and MAPK signaling in response to insulin, insulin-like growth factor (IGF) and EGF [107–109]. Other areas of phosphorylation include S305 in hinge domain by protein-kinase A (PKA), which has been shown to promote receptor activity that is refractory to tamoxifen [110] partly through expression of oncogene c-MYC [111].

In addition to aberrant cellular signaling through growth factors, paracrine signaling, particularly through cytokines is also a major source of phosphorylation. The tumor microenvironment which include a variety of cell types has been shown to impact ERα expression, activity, drug resistance and tumor progression [112–115]. Stromal cells may increase tumor size and metastatic potential through adipokine leptin and interleukin-6 (IL-6) [116]. Other cells such as macrophages, may induce phosphorylation of S305 through tumor necrosis factor α (TNFα) and IL-1β. Upon cytokine-induced S305 phosphorylation of ER, a subset of estradiol-induced genes are targeted, some of which have previously been shown in MCF7 models to induce extravasation in the absence of estradiol. Furthermore, MCF7 cells can also become resistant to tamoxifen in the presence of cytokines like TNFα [101, 117]. Alternatively, other cytokines such as transforming growth factor β (TGFβ) may impact ERα activity in a SMAD4-dependant manner positively by repressing tumor growth in early stages of disease and then negatively by promoting metastasis and epithelial-to-mesenchymal transition in later stages of disease [118–120].

10.2.5 Downstream Redirection via Transcription Factors

ERα binding to chromatin is responsible for transcriptional modulation of target genes [99]. This activity is part of a larger transcriptional complex that includes transcription factors such as FOXA1, PBX1, GATA3 and AP-2γ that can directly impact ERα binding to chromatin [28, 121–123]. For example, there is data demonstrating FOXA1 and PBX1 can function in a so-called pioneer mode to modulate chromatin accessibility, while TNFα re-directs ERα to NF-κB binding site such as at *BIRC3* [124, 125]. In 2012, Ross-Innes et al. reported FoxA1-mediated reprogramming of ER binding on a rapid time scale as mechanism of generating distinct binding combinations with cis-regulatory elements that was associated with poor clinical outcomes [126]. To further complicate the matter, cytokines and growth factors can also influence ERα binding to transcription factors thereby altering transcriptional activity.

10.3 Endocrine Treatment & Resistance in Breast Cancer

More than 70% of all breast cancer diagnoses are found to express ERα [127]. These luminal-type cancers are driven by pro-oncogenic signaling from ERα [88, 89]. Pharmacologic inhibition of this pathway can be divided into 3 categories: receptor inhibition [128], ligand reduction [129] and downstream target/co-stimulatory factor inhibition. The latter, which includes CDK4/6 and mTOR inhibition, is beyond the scope of this chapter [130, 131].

10.3.1 Overview of Therapies Targeting ER

Receptor Inhibition: Mixed – Selective Estrogen Receptor Modulator
Tamoxifen was discovered in 1960s by a chemist named Dora Richardson in a pharmaceutical lab (now part of AstraZeneca) in an effort to synthesize female contraceptives. It is termed a selective estrogen receptor modulator (SERM) because of its mixed antagonist/agonist properties [132]. Specifically, it inhibits ERα-expressing breast cancer cells but sustains and promotes osteoblasts and cells lining the endometrium. The agonist effect of tamoxifen on osteoblasts is beneficial in terms of bone density promotion, however this effect on endometrial lining results in an undesirable risk of endometrial hyperplasia/carcinoma in the uterus. In 1977, the United States Federal Drug Administration Authority (FDA) first approved its use for advanced breast cancer [133]. Numerous clinical trials demonstrated benefit, expanding use to the curative setting while specifying it to those expressing ERα. Tamoxifen, like other SERMS, raloxifene and toremifene, competitively binds the ligand-binding pocket within the LBD of ERα, thereby stabilizing structural conformations that alter recruitment of cofactors to AF-2 cleft from coactivators to corepressors such as silencing mediator of retinoid acid or thyroid-hormone receptor (NCOR2/SMRT) or nuclear receptor

co-repressor (NCOR1) resulting in an altered transcriptional activation and overall inhibition [134]. Due to pharmacologic activity at the level of the receptor, SERMs have the distinct advantage of clinical activity in both menstruating and post-menopausal women [135].

Receptor Inhibition: Complete – Selective Estrogen Receptor Degrader

Fulvestrant, a 7α-alkylsulphinyl analogue of 17β-estradiol, was derived through a coordinated drug discovery effort to uncover a pure antiestrogen. Its binding affinity to ERα is reported at 89% of estradiol compared to tamoxifen at only 2.5%. Once bound, Fulvestrant impairs receptor dimerization and nuclear translocation [128]. It also disables both AF1 and AF2 rendering any fulvestrant-ER complexes that enter the nucleus transcriptionally inactive. Lastly, fulvestrant-bound ERα is more hydrophobic making it unstable which accelerates ubiquitination and proteasome-mediated degradation [38]. Clinically, this agent is primarily used as a 2nd or 3rd line endocrine agent. Two of its major limitations are administration and bioavailability. It requires two 250 mg/5 ml intramuscular injections in a proprietary formulation of castor oils and alcohols which are slowly absorbed to maximal levels in anywhere between 2 and 19 days [136].

Ligand Reduction – Aromatase Inhibition

Third generation aromatase inhibitors primarily consist of two non-steroidal agents (anastrazole, letrozole) and one steroidal agent (exemestane). They inhibit a key enzyme, aromatase, needed for the conversion of androstenedione and testosterone to estrone and estradiol [129]. For post-menopausal women, aromatase inhibition is generally considered the preferred first-line endocrine agent. However, in premenopausal women, these agents are unable to effectively inhibit ovarian aromatase due to its abundance and therefore requires ovarian suppression through gonadotropin releasing hormone analogues or bilateral oophorectomy [44].

10.3.2 Mechanisms of Endocrine Resistance

Clinically, endocrine resistance in HR+ breast cancer is first distinguished as either intrinsic (also known as *de novo*) or acquired [137, 138]. Intrinsic, a rarer scenario that is generally defined by clinical recurrence within 2 years of starting endocrine therapy in adjuvant setting or 6 months in metastatic setting, is usually due to a factor already present [139], such as insufficient drug uptake, failure of metabolic activation, or lack of reliance on ER pathway [137]. Meanwhile, the more common scenario is an acquired resistance that develops over time. There are 3 major categories of acquired resistance: somatic alterations, epigenetic or non-genetic and tumor microenvironment. Somatic alterations are the most extensively characterized, likely due to rapid advances in sequencing over the last 2 decades and have revealed important somatic alterations or molecular signatures contributing to resistance in breast cancer tissue prior to endocrine treatment [138, 140–143].

Somatic alterations can be specific to *ESR1* (See Sect. 10.4.1) or extend to genes involved in regulatory control of ERα or cross talk with signaling pathways through receptor tyrosine kinases [138]. According to some studies, *ESR1* alterations accounts for approximately 30% of endocrine resistance [143, 144]. Alterations in other genes include those encoding growth factors such as ERBB2, EGFR, FGFR1, cooperative signaling pathway factors such as PIK3CA, PTEN, AKT, NF1, KRAS, BRAF, MAP2K1 and transcriptional regulators such as MYC, FOXA1, CTCF and TBX3. It has also been shown by Razavi et al. that alterations in *ESR1* were mutually exclusive from those involving transcriptional regulators and MAPK pathway components [143].

Epigenetic mechanisms include either reprogramming through post-translational modifications and DNA methylation, or ERα-regulated transcriptional programs, or parallel signaling pathways that promote cell-cycle progression or survival independent of ERα pathway [138, 145]. Increased histone H3 lysine 4 (H3K4) demethylase KDM5B activity and YY1 transcription factor activity have been implicated in acquired endocrine resistance [146, 147]. Aberrant activity of cofactors such as HOXB7, RUNX2 or FOXA1 can drive endocrine resistance through transcriptional rewiring [148–150]. Non-genetic mechanisms include compensatory cross-talk between signaling pathways or feedback loops such as C-terminal Src kinase (CSK) inhibiting oncogenic SRC family tyrosine kinases such as PAK2 until disruption of ERα binding downregulates CSK transcription [151]. Other non-genetic mechanisms include epithelial-to-mesenchymal transition and metabolic reprogramming [152, 153].

Lastly, mechanisms of resistance pertaining to the tumor microenvironment are least understood and characterized [138]. This includes microenvironmental factors such as hypoxia [154–156], cancer-associated fibroblasts [113, 157, 158], collagen [159, 160], inflammatory cytokines [101, 161], and immune components [162].

10.4 *ESR1* Mutations in Breast Cancer

In the 1994, Karnik et al. evaluated tamoxifen-resistant breast cancer tumors and found a mutation in exon 6 of 2 patients, hypothesizing the possibility of a constitutively active ERα as a possible mechanism of resistance [163]. In 1997, Zhang et al's report of a Y537N *ESR1* mutation in a clinical sample of metastatic breast cancer corroborated this hypothesis and proposed LBD mutations induced ERα conformational changes as a mechanism of ERα ligand independence [164] which subsequently has been validated by molecular dynamic simulations of other LBD mutations [165]. In the last decade, a number of larger scale studies utilizing deep sequencing have identified recurrent *ESR1* mutations in recurrent/metastatic HR+ breast cancer. The incidence rate of *ESR1* mutations in primary untreated tissue is low suggesting an acquired phenomenon that is not a primary driver of carcinogenesis but rather treatment resistance [143, 144, 165, 166]. Data from 962 primary samples in TCGA indicated *ESR1* mutation in only 0.5% of cases and amplification in only 2.6% of cases. In contrast, incidence of *ESR1* mutations in breast cancer previously treated with hormone therapy rises to 20–40% of cases [167]. The majority of these datasets note aromatase inhibitors as primary exposure, data on SERMs and SERDs is still unclear.

10.4.1 Types of Genomic Alterations

As previously discussed, primary components of ERα that the *ESR1* gene encodes is AF1, DBD, Hinge, AF2 & LBD. Several *ESR-1* related genomic abnormalities have been described including missense point mutations, copy number changes and genomic rearrangements resulting in fusion proteins [168] (Fig. 10.3a).

Data last accessed March 11, 2019 from COSMIC (https://cancer.sanger.ac.uk/cosmic) and cBioPortal reported 62 unique missense mutations of which 47 are in the LBD (https://www.cbioportal.org) [169]. The majority involving Y537 and D538 residues are in helix 12 of the LBD. As aforementioned, H12 is a key structural component of AF2 and regulator of agonist vs antagonist conformation. Other hotspot mutation regions include but are not limited to S463, L536, P535 and E380Q [168] (Fig. 10.3b). Early studies evaluating *ESR1* amplification via SNP microarray and FISH suggested high rates ~20%, however more recent biochemical and next generation sequencing studies report an *ESR1* amplification rate of ~2% in both primary and metastatic breast cancer tumors, suggesting a lesser role in resistance [170–173]. Lastly, a few genomic rearrangements involving *ESR1* that generate fusion proteins have been described and associated with endocrine treatment resistance (Fig. 10.3c). These include but are not limited to *ESR1-CCDC170* where the N-terminally truncated forms fuse with *ESR1* constitutive promoter, *ESR1-YAP1* and *ESR1-PCDH11X* where N-terminal end of ERα is fused in frame at C-terminus of *YAP1*and *PCDH11X* respectively [174, 175]. Functional

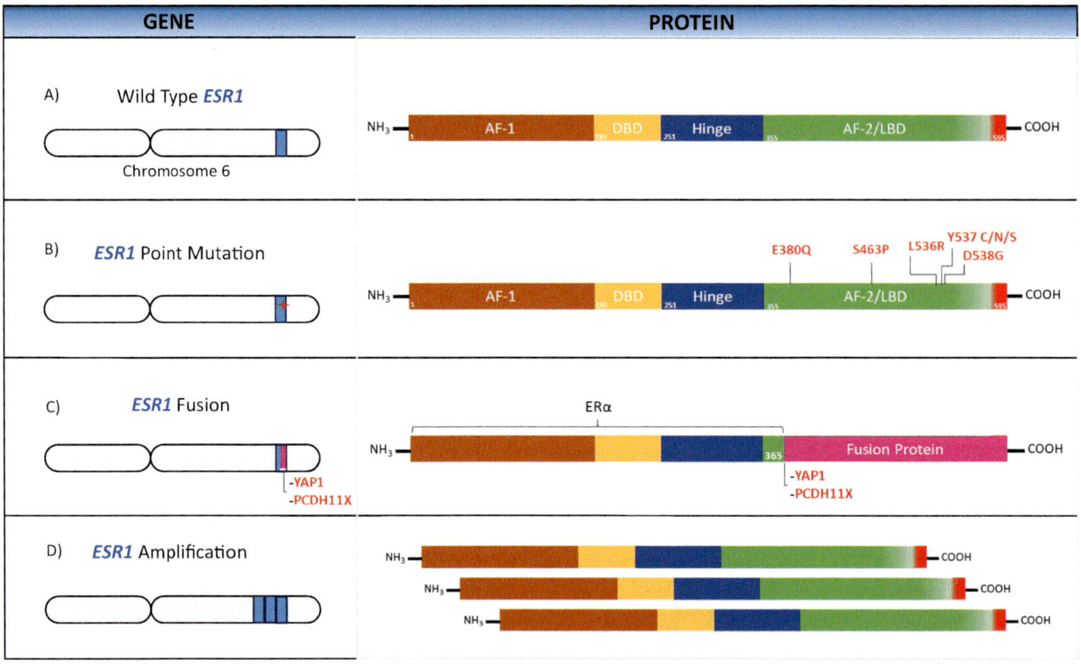

Fig. 10.3 **Types of *ESR1* alterations.** (**a**) Wild-type *ESR1* on chromosome 6 with corresponding unfolded ERα protein. (**b**) Mutant *ESR1* with a few commonly occurring *ESR1* point mutations [Y537 C/N/S, D538G, S436P, E380Q and L536R (**c**) *ESR1* fusion with *ESR1* (blue) – Fusion gene (pink) [i.e. *YAP-1* or *PCDH11X*] with corresponding fusion protein; Note numbers at bottom of protein schematic represent amino acid position. (**d**) *ESR1* amplification with multiple copies on chromosome 6 but otherwise normal ERα proteins

characterization by Lei et al. of latter two *ESR1* fusion proteins demonstrate a capacity to drive ligand-independent growth, transcriptional reprogramming that promote epithelial-mesenchymal transition, metastasis and endocrine therapy resistance [153].

10.4.2 *ESR1* Ligand Binding Domain Mutations

Between 2013 to 2014, five studies described mutations in LBD of *ESR1* from clinical samples through next-generation sequencing [144, 165, 175–177]. The first study [175] identified *ESR1* LBD mutations in 3 out of 7 patient-derived xenograft models of metastatic ER+ breast cancer which were confirmed to be present in originating tumors specimens [163]. In a second study, HR+ tumor samples from two patient cohorts were subjected to targeted next-generation sequencing [165]. These two cohorts included one cohort of 38 patients and a second cohort of 227 patients that participated in the BOLERO-2 clinical trial including 44 HR+ metastatic tumors and 181 primary treatmen- naïve tumors. BOLERO-2 was a phase III trial of metastatic HR+ breast cancer patients who had progressed on a non-steroidal aromatase inhibitor that were randomized to a steroidal aromatase inhibitor, exemestane or exemestane plus everolimus, a mammalian target of rapamycin complex 1 (MTORC1) inhibitor. Among the metastatic HR+ patients, 17.5% were found to have an *ESR1*LBD mutation. In contrast, *ESR1* LBD mutations were detected in only 3% of the primary HR+ primary tumors [131]. Another relatively large study of clinical samples with targeted next-generation sequencing of 208 breast cancer specimens consisted of 115 ER- tumors 76 ER+ metastatic tumors and 58 ER+ primary tumors. *ESR1* mutations were identified in 14.5% of the ER+ metastatic tumors and none were detected in ER+ primary tumors or ER- tumors. Furthermore, in this study there was a significant correlation between the number of prior endocrine treatments and the prevalence of the *ESR1* mutations. The *ESR1* mutations detected were in biopsies from different organ sites, suggesting no specific organotropism [144].

10.4.3 Endocrine Resistance and Cellular Phenotypes

Of the known genomic alterations in *ESR1* in breast cancer, missense mutations in LBD are the most common and arise following exposure to endocrine treatment [143, 144, 165]. This observation of increased *ESR1* mutations in the LBD that alter receptor activity in clinically recurrent samples suggested a mechanism of resistance that has been extensively studied in the preclinical setting [164, 178, 179]. Several breast cancer cell line studies evaluating the *ESR1* mutations involving the LBD, notably Y537S and D538G, have demonstrated ligand-independent transcriptional activity indicative of resistance to aromatase inhibitors (Fig. 10.4). In addition, cell lines expressing the Y537S or D538G mutation displayed decreased sensitivity to ER antagonists such as tamoxifen and fulvestrant, suggesting relative resistance to these endocrine treatments [104, 144, 165, 177].

Molecular modeling estimate that these mutations favor an apo-receptor conformation that is constitutively active despite antagonist binding. Nettles et al. [180] and Fanning et al. [181] solved the crystal structure of the mutant LBD and performed biophysical studies that demonstrate the Y537S and to a lesser degree D538G can stabilize H12 in the agonist conformation, similar to wild-type ligand-bound ERα (Fig. 10.2d). This agonist conformation is believed to aid in dimerization and prevention of proteosomal degradation of unbound mutant ERα, further supporting relative resistance to fulvestrant [144]. In addition, these studies showed the ERα LBD mutations confer reduced affinity to estrogen and ERα antagonists that bind to the LBD, also explaining endocrine resistance. Additionally, the ERα LBD mutations possess the capacity to recruit coregulators in a ligand-independent manner as demonstrated by a nuclear receptor-coregulator interaction microarray assay [168]. Consistent with these studies, co-immunoprecipitation protein studies have demonstrated that in the absence of ligand mutant ERα binds to NCOA1/SRC1 and NCOA3/SRC3, key co-activators of ERα. Other potential mechanisms of constitutive activity and endocrine resistance is ligand-independent phosphorylation at S118 and S167 [165]. In addition, Y537 itself is a phosphorylation site, however, the impact of *ESR1* LBD mutations on this phosphorylation site and downstream functional consequences to the ERα LBD remains unclear [97].

Several groups have studied the transcriptome of distinct *ESR1* mutations and discovered these mutations induce multiple genes that are not induced by wild-type ERα upon estrogen stimulation [97, 104, 182, 183]. One study revealed transcription of these mutant alleles is associated with unique ERα chromatin recruitment/binding that contribute to neomorphic properties. The unique transcriptome of mutant ERα includes the upregulation of genes that are involved in epithelial-mesenchymal transition and metastases. The mechanisms of differential ER binding for mutant ERα variants are still unclear. One possible explanation supported by this study was unique ER-cofactor binding in the presence of *ESR1* mutations. However, additional studies are needed to better understand underlying mechanisms. Follow-up *in-vivo* studies provided functional evidence for the metastatic phenotypes of the Y537S and D538G mutations [97]. A number of studies have now shown that the Y537S and D538G mutations differ in their transcriptional effects, with the Y537S displaying a larger number of uniquely regulated genes. Other phenotypes, such as the constitutive transcription are also more profound in the Y537S mutation.

The Y537S and D538G allelic mutations are the most prevalent LBD mutations. However, other recurrent LBD mutations, namely the E380Q, S436P, L536H, Y536C, Y537N and Y537C were identified in clinical samples of ER+ metastatic breast cancer at a lower frequency. Other less frequent mutations include

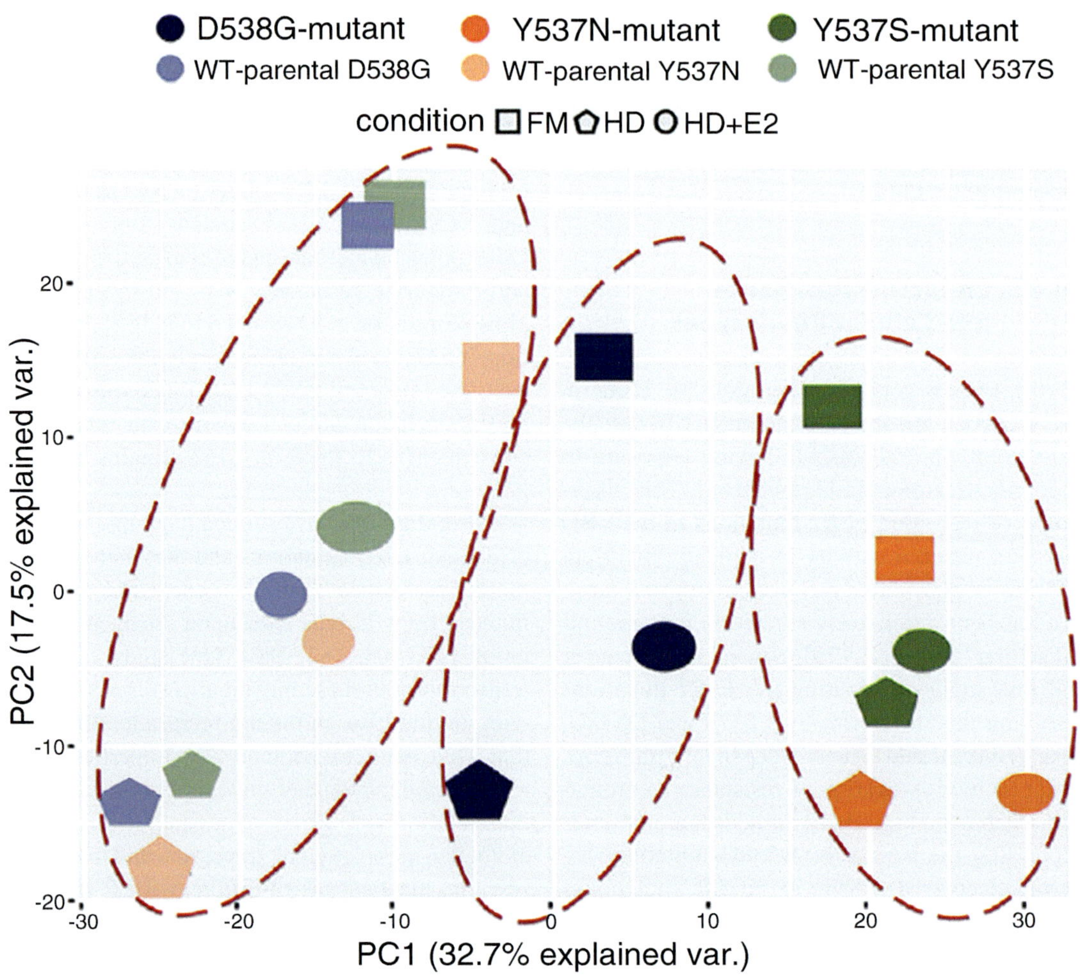

Fig. 10.4 Distinct transcriptional programs for wild-type and mutant *ESR1*. Principal-component analysis of transcriptomes from 6 cell lines (colors) under multiple treatment conditions [shapes] demonstrates distinct transcriptional programs. D538G-mutant (dark blue) and corresponding wild-type (WT) parental line (light blue). Y537N-mutant (orange) and corresponding WT parental line (light orange). Y537S-mutant (green) and corresponding WT parental line (light green). Treatment conditions include full medium [FM = square], hormone-depleted [HD = pentagon] and HD + estradiol [HD + E2 = circle]. (Adapted from Fig. 10.1b. In Jeselsohn et al. [97])

among others S432L, S461V, L466Q, V478L, V534E [144, 165, 175]. Cell lines modeling most of these mutations revealed ligand-dependent constitutive activity, with the exception of the S432L and V534E [165]. Additionally, the Y537S appeared to have the greatest effect on ligand-independent co-activator binding, phosphorylation and transcription compared to other mutations at the same 537 amino acid (Y537C, Y537N and Y537D) [104].

Earlier *in vitro* mutagenesis experiments identified LBD mutations that promote agonistic activity with tamoxifen or fulvestrant treatment, including L540Q, L543A/L544A [184, 185]. These variants, called "inversion" mutations, were found to have low activity in the absence of ligand and therefore are distinct from the recurrent constitutively active LBD mutations (Y537 and D538 alleles) that are found at a high frequency in patients with metastatic disease after treatment with an aromatase inhibitor. A more recent study, tested a panel of LBD mutations including V534N, L539D, M543K and L544D/E and showed that these mutations increase ERα transcriptional activity using a reporter assay in the presence of tamoxifen, fulvestrant and newer SERDS (GDC-0810 and GDC-0927). These mutations also have lower co-factor binding in the presence and absence of ligand [186]. At present, these mutations are infrequently observed however, their frequency may increase as newer more effective SERDs become available.

10.4.4 cfDNA Analysis of Clinical Trials

Discovery of *ESR1* mutations as a key driver of endocrine resistance in HR+ breast cancer has led to several retrospective studies evaluating for presence of these mutations in cell free DNA (cfDNA) [166, 187–191]. Plasma samples from the FERGI study were analyzed for 12 *ESR1* LBD mutations (E380Q, S436P, V534E, P535H, L536P, L536H, Y536S, Y536N, Y537C, D538G) [187]. FERGI was a phase II study of metastatic HR+ breast cancer patients who progressed on an aromatase inhibitor that were randomized to fulvestrant plus pictilisib, a pan-PI3Kinase inhibitor, versus fulvestrant plus placebo [192]. Analysis of 207 baseline plasma samples revealed *ESR1* mutations in 37% of patients who were enriched in luminal A type breast cancers and PIK3CA mutations. There was no significant association between the presence of a plasma *ESR1* mutation and progression-free survival with fulvestrant treatment or the combination of fulvestrant and pictilisib. Serial cfDNA testing in 71 of these patients also showed that a decreasing mutant *ESR1* allele fraction was associated with clinical response [187]. In previously referenced BOLERO-2 study-[131], baseline plasma samples were analyzed for Y537S and D538G mutations and these mutations were detected in 28.8% of 541 patients. Among these patients, 30 exhibited polyclonal Y537S and D538G mutations. Importantly, patients with either one of these mutations or both mutations had decreased overall survival, indicating that these mutations are prognostic of overall survival in metastatic HR+ breast cancer. Worse outcomes were seen in patients that had a Y537S mutation compared to a D538G mutation [193].

SoFEA was a phase III randomized study in which patients with metastatic HR+ breast cancer who had demonstrated prior sensitivity to a non-steroidal aromatase inhibitor were randomized to fulvestrant plus exemestane, fulvestrant plus placebo or exemestane [194]. Baseline plasma analysis for 7 *ESR1* LBD mutations (E380Q, L536R, Y537C, D538G, S436P, Y537N, Y537S) in a subset of 161 patients revealed plasma *ESR1* mutations in 39.1% of the patients. These patients had decreased benefit from exemestane when compared to a fulvestrant containing regimen (combining the two fulvestrant arms together). In contrast, there was no significant difference between exemestane and fulvestrant arms in patients with no plasma *ESR1* mutation [190]. Similarly, EFECT was a Phase III study [188] in which patients with advanced HR+ breast cancer who progressed on a non-steroidal aromatase inhibitor were randomized to exemestane versus fulvestrant [195]. The combined analysis of the SOFEA and EFECT studies included baseline cfDNA *ESR1* mutation data from 383 patients. Of these patients, 30% were found to have an *ESR1* mutation. The presence of an *ESR1* mutation was associated with significantly inferior outcomes in patients treated with exemestane versus fulvestrant, suggesting *ESR1* mutations are likely predictive of decreased benefit from an aromatase inhibitor [188].

In addition, pre-clinical data suggests *ESR1* mutations may also confer relative resistance to tamoxifen and fulvestrant [144, 165, 175]. However, clinical data to determine if the presence of the *ESR1* mutations associates with clinical benefit or lack thereof from tamoxifen or fulvestrant is limited [191]. PALOMA3 is a Phase III study in which patients who had progressed on a prior aromatase inhibitor were randomized to fulvetsrant plus palbociclib and fulvetsrant plus placebo [196]. An analysis of baseline cfDNA from 151 patients in the fulvestrant plus placebo arm only showed *ESR1* mutation present in 25% of the patients. Furthermore, patients with a mutation had decreased progression-free survival compared to patients without a mutation while on fulvestrant treatment, suggesting the *ESR1* mutations were associated with decreased benefit from fulvestrant compared to patients with wild-type *ESR1*. Overall, this study is retrospective and does not include a multivariate analysis, which limits interpretation of results [191].

Another retrospective analyses of baseline cfDNA samples from 360 patients on the PALOMA3 trial utilizing drop digital PCR analysis demonstrated *ESR1* mutations in 25.3% of patients previously treated with an aromatase inhibitor. Regardless of *ESR1* mutation status, patients had improved progression-free survival when treated with fulvestrant + palbociclib compared to fulvestrant plus placebo [190]. Subsequent retrospective cfDNA analysis of PALOMA-3 reveals enrichment of Y537S in both treatment arms to suggest a relative resistance to fulvestrant-based therapies [166].

10.4.5 Preclinical Evaluation of Therapeutic Vulnerabilities

Poor prognosis and treatment resistance to *ESR1* have spurred numerous preclinical studies aimed at identifying and targeting therapeutic vulnerabilities to *ESR1* mutations. Most leverage genome wide CRISPR knockout screens and transcriptomic analyses to identify potential ERα coregulators, kinases, growth factors, downstream receptors and epigenetic modifying proteins essential for *ESR1* mutant tumor growth [97, 175, 197].

NCOA3/SRC3 is a well-known co-activator in breast cancer [198]. In 2018, Gates et al., performed mass spectrometry based proteomic profiling of wild-type ERα, Y537S and D538G protein complexes to reveal NCOA3/SRC3 as potential target. In vivo pharmacologic inhibition of coactivator NCOA3/SRC3 in combination with endocrine therapy synergistically reduced transcriptional activity and growth of mutant expressing ERα, specifically Y537S [199].

CDK7 phosphorylates ERα at S118 similar to EGF which impacts downstream chromatin binding. In 2017, Harrod et al., demonstrated in vitro efficacy of CDK7 inhibitor THZ1 in MCF-7 cells expressing Y537S. Combined with fulvestrant, THZ1 reduced S118 phosphorylation, ERα-mediated gene expression and growth [104]. Our group confirmed these findings in vivo and reported on distinct allele-specific (ERα Y537S and D538G) cistromes and transcriptomes [97]. Similarly, Scott et al., noted hyperphosphorylation of ERα S294 in MCF-7 *ESR1* mutant Y537 and D538G cells compared to wild-type. CDK2 was discovered to phosphorylate ERα S294 resulting in ligand-independent ERα-mediated transcription. In vivo CDK2 inhibition with dinaciclib in combination with tamoxifen resulted in tumor regression of Y537S xenograft models [200].

In 2016, Mao et al., reported on constitutive hyperactivation of unfolded protein response (UPR) in *ESR1* Y537S and D538G mutant cells resulting in endocrine-resistant phenotype. A potent noncompetive ERα biomodulator, BHPI, previously discovered by their group [201], that activates UPR, was used to treat T47D *ESR1* wild-type and mutant cells revealing decreased estrogen stimulated growth. Similarly, treatment with BHPI also reduced progestin-stimulated growth [202].

Targeting stemness is also garnering significant interest. *ESR1* mutants demonstrate increased stem cell-like phenotype (CD44+/CD24- cells) and enhanced Notch signaling compared to wild-type. Inhibition of Notch by Geelsomini et al. using RO4929097 reduces mammosphere formation efficiency. However, other inhibitors of stem pathways such as Wnt/B-cantenin and sonic hedgehog have not been effective [203].

Targeting histone deacetylase (HDAC) is another attractive target due to multiple epigenetic mechanisms of resistance previously described (see Sect. 10.3.2), elevated HDAC levels in breast cancer and supportive in vivo preclinical data [204]. ENCORE301 is a Phase 2 study evaluating exemestane +/− entinostat (HDAC inhibitor) which revealed PFS 4.3 vs 2.3, OS 28.1 vs 19.8 [205]. However, disappointingly, E2112, a Phase 3 randomized placebo control study evaluating entinostat with exemestane versus placebo with exemestane did not show improved overall survival [206].

Finally, targeting the tumor immune microenvironment, cancer-associated fibroblasts and other stromal factors is still in its infancy but certainly an area of significant interest given success of anti-CTLA-4 and anti-PD-1 in other malignancies [138, 207]. The same has not been observed in breast cancer [208, 209] but the possibility remains that a deeper understanding of tumor-TME interactions and impact of endocrine therapy on immune function may aid in our identification of novel targets and therapeutic strategies.

10.4.6 Future Directions for Treatment

The next generation of endocrine therapy for breast cancer is heavily focused on ERα degradation and optimal sequencing of combinations with CDK4/6 and mTOR/PI3K/AKT pathway inhibitors [76, 210]. Fulvestrant, the only FDA approved SERD, is limited by a formulation that requires intramuscular injection and variable bioavailability that may limit maximal ERα receptor binding and degradation. Newer SERDs that offer oral administration are designed to optimize ERα receptor binding over time through improved absorption and pharmacokinetic properties. This is particularly attractive for *ESR1* mutations given constitutive ERα activity and relative clinical resistance to fulvestrant. Another similar but mechanistically different class of SERDs are proteolysis targeting chimeras (PROTACs) which are designed to leverage cellular ubiquitination machinery for enhanced degradation. Multiple SERDs and PROTACs are currently under development and clinical trial testing [210]. In 2021, the Menarini Group/Radius Heath® reported improved PFS and tolerability in their phase 3 study (EMERALD trial (NCT03778931)) evaluating their oral SERD, elacestrant vs either AI or fulvestrant in advanced HR+ breast cancer following progression on a CDK4/6 inhibitor in combination with either AI or fulvestrant. Of the 466 patients treated, 47% had *ESR1* mutations [211, 212]. Other SERDs in clinical development include amcenestrant [213], giredestrant [214], camizestrant [215], LY3484356 [216], ZN-c5 [217] and rintodestrant [218]. Many of these are also exploring first-line use in combination with CDK-4/6 inhibitor for advanced HR+ breast cancer as well single-agent use in early-stage breast cancer [210].

In addition to our evaluation of novel ERα targeting agents and drug combinations, future clinical studies are increasingly focused on developing and incorporating biomarkers, such as *ESR1* status to rationally tailor or personalize treatment regimens. Although prognostic, it is unclear if *ESR1* status will be predictive for any of the emerging class of SERDs. Studies such as SERENA-6 by AstraZeneca seek to leverage cfDNA technology along with biologic rationale of *ESR1* mutations to inform treatment selection. This is a phase 3 evaluating AZD9833 in combination with a CDK4/6 inhibitor in advanced HR+ breast cancer using detectable *ESR1* mutation on cfDNA before clinical progression to inform treatment selection [219]. An improved understanding of endocrine resistance and therapeutic agents is poised to fuel the next generation of advancements in endocrine therapy for breast cancer.

References

1. Jensen EV (1962) On the mechanism of estrogen action. Perspect Biol Med 6:47–59. https://doi.org/10.1353/pbm.1963.0005
2. Jensen EV, Jacobson HI, Walf AA, Frye CA (2010) Estrogen action: a historic perspective on the implications of considering alternative approaches. Physiol Behav 99:151–162. https://doi.org/10.1016/j.physbeh.2009.08.013
3. Kuiper GG, Enmark E, Pelto-Huikko M, Nilsson S, Gustafsson JA (1996) Cloning of a novel receptor expressed in rat prostate and ovary. Proc Natl Acad Sci U S A 93:5925–5930. https://doi.org/10.1073/pnas.93.12.5925

4. Mosselman S, Polman J, Dijkema R (1996) ER beta: identification and characterization of a novel human estrogen receptor. FEBS Lett 392:49–53. https://doi.org/10.1016/0014-5793(96)00782-x
5. Sever R, Glass CK (2013) Signaling by nuclear receptors. Cold Spring Harb Perspect Biol 5:a016709. https://doi.org/10.1101/cshperspect.a016709
6. Penvose A, Keenan JL, Bray D, Ramlall V, Siggers T (2019) Comprehensive study of nuclear receptor DNA binding provides a revised framework for understanding receptor specificity. Nat Commun 10:2514. https://doi.org/10.1038/s41467-019-10264-3
7. Bain DL, Heneghan AF, Connaghan-Jones KD, Miura MT (2007) Nuclear receptor structure: implications for function. Annu Rev Physiol 69:201–220. https://doi.org/10.1146/annurev.physiol.69.031905.160308
8. McKenna NJ, O'Malley BW (2002) Combinatorial control of gene expression by nuclear receptors and coregulators. Cell 108:465–474. https://doi.org/10.1016/s0092-8674(02)00641-4
9. Green S et al (1986) Human oestrogen receptor cDNA: sequence, expression and homology to v-erb-A. Nature 320:134–139. https://doi.org/10.1038/320134a0
10. Walter P et al (1985) Cloning of the human estrogen receptor cDNA. Proc Natl Acad Sci U S A 82:7889–7893. https://doi.org/10.1073/pnas.82.23.7889
11. Ogawa S et al (1998) The complete primary structure of human estrogen receptor β (hERβ) and its Heterodimerization with ER αin Vivoandin vitro. Biochem Biophys Res Commun 243:122–126. https://doi.org/10.1006/bbrc.1997.7893
12. Sand P, Luckhaus C, Schlurmann K, Götz M, Deckert J (2002) Untangling the human estrogen receptor gene structure. J Neural Transm 109:567–583. https://doi.org/10.1007/s007020200047
13. Kumar V et al (1987) Functional domains of the human estrogen receptor. Cell 51:941–951. https://doi.org/10.1016/0092-8674(87)90581-2
14. Bunce CM, Campbell MJ, Nilsson S, Gustafsson J-Å (2010) Estrogen receptors: their actions and functional roles in health and disease. In: Bunce CM, Campbell MJ (eds) Nuclear receptors. Springer, Dordrecht, pp 91–141. https://doi.org/10.1007/978-90-481-3303-1_5
15. Huang W et al (2018) Multidomain architecture of estrogen receptor reveals interfacial cross-talk between its DNA-binding and ligand-binding domains. Nat Commun 9:3520. https://doi.org/10.1038/s41467-018-06034-2
16. Métivier R et al (2003) Estrogen receptor-alpha directs ordered, cyclical, and combinatorial recruitment of cofactors on a natural target promoter. Cell 115:751–763. https://doi.org/10.1016/s0092-8674(03)00934-6
17. Katzenellenbogen BS (1996) Estrogen receptors: bioactivities and interactions with cell signaling pathways. Biol Reprod 54:287–293. https://doi.org/10.1095/biolreprod54.2.287
18. Bunone G, Briand PA, Miksicek RJ, Picard D (1996) Activation of the unliganded estrogen receptor by EGF involves the MAP kinase pathway and direct phosphorylation. EMBO J 15:2174–2183
19. Kato S et al (1995) Activation of the estrogen receptor through phosphorylation by mitogen-activated protein kinase. Science (New York, NY) 270:1491–1494. https://doi.org/10.1126/science.270.5241.1491
20. Cenni B, Picard D (1999) Ligand-independent activation of steroid receptors: new roles for old players. Trends Endocrinol Metab 10:41–46. https://doi.org/10.1016/s1043-2760(98)00121-0
21. Lavery DN, McEwan IJ (2005) Structure and function of steroid receptor AF1 transactivation domains: induction of active conformations. Biochem J 391:449–464. https://doi.org/10.1042/BJ20050872
22. Ignar-Trowbridge DM, Pimentel M, Teng CT, Korach KS, McLachlan JA (1995) Cross talk between peptide growth factor and estrogen receptor signaling systems. Environ Health Perspect 103(Suppl 7):35–38. https://doi.org/10.1289/ehp.95103s735
23. Métivier RI, Petit FG, Valotaire Y, Pakdel F (2000) Function of N-terminal transactivation domain of the estrogen receptor requires a potential α-helical structure and is negatively regulated by the a domain. Mol Endocrinol 14:1849–1871. https://doi.org/10.1210/mend.14.11.0546
24. Schwabe JW, Chapman L, Finch JT, Rhodes D, Neuhaus D (1993) DNA recognition by the oestrogen receptor: from solution to the crystal. Structure (London, England: 1993) 1:187–204. https://doi.org/10.1016/0969-2126(93)90020-h
25. Carroll JS et al (2005) Chromosome-wide mapping of estrogen receptor binding reveals long-range regulation requiring the forkhead protein FoxA1. Cell 122:33–43. https://doi.org/10.1016/j.cell.2005.05.008
26. Laganière J et al (2005) Location analysis of estrogen receptor α target promoters reveals that FOXA1 defines a domain of the estrogen response. Proc Natl Acad Sci U S A 102:11651–11656. https://doi.org/10.1073/pnas.0505575102
27. Lupien M et al (2008) FoxA1 translates epigenetic signatures into enhancer-driven lineage-specific transcription. Cell 132:958–970. https://doi.org/10.1016/j.cell.2008.01.018
28. Hurtado A, Holmes KA, Ross-Innes CS, Schmidt D, Carroll JS (2011) FOXA1 is a key determinant of estrogen receptor function and endocrine response. Nat Genet 43:27–33. https://doi.org/10.1038/ng.730
29. Wang C et al (2001) Direct acetylation of the estrogen receptor alpha hinge region by p300 regulates transactivation and hormone sensitivity. J Biol Chem 276:18375–18383. https://doi.org/10.1074/jbc.M100800200
30. Kim MY, Woo EM, Chong YTE, Homenko DR, Kraus WL (2006) Acetylation of estrogen receptor

alpha by p300 at lysines 266 and 268 enhances the deoxyribonucleic acid binding and transactivation activities of the receptor. Mol Endocrinol 20:1479–1493. https://doi.org/10.1210/me.2005-0531
31. Sentis S, Le Romancer M, Bianchin C, Rostan MC, Corbo L (2005) Sumoylation of the estrogen receptor alpha hinge region regulates its transcriptional activity. Mol Endocrinol 19:2671–2684. https://doi.org/10.1210/me.2005-0042
32. Berry NB, Fan M, Nephew KP (2008) Estrogen receptor-alpha hinge-region lysines 302 and 303 regulate receptor degradation by the proteasome. Mol Endocrinol 22:1535–1551. https://doi.org/10.1210/me.2007-0449
33. Moras D, Gronemeyer H (1998) The nuclear receptor ligand-binding domain: structure and function. Curr Opin Cell Biol 10:384–391. https://doi.org/10.1016/s0955-0674(98)80015-x
34. Brzozowski AM et al (1997) Molecular basis of agonism and antagonism in the oestrogen receptor. Nature 389:753–758. https://doi.org/10.1038/39645
35. Shiau AK et al (1998) The structural basis of estrogen receptor/coactivator recognition and the antagonism of this interaction by tamoxifen. Cell 95:927–937. https://doi.org/10.1016/s0092-8674(00)81717-1
36. Pike AC et al (1999) Structure of the ligand-binding domain of oestrogen receptor beta in the presence of a partial agonist and a full antagonist. EMBO J 18:4608–4618. https://doi.org/10.1093/emboj/18.17.4608
37. Shang Y, Hu X, DiRenzo J, Lazar MA, Brown M (2000) Cofactor dynamics and sufficiency in estrogen receptor-regulated transcription. Cell 103:843–852. https://doi.org/10.1016/s0092-8674(00)00188-4
38. Pike AC et al (2001) Structural insights into the mode of action of a pure antiestrogen. Structure (London, England: 1993) 9:145–153. https://doi.org/10.1016/s0969-2126(01)00568-8
39. Valentine JE, Kalkhoven E, White R, Hoare S, Parker MG (2000) Mutations in the estrogen receptor ligand binding domain discriminate between hormone-dependent transactivation and transrepression. J Biol Chem 275:25322–25329. https://doi.org/10.1074/jbc.M002497200
40. Tamrazi A, Carlson KE, Daniels JR, Hurth KM, Katzenellenbogen JA (2002) Estrogen receptor dimerization: ligand binding regulates dimer affinity and DimerDissociation rate. Mol Endocrinol 16:2706–2719. https://doi.org/10.1210/me.2002-0250
41. Tamrazi A, Carlson KE, Rodriguez AL, Katzenellenbogen JA (2005) Coactivator proteins as determinants of estrogen receptor structure and function: spectroscopic evidence for a novel coactivator-stabilized receptor conformation. Mol Endocrinol 19:1516–1528. https://doi.org/10.1210/me.2004-0458
42. Montano MM, Müller V, Trobaugh A, Katzenellenbogen BS (1995) The carboxy-terminal F domain of the human estrogen receptor: role in the transcriptional activity of the receptor and the effectiveness of antiestrogens as estrogen antagonists. Mol Endocrinol 9:814–825. https://doi.org/10.1210/mend.9.7.7476965
43. Kim K, Thu N, Saville B, Safe S (2003) Domains of estrogen receptor alpha (ERalpha) required for ERalpha/Sp1-mediated activation of GC-rich promoters by estrogens and antiestrogens in breast cancer cells. Mol Endocrinol 17:804–817. https://doi.org/10.1210/me.2002-0406
44. Gruber CJ, Tschugguel W, Schneeberger C, Huber JC (2002) Production and actions of estrogens. N Engl J Med 346:340–352. https://doi.org/10.1056/NEJMra000471
45. Webb P et al (1999) The estrogen receptor enhances AP-1 activity by two distinct mechanisms with different requirements for receptor transactivation functions. Mol Endocrinol 13:1672–1685. https://doi.org/10.1210/mend.13.10.0357
46. Nagy L, Schwabe JW (2004) Mechanism of the nuclear receptor molecular switch. Trends Biochem Sci 29:317–324. https://doi.org/10.1016/j.tibs.2004.04.006
47. Wärnmark A et al (2002) Interaction of transcriptional intermediary factor 2 nuclear receptor box peptides with the coactivator binding site of estrogen receptor alpha. J Biol Chem 277:21862–21868. https://doi.org/10.1074/jbc.M200764200
48. Pike AC (2006) Lessons learnt from structural studies of the oestrogen receptor. Best Pract Res Clin Endocrinol Metab 20:1–14. https://doi.org/10.1016/j.beem.2005.09.002
49. Wu YL et al (2005) Structural basis for an unexpected mode of SERM-mediated ER antagonism. Mol Cell 18:413–424. https://doi.org/10.1016/j.molcel.2005.04.014
50. Hu X, Lazar MA (1999) The CoRNR motif controls the recruitment of corepressors by nuclear hormone receptors. Nature 402:93–96. https://doi.org/10.1038/47069
51. Chang CY et al (1999) Dissection of the LXXLL nuclear receptor-coactivator interaction motif using combinatorial peptide libraries: discovery of peptide antagonists of estrogen receptors alpha and beta. Mol Cell Biol 19:8226–8239. https://doi.org/10.1128/MCB.19.12.8226
52. Kong EH et al (2005) Delineation of a unique protein-protein interaction site on the surface of the estrogen receptor. Proc Natl Acad Sci U S A 102:3593–3598. https://doi.org/10.1073/pnas.0407189102
53. Webb P et al (1998) Estrogen receptor activation function 1 works by binding p160 coactivator proteins. Mol Endocrinol 12:1605–1618. https://doi.org/10.1210/mend.12.10.0185
54. Métivier RI, Penot G, Flouriot G, Pakdel F (2001) Synergism between ERα transactivation function 1 (AF-1) and AF-2 mediated by steroid receptor coactivator Protein-1: requirement for the AF-1 α-helical Core and for a direct interaction between the N- and C-terminal domains. Mol Endocrinol 15:1953–1970. https://doi.org/10.1210/mend.15.11.0727

55. Onate SA et al (1998) The steroid receptor coactivator-1 contains multiple receptor interacting and activation domains that cooperatively enhance the activation function 1 (AF1) and AF2 domains of steroid receptors. J Biol Chem 273:12101–12108. https://doi.org/10.1074/jbc.273.20.12101
56. Benecke A, Chambon P, Gronemeyer H (2000) Synergy between estrogen receptor alpha activation functions AF1 and AF2 mediated by transcription intermediary factor TIF2. EMBO Rep 1:151–157. https://doi.org/10.1093/embo-reports/kvd028
57. Cheskis BJ, Karathanasis S, Lyttle CR (1997) Estrogen receptor ligands modulate its interaction with DNA. J Biol Chem 272:11384–11391. https://doi.org/10.1074/jbc.272.17.11384
58. Krieg AJ, Krieg SA, Ahn BS, Shapiro DJ (2004) Interplay between estrogen response element sequence and ligands controls in vivo binding of estrogen receptor to regulated genes. J Biol Chem 279:5025–5034. https://doi.org/10.1074/jbc.M307076200
59. Treviño LS, Weigel NL (2013) Phosphorylation: a fundamental regulator of steroid receptor action. Trends Endocrinol Metab 24:515–524. https://doi.org/10.1016/j.tem.2013.05.008
60. Shao D, Lazar MA (1999) Modulating nuclear receptor function: may the phos be with you. J Clin Invest 103:1617–1618. https://doi.org/10.1172/JCI7421
61. Smith CL, Conneely OM, O'Malley BW (1993) Modulation of the ligand-independent activation of the human estrogen receptor by hormone and antihormone. Proc Natl Acad Sci U S A 90:6120–6124. https://doi.org/10.1073/pnas.90.13.6120
62. Ignar-Trowbridge DM et al (1993) Peptide growth factors elicit estrogen receptor-dependent transcriptional activation of an estrogen-responsive element. Mol Endocrinol 7:992–998. https://doi.org/10.1210/mend.7.8.8232319
63. Newton CJ et al (1994) The unliganded estrogen receptor (ER) transduces growth factor signals. J Steroid Biochem Mol Biol 48:481–486. https://doi.org/10.1016/0960-0760(94)90197-x
64. Pietras RJ et al (1995) HER-2 tyrosine kinase pathway targets estrogen receptor and promotes hormone-independent growth in human breast cancer cells. Oncogene 10:2435–2446
65. Gruber CJ, Gruber DM, Gruber IM, Wieser F, Huber JC (2004) Anatomy of the estrogen response element. Trends Endocrinol Metab 15:73–78. https://doi.org/10.1016/j.tem.2004.01.008
66. Zilliacus J, Wright AP, Carlstedt-Duke J, Gustafsson JA (1995) Structural determinants of DNA-binding specificity by steroid receptors. Mol Endocrinol 9:389–400. https://doi.org/10.1210/mend.9.4.7659083
67. Klinge CM (2001) Estrogen receptor interaction with estrogen response elements. Nucleic Acids Res 29:2905–2919. https://doi.org/10.1093/nar/29.14.2905
68. Anolik JH, Klinge CM, Hilf R, Bambara RA (1995) Cooperative binding of estrogen receptor to DNA depends on spacing of binding sites, flanking sequence, and ligand. Biochemistry 34:2511–2520. https://doi.org/10.1021/bi00008a015
69. Yi P et al (2002) The effects of estrogen-responsive element- and ligand-induced structural changes on the recruitment of cofactors and transcriptional responses by ER alpha and ER beta. Mol Endocrinol 16:674–693. https://doi.org/10.1210/mend.16.4.0810
70. Peale FV Jr, Ludwig LB, Zain S, Hilf R, Bambara RA (1988) Properties of a high-affinity DNA binding site for estrogen receptor. Proc Natl Acad Sci U S A 85:1038–1042. https://doi.org/10.1073/pnas.85.4.1038
71. Klein-Hitpass L, Ryffel GU, Heitlinger E, Cato AC (1988) A 13 bp palindrome is a functional estrogen responsive element and interacts specifically with estrogen receptor. Nucleic Acids Res 16:647–663. https://doi.org/10.1093/nar/16.2.647
72. Driscoll MD et al (1998) Sequence requirements for estrogen receptor binding to estrogen response elements. J Biol Chem 273:29321–29330. https://doi.org/10.1074/jbc.273.45.29321
73. Palaniappan M et al (2019) The genomic landscape of estrogen receptor α binding sites in mouse mammary gland. PLoS One 14:e0220311–e0220311. https://doi.org/10.1371/journal.pone.0220311
74. Grober OM et al (2011) Global analysis of estrogen receptor beta binding to breast cancer cell genome reveals an extensive interplay with estrogen receptor alpha for target gene regulation. BMC Genomics 12:36. https://doi.org/10.1186/1471-2164-12-36
75. Lee H-R, Kim T-H, Choi K-C (2012) Functions and physiological roles of two types of estrogen receptors, ERα and ERβ, identified by estrogen receptor knockout mouse. Lab Anim Res 28:71–76. https://doi.org/10.5625/lar.2012.28.2.71
76. Chi D et al (2019) Estrogen receptor signaling is reprogrammed during breast tumorigenesis. Proc Natl Acad Sci 116:11437–11443. https://doi.org/10.1073/pnas.1819155116
77. Hammond ME et al (2010) American Society of Clinical Oncology/College Of American Pathologists guideline recommendations for immunohistochemical testing of estrogen and progesterone receptors in breast cancer. J Clin Oncol 28:2784–2795. https://doi.org/10.1200/jco.2009.25.6529
78. Barnes DM, Harris WH, Smith P, Millis RR, Rubens RD (1996) Immunohistochemical determination of oestrogen receptor: comparison of different methods of assessment of staining and correlation with clinical outcome of breast cancer patients. Br J Cancer 74:1445–1451. https://doi.org/10.1038/bjc.1996.563
79. Harvey JM, Clark GM, Osborne CK, Allred DC (1999) Estrogen receptor status by immunohistochemistry is superior to the ligand-binding assay

for predicting response to adjuvant endocrine therapy in breast cancer. J Clin Oncol 17:1474–1474. https://doi.org/10.1200/jco.1999.17.5.1474
80. Elledge RM et al (2000) Estrogen receptor (ER) and progesterone receptor (PgR), by ligand-binding assay compared with ER, PgR and pS2, by immuno-histochemistry in predicting response to tamoxifen in metastatic breast cancer: A Southwest Oncology Group study. Int J Cancer 89:111–117. https://doi.org/10.1002/(SICI)1097--0215(20000320)89:2<111::AID-IJC2>3.0.CO;2-W
81. Baum M et al (1983) Improved survival among patients treated with adjuvant tamoxifen after mastectomy for early breast cancer. Lancet (London, England) 2:450. https://doi.org/10.1016/s0140-6736(83)90406-3
82. (1987) Adjuvant tamoxifen in the management of operable breast cancer: the Scottish trial. Report from the breast cancer trials committee, Scottish cancer trials office (MRC), Edinburgh. Lancet (London, England) 2:171–175
83. Fisher B et al (2001) Tamoxifen and chemotherapy for axillary node-negative, estrogen receptor-negative breast cancer: findings from National Surgical Adjuvant Breast and Bowel Project B-23. J Clin Oncol 19:931–942. https://doi.org/10.1200/jco.2001.19.4.931
84. Mohsin SK et al (2004) Progesterone receptor by immunohistochemistry and clinical outcome in breast cancer: a validation study. Mod Pathol 17:1545–1554. https://doi.org/10.1038/modpathol.3800229
85. Lindström LS et al (2018) Intratumor heterogeneity of the estrogen receptor and the long-term risk of fatal breast cancer. J Natl Cancer Inst 110:726–733. https://doi.org/10.1093/jnci/djx270
86. Curigliano G et al (2017) De-escalating and escalating treatments for early-stage breast cancer: the St. Gallen international expert consensus conference on the primary therapy of early breast cancer 2017. Ann Oncol 28:1700–1712. https://doi.org/10.1093/annonc/mdx308
87. Goldhirsch A et al (2003) Meeting highlights: updated international expert consensus on the primary therapy of early breast cancer. J Clin Oncol 21:3357–3365. https://doi.org/10.1200/jco.2003.04.576
88. Perou CM et al (2000) Molecular portraits of human breast tumours. Nature 406:747–752. https://doi.org/10.1038/35021093
89. Sørlie T et al (2001) Gene expression patterns of breast carcinomas distinguish tumor subclasses with clinical implications. Proc Natl Acad Sci 98:10869–10874. https://doi.org/10.1073/pnas.191367098
90. Yanagawa M et al (2012) Luminal A and luminal B (HER2 negative) subtypes of breast cancer consist of a mixture of tumors with different genotype. BMC Res Notes 5:376. https://doi.org/10.1186/1756-0500-5-376
91. Goldhirsch A et al (2011) Strategies for subtypes--dealing with the diversity of breast cancer: highlights of the St. Gallen international expert consensus on the primary therapy of early breast cancer 2011. Ann Oncol 22:1736–1747. https://doi.org/10.1093/annonc/mdr304
92. Sparano JA et al (2015) Prospective validation of a 21-gene expression assay in breast cancer. N Engl J Med 373:2005–2014. https://doi.org/10.1056/NEJMoa1510764
93. Knauer M et al (2010) The predictive value of the 70-gene signature for adjuvant chemotherapy in early breast cancer. Breast Cancer Res Treat 120:655–661. https://doi.org/10.1007/s10549-010-0814-2
94. Sparano JA et al (2019) Clinical and genomic risk to guide the use of adjuvant therapy for breast cancer. N Engl J Med 380:2395–2405. https://doi.org/10.1056/NEJMoa1904819
95. Paik S et al (2004) A multigene assay to predict recurrence of tamoxifen-treated, node-negative breast cancer. New Eng J Med 351:2817–2826. https://doi.org/10.1056/NEJMoa041588
96. Sparano JA et al (2018) Adjuvant chemotherapy guided by a 21-gene expression assay in breast cancer. N Engl J Med 379:111–121. https://doi.org/10.1056/NEJMoa1804710
97. Jeselsohn R et al (2018) Allele-specific chromatin recruitment and therapeutic vulnerabilities of ESR1 activating mutations. Cancer Cell 33:173–186 e175. https://doi.org/10.1016/j.ccell.2018.01.004
98. Green KA, Carroll JS (2007) Oestrogen-receptor-mediated transcription and the influence of co-factors and chromatin state. Nat Rev Cancer 7:713–722. https://doi.org/10.1038/nrc2211
99. Siersbæk R, Kumar S, Carroll JS (2018) Signaling pathways and steroid receptors modulating estrogen receptor α function in breast cancer. Genes Dev 32:1141–1154. https://doi.org/10.1101/gad.316646.118
100. Anbalagan M, Rowan BG (2015) Estrogen receptor alpha phosphorylation and its functional impact in human breast cancer. Mol Cell Endocrinol 418(Pt 3):264–272. https://doi.org/10.1016/j.mce.2015.01.016
101. Stender JD et al (2017) Structural and molecular mechanisms of cytokine-mediated endocrine resistance in human breast cancer cells. Mol Cell 65:1122–1135.e1125. https://doi.org/10.1016/j.molcel.2017.02.008
102. Le Goff P, Montano MM, Schodin DJ, Katzenellenbogen BS (1994) Phosphorylation of the human estrogen receptor. Identification of hormone-regulated sites and examination of their influence on transcriptional activity. J Biol Chem 269:4458–4466
103. Chen D et al (2002) Phosphorylation of human estrogen receptor α at serine 118 by two distinct signal transduction pathways revealed by phosphorylation-

specific antisera. Oncogene 21:4921–4931. https://doi.org/10.1038/sj.onc.1205420
104. Harrod A et al (2017) Genomic modelling of the ESR1 Y537S mutation for evaluating function and new therapeutic approaches for metastatic breast cancer. Oncogene 36:2286–2296. https://doi.org/10.1038/onc.2016.382
105. Lupien M et al (2010) Growth factor stimulation induces a distinct ER(alpha) cistrome underlying breast cancer endocrine resistance. Genes Dev 24:2219–2227. https://doi.org/10.1101/gad.1944810
106. Magnani L et al (2015) The pioneer factor PBX1 is a novel driver of metastatic progression in ERα-positive breast cancer. Oncotarget 6:21878–21891. https://doi.org/10.18632/oncotarget.4243
107. Joel PB et al (1998) pp90rsk1 regulates estrogen receptor-mediated transcription through phosphorylation of Ser-167. Mol Cell Biol 18:1978–1984. https://doi.org/10.1128/mcb.18.4.1978
108. Becker MA, Ibrahim YH, Cui X, Lee AV, Yee D (2011) The IGF pathway regulates ERα through a S6K1-dependent mechanism in breast cancer cells. Mol Endocrinol 25:516–528. https://doi.org/10.1210/me.2010-0373
109. Yamnik RL, Holz MK (2010) mTOR/S6K1 and MAPK/RSK signaling pathways coordinately regulate estrogen receptor alpha serine 167 phosphorylation. FEBS Lett 584:124–128. https://doi.org/10.1016/j.febslet.2009.11.041
110. Michalides R et al (2004) Tamoxifen resistance by a conformational arrest of the estrogen receptor alpha after PKA activation in breast cancer. Cancer Cell 5:597–605. https://doi.org/10.1016/j.ccr.2004.05.016
111. de Leeuw R et al (2013) PKA phosphorylation redirects ERα to promoters of a unique gene set to induce tamoxifen resistance. Oncogene 32:3543–3551. https://doi.org/10.1038/onc.2012.361
112. Finak G et al (2008) Stromal gene expression predicts clinical outcome in breast cancer. Nat Med 14:518–527. https://doi.org/10.1038/nm1764
113. Brechbuhl HM et al (2017) Fibroblast subtypes regulate responsiveness of luminal breast cancer to estrogen. Clin Cancer Res 23:1710–1721. https://doi.org/10.1158/1078-0432.ccr-15-2851
114. Huang J et al (2017) Downregulation of estrogen receptor and modulation of growth of breast cancer cell lines mediated by paracrine stromal cell signals. Breast Cancer Res Treat 161:229–243. https://doi.org/10.1007/s10549-016-4052-0
115. Morgan MM et al (2018) Mammary fibroblasts reduce apoptosis and speed estrogen-induced hyperplasia in an organotypic MCF7-derived duct model. Sci Rep 8:7139. https://doi.org/10.1038/s41598-018-25461-1
116. Madeddu C et al (2014) Role of inflammation and oxidative stress in post-menopausal oestrogen-dependent breast cancer. J Cell Mol Med 18:2519–2529. https://doi.org/10.1111/jcmm.12413
117. Castellaro AM, Rodriguez-Baili MC, Di Tada CE, Gil GA (2019) Tumor-associated macrophages induce endocrine therapy resistance in ER+ breast cancer cells. Cancers (Basel) 11:189. https://doi.org/10.3390/cancers11020189
118. Wu L et al (2003) Smad4 as a transcription corepressor for estrogen receptor alpha. J Biol Chem 278:15192–15200. https://doi.org/10.1074/jbc.M212332200
119. Ren Y et al (2009) Dual effects of TGF-beta on ERalpha-mediated estrogenic transcriptional activity in breast cancer. Mol Cancer 8:111. https://doi.org/10.1186/1476-4598-8-111
120. Deckers M et al (2006) The tumor suppressor Smad4 is required for transforming growth factor beta-induced epithelial to mesenchymal transition and bone metastasis of breast cancer cells. Cancer Res 66:2202–2209. https://doi.org/10.1158/0008-5472.can-05-3560
121. Magnani L, Ballantyne EB, Zhang X, Lupien M (2011) PBX1 genomic pioneer function drives ERα signaling underlying progression in breast cancer. PLoS Genet 7:e1002368. https://doi.org/10.1371/journal.pgen.1002368
122. Theodorou V, Stark R, Menon S, Carroll JS (2013) GATA3 acts upstream of FOXA1 in mediating ESR1 binding by shaping enhancer accessibility. Genome Res 23:12–22. https://doi.org/10.1101/gr.139469.112
123. Tan SK et al (2011) AP-2γ regulates oestrogen receptor-mediated long-range chromatin interaction and gene transcription. EMBO J 30:2569–2581. https://doi.org/10.1038/emboj.2011.151
124. Pradhan M, Baumgarten SC, Bembinster LA, Frasor J (2012) CBP mediates NF-κB-dependent histone acetylation and estrogen receptor recruitment to an estrogen response element in the BIRC3 promoter. Mol Cell Biol 32:569–575. https://doi.org/10.1128/MCB.05869-11
125. Franco HL, Nagari A, Kraus WL (2015) TNFα signaling exposes latent estrogen receptor binding sites to alter the breast cancer cell transcriptome. Mol Cell 58:21–34. https://doi.org/10.1016/j.molcel.2015.02.001
126. Ross-Innes CS et al (2012) Differential oestrogen receptor binding is associated with clinical outcome in breast cancer. Nature 481:389–393. https://doi.org/10.1038/nature10730
127. Surveillance E, and End Results (SEER) Program (www.seer.cancer.gov) SEER*Stat Database: Incidence – SEER Research Data, 9 Registries, Nov 2020 Sub (1975–2018) – Linked To County Attributes – Time Dependent (1990–2018) Income/Rurality, 1969–2019 Counties, National Cancer Institute, DCCPS, Surveillance Research Program, released April 2021, based on the November 2020 submission
128. Nathan MR, Schmid P (2017) A review of Fulvestrant in breast cancer. Oncol Ther 5:17–29. https://doi.org/10.1007/s40487-017-0046-2

129. Brodie A (2002) Aromatase inhibitors in breast cancer. Trends Endocrinol Metab 13:61–65. https://doi.org/10.1016/s1043-2760(01)00529-x
130. Klein ME, Kovatcheva M, Davis LE, Tap WD, Koff A (2018) CDK4/6 inhibitors: the mechanism of action may not be as simple as once thought. Cancer Cell 34:9–20. https://doi.org/10.1016/j.ccell.2018.03.023
131. Baselga J et al (2011) Everolimus in postmenopausal hormone-receptor–positive advanced breast cancer. N Engl J Med 366:520–529. https://doi.org/10.1056/NEJMoa1109653
132. Quirke VM (2017) Tamoxifen from failed contraceptive pill to best-selling breast cancer medicine: a case-study in pharmaceutical innovation. Front Pharmacol 8:620–620. https://doi.org/10.3389/fphar.2017.00620
133. Jordan VC (2003) Tamoxifen: a most unlikely pioneering medicine. Nat Rev Drug Discov 2:205–213. https://doi.org/10.1038/nrd1031
134. Smith CL, Nawaz Z, O'Malley BW (1997) Coactivator and corepressor regulation of the agonist/antagonist activity of the mixed antiestrogen, 4-hydroxytamoxifen. Mol Endocrinol 11:657–666. https://doi.org/10.1210/mend.11.6.0009
135. Sunderland MC, Osborne CK (1991) Tamoxifen in premenopausal patients with metastatic breast cancer: a review. J Clin Oncol 9:1283–1297. https://doi.org/10.1200/jco.1991.9.7.1283
136. Robertson JF, Harrison M (2004) Fulvestrant: pharmacokinetics and pharmacology. Br J Cancer 90(Suppl 1):S7–S10. https://doi.org/10.1038/sj.bjc.6601630
137. Musgrove EA, Sutherland RL (2009) Biological determinants of endocrine resistance in breast cancer. Nat Rev Cancer 9:631–643. https://doi.org/10.1038/nrc2713
138. Hanker AB, Sudhan DR, Arteaga CL (2020) Overcoming endocrine resistance in breast cancer. Cancer Cell 37:496–513. https://doi.org/10.1016/j.ccell.2020.03.009
139. Finn RS et al (2016) Palbociclib and Letrozole in advanced breast cancer. N Engl J Med 375:1925–1936. https://doi.org/10.1056/NEJMoa1607303
140. Bertucci F et al (2019) Genomic characterization of metastatic breast cancers. Nature 569:560–564. https://doi.org/10.1038/s41586-019-1056-z
141. Griffith OL et al (2018) The prognostic effects of somatic mutations in ER-positive breast cancer. Nat Commun 9:3476. https://doi.org/10.1038/s41467-018-05914-x
142. Angus L et al (2019) The genomic landscape of metastatic breast cancer highlights changes in mutation and signature frequencies. Nat Genet 51:1450–1458. https://doi.org/10.1038/s41588-019-0507-7
143. Razavi P et al (2018) The genomic landscape of endocrine-resistant advanced breast cancers. Cancer Cell 34:427–438.e426. https://doi.org/10.1016/j.ccell.2018.08.008
144. Jeselsohn R et al (2014) Emergence of constitutively active estrogen receptor-alpha mutations in pretreated advanced estrogen receptor-positive breast cancer. Clin Cancer Res 20:1757–1767. https://doi.org/10.1158/1078-0432.ccr-13-2332
145. Patten DK et al (2018) Enhancer mapping uncovers phenotypic heterogeneity and evolution in patients with luminal breast cancer. Nat Med 24:1469–1480. https://doi.org/10.1038/s41591-018-0091-x
146. Hinohara K et al (2018) KDM5 histone demethylase activity links cellular transcriptomic heterogeneity to therapeutic resistance. Cancer Cell 34:939–953.e939. https://doi.org/10.1016/j.ccell.2018.10.014
147. Gala K et al (2018) KMT2C mediates the estrogen dependence of breast cancer through regulation of ERα enhancer function. Oncogene 37:4692–4710. https://doi.org/10.1038/s41388-018-0273-5
148. Jin K et al (2015) HOXB7 is an ERα cofactor in the activation of HER2 and multiple ER target genes leading to endocrine resistance. Cancer Discov 5:944–959. https://doi.org/10.1158/2159-8290.cd-15-0090
149. Jeselsohn R et al (2017) Embryonic transcription factor SOX9 drives breast cancer endocrine resistance. Proc Natl Acad Sci 114:E4482–E4491. https://doi.org/10.1073/pnas.1620993114
150. Fu X et al (2016) FOXA1 overexpression mediates endocrine resistance by altering the ER transcriptome and IL-8 expression in ER-positive breast cancer. Proc Natl Acad Sci 113:E6600–E6609. https://doi.org/10.1073/pnas.1612835113
151. Xiao T et al (2018) Estrogen-regulated feedback loop limits the efficacy of estrogen receptor–targeted breast cancer therapy. Proc Natl Acad Sci 115:7869–7878. https://doi.org/10.1073/pnas.1722617115
152. Hong SP et al (2019) Single-cell transcriptomics reveals multi-step adaptations to endocrine therapy. Nat Commun 10:3840. https://doi.org/10.1038/s41467-019-11721-9
153. Lei JT et al (2018) Functional annotation of ESR1 gene fusions in estrogen receptor-positive breast cancer. Cell Rep 24:1434–1444.e1437. https://doi.org/10.1016/j.celrep.2018.07.009
154. Generali D et al (2006) Hypoxia-inducible factor-1alpha expression predicts a poor response to primary chemoendocrine therapy and disease-free survival in primary human breast cancer. Clin Cancer Res 12:4562–4568. https://doi.org/10.1158/1078-0432.ccr-05-2690
155. Yang J et al (2015) Estrogen receptor-α directly regulates the hypoxia-inducible factor 1 pathway associated with antiestrogen response in breast cancer. Proc Natl Acad Sci U S A 112:15172–15177. https://doi.org/10.1073/pnas.1422015112
156. Todd VM et al (2021) Hypoxia inducible factor signaling in breast tumors controls spontaneous tumor dissemination in a site-specific manner. Commun Biol 4:1122. https://doi.org/10.1038/s42003-021-02648-3

157. Houthuijzen JM, Jonkers J (2018) Cancer-associated fibroblasts as key regulators of the breast cancer tumor microenvironment. Cancer Metastasis Rev 37:577–597. https://doi.org/10.1007/s10555-018-9768-3
158. Louault K et al (2019) Interactions between cancer-associated fibroblasts and tumor cells promote MCL-1 dependency in estrogen receptor-positive breast cancers. Oncogene 38:3261–3273. https://doi.org/10.1038/s41388-018-0635-z
159. Jallow F et al (2019) Dynamic interactions between the extracellular matrix and estrogen activity in progression of ER+ breast cancer. Oncogene 38:6913–6925. https://doi.org/10.1038/s41388-019-0941-0
160. Reyes-Ramos AM et al (2021) Collagen I fibrous substrates modulate the proliferation and Secretome of estrogen receptor-positive breast tumor cells in a hormone-restricted microenvironment. ACS Biomater Sci Eng 7:2430–2443. https://doi.org/10.1021/acsbiomaterials.0c01803
161. Joffroy CM et al (2010) Antiestrogens induce transforming growth factor beta-mediated immunosuppression in breast cancer. Cancer Res 70:1314–1322. https://doi.org/10.1158/0008-5472.can-09-3292
162. Anurag M et al (2019) Immune checkpoint profiles in luminal B breast cancer (Alliance). J Natl Cancer Inst 112:737–746. https://doi.org/10.1093/jnci/djz213
163. Karnik PS, Kulkarni S, Liu XP, Budd GT, Bukowski RM (1994) Estrogen receptor mutations in tamoxifen-resistant breast cancer. Cancer Res 54:349–353
164. Zhang QX, Borg A, Wolf DM, Oesterreich S, Fuqua SA (1997) An estrogen receptor mutant with strong hormone-independent activity from a metastatic breast cancer. Cancer Res 57:1244–1249
165. Toy W et al (2013) ESR1 ligand-binding domain mutations in hormone-resistant breast cancer. Nat Genet 45:1439–1445. https://doi.org/10.1038/ng.2822
166. O'Leary B et al (2018) The genetic landscape and clonal evolution of breast cancer resistance to Palbociclib plus Fulvestrant in the PALOMA-3 trial. Cancer Discov 8:1390–1403. https://doi.org/10.1158/2159-8290.cd-18-0264
167. Koboldt DC et al (2012) Comprehensive molecular portraits of human breast tumours. Nature 490:61–70. https://doi.org/10.1038/nature11412
168. Jeselsohn R, Buchwalter G, De Angelis C, Brown M, Schiff R (2015) ESR1 mutations-a mechanism for acquired endocrine resistance in breast cancer. Nat Rev Clin Oncol 12:573–583. https://doi.org/10.1038/nrclinonc.2015.117
169. Dustin D, Gu G, Fuqua SAW (2019) ESR1 mutations in breast cancer. Cancer 125:3714–3728. https://doi.org/10.1002/cncr.32345
170. Holst F et al (2007) Estrogen receptor alpha (ESR1) gene amplification is frequent in breast cancer. Nat Genet 39:655–660. https://doi.org/10.1038/ng2006
171. Tomita S et al (2009) Estrogen receptor alpha gene ESR1 amplification may predict endocrine therapy responsiveness in breast cancer patients. Cancer Sci 100:1012–1017. https://doi.org/10.1111/j.1349-7006.2009.01145.x
172. Reis-Filho JS et al (2008) ESR1 gene amplification in breast cancer: a common phenomenon? Nat Genet 40:809–810; author reply 810-802. https://doi.org/10.1038/ng0708-809b
173. Ooi A et al (2012) Gene amplification of ESR1 in breast cancers--fact or fiction? A fluorescence in situ hybridization and multiplex ligation-dependent probe amplification study. J Pathol 227:8–16. https://doi.org/10.1002/path.3974
174. Veeraraghavan J et al (2014) Recurrent ESR1–CCDC170 rearrangements in an aggressive subset of oestrogen receptor-positive breast cancers. Nat Commun 5:4577. https://doi.org/10.1038/ncomms5577
175. Li S et al (2013) Endocrine-therapy-resistant ESR1 variants revealed by genomic characterization of breast-cancer-derived xenografts. Cell Rep 4:1116–1130. https://doi.org/10.1016/j.celrep.2013.08.022
176. Robinson DR et al (2013) Activating ESR1 mutations in hormone-resistant metastatic breast cancer. Nat Genet 45:1446–1451. https://doi.org/10.1038/ng.2823
177. Merenbakh-Lamin K et al (2013) D538G mutation in estrogen receptor-α: a novel mechanism for acquired endocrine resistance in breast cancer. Cancer Res 73:6856–6864. https://doi.org/10.1158/0008-5472.can-13-1197
178. Carlson KE, Choi I, Gee A, Katzenellenbogen BS, Katzenellenbogen JA (1997) Altered ligand binding properties and enhanced stability of a constitutively active estrogen receptor: evidence that an open pocket conformation is required for ligand interaction. Biochemistry 36:14897–14905. https://doi.org/10.1021/bi971746l
179. Lazennec G, Ediger TR, Petz LN, Nardulli AM, Katzenellenbogen BS (1997) Mechanistic aspects of estrogen receptor activation probed with constitutively active estrogen receptors: correlations with DNA and coregulator interactions and receptor conformational changes. Mol Endocrinol 11:1375–1386. https://doi.org/10.1210/mend.11.9.9983
180. Nettles KW et al (2008) NFκB selectivity of estrogen receptor ligands revealed by comparative crystallographic analyses. Nat Chem Biol 4:241–247. https://doi.org/10.1038/nchembio.76
181. Fanning SW et al (2016) Estrogen receptor alpha somatic mutations Y537S and D538G confer breast cancer endocrine resistance by stabilizing the activating function-2 binding conformation. elife 5. https://doi.org/10.7554/eLife.12792
182. Bahreini A et al (2017) Mutation site and context dependent effects of ESR1 mutation in genome-edited breast cancer cell models. Breast Cancer Res 19:60. https://doi.org/10.1186/s13058-017-0851-4

183. Martin L-A et al (2017) Discovery of naturally occurring ESR1 mutations in breast cancer cell lines modelling endocrine resistance. Nat Commun 8:1865. https://doi.org/10.1038/s41467-017-01864-y
184. Arao Y, Hamilton KJ, Coons LA, Korach KS (2013) Estrogen receptor α L543A,L544A mutation changes antagonists to agonists, correlating with the ligand binding domain dimerization associated with DNA binding activity. J Biol Chem 288:21105–21116. https://doi.org/10.1074/jbc.M113.463455
185. Katzenellenbogen JA, Mayne CG, Katzenellenbogen BS, Greene GL, Chandarlapaty S (2018) Structural underpinnings of oestrogen receptor mutations in endocrine therapy resistance. Nat Rev Cancer 18:377–388. https://doi.org/10.1038/s41568-018-0001-z
186. Guan J et al (2019) Therapeutic ligands antagonize estrogen receptor function by impairing its mobility. Cell 178:949–963.e918. https://doi.org/10.1016/j.cell.2019.06.026
187. Spoerke JM et al (2016) Heterogeneity and clinical significance of ESR1 mutations in ER-positive metastatic breast cancer patients receiving fulvestrant. Nat Commun 7:11579. https://doi.org/10.1038/ncomms11579
188. Turner NC et al (2020) ESR1 mutations and overall survival on Fulvestrant versus Exemestane in advanced hormone receptor–positive breast cancer: a combined analysis of the phase III SoFEA and EFECT trials. Clin Cancer Res 26:5172–5177. https://doi.org/10.1158/1078-0432.ccr-20-0224
189. Schiavon G et al (2015) Analysis of ESR1 mutation in circulating tumor DNA demonstrates evolution during therapy for metastatic breast cancer. Sci Transl Med 7:313ra182. https://doi.org/10.1126/scitranslmed.aac7551
190. Fribbens C et al (2016) Plasma ESR1 mutations and the treatment of estrogen receptor-positive advanced breast cancer. J Clin Oncol 34:2961–2968. https://doi.org/10.1200/jco.2016.67.3061
191. O'Leary B et al (2018) Early circulating tumor DNA dynamics and clonal selection with palbociclib and fulvestrant for breast cancer. Nat Commun 9:896. https://doi.org/10.1038/s41467-018-03215-x
192. Krop IE et al (2016) Pictilisib for oestrogen receptor-positive, aromatase inhibitor-resistant, advanced or metastatic breast cancer (FERGI): a randomised, double-blind, placebo-controlled, phase 2 trial. Lancet Oncol 17:811–821. https://doi.org/10.1016/s1470-2045(16)00106-6
193. Chandarlapaty S et al (2016) Prevalence of ESR1 mutations in cell-free DNA and outcomes in metastatic breast cancer: a secondary analysis of the BOLERO-2 clinical trial. JAMA Oncol 2:1310–1315. https://doi.org/10.1001/jamaoncol.2016.1279
194. Johnston SR et al (2013) Fulvestrant plus anastrozole or placebo versus exemestane alone after progression on non-steroidal aromatase inhibitors in postmenopausal patients with hormone-receptor-positive locally advanced or metastatic breast cancer (SoFEA): a composite, multicentre, phase 3 randomised trial. Lancet Oncol 14:989–998. https://doi.org/10.1016/s1470-2045(13)70322-x
195. Chia S et al (2008) Double-blind, randomized placebo controlled trial of fulvestrant compared with exemestane after prior nonsteroidal aromatase inhibitor therapy in postmenopausal women with hormone receptor positive, advanced breast cancer: results from EFECT. J Clin Oncol 26:1664–1670. https://doi.org/10.1200/jco.2007.13.5822
196. Cristofanilli M et al (2016) Fulvestrant plus palbociclib versus fulvestrant plus placebo for treatment of hormone-receptor-positive, HER2-negative metastatic breast cancer that progressed on previous endocrine therapy (PALOMA-3): final analysis of the multicentre, double-blind, phase 3 randomised controlled trial. Lancet Oncol 17:425–439. https://doi.org/10.1016/s1470-2045(15)00613-0
197. Wang T, Wei JJ, Sabatini DM, Lander ES (2014) Genetic screens in human cells using the CRISPR-Cas9 system. Science (New York, NY) 343:80–84. https://doi.org/10.1126/science.1246981
198. Gojis O et al (2010) The role of SRC-3 in human breast cancer. Nat Rev Clin Oncol 7:83–89. https://doi.org/10.1038/nrclinonc.2009.219
199. Gates LA et al (2018) Proteomic profiling identifies key coactivators utilized by mutant ERα proteins as potential new therapeutic targets. Oncogene 37:4581–4598. https://doi.org/10.1038/s41388-018-0284-2
200. Scott GK et al (2016) ERpS294 is a biomarker of ligand or mutational ERα activation and a breast cancer target for CDK2 inhibition. Oncotarget 8:83432–83445. https://doi.org/10.18632/oncotarget.12735
201. Andruska ND et al (2015) Estrogen receptor α inhibitor activates the unfolded protein response, blocks protein synthesis, and induces tumor regression. Proc Natl Acad Sci U S A 112:4737–4742. https://doi.org/10.1073/pnas.1403685112
202. Mao C, Livezey M, Kim JE, Shapiro DJ (2016) Antiestrogen resistant cell lines expressing estrogen receptor α mutations upregulate the unfolded protein response and are killed by BHPI. Sci Rep 6:34753. https://doi.org/10.1038/srep34753
203. Gelsomino L et al (2018) Mutations in the estrogen receptor alpha hormone binding domain promote stem cell phenotype through notch activation in breast cancer cell lines. Cancer Lett 428:12–20. https://doi.org/10.1016/j.canlet.2018.04.023
204. Ladd B et al (2016) Effective combination therapies in preclinical endocrine resistant breast cancer models harboring ER mutations. Oncotarget 7:54120–54136
205. Yardley DA et al (2013) Randomized phase II, double-blind, placebo-controlled study of exemes-

tane with or without entinostat in postmenopausal women with locally recurrent or metastatic estrogen receptor-positive breast cancer progressing on treatment with a nonsteroidal aromatase inhibitor. J Clin Oncol 31:2128–2135. https://doi.org/10.1200/jco.2012.43.7251

206. Connolly RM et al (2021) E2112: randomized phase III trial of endocrine therapy plus Entinostat or placebo in hormone receptor-positive advanced breast cancer. A trial of the ECOG-ACRIN cancer research group. J Clin Oncol 39:3171–3181. https://doi.org/10.1200/jco.21.00944
207. Waldman AD, Fritz JM, Lenardo MJ (2020) A guide to cancer immunotherapy: from T cell basic science to clinical practice. Nat Rev Immunol 20:651–668. https://doi.org/10.1038/s41577-020-0306-5
208. Loi S et al (2013) Prognostic and predictive value of tumor-infiltrating lymphocytes in a phase III randomized adjuvant breast cancer trial in node-positive breast cancer comparing the addition of docetaxel to doxorubicin with doxorubicin-based chemotherapy: BIG 02-98. J Clin Oncol 31:860–867. https://doi.org/10.1200/jco.2011.41.0902
209. Rugo HS et al (2018) Safety and antitumor activity of Pembrolizumab in patients with estrogen receptor-positive/human epidermal growth factor receptor 2-negative advanced breast cancer. Clin Cancer Res 24:2804–2811. https://doi.org/10.1158/1078-0432.ccr-17-3452
210. Jeselsohn RM (2021) The evolving use of SERDs in estrogen receptor-positive, HER2-negative metastatic breast cancer. Clin Adv Hematol Oncol H&O 19:428–431
211. Menarini Group and Radius Health, I (2021) Menarini Group and Radius Health announce positive phase 3 topline results from the EMERALD trial evaluating elacestrant in breast cancer. News release, https://bit.ly/3E0Nmby
212. Bardia A et al (2019) EMERALD: phase III trial of elacestrant (RAD1901) vs endocrine therapy for previously treated ER+ advanced breast cancer. Future Oncol (London, England) 15:3209–3218. https://doi.org/10.2217/fon-2019-0370
213. Bardia A et al (2021) Phase I study of Elacestrant (RAD1901), a novel selective estrogen receptor degrader, in ER-positive, HER2-negative advanced breast cancer. J Clin Oncol 39:1360–1370. https://doi.org/10.1200/jco.20.02272
214. Jhaveri KL et al (2021) Safety and activity of single-agent giredestrant (GDC-9545) from a phase Ia/b study in patients (pts) with estrogen receptor-positive (ER+), HER2-negative locally advanced/metastatic breast cancer (LA/mBC). J Clin Oncol 39:1017–1017. https://doi.org/10.1200/JCO.2021.39.15_suppl.1017
215. Im S-A et al (2021) SERENA-4: a phase 3 comparison of AZD9833 (camizestrant) plus palbociclib, versus anastrozole plus palbociclib, for patients with ER-positive, HER2-negative advanced breast cancer who have not previously received systemic treatment for advanced disease. J Clin Oncol 39:TPS1101–TPS1101. https://doi.org/10.1200/JCO.2021.39.15_suppl.TPS1101
216. Jhaveri KL et al (2021) A first-in-human phase 1a/b trial of LY3484356, an oral selective estrogen receptor (ER) degrader (SERD) in ER+ advanced breast cancer (aBC) and endometrial endometrioid cancer (EEC): results from the EMBER study. J Clin Oncol 39:1050–1050. https://doi.org/10.1200/JCO.2021.39.15_suppl.1050
217. Samatar AA et al (2020) Abstract 4373: discovery of ZN-c5, a novel potent and oral selective estrogen receptor degrader. Cancer Res 80:4373–4373. https://doi.org/10.1158/1538-7445.am2020-4373
218. Maglakelidze M et al (2021) Rintodestrant (G1T48), an oral selective estrogen receptor degrader, in combination with palbociclib for ER+/HER2– advanced breast cancer: phase 1 results. J Clin Oncol 39:1063–1063. https://doi.org/10.1200/JCO.2021.39.15_suppl.1063
219. National Library of Medicine (U.S.) (2021) J.-. Phase III study to assess AZD9833+ CDK4/6 inhibitor in HR+/HER2-MBC with detectable ESR1m before progression (SERENA-6). https://ClinicalTrials.gov/show/NCT04964934

11 AR Structural Variants and Prostate Cancer

Laura Cato and Maysoun Shomali

Abstract

Therapeutic interventions for advanced castration-resistant prostate cancer (CRPC) are focused on inhibiting the androgen receptor (AR) through targeting of its C-terminal ligand binding domain (LBD). However, a significant subset of CRPC patients demonstrate primary resistance to androgen deprivation and anti-androgen therapies, suggesting that other targets, outside of the AR, might be pertinent to the cancer progression. One explanation is the expression of androgen receptor splice variants (AR-Vs). So far, more than 20 AR-Vs have been identified from both prostate cancer cell lines and prostate cancer tissue biopsies. Most of the AR-Vs have a conserved N-terminal domain, but lack the LBD, yet retain the ability to bind DNA and activate downstream signaling. Although it remains unclear whether AR-Vs are principal divers or mere bystanders of CRPC progression, inhibiting AR-Vs, through drugs that target the AR transactivation function outside of the LBD, has been a major emphasis for next generation therapeutics in prostate cancer. This book chapter is dedicated to the role of AR variants and their clinical importance. We will review the initial discovery of AR-Vs, their regulation and prevalence, as well as their biological function in prostate cancer. We will provide an overview of the role of AR-Vs in the development of metastatic CRPC and in promoting clinical treatment failures. Lastly, we will present an introduction to the therapeutic approaches towards developing AR-V-targeted therapies including the continuing progress, the old challenges, and the new prospects.

Keywords

Androgen receptor (AR) · AR variants · Prostate cancer (PCa) · Small molecule inhibitors · Degraders

L. Cato (✉) · M. Shomali
Sanofi, Research and Development, Cambridge, MA, USA
e-mail: Laura.Cato@sanofi.com

11.1 The Androgen Receptor and Its Role in Prostate Cancer

Prostate carcinoma is one of the leading causes of cancer death in men worldwide and the fourth most common cancer overall [1, 2]. The underlying biological mechanisms contributing to the initiation and progression of this disease can be attributed to alterations in hormonal factors and particularly changes to circulating androgens. Androgens bind and control the function of the

M. J. Campbell, C. L. Bevan (eds.), *Nuclear Receptors in Human Health and Disease*, Advances in Experimental Medicine and Biology 1390, https://doi.org/10.1007/978-3-031-11836-4_11

androgen receptor (AR), a hormone-regulated transcription factor which serves as a major modulator of normal prostate and prostatic tumor growth. In its non-hormone bound state, the majority of AR is localized in the cytoplasm in complex with molecular chaperones such as Hsp70, Hsp90 and immunophilins [3]. However, in the presence of androgens, the receptor undergoes conformational changes (i.e., phosphorylation), dimerizes and translocates to the nucleus. Here it recognizes and binds specific DNA sequences, known as AR response elements (AREs), to regulate the expression of genes involved in prostate cancer (PCa) cell growth.

The AR, like other members of the nuclear hormone receptor (NHR) family, consists of four distinct functional domains (encoded by eight exons): A N-terminal regulator domain (activation function 1, AF1; encoded by exon 1), a highly conserved DNA-binding domain (DBD; encoded by exons 2 and 3), a hinge region (encoded by exon 4) and a carboxy-terminal ligand-binding domain (LBD; encoded by exons 4–8) (Fig. 11.1). The domain containing the LBD is also referred to as activation function 2 (AF2) and is made up of 12 helical structures, although unlike the LBD of other NHRs helix 2 is missing [4]. Upon hormone binding, the 12th helix is reorganized to an agonist conformation generating a hydrophobic surface termed the "coactivator pocket" for tissue-specific cofactor binding. Coactivators and corepressors bind to this hormone-induced conformation via a conserved "LXXLL" sequence to enhance or repress the transcriptional activity of the receptor [5–7]. Due to this function, it was long hypothesized that the AF2 is the critical domain for the transactivation function of the receptor [8, 9]. Although this is indeed the case for other NHRs (e.g. estrogen receptor (ER) [10]), most of the transactivation function of the AR lies within its N-terminal AF1 domain, subdivided into transactivation units $\tau1$ (amino acids 100–360) and $\tau5$ (amino acids 360–528) [11]. This domain presents a site for AR interaction with cooperating transcription factors and coregulator proteins [12].

All clinically approved AR targeting agents for PCa therapy function by competing with the endogenous androgens for AR LBD binding, thereby inhibiting AR-driven oncogenic gene expression [13–15]. Targeting the AR N-terminus has been historically challenging, due to a lack of enzymatic activity or rigid binding clefts in this domain [16]. However, recent therapeutic advances have been made in targeting the AR N-termini and this is covered in detail later in this chapter. Despite targeting the AR AF2 domain, endocrine therapies in PCa remain largely successful, but their effectiveness is challenged by intrinsic and acquired resistance as the disease progresses. Typically, after an initial response to anti-androgen or androgen-ablation therapies tumors will progress, usually in concert with an increase in serum prostate-specific antigen (PSA). Secondary responses to block adrenal androgens may show initial promise, but are rarely successful at significantly shrinking the existing tumor mass or healing bone metastases. This stage of hormone ablation unresponsive prostate carcinoma remains incurable to date and is referred to as castrate-resistant prostate cancer (CRPC) [17–19].

CRPC is brought about by an array of mechanisms, most of which involve alterations of the AR itself. Proposed mechanisms include the development of AR hypersensitivity to residual androgens, or the acquired ability of AR to become activated in a ligand-independent manner through genetic alterations, such as mutations or changes to tyrosine phosphorylation of the receptor [20, 21]. More recently, the occurrence of AR variants (AR-Vs) that lack some or most of the C-terminal LBD of the full-length AR (AR-FL) has been linked with CRPC development and progression [22–26]. Although their exact significance in PCa and CRPC remains unresolved, it is widely appreciated that AR-Vs are constitutively active and demonstrate resistance to androgen ablation therapy [27]. This chapter will focus on the AR-Vs, their discovery and regulation, as well as their significance in PCa. We will also highlight therapeutic advances in targeting AR-Vs including recent progress and ongoing challenges.

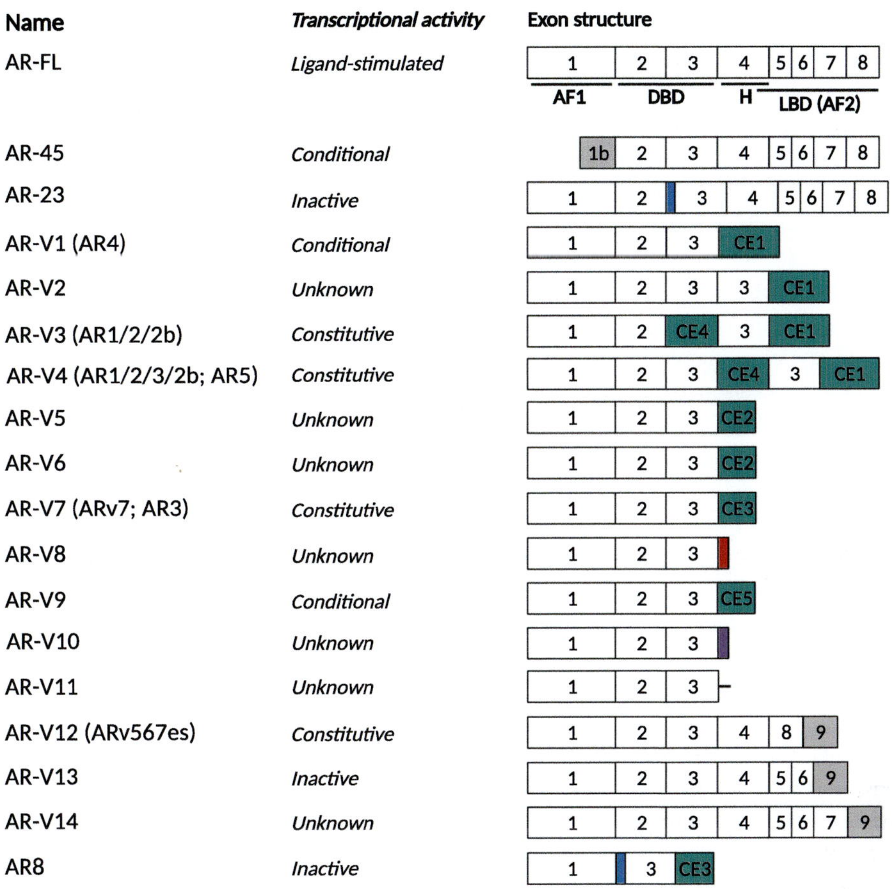

Fig. 11.1 Schematic of AR-FL and the most prominent AR-Vs in PCa. Image depicting the exon structures and functional activities of AR-FL and its variants was adapted from Zhu and Luo (2020) [53]. Large exon insertions and cryptic exons (CE) are shown in grey and turquoise respectively. Small nucleotide insertions are shown in blue, red and purple. The locations of the functional domains of AR-FL are shown. *AF1* and *AF2* Activation function 1 and 2, *DBD* DNA-binding domain, *H* Hinge domain, *LBD* Ligand-binding domain. Images are not drawn to scale

11.2 Discovery of the AR-Vs

The first discovery of AR-Vs dates to the early 2000s, with the identification of multiple 80–90 kDa molecular weight bands detectable by AR immunoblot in the CWR22Rv1 prostate cell line model [28]. Although initially presumed to be proteolytic cleavage bands of the AR-FL [29], it is now clear that this was the earliest identification of AR-Vs in PCa. Characterization of the first bona fide AR-Vs followed shortly thereafter. These variants contained intact LBDs, but had other notable disruptions to their protein structure. The first variant of this kind to be described was AR45. AR variant AR45 (named for its protein size of 45 kDa) is highly expressed in skeletal and cardiac muscle, and although contradicting literature surrounding its biological function exists, it has largely been described as exerting a dominant negative role on AR-FL, at least in the heart [30, 31]. Subsequently, AR variant AR23 was discovered. The transcriptionally inactive AR23 has a set of unique 23 amino acids inserted in the DBD, which interfere with its ability to bind to DNA in chromatin, hence rendering it inactive [32, 33]. AR23 is exclusively cytoplasmic and unable to translocate to the nucleus. Given the lack of AR23 transcriptional activity and lack of endogenous expression in PCa cell lines, it is unlikely that this variant is of major clinical significance in PCa.

Since these early discoveries, more than 20 other AR-Vs have been described in PCa cell lines, xenografts, and patient tumors [34] (Fig. 11.1). Most of the AR-Vs share common structural features: conserved DBDs and N-terminal domains, but they often diverge in the length and composition of their C-terminal AR LBDs [35]. One consequence of this is that AR-Vs may lack the consensus nuclear localization signal (NLS), a small amino acid sequence located within the hinge region and LBD of the receptor. Since the NLS is essential for the correct nuclear transport of the receptor, this, at least in part, explains why several AR-Vs are exclusively cytoplasmic and do not possess the ability to translocate to the nucleus by themselves [36, 37].

Other features that distinguish the different AR-Vs from one another include their relative expression levels (in cancer cells and clinical PCa samples) and their varying abilities to function as transcription factors. Of the AR-Vs found in PCa some have no known function, while others appear to enhance AR-FL activity or can regulate their own transcriptome [23–25, 38, 39] (Fig. 11.1). For example, AR-Vs with an LBD lacking exons 7 and/or 8 (e.g., AR-V13 or AR-V14) demonstrate no activity, while variants lacking exons 5, 6 and 7 are constitutively active (ARv567es or ARv7) [24, 40]. It is noteworthy that even amongst the constitutively active AR-Vs differing levels of activity exist, suggesting that other factors might differentially regulate their functionality.

One of the most abundant and best-studied variants lacking the entire LBD is AR variant 7 (ARv7, or also known as AR3). This AR isoform arises through inclusion of cryptic exon 3 (CE3), resulting in a partial hinge region and unique 16 amino acid C-terminal tail. However, ARv7 contains a conserved N-terminal AF1 domain and central DBD. Due to these structural features ARv7, unlike AR-FL, is continuously localized to the nucleus and is constitutively active [25]. It is notable that ARv7 is one of the best characterized AR-Vs. This is particularly the case with regards to assessing its abundance in clinical patient samples and xenograft models [41], as it is one of the few AR-Vs for which a specific and selective antibody exists (targeting its unique CE3). In fact, ARv7 levels are increased (~20-fold) in metastases of CRPC patients compared to primary disease [42, 43], and its expression can be observed in over 80% of CRPC patient samples in the Stand Up To Cancer (SU2C) initiative PCa cohort [26]. Its expression is also correlated with androgen-independent proliferation in cell line models [22]. Many studies have confirmed this correlation of ARv7 status and disease progression [44–46]. Moreover, detection of ARv7 in patients seems to predict treatment response to anti-hormonal therapies [27].

Other variants exist that arise from cryptic exon splicing. For example, AR-V3 arises through a cryptic exon insertion between exon 2 and 3 and is highly detectable in circulating tumor DNA from CRPC patients [47]. AR-V9 shares a common 3′ terminal cryptic exon with ARv7, is conditionally active and promotes androgen-independent AR transcriptional activity [48]. Similarly, AR-V1 arises from splicing of the cryptic exon 3. However, unlike other AR-Vs, AR-V1 does not regulate gene expression by itself but heterodimerizes with AR-FL (and other NHRs) and reduces their ability to translocate to the nucleus, creating a dominate negative effect [37]. AR-V1 is detectable in non-malignant prostate tissue and primary prostate tumors, but also in bone metastasis. In fact, expression of AR-V1 in advanced disease and metastases seems to be elevated compared to primary tumors and this seems to correlate with therapy response [25]. In agreement, high AR-V1 expression has been demonstrated in a VCaP xenograft model following castration [49].

Significant literature also exists on the AR variant ARv567es (also known as AR-V12), which arises through an intragenic re-arrangement of the *AR* gene (namely deletion of exons 5, 6 and 7) [50]. Despite containing the sequence for exon 8, a frame shift that occurs as a consequence of exon skipping results in a premature stop codon [24]. Like ARv7, ARv567es is constitutively active due to the lack of an LBD and can enhance the activity of AR-FL [24]. Although the presence of ARv567es has been reported in tissue from healthy men treated with anti-androgens as part of a male contraceptive study [24], its expression, like that of ARv7, is primarily associated with CRPC and its metastases.

Perhaps one of the most unique of the AR-Vs discovered to date is AR8. AR8 lacks a DBD and is exclusively associated with the plasma membrane. It remains relatively unclear what role AR8 plays for disease progression and in the development of therapy resistance in PCa. It does not seem to function as an active transcription factor nor interact with AR-FL [51]. Instead, it seems to be involved in intercellular signaling pathways by binding to EGFR.

11.3 Regulation of AR-V Expression

It has long been hypothesized why AR-Vs exist and how their expression and prevalence is regulated. Two major mechanisms driving AR-V expression have been reported: alterations in splicing (including altered splicing factor expression) and alterations of the *AR* gene locus itself [52, 53]. Additionally, emerging evidence suggests that long non-coding RNAs (lncRNAs), micro RNAs (miRNAs) or even circular RNAs (circRNAs) may be involved in the expression of AR-Vs [54–58]. Transcription of AR-Vs may also be the result of treatment response and can be regulated through calpain-mediated cleavage of AR-FL [59]. Regardless of the mechanisms involved, AR-V expression is dynamic and can be rapidly suppressed by androgens or induced through AR inhibition (e.g., androgen deprivation, anti-androgens or via genetic tools) [60, 61].

Regulation of AR-V expression through RNA splicing is thought to be an adaptive response to hormone-deprivation therapy, and specific components of the spliceosome that generate the constitutively active AR-Vs can be regulated directly by AR itself. For example, the splicing factor hnRNPA1 plays a central role in generating ARv7 and is highly overexpressed in prostate tumors compared to the benign prostate [62]. Similarly, Jiménez-Vacas et al. observed that multiple components of the spliceosome machinery are profoundly dysregulated in clinically localized PCa compared to adjacent non-tumor tissue. SiRNA-mediated knockdown of spliceosome factors SNRNP200, SRSF3 and SRPM1 in PCa cell lines did not only lead to a decrease in AR-V expression, but also resulted in the inhibition of cell proliferation and migration [63]. These data indicate that the generation of AR-Vs are not simply random splicing events, but are likely specific tumor survival events in response to disease progression. In fact, at least in cell line models, AR-Vs expression occurs concurrently with the expression, activity and recruitment of the splicing machinery that regulates them [64].

Alterations of the *AR* gene have also been reported in ~60% of all CRPC cases [52]. In addition, AR-V formation as a result of select amplifications and genomic rearrangements has been identified through the study of PCa cell lines and xenograft models (e.g. CWR22Rv1, LuCaP and VCaP) [50]. For example, the castration resistant CWR22Rv1 cell line expresses AR-Vs as a result of a tandem duplication of a 35-kb DNA sequence including exon 3 of the AR gene [65]. Similarly, an in-frame deletion/inversion of exon 5, 6 and 7 is responsible for the expression of ARv567es in the patient-derived xenograft (PDX) LuCaP 86.2 and LuCAP136 models [66, 67], and in human patient samples [52]. Moreover, in the CRPC CWR-R1 model a ~ 48 kb intragenic deletion within the AR intron 1 was identified in a subset of cells with enhanced expression of ARv7 [66]. However, not all cell line models display such *AR* gene alterations. The LNCaP95 long-term androgen-depleted cell line for example, which is derived from LNCaP, demonstrates high AR-V (specifically ARv7) expression, without an apparent intragenic *AR* gene rearrangement [68]. This clearly highlights that different AR-Vs may rely on varying mechanisms for their generation, which may additionally be governed by underlying cell line- and tissue/tumor-specific factors.

11.4 Biological Activity and Function of AR-Vs

It remains unclear whether AR-Vs harbor the ability to regulate unique transcriptomes, distinct from that of the AR-FL, or if they require AR-FL for their transcriptional cooperativity [24, 42, 50]. Multiple studies have demonstrated that AR-Vs can dimerize with one another, as well as the full-length receptor [42, 60, 69, 70]. In fact, it was shown that AR-Vs can enhance AR-FL function by facilitating the nuclear transport of the receptor, particularly in the absence of androgens or the presence of AR antagonists [71]. It is therefore apparent that AR-V/AR-FL dimerization can result in the nuclear localization of the heterodimer, subsequent binding to AREs and, as a result, enhancement of AR-FL activity. For example, ARv7 and AR-FL are mutually dependent on each other when occupying the *PSA* promoter and other canonical AR-regulated genes. Moreover, a chromatin immunoprecipitation-sequencing (ChIP-seq) study in a cell line engineered to exclusively express AR-Vs revealed that the majority of ARv567es binding sites were identical to those identified for AR-FL in the (wild-type) parental line [72]. Although unique AR-V binding was initially identified, these sites were considered false positives, as they were localized to repetitive DNA sequences and could not be verified in a subsequent experiment. The study concluded that AR-Vs do not bind to unique DNA loci, but instead bind canonical ARE sequences in an AR-FL-dependent manner. In agreement, using ChIP-seq we demonstrated co-dependent binding of ARv7 with AR-FL in the CRPC cell line LNCaP95 [73]. Although some exclusive ARv7 binding sites could be identified, particularly in the absence of androgens, we failed to validate them using shRNA-mediated silencing of ARv7, suggesting that such binding sites are likely experimental artifacts rather than bona fide ARv7 binding sites. Moreover, shRNA-mediated knockdown of the full-length receptor led to a decrease in binding of ARv7 and vice versa, indicating that their ability to bind chromatin are inherently linked. This is in agreement with data from Cai et al., who demonstrated that ARv7 binding sites, although enriched under hormone-starvation conditions, largely overlapped those of AR-FL upon DHT induction [74].

Although these data may indicate that AR-Vs require the presence of AR-FL for transcriptional activity, many studies corroborate that AR-Vs can also act independently of AR-FL [24, 73]. AR-Vs seem to possess the ability to homodimerize, independently bind to chromatin and independently regulate their transcriptome [69, 75–77]. For example, ARv7, but not AR-FL, was shown to bind the ARE sequence within the promoter region of *AKT1* in both 22Rv1 and CWR-R1 cells [23]. ARv7, as well as other variants are also involved in the regulation of *UBE2C*, an enzyme important for the ubiquitin-mediated turnover of other proteins [60]. Compellingly, *UBE2C* is significantly upregulated in CRPC, when AR-Vs are highly expressed [14, 60, 78]. Other reported AR-V-regulated genes include genes involved in cell cycle progression, genes associated with mitotic and anti-apoptotic pathways, and genes involved in DNA damage response [60, 73, 79]. Comparable to genes regulated by AR-FL, unique AR-V target genes are described to be biphasic and demonstrate increased expression under low androgen levels, but decreased expression under high androgen and AR levels [50].

The divergent transcriptional properties of AR-Vs compared to AR-FL (yet shared binding sites in chromatin) can be, in part, explained by their structural differences, resulting in alterations in their chromatin binding behaviors (i.e., residence times) and hence distinct transactivation function [36]. Additional contributing factors may be the differential interactions between the receptors and transcriptional coregulators and cooperating transcription factors. For example, we have demonstrated that ARv7 preferentially associates with the transcriptional corepressors NCOR1 and 2, resulting in an overall transcriptional repressive function of ARv7, while AR-FL interacts with both coactivators and compressors, and on balance is associated with a slightly more transcriptional transactivation [73]. Others have demonstrated the necessity of the transcription factor HOXB13 for ARv7 chromatin binding and ARv7-

mediated gene expression [77, 80]. It is important to note that factors governing the differences in transcriptomes between AR-V and AR-FL may be cell-context-dependent, which likely reflects the heterogeneity of AR-V-driven PCa, as reported by others [81].

11.5 AR-V Prevalence and Significance in Disease

Although high AR-V levels on a whole are associated with disease recurrence, it remains uncertain whether AR-Vs are responsible for CRPC progression or are epiphenomena caused by castration or androgen-deprivation. AR-V expression in tissue and cell lines of primary disease, although inherently linked to AR-FL expression, is only approximately 0.1–2.5% of the levels of the full-length receptor [38]. However surprisingly, ARv7 expression has been reported in 80% of benign and hypoplasia prostate tissues from men with no evidence of cancer [42, 82]. Given these results it is conceivable that initially (in normal tissue or primary disease) AR-Vs may only exist at low levels or in a subset of cancer cells, but in response to castration and/or AR inhibition through targeted therapies, AR-V expression may emerge and promote tumor progression. This hypothesis is supported by data from numerous studies which report AR-Vs as a driver of PCa progression [40, 42]. Overexpression of several AR-Vs, and specifically ARv7, has been demonstrated in 20–40% of CRPC patients [25, 47, 83, 84], although Ma et al. recently postulated that ARv7 expression in CRPC may even be higher (due to underestimations by previous detection methods) [85]. In agreement, human xenograft mouse models expressing ARv7, display significantly higher tumor burdens than the same model lacking AR-Vs [23, 38]. In clinical patient samples, high AR-V expression has been shown in bone metastasis and metastatic CRPC tissue [61], and is associated with a shorter time to death in men with metastatic CRPC [42]. However, not all AR-V expression correlates with disease progression. Bernemann et al., recently suggested that expression of ARv567es is not correlative with late stage PCa and clinical treatment response [86]. It is conceivable that variant-specific differences exist or may not have been fully captured due to a lack of available tools to study them.

It is remarkable that the variants most prevalent in human CRPC tissue are those most consistently found following androgen deprivation *in vitro* [22, 61]. This suggests that at least one mechanism by which AR-Vs may exert their function in disease progression is linked to their upregulation in response to PCa therapies and treatment response. Watson et al. reported that ARv7 mRNA expression is rapidly upregulated (in a reversible manner) following androgen deprivation [38]. Similarly, significantly increased AR-V expression was observed in tumors from xenograft models following castration [24, 25, 49], or treatment with abiraterone acetate, a clinically approved inhibitor of testosterone synthesis used in the treatment of CRPC [87, 88]. Increases in ARv7 expression also seem to correlate with response to the AR antagonist enzalutamide, and high levels of AR-Vs are detectable in PCa cell lines after long-term culture conditions with enzalutamide [89]. Moreover, a high abundance in AR-Vs is associated with enzalutamide resistance in xenograft models [71, 72], and is observable in patients with metastatic CRPC resistant to enzalutamide [90]. This work has been corroborated by Zhu et al. who recently demonstrated that enzalutamide-resistance is consistently correlated with enhanced ARv7, but not AR-FL expression [91]. Although it is unclear whether AR-V expression is associated with response or resistance to taxanes (i.e., docetaxel) [92–94], combined these data suggest that AR-V expression can be associated with resistance to second-generation PCa therapies. In fact, multiple studies have demonstrated the use of at least ARv7 as a prognostic biomarker in the context of AR-targeting therapies [27, 44, 95]. In these studies, ARv7 expression was associated with reduced overall survival (OS), progression-free survival (PFS) and PSA response. This suggests a potential causative role of AR-Vs in the continued failure of therapeutic intervention in PCa.

11.6 Therapeutically Targeting AR-Vs

Many currently available therapies in PCa seem to be effective at slowing the progression of the disease by inhibiting the production or the binding of AR LBD-activating ligands. These include suppression of gonadal steroid synthesis by GnRH agonists (leuprolide or goserelin) or inhibition of androgen biosynthesis (abiraterone acetate), which blocks the production of the AR activating ligand testosterone [96–98]. Other therapies prevent testosterone conversion to DHT, the higher affinity ligand of AR, through inhibition of the enzyme 5-α reductase that converts it (finasteride or dutasteride) [99–104]. Moreover, other antiandrogen therapies (bicalutamide, nilutamide, flutamide, enzalutamide, apalutamide, and darolutamide) are competitive inhibitors of AR ligand binding itself [105–111].

As highlighted earlier in the chapter, multiple studies have suggested that AR-Vs can function independently of AR-FL and can contribute, or even drive CRPC. From a clinical perspective these findings could be capitalized on by generating AR-V-selective therapies. Multiple approaches to target AR-Vs exist. For example, AR-V expression may be abrogated by targeting the spliceosome machinery that is responsible for generating it. It is also conceivable that abrogation of cofactors that regulate AR-Vs through interaction with their NTD may pose another avenue for therapeutic intervention. However, in the interest of the readership, we will be focusing this portion of the chapter on the direct AR-V targeting mechanisms, with particular emphasis on more developed therapeutics and those currently in clinical trials (Fig. 11.2). Moreover, some of the newer classes of AR targeting agents are being explored in combination with traditional

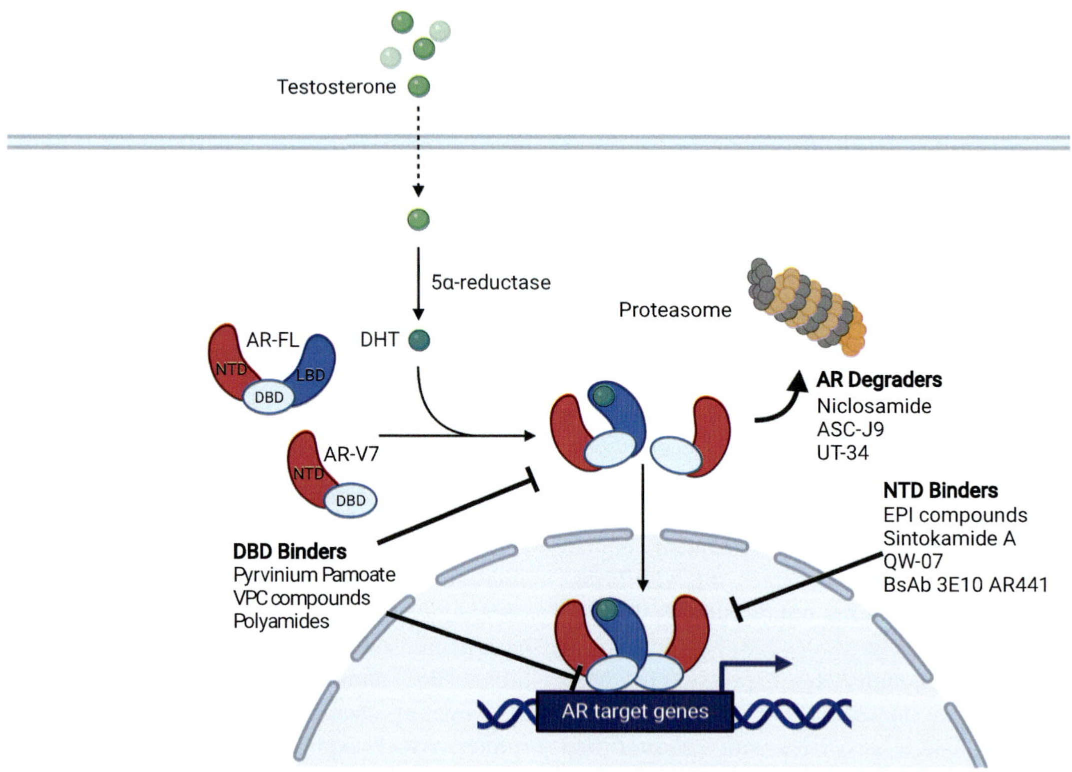

Fig. 11.2 Novel AR targeting agents and their functionality in PCa. Emerging new therapies can bind to the AR (AR-FL and/ or AR-Vs) at its NTD or DBD. Targeting the NTD blocks the receptor transactivation function and correct nuclear transport. In comparison, inhibition of the DBD prevents the AR interaction with DNA in chromatin and potentially receptor dimerization. Moreover, AR degradation presents an additional opportunity for therapeutic intervention of the receptor function. This involves reduction of AR protein levels in a proteasome-dependent manner. Examples of compounds for each category that are under current biological investigation or in clinical trials are listed

(androgen competitive) AR inhibitors and results, as apparent, are described. It is of note that most inhibitory agents directed at AR-Vs likely also abrogate AR-FL function. This is due to their conserved structural features, particularly for the NTD and DBD. Given their dual targeting capacity we will refer to such agents as "AR" inhibitors or degraders, encompassing the ability to alter both AR-V and AR-FL functionality.

11.7 Inhibition of the AR N-Terminal Domain

The NTD, as described earlier in this chapter, is critical for the interaction of AR with its cofactors, and consequently for transcriptional activity [112]. Inhibiting this domain of the protein blocks the AR activity regardless of the presence or absence of the LBD. An apparent consequence of this is the alteration of the correct receptor folding and the aberration of efficient nuclear transport of the receptor (Fig. 11.2). Given these features, the NTD presents as an attractive domain for AR inhibition. However, the identification of NTD targeting agents has been challenging due to its intrinsic disorder, and the lack of available structure determination of this domain. Nonetheless, some recent advances have been made and are outlined below.

Epoxide-based EPI compounds are small molecule inhibitors (Fig. 11.3), which block AR transcriptional activity [113]. The earliest of its kind, EPI-001, a mixture of 4 stereoisomers, was shown to bind to the AR NTD covalently and irreversibly by nuclear magnetic resonance (NMR) [114]. This molecule was reported to block the N/C inter-domain interaction of AR but did not block ligand binding to the LBD. Furthermore, the authors demonstrated that EPI-001 blocks the interaction of cofactors, primarily RAP74 and CBP, to the NTD [113]. This results in the effective inhibition of both ligand-dependent and -independent transactivation of the AR, as well as the inhibition of constitutively active AR-Vs. However, this molecule lacked the efficacy to become a clinical candidate and hence the more potent stereoisomer EPI-002 was developed into the pro-drug, EPI-506 [115]. In a Phase I study in men, EPI-506 was well-tolerated, but demonstrated poor bioavailability and high pill burden, and as a result, the study was terminated [116]. A successor of EPI-506, EPI-7386, has improved stability and is currently in Phase I/II clinical testing as a monotherapy and in combination with enzalutamide [113, 117].

Sintokamides are natural compounds that antagonize AR and are reported to bind the AR NTD [118] (Fig. 11.3). Like the EPI series, these molecules also inhibit the transactivation function of the AR. *In vivo*, sintokamide A (SINT1) caused regression of tumors in CRPC xenograft models and reduced the expression of the canonical AR target gene *PSA*. Interestingly, combination studies with EPI-002 and SINT1 demonstrated an additive effect, suggesting that these two molecules bind distinct regions of the NTD [118]. Subsequent binding studies have shown that SINT1 binds to the AF1 domain, but closer to the N-terminus and potentially within ι1, while EPI-002 binds ι5 [119]. Another peptidomimetic, D2 (Fig. 11.3), disrupts the interaction between AR and the coregulator PELP1, by mimicking the LXXLL motif in the AF2 domain of the AR C-terminus [120]. Inhibition of this interaction blocks the nuclear translocation of the AR and inhibits the growth of PCa cells [121].

Based on the above-described compounds it is apparent that most AR NTD inhibitors concomitantly affect the receptor transactivation function. In fact, the N-terminal small molecule inhibitor QW-07 (Fig. 11.3), was recently identified in a phenotypic screen assessing the NTD transcriptional activity [122]. Although QW-07 is believed to directly interact with the AR NTD, further work is required to refine the precise binding region [122, 123]. Regardless of its site of association, QW-07 was shown to block the transcriptional activity of AR-FL and AR-Vs by blocking their interaction with DNA and regulating cofactors. Additionally, treatment with QW-07 inhibited PCa cell growth *in vitro* and inhibited tumor growth *in vivo* [122]. QW-07 seems to show more potent inhibition of transcriptional activity and stronger inhibition of proliferation in PCa cell lines than EPI-001 [122]. However, to date there is no reported clinical data on QW-07.

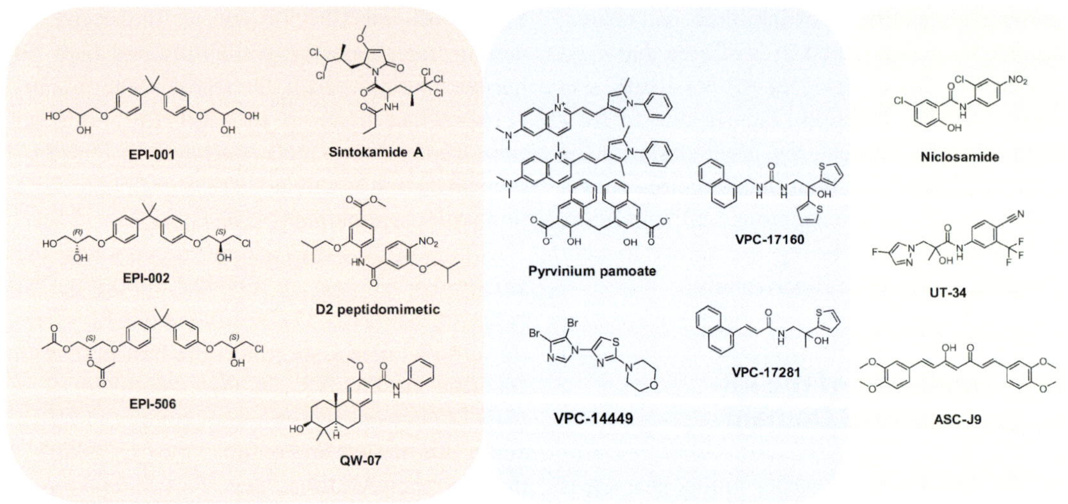

Fig. 11.3 Chemical structures of AR targeting agents that bind outside of the LBD. Chemical structures, as indicated, of the NTD-targeting small molecules EPI-001, EPI-002, EPI-506, Sintokamide A, D2 peptidomimetic and QW-07 (pink-shaded box); the DBD-targeting small molecules Pyrvinium pamoate, VPC-14449, VPC-17160 and VPC-17281 (blue-shaded box); and the AR degraders Niclosamide, UT-34 and ASC-J9 (yellow-shaded box). The structures of additional compounds described in this chapter have not been disclosed yet

An alternative modality for targeting the NTD is the use of bispecific antibodies (BsAb) to block AR signaling. BsAb can target multiple antigens (i.e., immune and tumor cells) simultaneously and by doing so are associated with reduced adverse reactions and drug resistance than traditional therapies. 340-AR441 is a BsAb engineered based on the monoclonal antibody AR441 which recognizes an epitope in the AR NTD, and is capable of penetrating cells [123]. Although this antibody demonstrated highly specific and potent *in vitro* activity in PCa cell lines, no *in vivo* data has been reported as of yet [123]. Next generation BsAb are much anticipated.

11.8 Inhibition of the AR-DBD

The conserved DBD presents another avenue in targeting AR-Vs by preventing their interaction with DNA in chromatin [124]. The crystal structure of the AR-DBD (in complex with DNA) has been previously solved, enabling the use of docking programs to identify future candidate inhibitors. Binders of the DBD can abolish AR transcriptional activity by several different mechanisms, including inhibition of DNA binding at AREs, prevention of AR dimerization, or inhibition of AR-V nuclear localization.

Pyrvinium pamoate (PP), a cyanine dye derived from quinoline and an anthelmintic drug, is a non-competitive inhibitor of the AR (Fig. 11.3). Treatment with PP demonstrated transcriptional inhibition of AR-FL and AR constructs lacking the LBD or NTD *in vitro* [125]. Furthermore, the combined treatment of PP with the LBD-inhibitors hydroxyflutamide and bicalutamide synergistically disrupted AR activity, leading to decreases in cell growth [125, 126]. Comparable results were observed *in vivo* where PP treatment alone reduced tumor weight, but more significantly when used in combination with an LBD-targeting inhibitor [125, 126]. Modelling predicted that PP binds at the interface of the DBD of the AR dimer and the minor groove of the AR response element [126]. However this view has been challenged by the fact that in a subsequent study Pal et al. demonstrated a lack of significant interaction between P24, a soluble derivative of PP, and the AR DBD-DNA complex by NMR [127]. Moreover, besides the AR, PP also reduces ERα and glucocorticoid receptor (GR) signaling [126], presumably through direct binding with these NHRs. Additionally, PP seems to inhibit the interaction between AR and several splicing factors, such as DDX17 [127]. Combined these data point towards an indirect, rather than a direct method of action for PP in modulating AR function. Further studies are required for a more comprehensive understanding of this molecule.

Using an *in silico* drug design approach has enabled the identification of a surface-exposed pocket (including residues Ser-579 and Lys-610) on the AR-DBD, involved in the head-to-tail dimerization of the receptor [124, 128, 129]. This site was postulated to be a good alternative site for AR inhibition and subsequent computer-aided high throughput screening of it led to the discovery of the VPC molecules (Fig. 11.3). The earliest lead compound, VPC-14449, did not impede the nuclear localization of the AR, but blocked the transcriptional activity of both AR-FL and AR-Vs within the nucleus. Here it reportedly reduced the interaction of the receptor with chromatin, resulting in reduced (*PSA*) gene expression and inhibition of tumor growth [129]. The follow-up molecule VPC-17005, disrupted the AR-DBD dimerization interface and consequently lead to reduced transcriptional activity of the receptor and repressed PCa cell growth [124]. However, VPC-17005 demonstrated poor metabolic stability and was not suitable for *in vivo* studies [130]. Hence, further optimization was required to generate the chemically diverse VPC-17160 and VPC-17281 [131]. Both newer molecules demonstrate improved potency in inhibiting transcription and cell proliferation compared to VPC-17005, at least *in vitro* in ARv7-dependent and CRPC cell lines. In addition, VPC-17281 demonstrates improved metabolic stability, but surprisingly also reduces growth of the AR-negative PC3 cell line. This suggests potential off-target activities of at least this VPC molecule. Combined these results indicate that although advances on the VPC molecules have clearly been made, further optimization of these compounds are necessary to improve their potency and drug-like properties.

Another method of inhibiting the DBD involves the targeting of polyamides to the ARE consensus sequence to block AR interaction on the DNA and thus indirectly alter the receptor activity [132, 133]. This targeting approach has been corroborated by experimental data that suggests that polyamides can bind to ARE sequences and reduce the gene expression of select AR target genes, such as *FKBP5* and *PSA*, resulting in the inhibition of PCa cell proliferation [132, 133]. In *in vivo* studies, subcutaneous dosing of the polyamide ARE-1 reduced xenograft prostate tumor sizes by 40% [133]. Although promising on face value, a limitation of polyamides as a therapeutic agent may be their poor cellular permeability and diffusion into tumor tissue which may result in poor efficacy as a single agent.

11.9 Degradation of AR

A more recent and potentially more successful approach in targeting AR-Vs is induced protein degradation. A variety of drugs have been investigated that degrade AR-FL and AR-V proteins, specifically ARv7, resulting in the inhibition of its oncogenic signaling.

The recent discovery of proteolysis-targeting chimeras (PROTACs) has triggered an interest in identifying molecules to degrade the AR. AR PROTAC molecules combine known binders of AR with known binders to an E3 ubiquitin ligase separated by a linker molecule [134–136]. This promotes a special proximity between the AR and the E3 ligase, resulting in receptor ubiquitination and ultimately degradation, in a proteasome-dependent manner. Utilizing this technology, ARV-110 was discovered and is currently in clinical trials for advanced CRPC [137]. One of the liabilities of current AR PROTACs, including ARV-110, is their sole affinity to the AR LBD. At face value this would support a scenario in which only AR-FL and not AR-Vs would get degraded. However, it was hypothesized that due to the ability of AR-Vs to heterodimerize with AR-FL, the former could concomitantly be degraded. Unfortunately, preclinical studies assessing AR degraders in AR-V-expressing cell lines, such as 22Rv1, CWR-R1 or VCAP, could not identify a reduction of AR-V protein levels [138]. Presumably a direct interaction between ARV-110 and AR-Vs is required for degradation.

Others have reported AR degradation by mechanisms outside of the use of PROTAC molecules. Niclosamide, an anthelmintic drug (Fig. 11.3), seemingly degrades ARv7 selectively via the proteasome-dependent pathway [139, 140]. The exact mechanism and binding site for niclosamide is not known, however treatment with niclosamide reduces ARv7 protein levels and thus inhibits its chromatin recruitment and expression of the *PSA* gene. While niclosamide reduced tumor growth in enzalutamide-resistant C4-2b cells and CWR22Rv1 cells, the combination of niclosamide with either abiraterone or enzalutamide had a more profound effect on tumor growth inhibition than treatment of either compound alone [141]. Additionally, niclosamide was shown to inhibit androgen-independent AR activation by blocking the interleukin-6-STAT3 AR pathway. Constitutively active STAT3 is a part of the positive autocrine IL6 loop and STAT3 activation is thought to be a mechanism of enzalutamide resistance [142]. These properties imply that niclosamide may be a suitable candidate for the treatment of metastatic and treatment-resistant CRPC. However, further differentiation of the impact of niclosamide on androgen-dependent versus -independent pathways will be necessary for the therapeutic advances of this compound. The outcome of ongoing clinical trials combining niclosamide with abiraterone acetate (NCT02807805) or enzalutamide (NCT03123978) in men with CRPC are eagerly awaited.

Selective androgen receptor degraders (SARDs) are a newly discovered group of molecules that interact with both the AR AF1 and the carboxy terminal of the AR LBD. Mediated through these interactions, SARDs reportedly degrade both AR-FL and AR-Vs, via the ubiquitin proteasome pathway. UT-34 is the most advanced of these compounds [143] (Fig. 11.3). In *in vivo* studies, treatment with UT-34 decreased the growth of enzalutamide-resistant, castration-resistant xenograft models [144], implying great promise for this therapeutic approach in the treatment of CRPC patients. Although work from steady-state fluorescence emission spectra suggests that UT-34 binds to the NTD, thus making it a good therapeutic agent for both AR-FL and AR-Vs, more work is needed to determine the exact binding region of this molecule [143].

More recently the nonsteroidal anti-androgen ASC-J9 (dimethylcurcumin; Fig. 11.3) was reported to inhibit PCa cell proliferation *in vitro* and tumor growth *in vivo*, by reducing both AR-FL and AR-V protein levels [139, 140]. This reported degradation is thought to be facilitated by the compound-mediated interaction between AR and MDM2, an E3 ubiquitin ligase involved in AR stability [140]. However, ASC-J9 is believed to also have many AR-independent anticancer activities, including induction of apoptosis through NF-κB signaling and inhibition of ATF3/PTK2 signaling and STAT3 sumoylation [145, 146]. Therefor the specificity and selectivity of ASC-J9 is questionable.

Given the above data it is conceivable that a degradation approach may be the best approach going forward to directly target AR-Vs (and AR-FL) and ablate their oncogenic signaling. Building on the principle of PROTACs, a ligand with affinity for the NTD or DBD could be used to degrade both receptors and may be the most effective way to combat AR-dependent CRPC. Since at least one AR PROTAC has already entered the clinic [137], this seems an achievable, yet challenging, feat to be tackled.

11.10 Concluding Remarks

The wealth of knowledge now available on the biology and function of AR-Vs in PCa are a true tribute to the recent advances in the field of AR biology. Based on the data presented in this chapter it is apparent that knowledge gained from *in vitro*, *in vivo* and clinical studies demonstrates the importance of AR-Vs in PCa, CRPC development and poor treatment response, but they also highlight the urgent need for further research on them to improve patient response. Despite the depth of work that has been done, there is a noticeable lack of FDA-approved drugs that inhibit or degrade AR-Vs. While many anti-AR-V drugs are currently under investigation, the efficacy of treating PCa with such approaches remains to be determined. Novel, selective and more potent drugs targeting AR-Vs are warranted.

Acknowledgments Chemical structures were made with ChemDraw. Figures were made using BioRender.com

References

1. Siegel RL et al (2021) Cancer statistics, 2021. CA Cancer J Clin 71(1):7–33
2. Sung H et al (2021) Global cancer statistics 2020: GLOBOCAN estimates of incidence and mortality worldwide for 36 cancers in 185 countries. CA Cancer J Clin 71(3):209–249
3. Cato L et al (2014) Control of steroid receptor dynamics and function by genomic actions of the cochaperones p23 and bag-1L. Nucl Recept Signal 12:e005
4. Estebanez-Perpina E, Bevan CL, McEwan IJ (2021) Eighty years of targeting androgen receptor activity in prostate cancer: the fight goes on. Cancers (Basel) 13(3)
5. Darimont BD et al (1998) Structure and specificity of nuclear receptor-coactivator interactions. Genes Dev 12(21):3343–3356
6. Heery DM et al (1997) A signature motif in transcriptional co-activators mediates binding to nuclear receptors. Nature 387(6634):733–736
7. Nolte RT et al (1998) Ligand binding and co-activator assembly of the peroxisome proliferator-activated receptor-gamma. Nature 395(6698):137–143
8. Glass CK, Rosenfeld MG (2000) The coregulator exchange in transcriptional functions of nuclear receptors. Genes Dev 14(2):121–141
9. Moras D, Gronemeyer H (1998) The nuclear receptor ligand-binding domain: structure and function. Curr Opin Cell Biol 10(3):384–391
10. Tora L et al (1989) The human estrogen receptor has two independent nonacidic transcriptional activation functions. Cell 59(3):477–487
11. Claessens F et al (2008) Diverse roles of androgen receptor (AR) domains in AR-mediated signaling. Nucl Recept Signal 6:e008
12. Bevan CL et al (1999) The AF1 and AF2 domains of the androgen receptor interact with distinct regions of SRC1. Mol Cell Biol 19(12):8383–8392
13. Matsumoto H et al (2013) Cotargeting androgen receptor and clusterin delays castrate-resistant prostate cancer progression by inhibiting adaptive stress response and AR stability. Cancer Res 73(16):5206–5217
14. Wang Q et al (2009) Androgen receptor regulates a distinct transcription program in androgen-independent prostate cancer. Cell 138(2):245–256
15. Fujii S, Kagechika H (2019) Androgen receptor modulators: a review of recent patents and reports (2012-2018). Expert Opin Ther Pat 29(6):439–453
16. Lavery DN, McEwan IJ (2008) Structural characterization of the native NH2-terminal transactivation domain of the human androgen receptor: a collapsed disordered conformation underlies structural plasticity and protein-induced folding. Biochemistry 47(11):3360–3369
17. Culig Z et al (2002) Androgen receptors in prostate cancer. Endocr Relat Cancer 9(3):155–170
18. Galletti G et al (2017) Mechanisms of resistance to systemic therapy in metastatic castration-resistant prostate cancer. Cancer Treat Rev 57:16–27
19. Zegarra-Moro OL et al (2002) Disruption of androgen receptor function inhibits proliferation of androgen-refractory prostate cancer cells. Cancer Res 62(4):1008–1013
20. Taplin ME et al (1995) Mutation of the androgen-receptor gene in metastatic androgen-independent prostate cancer. N Engl J Med 332(21):1393–1398
21. Guo Z et al (2006) Regulation of androgen receptor activity by tyrosine phosphorylation. Cancer Cell 10(4):309–319
22. Dehm SM et al (2008) Splicing of a novel androgen receptor exon generates a constitutively active androgen receptor that mediates prostate cancer therapy resistance. Cancer Res 68(13):5469–5477
23. Guo Z et al (2009) A novel androgen receptor splice variant is up-regulated during prostate cancer progression and promotes androgen depletion-resistant growth. Cancer Res 69(6):2305–2313
24. Sun S et al (2010) Castration resistance in human prostate cancer is conferred by a frequently occurring androgen receptor splice variant. J Clin Invest 120(8):2715–2730
25. Hu R et al (2009) Ligand-independent androgen receptor variants derived from splicing of cryptic exons signify hormone-refractory prostate cancer. Cancer Res 69(1):16–22
26. Robinson D et al (2015) Integrative clinical genomics of advanced prostate cancer. Cell 161(5):1215–1228
27. Antonarakis ES et al (2014) AR-V7 and resistance to enzalutamide and abiraterone in prostate cancer. N Engl J Med 371(11):1028–1038
28. Tepper CG et al (2002) Characterization of a novel androgen receptor mutation in a relapsed CWR22 prostate cancer xenograft and cell line. Cancer Res 62(22):6606–6614
29. Gregory CW, He B, Wilson EM (2001) The putative androgen receptor-a form results from in vitro proteolysis. J Mol Endocrinol 27(3):309–319
30. Weiss B, Faus H, Haendler B (2007) Phylogenetic conservation of the androgen receptor AR45 variant form in placental mammals. Gene 399(2):105–111
31. Ahrens-Fath I et al (2005) Androgen receptor function is modulated by the tissue-specific AR45 variant. FEBS J 272(1):74–84
32. Jagla M et al (2007) A splicing variant of the androgen receptor detected in a metastatic prostate cancer exhibits exclusively cytoplasmic actions. Endocrinology 148(9):4334–4343

33. Wadosky KM, Koochekpour S (2016) Molecular mechanisms underlying resistance to androgen deprivation therapy in prostate cancer. Oncotarget 7(39):64447–64470
34. Lu C, Luo J (2013) Decoding the androgen receptor splice variants. Transl Androl Urol 2(3):178–186
35. Ware KE et al (2014) Biologic and clinical significance of androgen receptor variants in castration resistant prostate cancer. Endocr Relat Cancer 21(4):T87–T103
36. Chan SC, Li Y, Dehm SM (2012) Androgen receptor splice variants activate androgen receptor target genes and support aberrant prostate cancer cell growth independent of canonical androgen receptor nuclear localization signal. J Biol Chem 287(23):19736–19749
37. Zhan Y et al (2017) Interplay between cytoplasmic and nuclear androgen receptor splice variants mediates castration resistance. Mol Cancer Res 15(1):59–68
38. Watson PA et al (2010) Constitutively active androgen receptor splice variants expressed in castration-resistant prostate cancer require full-length androgen receptor. Proc Natl Acad Sci U S A 107(39):16759–16765
39. Marcias G et al (2010) Identification of novel truncated androgen receptor (AR) mutants including unreported pre-mRNA splicing variants in the 22Rv1 hormone-refractory prostate cancer (PCa) cell line. Hum Mutat 31(1):74–80
40. Hu R, Isaacs WB, Luo J (2011) A snapshot of the expression signature of androgen receptor splicing variants and their distinctive transcriptional activities. Prostate 71(15):1656–1667
41. Lu C et al (2020) Androgen receptor variant-driven prostate cancer II: advances in laboratory investigations. Prostate Cancer Prostatic Dis 23(3):381–397
42. Hornberg E et al (2011) Expression of androgen receptor splice variants in prostate cancer bone metastases is associated with castration-resistance and short survival. PLoS One 6(4):e19059
43. Qu Y et al (2015) Constitutively active AR-V7 plays an essential role in the development and progression of castration-resistant prostate cancer. Sci Rep 5:7654
44. Antonarakis ES et al (2017) Clinical significance of androgen receptor splice Variant-7 mRNA detection in circulating tumor cells of men with metastatic castration-resistant prostate cancer treated with first- and second-line Abiraterone and Enzalutamide. J Clin Oncol 35(19):2149–2156
45. Scher HI et al (2016) Association of AR-V7 on circulating tumor cells as a treatment-specific biomarker with outcomes and survival in castration-resistant prostate cancer. JAMA Oncol 2(11):1441–1449
46. Welti J et al (2016) Analytical validation and clinical qualification of a new Immunohistochemical assay for androgen receptor splice Variant-7 protein expression in metastatic castration-resistant prostate cancer. Eur Urol 70(4):599–608
47. De Laere B et al (2017) Comprehensive profiling of the androgen receptor in liquid biopsies from castration-resistant prostate cancer reveals novel intra-AR structural variation and splice variant expression patterns. Eur Urol 72(2):192–200
48. Kohli M et al (2017) Androgen receptor variant AR-V9 is coexpressed with AR-V7 in prostate cancer metastases and predicts abiraterone resistance. Clin Cancer Res 23(16):4704–4715
49. Knuuttila M et al (2014) Castration induces up-regulation of intratumoral androgen biosynthesis and androgen receptor expression in an orthotopic VCaP human prostate cancer xenograft model. Am J Pathol 184(8):2163–2173
50. Li Y et al (2013) Androgen receptor splice variants mediate enzalutamide resistance in castration-resistant prostate cancer cell lines. Cancer Res 73(2):483–489
51. Yang X et al (2011) Novel membrane-associated androgen receptor splice variant potentiates proliferative and survival responses in prostate cancer cells. J Biol Chem 286(41):36152–36160
52. Henzler C et al (2016) Truncation and constitutive activation of the androgen receptor by diverse genomic rearrangements in prostate cancer. Nat Commun 7:13668
53. Zhu Y, Luo J (2020) Regulation of androgen receptor variants in prostate cancer. Asian J Urol 7(3):251–257
54. Kumar B et al (2016) Identification of miR-30b-3p and miR-30d-5p as direct regulators of androgen receptor signaling in prostate cancer by complementary functional microRNA library screening. Oncotarget 7(45):72593–72607
55. Greene J et al (2019) Circular RNAs are differentially expressed in prostate cancer and are potentially associated with resistance to enzalutamide. Sci Rep 9(1):10739
56. Zhang A et al (2016) Long non-coding RNA: a newly deciphered "code" in prostate cancer. Cancer Lett 375(2):323–330
57. Takayama KI et al (2020) Identification of long non-coding RNAs in advanced prostate cancer associated with androgen receptor splicing factors. Commun Biol 3(1):393
58. Fletcher CE et al (2019) Androgen receptor-modulatory microRNAs provide insight into therapy resistance and therapeutic targets in advanced prostate cancer. Oncogene 38(28):5700–5724
59. Libertini SJ et al (2007) Evidence for calpain-mediated androgen receptor cleavage as a mechanism for androgen independence. Cancer Res 67(19):9001–9005
60. Hu R et al (2012) Distinct transcriptional programs mediated by the ligand-dependent full-length andro-

gen receptor and its splice variants in castration-resistant prostate cancer. Cancer Res 72(14):3457–3462
61. Zhang X et al (2011) Androgen receptor variants occur frequently in castration resistant prostate cancer metastases. PLoS One 6(11):e27970
62. Nadiminty N et al (2015) NF-kappaB2/p52:c-Myc:hnRNPA1 pathway regulates expression of androgen receptor splice variants and Enzalutamide sensitivity in prostate cancer. Mol Cancer Ther 14(8):1884–1895
63. Jimenez-Vacas JM et al (2020) Dysregulation of the splicing machinery is directly associated to aggressiveness of prostate cancer. EBioMedicine 51:102547
64. Liu LL et al (2014) Mechanisms of the androgen receptor splicing in prostate cancer cells. Oncogene 33(24):3140–3150
65. Li Y et al (2011) Intragenic rearrangement and altered RNA splicing of the androgen receptor in a cell-based model of prostate cancer progression. Cancer Res 71(6):2108–2117
66. Li Y et al (2012) AR intragenic deletions linked to androgen receptor splice variant expression and activity in models of prostate cancer progression. Oncogene 31(45):4759–4767
67. Nyquist MD et al (2013) TALEN-engineered AR gene rearrangements reveal endocrine uncoupling of androgen receptor in prostate cancer. Proc Natl Acad Sci U S A 110(43):17492–17497
68. van Bokhoven A et al (2003) Molecular characterization of human prostate carcinoma cell lines. Prostate 57(3):205–225
69. Xu D et al (2015) Androgen receptor splice variants dimerize to transactivate target genes. Cancer Res 75(17):3663–3671
70. Ozgun F et al (2021) DNA binding alters ARv7 dimer interactions. J Cell Sci 134(14)
71. Cao B et al (2014) Androgen receptor splice variants activating the full-length receptor in mediating resistance to androgen-directed therapy. Oncotarget 5(6):1646–1656
72. Chan SC et al (2015) Targeting chromatin binding regulation of constitutively active AR variants to overcome prostate cancer resistance to endocrine-based therapies. Nucleic Acids Res 43(12):5880–5897
73. Cato L et al (2019) ARv7 represses tumor-suppressor genes in castration-resistant prostate cancer. Cancer Cell 35(3):401-413 e6
74. Cai Z et al (2018) Androgen receptor: what we know and what we expect in castration-resistant prostate cancer. Int Urol Nephrol 50(10):1753–1764
75. He Y et al (2018) Androgen receptor splice variants bind to constitutively open chromatin and promote abiraterone-resistant growth of prostate cancer. Nucleic Acids Res 46(4):1895–1911
76. Lu J et al (2015) The cistrome and gene signature of androgen receptor splice variants in castration resistant prostate cancer cells. J Urol 193(2):690–698
77. Liang J et al (2021) Androgen receptor splice variant 7 functions independently of the full length receptor in prostate cancer cells. Cancer Lett 519:172–184
78. Cuzick J et al (2011) Prognostic value of an RNA expression signature derived from cell cycle proliferation genes in patients with prostate cancer: a retrospective study. Lancet Oncol 12(3):245–255
79. Kounatidou E et al (2019) A novel CRISPR-engineered prostate cancer cell line defines the AR-V transcriptome and identifies PARP inhibitor sensitivities. Nucleic Acids Res 47(11):5634–5647
80. Chen Z et al (2018) Diverse AR-V7 cistromes in castration-resistant prostate cancer are governed by HoxB13. Proc Natl Acad Sci U S A 115(26):6810–6815
81. Terada N et al (2017) Prognostic and predictive biomarkers in prostate cancer: latest evidence and clinical implications. Ther Adv Med Oncol 9(8):565–573
82. Hillebrand AC et al (2018) Androgen receptor isoforms expression in benign prostatic hyperplasia and primary prostate cancer. PLoS One 13(7):e0200613
83. Kallio HML et al (2018) Constitutively active androgen receptor splice variants AR-V3, AR-V7 and AR-V9 are co-expressed in castration-resistant prostate cancer metastases. Br J Cancer 119(3):347–356
84. Rizzo A et al (2021) Exploring the association between metastatic sites and androgen receptor splice variant 7 (AR-V7) in castration-resistant prostate cancer patients: a meta-analysis of prospective clinical trials. Pathol Res Pract 222:153440
85. Ma T et al (2021) Increased transcription and high translation efficiency lead to accumulation of androgen receptor splice variant after androgen deprivation therapy. Cancer Lett 504:37–48
86. Bernemann C et al (2019) Comparative analysis of AR variant AR-V567es mRNA detection systems reveals eminent variability and questions the role as a clinical biomarker in prostate cancer. Clin Cancer Res 25(13):3856–3864
87. Yu Z et al (2014) Rapid induction of androgen receptor splice variants by androgen deprivation in prostate cancer. Clin Cancer Res 20(6):1590–1600
88. Sharp A et al (2019) Androgen receptor splice variant-7 expression emerges with castration resistance in prostate cancer. J Clin Invest 129(1):192–208
89. Hoefer J et al (2016) Critical role of androgen receptor level in prostate cancer cell resistance to new generation antiandrogen enzalutamide. Oncotarget 7(37):59781–59794
90. Efstathiou E et al (2015) Molecular characterization of enzalutamide-treated bone metastatic castration-resistant prostate cancer. Eur Urol 67(1):53–60
91. Zhu Y et al (2020) Role of androgen receptor splice variant-7 (AR-V7) in prostate cancer resistance to 2nd-generation androgen receptor signaling inhibitors. Oncogene 39(45):6935–6949

92. Thadani-Mulero M et al (2014) Androgen receptor splice variants determine taxane sensitivity in prostate cancer. Cancer Res 74(8):2270–2282
93. Antonarakis ES et al (2015) Androgen receptor splice variant 7 and efficacy of Taxane chemotherapy in patients with metastatic castration-resistant prostate cancer. JAMA Oncol 1(5):582–591
94. Zhang G et al (2015) Androgen receptor splice variants circumvent AR blockade by microtubule-targeting agents. Oncotarget 6(27):23358–23371
95. Sharp A et al (2019) Clinical utility of circulating tumour cell androgen receptor splice Variant-7 status in metastatic castration-resistant prostate cancer. Eur Urol 76(5):676–685
96. Klotz L et al (2008) The efficacy and safety of degarelix: a 12-month, comparative, randomized, open-label, parallel-group phase III study in patients with prostate cancer. BJU Int 102(11):1531–1538
97. Pilepich MV et al (2005) Androgen suppression adjuvant to definitive radiotherapy in prostate carcinoma--long-term results of phase III RTOG 85-31. Int J Radiat Oncol Biol Phys 61(5):1285–1290
98. de Bono JS et al (2011) Abiraterone and increased survival in metastatic prostate cancer. N Engl J Med 364(21):1995–2005
99. Thompson CA (2003) Finasteride may prevent prostate cancer. Am J Health Syst Pharm 60(15):1511–1515
100. Thompson IM et al (2003) The influence of finasteride on the development of prostate cancer. N Engl J Med 349(3):215–224
101. Thompson IM et al (2003) Prevention of prostate cancer with finasteride: US/European perspective. Eur Urol 44(6):650–655
102. Andriole G et al (2004) Chemoprevention of prostate cancer in men at high risk: rationale and design of the reduction by dutasteride of prostate cancer events (REDUCE) trial. J Urol 172(4 Pt 1):1314–1317
103. Andriole GL et al (2010) Effect of dutasteride on the risk of prostate cancer. N Engl J Med 362(13):1192–1202
104. Clark RV et al (2004) Marked suppression of dihydrotestosterone in men with benign prostatic hyperplasia by dutasteride, a dual 5alpha-reductase inhibitor. J Clin Endocrinol Metab 89(5):2179–2184
105. Iversen P et al (2010) Antiandrogen monotherapy in patients with localized or locally advanced prostate cancer: final results from the bicalutamide early prostate cancer programme at a median follow-up of 9.7 years. BJU Int 105(8):1074–1081
106. Harris MG et al (1993) Nilutamide. A review of its pharmacodynamic and pharmacokinetic properties, and therapeutic efficacy in prostate cancer. Drugs Aging 3(1):9–25
107. Janknegt RA (1993) Total androgen blockade with the use of orchiectomy and nilutamide (Anandron) or placebo as treatment of metastatic prostate cancer. Anandron International Study Group. Cancer 72(12 Suppl):3874–3877
108. Beer TM et al (2014) Enzalutamide in metastatic prostate cancer before chemotherapy. N Engl J Med 371(5):424–433
109. Scher HI et al (2012) Increased survival with enzalutamide in prostate cancer after chemotherapy. N Engl J Med 367(13):1187–1197
110. Clegg NJ et al (2012) ARN-509: a novel antiandrogen for prostate cancer treatment. Cancer Res 72(6):1494–1503
111. Moilanen AM et al (2015) Discovery of ODM-201, a new-generation androgen receptor inhibitor targeting resistance mechanisms to androgen signaling-directed prostate cancer therapies. Sci Rep 5:12007
112. Yu X et al (2020) Structural insights of transcriptionally active, full-length androgen receptor Coactivator complexes. Mol Cell 79(5):812-823 e4
113. Andersen RJ et al (2010) Regression of castrate-recurrent prostate cancer by a small-molecule inhibitor of the amino-terminus domain of the androgen receptor. Cancer Cell 17(6):535–546
114. De Mol E et al (2016) EPI-001, a compound active against castration-resistant prostate cancer, targets transactivation unit 5 of the androgen receptor. ACS Chem Biol 11(9):2499–2505
115. Imamura Y et al (2016) An imaging agent to detect androgen receptor and its active splice variants in prostate cancer. JCI Insight 1(11)
116. Obst JK et al (2019) Revealing metabolic liabilities of ralaniten to enhance novel androgen receptor targeted therapies. ACS Pharmacol Transl Sci 2(6):453–467
117. Le Moigne G et al (2021) Noninvasive tidal volume measurements, using a time-of-flight camera, under high-flow nasal cannula-a physiological evaluation, in healthy volunteers. Crit Care Med
118. Banuelos CA et al (2016) Sintokamide a is a novel antagonist of androgen receptor that uniquely binds activation Function-1 in its amino-terminal domain. J Biol Chem 291(42):22231–22243
119. Yang YC et al (2016) Targeting androgen receptor activation function-1 with EPI to overcome resistance mechanisms in castration-resistant prostate cancer. Clin Cancer Res 22(17):4466–4477
120. Ravindranathan P et al (2013) Peptidomimetic targeting of critical androgen receptor-coregulator interactions in prostate cancer. Nat Commun 4:1923
121. Akram ON et al (2014) Tailoring peptidomimetics for targeting protein-protein interactions. Mol Cancer Res 12(7):967–978
122. Peng S et al (2020) Regression of castration-resistant prostate cancer by a novel compound QW07 targeting androgen receptor N-terminal domain. Cell Biol Toxicol 36(5):399–416
123. Goicochea NL et al (2017) Development of cell-penetrating bispecific antibodies targeting the N-terminal domain of androgen receptor for prostate cancer therapy†. Protein Eng Des Sel 30(12):785–793
124. Dalal K et al (2014) Selectively targeting the DNA-binding domain of the androgen receptor as a pro-

spective therapy for prostate cancer. J Biol Chem 289(38):26417–26429
125. Jones JO et al (2009) Non-competitive androgen receptor inhibition in vitro and in vivo. Proc Natl Acad Sci U S A 106(17):7233–7238
126. Lim M et al (2014) Ligand-independent and tissue-selective androgen receptor inhibition by pyrvinium. ACS Chem Biol 9(3):692–702
127. Pal SK et al (2019) Mechanistic investigation of the androgen receptor DNA-binding domain inhibitor Pyrvinium. ACS Omega 4(2):2472–2481
128. Li H et al (2014) Discovery of small-molecule inhibitors selectively targeting the DNA-binding domain of the human androgen receptor. J Med Chem 57(15):6458–6467
129. Dalal K et al (2017) Bypassing drug resistance mechanisms of prostate cancer with small molecules that target androgen receptor-chromatin interactions. Mol Cancer Ther 16(10):2281–2291
130. Dalal K et al (2018) Selectively targeting the dimerization interface of human androgen receptor with small-molecules to treat castration-resistant prostate cancer. Cancer Lett 437:35–43
131. Radaeva M et al (2021) Development of novel inhibitors targeting the D-Box of the DNA binding domain of androgen receptor. Int J Mol Sci 22(5)
132. Dervan PB, Kurmis AA, Finn PB (2018) Molecular recognition of DNA by py–im polyamides: from discovery to oncology. DNA-Target Mol Ther Agents 7:298
133. Kurmis AA, Dervan PB (2019) Sequence specific suppression of androgen receptor-DNA binding in vivo by a Py-Im polyamide. Nucleic Acids Res 47(8):3828–3835
134. Han X et al (2019) Discovery of ARD-69 as a highly potent proteolysis targeting chimera (PROTAC) degrader of androgen receptor (AR) for the treatment of prostate cancer. J Med Chem 62(2):941–964
135. Schneekloth AR et al (2008) Targeted intracellular protein degradation induced by a small molecule: en route to chemical proteomics. Bioorg Med Chem Lett 18(22):5904–5908
136. Salami J et al (2018) Androgen receptor degradation by the proteolysis-targeting chimera ARCC-4 outperforms enzalutamide in cellular models of prostate cancer drug resistance. Commun Biol 1:100
137. Petrylak DP et al (2020) First-in-human phase I study of ARV-110, an androgen receptor (AR) PROTAC degrader in patients (pts) with metastatic castrate-resistant prostate cancer (mCRPC) following enzalutamide (ENZ) and/or abiraterone (ABI). J Clin Oncol 38(15_suppl):3500
138. Kregel S et al (2020) Androgen receptor degraders overcome common resistance mechanisms developed during prostate cancer treatment. Neoplasia 22(2):111–119
139. Cheng MA et al (2018) Androgen receptor (AR) degradation enhancer ASC-J9((R)) in an FDA-approved formulated solution suppresses castration resistant prostate cancer cell growth. Cancer Lett 417:182–191
140. Lai KP et al (2013) New therapeutic approach to suppress castration-resistant prostate cancer using ASC-J9 via targeting androgen receptor in selective prostate cells. Am J Pathol 182(2):460–473
141. Liu C et al (2014) Niclosamide inhibits androgen receptor variants expression and overcomes enzalutamide resistance in castration-resistant prostate cancer. Clin Cancer Res 20(12):3198–3210
142. Lin TH et al (2013) Anti-androgen receptor ASC-J9 versus anti-androgens MDV3100 (Enzalutamide) or Casodex (Bicalutamide) leads to opposite effects on prostate cancer metastasis via differential modulation of macrophage infiltration and STAT3-CCL2 signaling. Cell Death Dis 4:e764
143. Ponnusamy S et al (2019) Orally bioavailable androgen receptor degrader, potential next-generation therapeutic for Enzalutamide-resistant prostate cancer. Clin Cancer Res 25(22):6764–6780
144. Ponnusamy S et al (2017) Novel selective agents for the degradation of androgen receptor variants to treat castration-resistant prostate cancer. Cancer Res 77(22):6282–6298
145. Lin W et al (2018) ASC-J9((R)) suppresses prostate cancer cell invasion via altering the sumoylation-phosphorylation of STAT3. Cancer Lett 425:21–30
146. Huang CP et al (2019) ASC-J9(R) increases the bladder cancer chemotherapy efficacy via altering the androgen receptor (AR) and NF-kappaB survival signals. J Exp Clin Cancer Res 38(1):275

ERβ and Inflammation

12

Linnea Hases, Amena Archer, and Cecilia Williams

Abstract

Estrogen, through the regulation of cytokine production, can act both as pro-inflammatory and anti-inflammatory signals dependent on the tissue context. In breast cancer cells, ERα is known to modulate inflammatory signaling through interaction with NFκB. Whether ERβ has a role in inflammation is less explored. Low levels of ERβ have been corroborated in several immune-related organs and, for example, in colonic epithelial cells. Specifically, an impact of ERβ on colitis and colitis-associated colorectal cancer (CRC) is experimentally supported, using ERβ-selective agonists, full-body ERβ knockout mice and, most recently, intestinal epithelial-specific knockout mice. An intricate crosstalk between ERβ and TNFα/NFκB signaling in the colon is supported, and ERβ activation appears to reduce macrophage infiltration also during high fat diet (HFD)-induced colon inflammation. Finally, the gut microbiota plays a fundamental role in the pathogenesis of colitis and ERβ has been indicated to modulate the microbiota diversity during colitis and colitis-induced CRC. ERβ is thus proposed to protect against colitis, by modulating NFκB signaling, immune cell infiltration, and/or microbiota composition. Selective activation of ERβ may therefore constitute a suitable preventative approach for the treatment of for example colitis-associated CRC.

Keywords

Estrogen receptor · Colon · Colorectal cancer · NFκB · Circadian clock · Gut microbiota

12.1 Colon Inflammation and Colorectal Cancer

Chronic inflammation promotes the development of colitis-associated CRC (CAC) [1], and inflammatory bowel disease (IBD) is a well-known risk factor for CRC. In fact, 20–30% of IBD patients will develop CAC during their lifetime [2]. The exact etiology of IBD is unknown, but it is understood to be multifactorial. It is driven by a disrupted immune system and/or gut microbiota, initiated by environmental factors in a genetically predisposed host [3]. IBD patients present an increased intestinal permeability with tight junction abnormalities [4–6], mucus abnormalities [7], and gut microbiota dysbiosis [8, 9]. Moreover,

L. Hases · A. Archer · C. Williams (✉)
SciLifeLab, Department of Protein Science, KTH Royal Institute of Technology, Solna, Sweden

Department of Biosciences and Nutrition, Karolinska Institutet, Huddinge, Sweden
e-mail: cecilia.williams@scilifelab.se

M. J. Campbell, C. L. Bevan (eds.), *Nuclear Receptors in Human Health and Disease*, Advances in Experimental Medicine and Biology 1390, https://doi.org/10.1007/978-3-031-11836-4_12

IBD patients show increased local and systemic levels of TNFα [10, 11].

The majority of CRC cases are sporadic and arise without any genetic predisposition [12]. These are associated with modifiable risk factors, several of which can lead to a low-grade chronic inflammation, such as intake of high-fat diet (HFD), obesity, low physical activity, and consumption of alcohol and cigarettes [13]. Also, alterations of the circadian rhythm (e.g., jet lag, "shift" work) has been reported to impact cancer development [14, 15]. Epidemiological studies support a strong association between a high body mass index (BMI) and an elevated CRC risk [16–18]. Obesity leads to a low-grade chronic inflammation, which can promote tumorigenesis [19]. HFD increases the risk, and the CRC incidence is increasing in young adults [20], which may be explained by an altered life-style. The frequency of obesity and IBD is increasing in parallel and 15–40% of IBD patients are obese, suggesting a shared environmental link between the conditions [21]. Colon is the first organ to respond to HFD [22], with increased intestinal permeability [23], inflammation [24, 25], stem cell activity [23, 26], and altered gut microbiota [27–30]. HFD-induced obesity increases CAC formation in mice [31, 32]. Pro-inflammatory adipokines produced by the adipose tissue may be the mechanism linking IBD and obesity [33], and circulating TNFα levels are found in obese individuals [34].

Another plausible mechanism linking the conditions is gut microbiota dysbiosis [33]. The gut microbiota presents vital benefits to the host. It produces short chain fatty acids (SCFAs) by fermentation of resistant dietary fibers and starches [35]. SCFAs, especially butyrate, are important energy sources for the intestinal epithelium that impacts epithelial cell differentiation, apoptosis, and proliferation [36]. Butyrate can also regulate gene expression by inhibiting histone deacetylases (HDACs), favoring an acetylated state, which remodels the chromatin toward an open and transcriptionally active state [37]. In addition, SCFAs regulate the immune system by increasing the production of mucus, strengthening the intestinal barrier, and inhibiting inflammatory responses [38]. Loss of SCFA-producing beneficial bacteria, outgrowth of opportunistic pathogens, and loss of overall bacterial diversity characterize gut dysbiosis. Resulting increase of gram-negative bacteria during dysbiosis can increase the levels of the endotoxin LPS, which can cause metabolic endotoxemia and obesity-related disorders [29]. Interaction of LPS with pattern recognition receptors can activate NFκB, which leads to increased expression of many pro-inflammatory cytokines [29]. Moreover, outgrowth of opportunistic pathogens that secrete enterotoxins can cause epithelial cell damage, intestinal permeability, and intestinal inflammation [39]. The gut microbiota is important to maintain intestinal homeostasis and dysbiosis can contribute to obesity, IBD, and CRC. Dysbiosis is thought to be one of the leading causes for CAC, by aggravating chronic inflammation. The release of toxic metabolites during dysbiosis may be associated with the neoplastic changes in the intestinal epithelium.

Interestingly, in obesity, IBD, and CRC, circadian clock dysregulation has been reported. The central circadian clock located in the suprachiasmatic nucleus in the anterior hypothalamus controls several physiological processes including feeding time and energy intake. The classical regulators include the transcriptional CLOCK complex (CLOCK, ARNTL/BMAL1, NPAS2), the core clock genes Period (PER1–3), and the cryptochrome circadian regulators (CRY1, CRY2). Peripheral tissues and cells have also their own autonomous circadian clocks, which play a central role in metabolic output. In the intestinal tract, the circadian clock is involved in motility, nutrient absorption, intestinal permeability, metabolism, and cell proliferation. Its alteration is associated with changed microbiota composition, increased gut inflammation, and CRC risks [40]. Furthermore, diet-induced obesity that leads to colonic inflammation with increased intestinal permeability and modification of the microbiota, alters also the intestinal circadian clock as observed in other organs such as the liver. Interestingly, estrogenic and sex-dependent mechanisms regulate the circadian clock [41], and proinflammatory NFκB has been implicated in the maintenance of the circadian rhythm in mouse [42].

12.1.1 Pro-Inflammatory Signaling

Chronic inflammation is a hallmark for CRC promotion and correlates with poor prognosis. The inflammation is driven by cytokines and chemokines produced both by the tumor cells themselves and by other cells in the tumor microenvironment. Pro-inflammatory cytokines, including TNFα, IL-1β, and IL-6, activate the oncogenic signaling pathways NFκB and STAT3 in the tumor cells, thereby enhancing tumor growth and invasiveness. TNFα is one of the main pro-inflammatory mediators that trigger inflammatory response. TNFα is involved in the pathogenesis of both obesity and IBD. TNFα is an early response pro-inflammatory cytokine and regulates multiple cellular processes, including proliferation, apoptosis, differentiation, and the production of inflammatory molecules, which contributes to epithelial cell injury [43]. TNFα promotes inflammation-associated cancer by activating NFκB [44]. NFκB induces several cytokines and oncogenes involved in tumorigenesis and is critical for induction of CAC [45–47]. Treatment with anti-inflammatory drugs (e.g., aspirin or non-steroid anti-inflammatory drugs, NSAIDs) reduces CRC incidence in clinical trials [48, 49], by inhibiting nuclear translocation of NFκB [50]. Anti-TNFα therapy has also been used in IBD patients; however patient response is low, and the patients that respond may lose response over time [51]. The pro-inflammatory cytokine IL-6 activates the oncogenic pathway STAT3. STAT3 drives several malignant pathways in CRC, including proliferation, anti-apoptosis, epithelial-to-mesenchymal transition (EMT), invasion and metastasis, and angiogenesis [52]. Currently, a number of clinical trials are investigating drugs targeting the IL-6/STAT3 pathway in CRC [52]. However, the risk-benefit balance of current treatments remains poor and there is a need for better preventative approaches to reduce CAC incidence.

12.2 Estrogen Signaling in Colon Inflammation and CAC

Estrogen can act both pro-inflammatory and anti-inflammatory, depending on the tissue context. Several studies have shown that estrogen receptors (ERs) can repress the transcriptional activity of NFκB [53]. This repression can occur via different mechanisms, including prevention of NFκB DNA binding, recruitment of corepressors into the ER-NFκB complex, competition of coactivators, or prevention of NFκB nuclear translocation [53]. Treatment with 17β-estradiol (E2) in the murine macrophage cell line RAW264.7 blocked LPS-induced nuclear translocation of NFκB and corresponding inflammatory signaling [54]. Furthermore, E2 inhibited inflammation in rat aortic smooth muscle cells by two mechanisms: directly inhibiting NFκB binding to inflammatory genes and promoting synthesis of the NFκB inhibitor IκBα [55]. Contrary, in breast cancer cells, estrogen via ERα can partially enhance inflammatory signaling through a positive crosstalk with NFκB [56].

There are clear sex differences in CRC incidence, survival, and tumor locations [57–59]. Sex differences are present both in incidence and onset of CRC, with men at higher risk [60]. Men with IBD possess a 60% higher risk to develop CRC compared to corresponding women [61]. Further, the overall survival for premenopausal CRC patients is better than for age-matched men, whereas the contrary is seen in postmenopausal patients [62]. Numerous epidemiological studies support a protective role of estrogen against CRC. Importantly, menopausal hormone therapy (MHT) [63, 64], endogenous estrogen [65], oral contraceptives [66, 67], and phytoestrogens [68] reduce the CRC incidence. MHT reduces the CRC incidence with around 20%, and estrogen-only therapy had the largest impact [63, 69–71]. Interestingly, while obesity is a risk factor for CRC in both sexes, this is stronger in men compared to women [72], Although the incidence of obesity is higher in women, women are less prone to develop insulin resistance and metabolic syndrome (MetS) than men [73]. Studies have linked hyperinsulinemia to enhanced CRC risks [74], highlighting the influence of obesity on CRC. Estrogen can attenuate the deleterious effects of obesity by improving insulin sensitivity, serum lipid profile, and fat distribution (reviewed in [75]). Absence of estrogen worsens the metabolic unbalance and can contribute to MetS, type-2 diabetes, and cardiovascular diseases. Indeed, women during menopausal transition are at higher risk to gain weight, get obese, and develop MetS, while MHT in patients with diabetes has been shown to reduce obesity, fasting glucose levels, and insulin resistance [76, 77].

Further, estrogen and the gut microbiota have synergistic actions that can impact obesity and cancer. The gut microbiota is in fact one of the major regulators of circulating estrogen, referred to as the estrobolome [78]. By producing the enzyme β-glucuronidase, microbiota metabolizes estrogen-like compounds to an active form by deconjugation [78]. Dysbiosis can thereby impact the circulating levels of estrogen. Genistein is a soy isoflavone metabolized by the gut microbiota to compounds with a structure similar to estrogen [78]. Genistein has been shown to exert antibacterial activities [79], and supplementation with genistein in postmenopausal women can induce growth of the beneficial soy-fermenting bacteria *Bifidobacterium* [80, 81]. A meta-analysis by Fang et al. revealed that dietary supplementation with phytoestrogens significantly reduced insulin resistance in postmenopausal women [82]. Intake of phytoestrogen has also been shown to reduce the risk of CRC [68]. There is also evidence that sex hormones themselves regulate the microbiota composition. The microbiota diversity is lower in postmenopausal women compared to premenopausal women [83]. Several human and animal studies support that sex impacts the microbiota composition, although the results are inconsistent. Clear sex differences in the microbiota composition were noted when 89 different inbred mouse strains were analyzed independently [84]. In addition, E2 treatment impacted the gut microbiota both in males and ovariectomized (OVX) females on a western diet, where it reduced the levels of the opportunistic pathogen *Proteobacteria* and LPS-related functional pathways [85]. Further, E2 treatment impacted the diversity of the gut microbiota in males with induced CAC [86].

The sex difference seen in CRC incidence is also evident in mice. In several studies, male mice have been shown to present a higher incidence of more and larger tumors compared to female mice during colitis-induced tumor formation (using carcinogen azoxymethane, AOM, and irritant dextran sodium sulfate, DSS) model [87–89]. Furthermore, studies have shown that E2 treatment protects against colitis and CAC by suppressing NFκB signaling. E2 treatment in males suppressed NFκB signaling and resulted in reduced tumor formation [87]. Moreover, OVX females presented increased colitis-induced tumor formation, which was reduced by E2 treatment [90]. The E2 treatment in OVX females suppressed the NFκB signaling consistent with the effects seen in males [90]. Interestingly, HFD-induced obesity has been shown to aggravate colitis-induced tumor formation by enhancing colonic cell proliferation [32]. We have shown that HFD feeding increased colonic cell proliferation specifically in males, which suggests that endogenous estrogen levels in females are protective [89]. In line with this, E2 treatment in males could oppose the HFD-induced proliferation [89]. Together, these data support that both endogenous and exogenous estrogens protect against CAC.

12.3 Expression Pattern and Role of ERβ in Colon

Multiple studies support that ERβ acts as a tumor suppressor in CRC. However, the expression and role of ERβ are debated, due to relative low mRNA levels in the colon and the use of unspecific antibodies in the literature. We have performed an antibody validation of 13 commercial or in-house anti-ERβ antibodies, including the most used ones [91]. Immunohistochemistry (IHC) analysis with extensive negative and positive controls, and validation of binding with immunoprecipitation (IP) followed by mass spectrometry (MS), revealed that only one antibody, PPZ0506, was specific for ERβ [91]. This antibody had not previously been used on clinical material. Using the PPZ0506 antibody, we performed protein expression profiling in 21 malignant and 44 normal human tissues. This revealed that ERβ protein expression could only be validated in a few normal tissues, including ovary, testis, rectum and colon, and lymphoid cells in for example the spleen and lymph nodes (Fig. 12.1a) [91]. Further, ERβ protein expression was found in only three cancer types examined: in a majority of granulosa cell tumors, and in rare cases of melanoma and thy-

A

TISSUE	RNA (FPKM)	IHC PPZ
Testis	5	
Adrenal gland	4	
Ovary	3	
Stomach	1	
Appendix	1	
Colon	1	
Rectum	1	
Urinary bladder	1	
Adipose tissue	1	
Lymph node	1	
Tonsil	1	
Spleen	1	
Liver	0	
Gallbladder	0	
Pancreas	0	
Salivary gland	0	
Esophagus	0	
Duodenum	0	
Small intestine	0	
Kidney	0	
Prostate	0	
Breast	0	
Endometrium	0	
Fallopian tube	0	
Placenta	0	
Skin	0	
Skeletal muscle	0	
Smooth muscle	0	
Bone marrow	0	
Cerebral cortex	0	
Thyroid gland	0	
Lung	0	
Heart muscle	0	

B

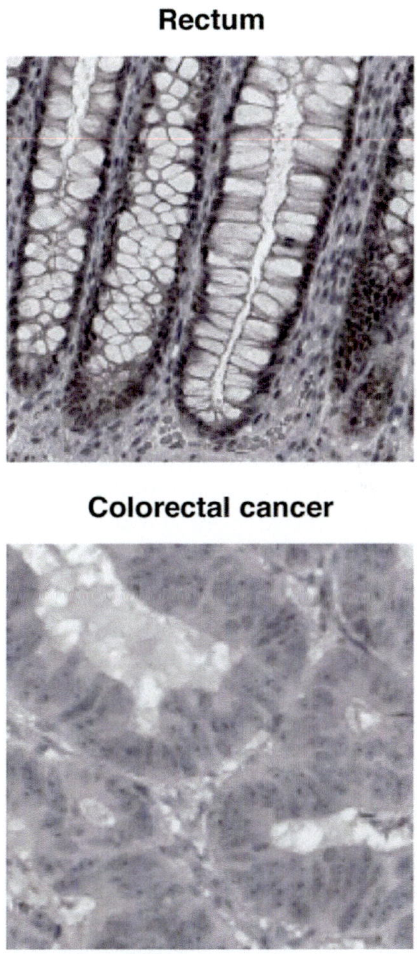

Fig. 12.1 ERβ protein expression obtained through IHC with PPZ0506 correlates well with transcript levels in a large panel of tissues. (**a**) mRNA and protein expression of ERβ in the Human Protein Atlas tissue panel. (**b**) IHC (PPZ0506) showing ERβ expression in intestinal epithelial cells in the rectum and lack of expression in CRC. (Note: This figure was adopted from Andersson et al. [91] and Hases et al. [89] both licensed under a Creative Commons Attribution 4.0 International License)

roid cancer [91]. This ERβ protein expression pattern correlated well to RNA-seq data (e.g., from the Genotype-Tissue Expression (GTEx) and Human Protein Atlas) but contradicted multiple studies using other antibodies.

Both ERα and ERβ have been reported to protect against obesogenic effects. While ERα clearly improves MetS, the role of ERβ is more controversial. Mice lacking ERα (αERKO) developed obesity, glucose intolerance, and insulin resistance [92, 93]. Although ERβ has been reported to protect against HFD-induced MetS, Foryst-Ludwig et al. demonstrated that while ERβ indeed protected against HFD-induced obesity it also presented pro-diabetogenic effects, as βERKO mice exhibited improved insulin sensitivity and glucose tolerance [94]. Moreover, we have noted that ERβ-selective activation with DPN impaired glucose levels in female mice, while no effect was seen in males [95]. While ERα improves the MetS, and in this respect would be beneficial against CRC, most of the estrogen-mediated CRC protective effects appear to be mediated via ERβ. Polymorphism in the ERβ promoter region is associated with CRC risk and survival [96, 97]. Data obtained from Passarelli et al. demonstrated that three out of 99 SNPs tested were statistically significant predictors of CRC-specific and overall survival, all three in the promoter region of *ESR2* [97]. Another study by Honma et al. demonstrated that cytosine-adenine repeat polymorphism (*ESR2* CA rs3223460) located within an intron of the ERβ gene correlated to CRC risk [96]. ERβ protein, but not ERα, is expressed at low levels in normal intestinal epithelial cells and loses its expression in CRC (Fig. 12.1b) [89]. When ERβ is exogenously expressed in human CRC cell lines, it mediates anti-proliferative and anti-tumorigenic roles, also in xenografts [98–100]. βERKO mice present impaired cell-to-cell junctions and cell architecture, suggesting that ERβ plays a role in the integrity of the intestinal barrier [101]. Moreover, multiple *in vivo* studies support a protective effect of ERβ against CRC. Female mice lacking ERβ (βERKO) presented increased tumor formation in the CAC model (AOM/DSS) [102], and treatment with a selective ERβ agonist reduced proliferation and tumor formation in the $Apc^{Min/+}$ model of intestinal tumorigenesis in both sexes [103]. In addition, ERβ activation with DPN repressed HFD-induced effects on the colon in both sexes (Fig. 12.2a) [95]. ERβ-selective activation also opposed the HFD-increased colonic infiltration of F4/80+ macrophages in both sexes (Fig. 12.2b) [95] along with colonic epithelial cell proliferation and modulation of circadian clock genes in males (Fig. 12.2c) [95]. Conclusively, we have recently demonstrated that the CAC-protective effect of ERβ is mediated by its expression in intestinal epithelial cells. Removing ERβ specifically from these cells ($ER\beta KO^{Vil}$) increased (AOM/DSS) tumor formation in a sex-dependent manner [89]. $ER\beta KO^{Vil}$ male mice developed significantly more tumors whereas $ER\beta KO^{Vil}$ female mice developed significantly larger tumors compared to their WT counterparts [89].

12.3.1 Intestinal Epithelial ERβ Regulates Core Clock Genes

Our use of intestinal ERβ knockout ($ER\beta KO^{Vil}$) mice also revealed that the regulation of core clock genes was impacted by intestinal epithelial ERβ in females [95]. Mechanistic studies revealed that ERβ binds to cis-regulatory chromatin regions of the core clock genes *Bmal1* and *Npas2* [95]. $ER\beta KO^{Vil}$ female mice presented a significant decrease in the expression of the core clock genes *Bmal1* and *Npas2* during HFD (Fig. 12.3) [95]. Since the circadian clock is important for intestinal homeostasis, and its impairment can impact cell proliferation, the immune system, intestinal permeability, and the microbiota composition [104–107], this may be a critical mechanism whereby ERβ impacts colon inflammation and cancer.

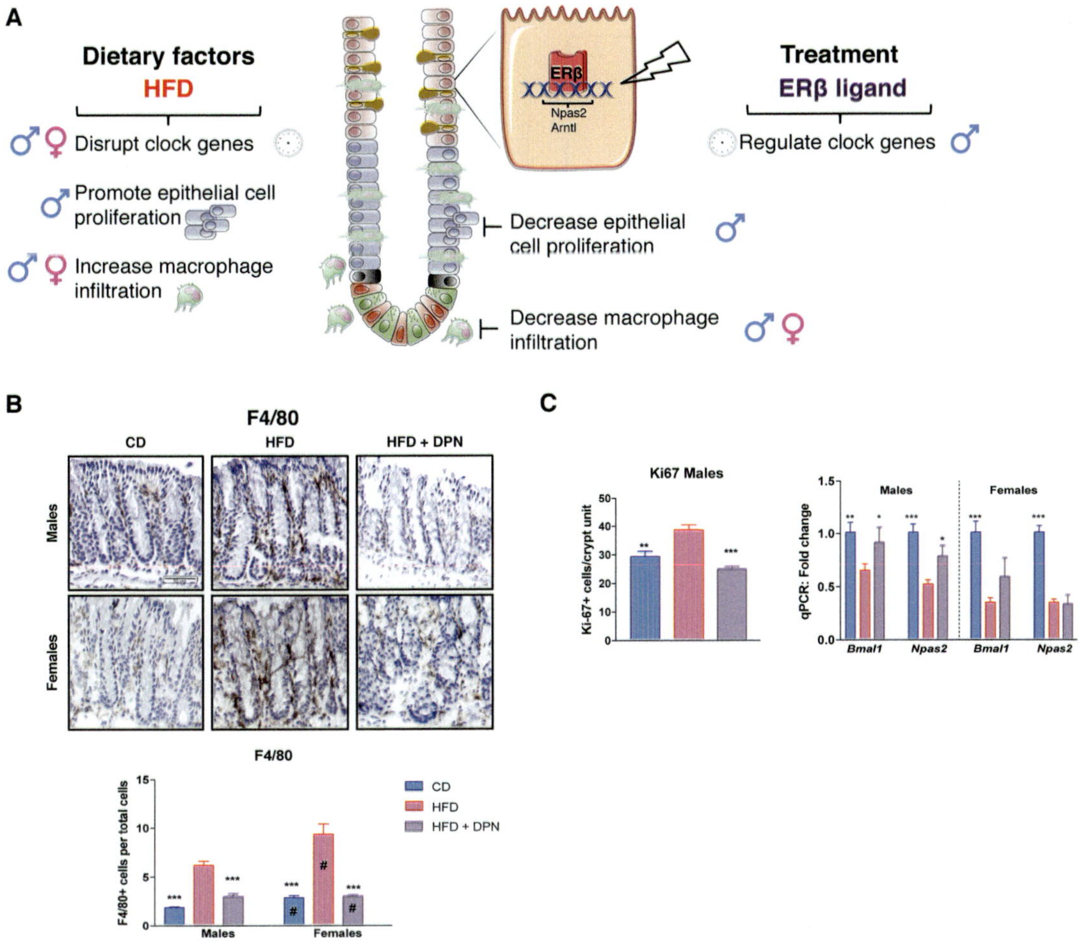

Fig. 12.2 Schematic illustration of the proposed estrogenic regulation of colon and its microenvironment during HFD-induced obesity and supportive data. (**a**) The proposed model for estrogenic regulation during HFD. (**b**) The HFD-induced macrophage infiltration was significantly counteracted by ERβ in both sexes (bottom left and middle). (**c**) Further, the HFD-induced cell proliferation and impaired colonic clock gene expression was significantly opposed by ERβ activation in males. (Note: This figure was modified from Hases et al. [95] licensed under a Creative Commons Attribution 4.0 International License)

12.3.2 Intestinal Epithelial ERβ Regulates NFκB Signaling and Gut Microbiota

Intestinal ERβ mediated protection against tumor formation, and its loss, resulted in increased expression of TNFα and NFκB target genes, including IL-6 [89]. Also *ex vivo,* intestinal organoids from ERβKO^{Vil} mice presented increased TNFα-induced epithelial cell damage, demonstrating a local regulation between intestinal epithelial ERβ and TNFα that is independent of the microbiota and the systemic immune response [89]. Mechanistic studies in human CRC cell lines revealed that ERβ could bind and regulate the expression of NFκB regulators, including *ATF3, BCL3,* and *BIRC3,* mimicking regulations observed *in vivo* [89]. The intestinal epithelial levels of ERβ are relatively low, but effects of its removal on CAC *in vivo* were relatively strong. Interestingly, using ERE-luciferase transactivation assay in a CRC cell line with exogenously expressed ERβ, we found that activation of ERβ was significantly increased by TNFα [89]. This suggests a protective feedback

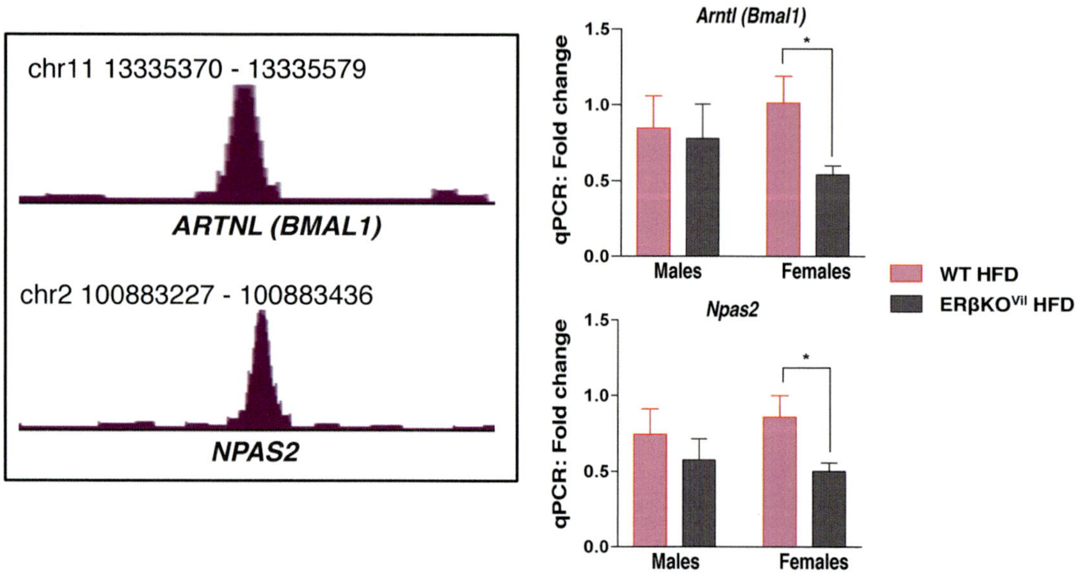

Fig. 12.3 Intestinal epithelial ERβ binds chromatin by and regulates core clock genes in females. ERβ binds to cis-regulatory chromatin regions of the core clock genes *BMAL1* and *NPAS2* in CRC cell lines with exogenously expressed ERβ. Intestinal epithelial ERβ regulates the expression of the two core clock genes in females fed HFD. (Note: This figure was modified from Hases et al. [95] licensed under a Creative Commons Attribution 4.0 International License)

mechanism in which the inflammatory state amplifies the activity of ERβ, which in turn inhibits the TNFα-induced NFκB inflammatory signaling and results in reduced tumorigenesis (Fig. 12.4). The expression of IL-6 was increased in ERβKO[Vil] mice during induced (AOM/DSS) tumor formation, and further studies are needed to explore the role of ERβ in the regulation of IL6/STAT3 pathway.

Although this shows that intestinal epithelial ERβ can attenuate TNFα-signaling locally in the colon, other mechanism may contribute. ERβ may impact the microbiota composition, which in turn can impact the tumor formation. ERβKO[Vil] mice presented a significant compositional dissimilarity compared to WT mice during DSS-induced colitis, as analyzed by 16S rRNA sequencing of stool samples [108]. This significant difference, however, disappeared during later stages of CAC, when the more acute DSS-induced inflammatory state had subsided. During DSS-induced colitis, ERβKO[Vil] mice presented an increased expression of many SCFA-producing bacteria and an increase in the gram-negative bacteria *Desulfovibrio* (phylum Proteobacteria) [108], which is a major producer of hydrogen sulfide (H_2S). This may cause increased SCFAs and H_2S levels in ERβKO[Vil] mice. Increased levels of the SCFA butyrate can inhibit intestinal stem cell proliferation [109], and may lead to an impaired ulcer healing after DSS treatment. Increased levels of H_2S can be detrimental to the intestinal epithelium by inhibiting butyrate oxidation, the major source of energy for the intestinal epithelium, and impair DNA repair [110]. Decreased butyrate oxidation may cause intestinal epithelial starvation and increased intestinal permeability [110]. Additionally, H_2S can reduce the disulfide bonds in the mucus, which can lead to an impaired intestinal barrier [111]. Further, the levels of the bacteria *Faecalibaculum* was reduced in ERβKO[Vil] mice [108]. *Faecalibaculum* has been shown to decrease tumor growth in both AOM/DSS-induced and $Apc^{Min/+}$ spontaneous tumor formation models [112]. Our study implicating a role for intestinal epithelial ERβ on the microbiome was a relatively small study, where limited sex and genotype stratifications and interactions were performed [108]. Larger studies, ideally

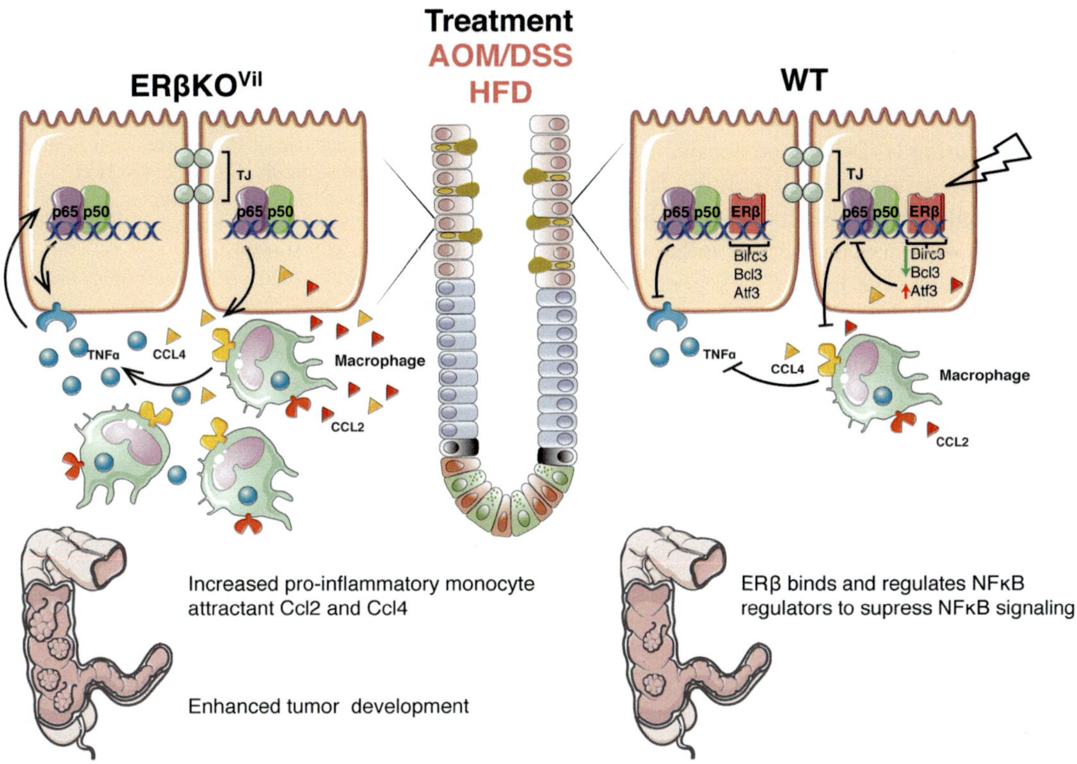

Fig. 12.4 A proposed model for how ERβ protects against colitis and CRC based on animal studies. Intestinal epithelial ERβ suppresses TNFα signaling by controlling NFκB regulators and diminishing NFκB-induced inflammatory signaling. The reduced inflammation leads to decreased levels of *Ccl2* and *Ccl4*, which can result in reduced infiltration of TNFα secreting pro-inflammatory macrophages, and subsequent reduced tumor formation and growth. (Note: This figure was adopted from Hases et al. [89] licensed under a Creative Commons Attribution 4.0 International License)

using whole genome metagenomic sequencing analyses, would be needed to corroborate and further characterize the role of intestinal epithelial ERβ in the regulation of the microbiota composition.

12.4 Concluding Remarks

CRC represents the third most deadly form of cancer in the Western world and the incidence is increasing among young adults. This highlights the need for better preventatives to reduce CRC mortality. Studies support that ERβ in intestinal epithelial cells protects against colon inflammation and CAC. Natural ERβ-selective agonists (phytoestrogens) in clinical trials have been proven to be safe [113, 114], and ERβ-selective agonists have been proposed as a promising approach to lower CRC incidence [115]. We have recently demonstrated that intestinal epithelial ERβ protects against induced tumor formation in a sex-dependent manner by modifying NFκB signaling. Further, selective activation of ERβ suppresses HFD-induced macrophage infiltration and downregulates core clock genes in the colon of both sexes and opposes HFD-induced epithelial cell proliferation in males. The regulation of the core clock genes *Bmal1* and *Npas2* was indeed regulated by intestinal epithelial ERβ in females. The protective effects of ERβ may in part be regulated by its modulation of the gut microbiota. Measuring the levels of microbiota metabolites would provide an improved understanding of the impact of the microbiota changes on the colon. Interestingly, the expression of colonic ERβ is reduced in antibiotic-treated mice [116], which suggest that the microbiota and its metabolites might regulate ERβ expression.

In addition to the circadian clock being dysregulated during HFD-induced obesity, it is also disrupted in induced (DSS) colitis. This suggests a common disease mechanism between HFD- and DSS-induced colon inflammation, which might involve ERβ. Clock gene disruption is an early event in IBD [117], and *Bmal1*-knockout mice present elevated severity of DSS-induced colitis [118]. We have demonstrated that intestinal ERβ regulate clock genes, and restoring the circadian clock by safe therapies, such as a selective agonist for ERβ, may be an ideal target for preventing CAC in for example IBD patients. Interestingly, the circadian clock regulates ERβ expression trough the conserved E-box motif located in the ERβ promoter region [119]. Further studies are needed to investigate the role of ERβ in the regulation of the colonic circadian rhythm and *vice versa*. Altogether, there is evidence that an ERβ selective agonist may constitute a novel chemopreventive approach against CAC.

Acknowledgements This work was supported by the Swedish Cancer Society (21 1632 Pj), Swedish Research Council (2017-01658), and Stockholm County Council (RS2021-0316).

References

1. Sokolova O, Naumann M (2019) Crosstalk between DNA damage and inflammation in the multiple steps of gastric carcinogenesis. Curr Top Microbiol Immunol 421:107–137
2. Choi PM, Zelig MP (1994) Similarity of colorectal cancer in Crohn's disease and ulcerative colitis: implications for carcinogenesis and prevention. Gut 35:950–954
3. Missaghi B, Barkema HW, Madsen KL, Ghosh S (2014) Perturbation of the human microbiome as a contributor to inflammatory bowel disease. Pathogens 3:510–527
4. Lee SH (2015) Intestinal permeability regulation by tight junction: implication on inflammatory bowel diseases. Intest Res 13:11–18
5. Salim SY, Söderholm JD (2011) Importance of disrupted intestinal barrier in inflammatory bowel diseases. Inflamm Bowel Dis 17:362–381
6. Zeissig S et al (2007) Changes in expression and distribution of claudin 2, 5 and 8 lead to discontinuous tight junctions and barrier dysfunction in active Crohn's disease. Gut 56:61–72
7. van der Post S et al (2019) Structural weakening of the colonic mucus barrier is an early event in ulcerative colitis pathogenesis. Gut 68:2142–2151
8. Ahmed I, Roy BC, Khan SA, Septer S, Umar S (2016) Microbiome, metabolome and inflammatory bowel disease. Microorganisms 4
9. Alam MT et al (2020) Microbial imbalance in inflammatory bowel disease patients at different taxonomic levels. Gut Pathog 12:1
10. Reimund JM et al (1996) Mucosal inflammatory cytokine production by intestinal biopsies in patients with ulcerative colitis and Crohn's disease. J Clin Immunol 16:144–150
11. Reimund JM et al (1996) Increased production of tumour necrosis factor-alpha interleukin-1 beta, and interleukin-6 by morphologically normal intestinal biopsies from patients with Crohn's disease. Gut 39:684–689
12. Müller MF, Ibrahim AE, Arends MJ (2016) Molecular pathological classification of colorectal cancer. Virchows Arch 469:125–134
13. Marley AR, Nan H (2016) Epidemiology of colorectal cancer. Int J Mol Epidemiol Genet 7:105–114
14. Parent M, El-Zein M, Rousseau MC, Pintos J, Siemiatycki J (2012) Night work and the risk of cancer among men. Am J Epidemiol 176:751–759
15. Lee Y (2021) Roles of circadian clocks in cancer pathogenesis and treatment. Exp Mol Med 53:1529–1538
16. Dai Z, Xu YC, Niu L (2007) Obesity and colorectal cancer risk: a meta-analysis of cohort studies. World J Gastroenterol 13:4199–4206
17. Harriss DJ et al (2009) Lifestyle factors and colorectal cancer risk (1): systematic review and meta-analysis of associations with body mass index. Color Dis Off J Assoc Coloproctol G B Irel 11:547–563
18. Polednak AP (2008) Estimating the number of U.S. incident cancers attributable to obesity and the impact on temporal trends in incidence rates for obesity-related cancers. Cancer Detect Prev 32:190–199
19. Yehuda-Shnaidman E, Schwartz B (2012) Mechanisms linking obesity, inflammation and altered metabolism to colon carcinogenesis. Obes Rev 13:1083–1095
20. Loomans-Kropp HA, Umar A (2019) Increasing incidence of colorectal cancer in young adults. J Cancer Epidemiol 2019:9841295
21. Singh S, Dulai PS, Zarrinpar A, Ramamoorthy S, Sandborn WJ (2017) Obesity in IBD: epidemiology, pathogenesis, disease course and treatment outcomes. Nat Rev Gastroenterol Hepatol 14:110–121
22. Kawano Y et al (2016) Colonic pro-inflammatory macrophages cause insulin resistance in an intestinal Ccl2/Ccr2-dependent manner. Cell Metab 24:295–310

23. Xie Y et al (2020) Impact of a high-fat diet on intestinal stem cells and epithelial barrier function in middle-aged female mice. Mol Med Rep 21:1133–1144
24. Liu Z et al (2012) Diet-induced obesity elevates colonic TNF-alpha in mice and is accompanied by an activation of Wnt signaling: a mechanism for obesity-associated colorectal cancer. J Nutr Biochem 23:1207–1213
25. Luck H et al (2015) Regulation of obesity-related insulin resistance with gut anti-inflammatory agents. Cell Metab 21:527–542
26. Beyaz S et al (2016) High-fat diet enhances stemness and tumorigenicity of intestinal progenitors. Nature 531:53–58
27. Cani PD et al (2008) Changes in gut microbiota control metabolic endotoxemia-induced inflammation in high-fat diet-induced obesity and diabetes in mice. Diabetes 57:1470–1481
28. Hildebrandt MA et al (2009) High-fat diet determines the composition of the murine gut microbiome independently of obesity. Gastroenterology 137:1716-1724.e1711-1712
29. Kim KA, Gu W, Lee IA, Joh EH, Kim DH (2012) High fat diet-induced gut microbiota exacerbates inflammation and obesity in mice via the TLR4 signaling pathway. PLoS One 7:e47713
30. Serino M et al (2012) Metabolic adaptation to a high-fat diet is associated with a change in the gut microbiota. Gut 61:543–553
31. Wunderlich CM et al (2018) Obesity exacerbates colitis-associated cancer via IL-6-regulated macrophage polarisation and CCL-20/CCR-6-mediated lymphocyte recruitment. Nat Commun 9:1646
32. Tuominen I et al (2013) Diet-induced obesity promotes colon tumor development in azoxymethane-treated mice. PLoS One 8:e60939
33. Harper JW, Zisman TL (2016) Interaction of obesity and inflammatory bowel disease. World J Gastroenterol 22:7868–7881
34. Rodriguez-Hernandez H, Simental-Mendia LE, Rodriguez-Ramirez G, Reyes-Romero MA (2013) Obesity and inflammation: epidemiology, risk factors, and markers of inflammation. Int J Endocrinol 2013:678159
35. Rowland I et al (2018) Gut microbiota functions: metabolism of nutrients and other food components. Eur J Nutr 57:1–24
36. Parada Venegas D et al (2019) Short chain fatty acids (SCFAs)-mediated gut epithelial and immune regulation and its relevance for inflammatory bowel diseases. Front Immunol 10:277
37. Candido EP, Reeves R, Davie JR (1978) Sodium butyrate inhibits histone deacetylation in cultured cells. Cell 14:105–113
38. Bilotta AJ, Cong Y (2019) Gut microbiota metabolite regulation of host defenses at mucosal surfaces: implication in precision medicine. Precis Clin Med 2:110–119
39. Lobionda S, Sittipo P, Kwon HY, Lee YK (2019) The role of gut microbiota in intestinal inflammation with respect to diet and extrinsic stressors. Microorganisms 7
40. Voigt RM, Forsyth CB, Keshavarzian A (2019) Circadian rhythms: a regulator of gastrointestinal health and dysfunction. Expert Rev Gastroenterol Hepatol 13:411–424
41. Yan L, Silver R (2016) Neuroendocrine underpinnings of sex differences in circadian timing systems. J Steroid Biochem Mol Biol 160:118–126
42. Hong HK et al (2018) Requirement for NF-κB in maintenance of molecular and behavioral circadian rhythms in mice. Genes Dev 32:1367–1379
43. Leppkes M, Roulis M, Neurath MF, Kollias G, Becker C (2014) Pleiotropic functions of TNF-α in the regulation of the intestinal epithelial response to inflammation. Int Immunol 26:509–515
44. Senftleben U, Karin M (2002) The IKK/NF-kappaB pathway. Crit Care Med 30:S18–s26
45. Baldwin AS (2001) Control of oncogenesis and cancer therapy resistance by the transcription factor NF-kappaB. J Clin Invest 107:241–246
46. Jana A et al (2017) NFkB is essential for activin-induced colorectal cancer migration via upregulation of PI3K-MDM2 pathway. Oncotarget 8:37377–37393
47. Schottelius AJ, Dinter H (2006) Cytokines, NF-kappaB, microenvironment, intestinal inflammation and cancer. Cancer Treat Res 130:67–87
48. Coyle C, Cafferty FH, Langley RE (2016) Aspirin and colorectal cancer prevention and treatment: is it for everyone? Curr Colorectal Cancer Rep 12:27–34
49. Zheng SL, Roddick AJ (2019) Association of Aspirin use for primary prevention with cardiovascular events and bleeding events: a systematic review and meta-analysis. JAMA 321:277–287
50. Nagaishi T et al (2016) Epithelial nuclear factor x03BA;B activation in inflammatory bowel diseases and colitis-associated carcinogenesis. Digestion 93:40–46
51. Singh S, George J, Boland BS, Vande Casteele N, Sandborn WJ (2018) Primary non-response to tumor necrosis factor antagonists is associated with inferior response to second-line biologics in patients with inflammatory bowel diseases: a systematic review and meta-analysis. J Crohns Colitis 12:635–643
52. Lin Y et al (2020) Progress in understanding the IL-6/STAT3 pathway in colorectal cancer. Onco Targets Ther 13:13023–13032
53. Kalaitzidis D, Gilmore TD (2005) Transcription factor cross-talk: the estrogen receptor and NF-kappaB. Trends Endocrinol Metab 16:46–52
54. Ghisletti S, Meda C, Maggi A, Vegeto E (2005) 17beta-estradiol inhibits inflammatory gene expression by controlling NF-kappaB intracellular localization. Mol Cell Biol 25:2957–2968

55. Xing D et al (2012) Estrogen modulates NFκB signaling by enhancing IκBα levels and blocking p65 binding at the promoters of inflammatory genes via estrogen receptor-β. PLoS One 7:e36890
56. Frasor J et al (2009) Positive cross-talk between estrogen receptor and NF-kappaB in breast cancer. Cancer Res 69:8918–8925
57. Kim SE et al (2015) Sex- and gender-specific disparities in colorectal cancer risk. World J Gastroenterol 21:5167–5175
58. Zheng D et al (2018) Regulation of sex hormone receptors in sexual dimorphism of human cancers. Cancer Lett 438:24–31
59. Hases L et al (2021) The importance of sex in the discovery of colorectal cancer prognostic biomarkers. Int J Mol Sci 22:1354
60. Brozek W, Kriwanek S, Bonner E, Peterlik M, Cross HS (2009) Mutual associations between malignancy, age, gender, and subsite incidence of colorectal cancer. Anticancer Res 29:3721–3726
61. Soderlund S et al (2010) Inflammatory bowel disease confers a lower risk of colorectal cancer to females than to males. Gastroenterology 138:1697–1703
62. Hendifar A et al (2009) Gender disparities in metastatic colorectal cancer survival. Clin Cancer Res 15:6391–6397
63. Grodstein F, Newcomb PA, Stampfer MJ (1999) Postmenopausal hormone therapy and the risk of colorectal cancer: a review and meta-analysis. Am J Med 106:574–582
64. Newcomb PA et al (2007) Estrogen plus progestin use, microsatellite instability, and the risk of colorectal cancer in women. Cancer Res 67:7534–7539
65. Murphy N et al (2015) A prospective evaluation of endogenous sex hormone levels and colorectal cancer risk in postmenopausal women. J Natl Cancer Inst 107
66. Fernandez E et al (2001) Oral contraceptives and colorectal cancer risk: a meta-analysis. Br J Cancer 84:722–727
67. Giardiello FM et al (2005) Oral contraceptives and polyp regression in familial adenomatous polyposis. Gastroenterology 128:1077–1080
68. Cotterchio M et al (2006) Dietary phytoestrogen intake is associated with reduced colorectal cancer risk. J Nutr 136:3046–3053
69. Botteri E et al (2017) Menopausal hormone therapy and colorectal cancer: a linkage between nationwide registries in Norway. BMJ Open 7:e017639
70. Lobo RA (2017) Hormone-replacement therapy: current thinking. Nat Rev Endocrinol 13:220–231
71. Liu Q et al (2021) Menopausal hormone therapies and risk of colorectal cancer: a Swedish matched-cohort study. Aliment Pharmacol Ther 53:1216–1225
72. Frezza EE, Wachtel MS, Chiriva-Internati M (2006) Influence of obesity on the risk of developing colon cancer. Gut 55:285–291
73. Ng M et al (2014) Global, regional, and national prevalence of overweight and obesity in children and adults during 1980-2013: a systematic analysis for the global burden of disease study 2013. Lancet 384:766–781
74. Kaaks R et al (2000) Serum C-peptide, insulin-like growth factor (IGF)-I, IGF-binding proteins, and colorectal cancer risk in women. J Natl Cancer Inst 92:1592–1600
75. De Paoli M, Zakharia A, Werstuck GH (2021) The role of estrogen in insulin resistance: a review of clinical and preclinical data. Am J Pathol 191:1490–1498
76. Salpeter SR et al (2006) Meta-analysis: effect of hormone-replacement therapy on components of the metabolic syndrome in postmenopausal women. Diabetes Obes Metab 8:538–554
77. Stachowiak G, Pertynski T, Pertynska-Marczewska M (2015) Metabolic disorders in menopause. Prz Menopauzalny 14:59–64
78. Baker JM, Al-Nakkash L, Herbst-Kralovetz MM (2017) Estrogen-gut microbiome axis: physiological and clinical implications. Maturitas 103:45–53
79. Hong H, Landauer MR, Foriska MA, Ledney GD (2006) Antibacterial activity of the soy isoflavone genistein. J Basic Microbiol 46:329–335
80. Clavel T et al (2005) Isoflavones and functional foods alter the dominant intestinal microbiota in postmenopausal women. J Nutr 135:2786–2792
81. Nakatsu CH et al (2014) Fecal bacterial community changes associated with isoflavone metabolites in postmenopausal women after soy bar consumption. PLoS One 9:e108924
82. Fang K et al (2016) Soy isoflavones and glucose metabolism in menopausal women: a systematic review and meta-analysis of randomized controlled trials. Mol Nutr Food Res 60:1602–1614
83. Zhao H et al (2019) Compositional and functional features of the female premenopausal and postmenopausal gut microbiota. FEBS Lett 593:2655–2664
84. Org E et al (2016) Sex differences and hormonal effects on gut microbiota composition in mice. Gut Microbes 7:313–322
85. Kaliannan K et al (2018) Estrogen-mediated gut microbiome alterations influence sexual dimorphism in metabolic syndrome in mice. Microbiome 6:205
86. Song CH et al (2020) 17β-estradiol supplementation changes gut microbiota diversity in intact and colorectal cancer-induced ICR male mice. Sci Rep 10:12283
87. Son HJ et al (2019) Effect of estradiol in an Azoxymethane/dextran sulfate sodium-treated mouse model of colorectal cancer: implication for sex difference in colorectal cancer development. Cancer Res Treat 51:632–648
88. Lee SM et al (2016) The effect of sex on the Azoxymethane/dextran sulfate sodium-treated mice model of colon cancer. J Cancer Prev 21:271–278

89. Hases L et al (2020) Intestinal estrogen receptor beta suppresses colon inflammation and tumorigenesis in both sexes. Cancer Lett 492:54–62
90. Song CH et al (2019) Effects of 17β-estradiol on colorectal cancer development after azoxymethane/dextran sulfate sodium treatment of ovariectomized mice. Biochem Pharmacol 164:139–151
91. Andersson S et al (2017) Insufficient antibody validation challenges oestrogen receptor beta research. Nat Commun 8:15840
92. Heine PA, Taylor JA, Iwamoto GA, Lubahn DB, Cooke PS (2000) Increased adipose tissue in male and female estrogen receptor-alpha knockout mice. Proc Natl Acad Sci U S A 97:12729–12734
93. Manrique C et al (2012) Loss of estrogen receptor α signaling leads to insulin resistance and obesity in young and adult female mice. Cardiorenal Med 2:200–210
94. Foryst-Ludwig A et al (2008) Metabolic actions of estrogen receptor beta (ERbeta) are mediated by a negative cross-talk with PPARgamma. PLoS Genet 4:e1000108
95. Hases L et al (2020) High-fat diet and estrogen impacts the colon and its transcriptome in a sex-dependent manner. Sci Rep 10:16160
96. Honma N et al (2013) Estrogen receptor-beta gene polymorphism and colorectal cancer risk: effect modified by body mass index and isoflavone intake. Int J Cancer 132:951–958
97. Passarelli MN et al (2013) Common single-nucleotide polymorphisms in the estrogen receptor beta promoter are associated with colorectal cancer survival in postmenopausal women. Cancer Res 73:767–775
98. Edvardsson K et al (2013) Estrogen receptor β expression induces changes in the microRNA pool in human colon cancer cells. Carcinogenesis 34:1431–1441
99. Nguyen Vu T et al (2016) Estrogen receptor beta reduces colon cancer metastasis through a novel miR-205 - PROX1 mechanism. Oncotarget 7:42159–42171
100. Hartman J et al (2009) Tumor repressive functions of estrogen receptor beta in SW480 colon cancer cells. Cancer Res 69:6100–6106
101. Wada-Hiraike O et al (2006) Role of estrogen receptor beta in colonic epithelium. Proc Natl Acad Sci U S A 103:2959–2964
102. Saleiro D et al (2012) Estrogen receptor-β protects against colitis-associated neoplasia in mice. Int J Cancer 131:2553–2561
103. Giroux V, Bernatchez G, Carrier JC (2011) Chemopreventive effect of ERβ-selective agonist on intestinal tumorigenesis in Apc(Min/+) mice. Mol Carcinog 50:359–369
104. Kyoko OO et al (2014) Expressions of tight junction proteins Occludin and Claudin-1 are under the circadian control in the mouse large intestine: implications in intestinal permeability and susceptibility to colitis. PLoS One 9:e98016
105. Deaver JA, Eum SY, Toborek M (2018) Circadian disruption changes gut microbiome taxa and functional gene composition. Front Microbiol 9:737
106. Pagel R et al (2017) Circadian rhythm disruption impairs tissue homeostasis and exacerbates chronic inflammation in the intestine. FASEB J 31:4707–4719
107. Voigt RM et al (2014) Circadian disorganization alters intestinal microbiota. PLoS One 9:e97500
108. Ibrahim A et al (2019) Colitis-induced colorectal cancer and intestinal epithelial estrogen receptor beta impact gut microbiota diversity. Int J Cancer 144:3086–3098
109. Kaiko GE et al (2016) The colonic crypt protects stem cells from microbiota-derived metabolites. Cell 165:1708–1720
110. Kushkevych I et al (2020) Recent advances in metabolic pathways of sulfate reduction in intestinal bacteria. Cells 9
111. Ijssennagger N et al (2015) Gut microbiota facilitates dietary heme-induced epithelial hyperproliferation by opening the mucus barrier in colon. Proc Natl Acad Sci U S A 112:10038–10043
112. Zagato E et al (2020) Endogenous murine microbiota member Faecalibaculum rodentium and its human homologue protect from intestinal tumour growth. Nat Microbiol 5:511–524
113. Aso T et al (2012) A natural S-equol supplement alleviates hot flushes and other menopausal symptoms in equol nonproducing postmenopausal Japanese women. J Womens Health (Larchmt) 21:92–100
114. Tagliaferri MA, Tagliaferri MC, Creasman JM, Koltun WD (2016) A selective estrogen receptor Beta agonist for the treatment of hot flushes: phase 2 clinical trial. J Altern Complement Med 22:722–728
115. Williams C, DiLeo A, Niv Y, Gustafsson J (2016) Estrogen receptor beta as target for colorectal cancer prevention. Cancer Lett 372:48–56
116. Mukherji A, Kobiita A, Ye T, Chambon P (2013) Homeostasis in intestinal epithelium is orchestrated by the circadian clock and microbiota cues transduced by TLRs. Cell 153:812–827
117. Weintraub Y et al (2020) Clock gene disruption is an initial manifestation of inflammatory bowel diseases. Clin Gastroenterol Hepatol 18:115-122.e111
118. Wang S et al (2018) REV-ERBα integrates colon clock with experimental colitis through regulation of NF-κB/NLRP3 axis. Nat Commun 9:4246
119. Cai W et al (2008) Expression levels of estrogen receptor beta are modulated by components of the molecular clock. Mol Cell Biol 28:784–793

Genomic Insights into Non-steroidal Nuclear Receptors in Prostate and Breast Cancer

13

Sajad A. Wani and Moray J. Campbell

Abstract

Alterations in transcriptional programs are a fundamental feature of prostate (PCa) and breast cancer (BrCa), and frequently target the actions of the principal steroidal nuclear receptors (NRs), namely the androgen receptor (AR) and the estrogen receptor alpha (ERα), respectively. Indeed, the functions of AR and ERα are central to both prostate and mammary gland biology. The genomic interactions of these NRs become highly distorted in part by changing how they functionally interact with a cohort of non-steroidal Type II NRs, which are by contrast relatively understudied compared to their steroidal cousins. For example, the AR cistrome overlaps with cistromes of different Type II NRs, which suggests a high potential for integrated NR functions to tailor transcriptional signals. Over recent years the cistromes of these Type II NRs, including HNF4s, RARs, PPARs and VDR, have been studied in PCa and BrCa revealing convergence and functional consequences, and are reviewed in the current chapter.

S. A. Wani · M. J. Campbell (✉)
Division of Pharmaceutics and Pharmacology, College of Pharmacy, The Ohio State University, Columbus, OH, USA
e-mail: campbell.1933@osu.edu

Keywords

Breast cancer · Prostate cancer · Non-steroidal nuclear receptors · Cistrome · Transcriptome · Bookmarking · Epigenetics

13.1 Nuclear Receptor Genomic Interactions Are Highly Integrated and Sense a Wide Variety of Inputs

The collective transcriptional actions of nuclear receptors (NRs) form a central conduit for hormonal, dietary and environmental compounds to signal to the genome. Specifically, NRs act as sensors that respond to both the presence and absence of a diverse array of ligands and in turn initiate and fine-tune transcriptional events. The impact of NR gene regulatory complexes is evident in development, metabolism, circadian rhythm and cell fate decisions including differentiation phenotypes. Reflecting this widespread importance, there is clear evidence for their disruption acting as disease drivers for various syndromes including cancer [1–5].

The classical sex steroids bind cognate receptors with high affinity; estradiol binds estrogen receptor, NR3A1/ERα, and dihydrotestosterone binds the androgen receptor, NR3C4/AR. Beyond these ligands seco-steroids, retinoid derivatives

M. J. Campbell, C. L. Bevan (eds.), *Nuclear Receptors in Human Health and Disease*, Advances in Experimental Medicine and Biology 1390, https://doi.org/10.1007/978-3-031-11836-4_13

and bioactive dietary-derived factors such as fatty acids, oxysterols, heme, and bile acids act as ligands and regulate the genomic interactions of a broader group of the NRs. More broadly, these integrated and environmentally-driven NR-genomic interactions are central to concepts such as nutrigenomics and provide the rationale for positioning a wider panel of NRs as promising therapeutic targets in cancer [6–8]. Finally, other NRs, without known ligands, have also been identified, known as orphan receptors [9]. Collectively, the interaction of all these NRs allows for the highly dexterous transcriptional outputs, underpinned by the dynamic and mobile NR-genomic interactions, known as NR cistromes. In turn, the NR cistrome gene-regulatory functions are regulated by NR-associated coregulators including coactivators, corepressors and other transcription factors (TFs) and thereby provide a further level of control to regulate transcription [10–13].

NRs are classified based on mode of action as Type I, II, III, or IV [14]. Steroid NRs are Type I and in the absence of ligand these receptors are often largely cytoplasmic associated with heat shock proteins. Ligand binding results in their dissociation from heat shock proteins and NR homo-dimerization and translocation to the nucleus. Type II NRs, in contrast, reside in the nucleus as heterodimers (for example with RXRs) and bound to genome even in the absence of ligands [15]. Types III and Type IV are orphan receptors, for which ligands are unknown, or possibly don't exist, and are also generally located in the nucleus and bind DNA as homodimers (Type III) or monomers (Type IV).

The impact of NRs is highly evident across many high-profile and impactful hormone-dependent cancers, including not only prostate cancer (PCa) and breast cancer (BrCa), but also other cancers including ovarian, endometrium, testis, thyroid, and pancreas. An appreciation of the relationship between steroids and cancers of the reproductive system was pioneered by the work of Sir George Beatson in the nineteenth century, who began to define the relationship between estrogen and BrCa risk [16]. Subsequently, in the 1940s this concept was echoed by the work of Dr. Charles Huggins and colleagues who established the endocrine synthesis of androgens and the relationship to PCa [17]. As a result, the genomic functions of AR and ERα in PCa and BrCa, respectively, are highly studied and these are well understood TFs. Additionally, there is a parallel and, in many cases, emerging appreciation of how these cancers are impacted by non-steroidal NRs, and the potential for the genomic cross-talk between steroidal and non-steroidal NRs. For example, there are physiological and gene regulatory studies that strongly support the concept that Type I and Type II NRs function in a range of cooperative and antagonist cross-talk signaling mechanisms, for example centered around AR [18–23], and ERα [24–29].

The focus of the current chapter is to summarize genomic insights into the Type II NRs in hormone-dependent cancer including the vitamin D receptor (NR1I1VDR), retinoic acid receptors (NR1B1/RARα, NR1B2/RARβ, and NR1B3/RARγ), and peroxisome proliferator-activated receptors (NR1C1/PPARα, NR1C2/PPARδ, and NR1C3/PPARγ) [9], summarized in Table 13.1. Clearly, orphan receptors, given they have no identified ligands, also fall under the classification of non-steroidal receptors. In parallel, the understanding of adopted nuclear orphans and orphan NRs is evolving, and reveal further insights into NR functions in terms of genomic distribution and cross-talk with signaling pathways including those that are key targets for pharmacological pathways [30].

13.2 Genomic Interactions of Non-steroidal Nuclear Receptors in PCa and BrCa

13.2.1 The Vitamin D Receptor

Supporting an anti-tumorigenic role for the VDR men whose prostate tumors have higher VDR expression have significantly lower prostate-specific antigen, lower Gleason score and less advanced tumor stage [31]. The circulating pre-hormone vitamin D_3 is the precursor to the active hormonecalcitriol(1alpha,25dihydroxyvitaminD3

Table 13.1 Nonsteroidal nuclear receptors

Receptor	Symbol	Ligands
TRα	NR1A1	Thyroxine (T4), triiodothyronine (T3)
TRβ	NR1A2	Thyroxine (T4), triiodothyronine (T3)
RARα	NR1B1	All-*trans* and 9-*cis* retinoic acid
RARβ	NR1B2	All-*trans* and 9-*cis* retinoic acid
RARγ	NR1B3	All-*trans* and 9-*cis* retinoic acid
PPARα	NR1C1	Fatty acids
PPARβ/δ	NR1C2	Fatty acids
PPARγ	NR1C3	Fatty acids
Rev-Erbα	NR1D1	Heme
Rev-Erbβ	NR1D2	Heme
RORα	NR1F1	Oxysterols
RORβ	NR1F2	Oxysterols
RORγ	NR1F3	Oxysterols
LXRβ	NR1H2	Oxysterols
LXRα	NR1H3	Oxysterols
FXR	NR1H4	Bile acids
VDR	NR1I1	Calcitriol (1′,25′-dihydroxy vitamin D_3)
PXR	NR1I2	Bile acids
CAR	NR1I3	Androstanol, androstenol
HNF-4α	NR2A1	Fatty acids
HNF-4γ	NR2A2	Fatty acids
RXRα	NR2B1	9-*cis*-retinoic acid
RXRβ	NR2B2	9-*cis*-retinoic acid
RXRγ	NR2B3	9-*cis*-retinoic acid
TR2	NR2C1	All-*trans* retinoic acid
TR4	NR2C2	All-*trans* retinoic acid
TLX	NR2E1	Not known
PNR	NR2E3	Benzimidazoles
COUP-TF I	NR2F1	Not known
COUP-TF II	NR2F2	Retinol/ATRA
EAR2	NR2F6	Not known

($1\alpha,25(OH)_2D_3$)) that binds to the VDR. Epidemiological approaches have identified relationships between low circulating vitamin D_3 and cancer incidence, and that $1\alpha,25(OH)_2D_3$ suppresses early prostate carcinogenesis by regulating genes involved in proliferation, differentiation and apoptosis [32]. Underscoring the potential importance of this signaling axis, genomic studies in murine VDR knockout cells as well as human studies have suggested that $1\alpha,25(OH)_2D_3$ can regulate as much as 3% of the mouse or human genome directly and/or indirectly [33].

Several studies have assessed the VDR cistrome in PCa [34, 35] by VDR chromatin immunoprecipitation sequencing (ChIP-Seq). Work by Fleet et al [34] identified binding at ~3400 protein-coding genes, ~680 long non-coding RNAs, and ~ 470 miRNAs. This included VDR-bound peaks at known VDR target genes including *CYP24A1* and *IGFBP3*. Peak distribution was evenly divided between intergenic and intronic regions, supporting both long-range and proximal regulation. These studies also suggested that $1\alpha,25(OH)_2D_3$ amplifies signals mediated through other TFs including NF-Kappa-B Inhibitor Alpha (NFKBIA) and FOXO1, and some peaks near immune response related genes (e.g., L1R2) hint towards VDR regulation of immune processes.

A further VDR-ChIP Seq study in non-malignant prostate cells (PrEc) [35] identified ~5000 VDR binding sites, again including well-known targets (e.g., *CYP24A1*) and, interestingly, ligand activation led to a significant decrease in the number of VDR-ChIP peaks, reflecting perhaps an active role for the basal VDR in gene expression. Sites with loss of peaks include aminoacyl tRNA synthetase genes, which in turn leads to decreased proliferation. VDR also binds near genes regulating neural differentiation, which raises a possibility that itmay also be linked to neuroendocrine trans differentiation in PCa.

Finally, a recent study from our lab [doi.org/10.1101/2022.01.31.478573] addressed VDR function in the context of PCa health disparities by examining a panel of European American (EA) (HPr1-AR and LNCaP) and African American (AA) cell lines (RC43N, RC43T, RC77N and RC77T). These analyses lent strong evidence to the concept that the VDR is a significantly more potent transcriptional regulator in AA than EA prostate cells, and that in PCa this signaling is distorted and suppressed. In non-malignant RC43N cells, VDR ChIP-Seq identified significant basal and 1α,25$(OH)_2D_3$ dependent VDR binding sites, with ~1300 in total associated with transcriptional responses enriched for circadian rhythm and inflammation networks. In parallel, 1α,25$(OH)_2D_3$-dependent ATAC-Seq also revealed the greatest impact on chromatin accessibility in RC43N cells, with significant gain of nucleosome-free regions at enhancers. By contrast, in malignant EA and AA cell models 1α,25$(OH)_2D_3$ led to a loss of VDR binding. Motif prediction identified a diverse set of enriched motifs within peaks, including the VDR motif and other NRs including the AR and RARs. The suppressed transcriptional responses in AA PCa cells associated with reduced expression of Bromodomain adjacent to zinc finger domain protein 1A (BAZ1A), a component of the human SWI/SNF complex, and restored expression of this protein led to significantly enhanced 1α,25$(OH)_2D_3$-regulated transcriptome.

There are also equally compelling epidemiological associations between vitamin D_3 and breast cancer incidence. For example, the Vdr −/− mouse [36, 37] displays a range of mammary gland phenotypes in terms of disrupted development of the gland, and then changing sensitives to the control of programmed cell death within epithelial cells. In parallel there are a wide range of pre-clinical studies which all support a potentially anti-tumorigenic role in BrCa [38].

Two studies have examined VDR genomic interactions which revealed that in MCF-7 BrCa cells, VDR has ~2300 VDR-binding sites in the absence of 1,25(OH)2D3, and ~7,400 sites following ligand stimulation (4 h). Out of these, ~700 sites remained unchanged in both presence and absence of ligand. A significant numbers of VDR-binding sites were detected in intergenic regions, and distal from promoters, and VDR-bound enhancers were enriched in apoptotic and metabolic pathways. In a series of comprehensive studies led by Kevin White and coworkers [39, 40] multi-cistrome analyses were undertaken for a range of more than 20 NRs including non-steroidal ones in BrCa cancer cell lines [39–41]. Within these studies VDR binding was analyzed in MCF-7 cells and also reported ~7000 binding regions, which were more distal to TSS regions than many of the other NRs, and in terms of network topology demonstrated lower interconnectedness compared to NRs such as the retinoic acid receptors. These workers were able to undertake integrative regions.

Together these data strongly support the VDR playing an important role in the biology of the prostate and mammary glands, and suggest disruption of VDR signalling is carcinogenic by disrupting a wide number of gene regulatory mechanisms including overlap with other NRs.

13.2.2 Retinoic Acid Receptors

The NR1B1/RARα represents one of the earliest examples of targeted cancer therapy, involving all-trans retinoic acid in acute promyelocytic leukemia [42, 43]. This was a major catalyst for the development of the field of differentiation therapy, whereby compounds such as retinoic acid would be in cancers to limit their proliferation and induce either differentiation or programmed cell death [44, 45]. In part, these actions were the motivations for cistromic studies on the VDR, RARs and multiple NRs in PCa, and BrCa [44, 45].

In the prostate, retinoic acid regulates normal differentiation and the Rarγ knockout mouse exhibits prostate metaplasia [46, 47], both suggesting the receptor plays a role in control of cell growth. Reflecting this, NR1B3/RARγ is commonly down-regulated in PCa[3], for example because of up-regulated miR-96-5p, and this leads to significant changes to AR signaling [48]. In a non-malignant prostate cell line, RWPE-1, under basal conditions the RARγ cistrome is ~1250 peaks and interestingly the addition of a RARγ-selective ligand (CD437) restricts the number of peaks to ~350, which are mostly shared with the basal state (only ~50 appear unique). These data also revealed that RARγ significantly enhanced AR function, and regulation of AR target genes, and that the RARγ cistrome significantly overlapped with AR binding at active enhancers. In turn, reduced expression genes that were annotated RARγ binding was associated with aggressive PCa [48].

In MCF-7 BrCa cells, RARα/γ and ERα form a genomic antagonism [40] in a so-called "Yin and Yang" manner to regulate proliferation and survival. These NRs balance expression of shared gene targets in part because RARs overlaps significantly with ERα binding in a genome wide fashion. These co-occupied regions are in the vicinity of genes for which estrogen and retinoic acid regulate antagonistically. The number of peaks in the presence of selective RARα (AM580) and RARγ (CD437) ligands was ~7300 for RARα and ~ 3200 for RARγ sites, and using a generous distance cut-off of 1 kb between the center of the peaks there was a significant overlap of sites; it is unclear how many of the peaks actually overlap as opposed to being closely adjacent. This therefore suggests convergence at the level of gene-regulatory actions rather than perhaps direct chromatin-accessibility [40]. Together, these data suggest significant genomic interactions between RARs and both AR and ERα in PCa and BrCa.

Interestingly, the related paralog, RARβ, appears to be a *bona fide* tumor suppressor in BrCa and PCa. For example, methylation patterns of the CpG islands associated with the RARβ promoter are exploited in algorithms to predict tumor grade and progression risks in these tumors [49–52]. Against this backdrop it is perhaps surprising that there are no cistrome data for this receptor in these cancers, although it has been undertaken in brain tissues [53].

13.2.3 RAR Related Orphan Receptor C

NR1F3/RORC encodes RORγ and is amplified and upregulated in metastatic recurrent PCa tumors following androgen deprivation therapy. It acts as an upstream regulator of AR and appears to drive AR expression, as well as to facilitate recruitment of coactivators such as Nuclear Receptor Coactivator 1 and 3 (NCOA1/3, SRC1/3). Furthermore, pharmacological targeting with an antagonist to RORγ reduces expression of AR as well as the oncogenic AR splice variant 7 and reduces AR genomic binding, and as a result reduced expression of various AR target genes. This regulation appears to be a targeted AR event, as inhibiting RORγ does not alter genome-wide histone modifications associated with chromatin accessibility [54].

Studies on RORγ in BrCa have suggested that its function is an essential activator of the cholesterol-biosynthesis program, as it binds to cholesterol-biosynthesis genes, and it facilitates the genomic recruitment of Sterol regulatory element-binding protein 2 (SREBP2) in Triple-negative BrCa [55]. From a genome-wide perspective there appear to be a massive number of RORγ binding sites in the HCC70 BrCa cell line, in excess of 30,000, and these are highly shared with SREBP2 binding sites. Again, similarly to PCa, a RORγ antagonist very potently inhibits BCa tumor growth in vitro and in xenografts [55]. Similarly, the related RORα is also a potential tumor suppressor and a therapeutic target for BrCa [56, 57] but as yet cistromic studies have not been undertaken and so the extent of genomic cooperation between these two receptors remains unknown. RORγ therefore plays a paramount role in regulating cholesterol-biosynthesis through its own genomic binding leads to the recruitment of SREBP2 at the gene targets to stimulate the cholesterol-biosynthesis.

13.2.4 Peroxisome Proliferator-Activated Receptors

PPARs regulate energy production, lipid metabolism, and inflammation [58]. In triple negative BrCa MDA-MB-231 cells, ChIP-Seq and transcriptomic analyses identified ~500 PPARδ peaks and, amongst these, the hormone *ANGPTL4* was a significant PPARδ target [59]. In another study, using a transformed variant of the non-malignant breast epithelial cell, MCF10A-NeuT cells, PPARγ binds to a large number of sites and regulates genes and notably EphA-Amphiregulin as well as genes involved in chemokine signaling [60]. Similarly, PPARα, PPARδ and PPARγ bind to ~2230, ~3250 and ~ 6300 genomic regions respectively in MCF7 cells with PPARγ binding as sites at a greater distal distance to TSS [39] than the other PPARs. Interestingly, the PPARδ cistrome shared a significant proportion (~70%) of its binding sites with RARα and RARγ, and in part this led to the concept of high occupancy target (HOT) regions in the genome. Specifically, these are regions that are significantly shared by multiple NRs and other TFs, and appear to be found disproportionately associated with genes associated with cancer development and progression. The functional significance of these sites is illustrated by shared PPARδ and RARs binding sites at target genes, which in turn are associated with poor prognosis in BrCa. More widely these genomic findings also support a concept of selectively targeting RARs and PPARδ to inhibit synergistically BrCa growth.

Set against these interesting data, to date there are no cistromic studies of PPARs in PCa. This is all the more striking given that there is a considerable literature on PPARs [61–65] and the PPAR coregulator PPARGC1α [66–68] playing significant roles in PCa carcinogenesis. Such studies would also be able to address the concept of HOT regions in PCa, and how these cistromic patterns impact AR signaling.

13.2.5 Hepatocyte Nuclear Factor 4 α and γ

In PCa, NR2A1/HNF4γ appears to function as a pioneer factor that generates and maintains enhancer landscape at lineage genes, for example those associated with neuronal lineages, and which impacts AR signaling in a more nuanced manner. For example, restoring HNF4γ expression reduces AR sensitivity towards androgen deprivation therapy [69], and increased HNF4γ expression does not alter the AR cistrome or AR signaling directly, but increased FOXA1 binding at a subset of HNF4γ sites. Approximately 35% of HNF4γ peaks share binding FOXA1, and a smaller proportion of HNF4γ peaks directly overlap with AR peaks. Therefore, HNF4γ binding sites appear to cooperate with FOXA1 to establish and maintain enhancers that facilitate lineage-specific transcriptomes in the prostate; this is potentially corrupted in PCa progression [69]. Similarly, NR2A2/HNF4α appears to exert a tumor suppressor function and has reduced expression in PCa tissues, cell lines, and xenografts of androgen deprivation therapy recurrent PCa [70] through epigenetic mechanisms. For example, HNF4α binds constitutively to binding sites in the promoter of *CDKN1A*, which guides AR to bind upon dihydrotestosterone stimulation. Indeed, the motifs of HNF4α are over-represented within unique AR-binding loci, and the cistrome shows significant overlap with AR-binding sites [71]. Again, given these potent cooperative actions between HNF4 receptors with a principal steroid hormone receptor, it is perhaps surprising that similar studies haven't yet been undertaken in BrCa.

13.2.6 COUP Transcription Factor I and II

NR2F1/COUP-TF I is one of the earliest cloned NRs, first being identified in the late 1980s [72], and subsequently led to the discovery of NR2F2/COUP-TF II [73]. Several studies [39, 74, 75] have analyzed the COUP-TF II cistrome in BrCa. High expression of COUP-TF II is related with better survival in ERα-positive BrCa patients but not in ERα-negative patients, and COUP-TF II cooperates with pioneer factors such as FOXA1 and GATA3 to promote ERα function [74, 75]. These findings suggest a cooperativity between ERα and COUP-TF-11, and although estradiol is not required for COUP-TF II binding, inhibition of COUP-TF II decreases ERα binding, chromatin accessibility (ATAC-Seq peaks were reduced by 70% after COUP-TF II depletion), and estradiol-dependent cell growth suggesting a protein-protein interaction. Together, these data suggest a complex interdependency between estradiol, ERα and COUP-TF II. In MCF-7 cells, approximately, 40% of ERα binding sites overlap with FOXA1, 60% with COUP-TF II and 70% with GATA3, and there is evidence for shared binding at super-enhancers on a wide-spread scale which directly leads to high *de novo* transcription. Indeed, this integration also impacts other NRs downstream, including RARβ [76]. These roles for COUP-TF II in regulating ERα-mediated transcription make it an interesting potential therapeutic target in BCa. In parallel studies COUP-TF I-specific agonists suppress metastasis supporting a wider role for COUP-TFs to interact with ERα and to regulate anticancer actions [77].

13.2.7 NUR77

NR4A1/NUR77 is an orphan NR that acts in a ligand-independent manner. In a recent study [78], NUR77 was reported to regulate immediate early genes, suppressing replication stress in BCa and acting as a master regulator through a transcriptional processing checkpoint. Genome-wide analyses revealed that NUR77 binds the gene body and 3' UTR of immediate early genes, inhibits transcriptional elongation, generating R-loops and accessible chromatin domains. Under stress, dissociation of NUR77 leads to a burst of expression of these transcriptionally poised genes thereby suggesting a role for NUR77 in governing transcriptional responses to chronic replication stress. Although there are no genome-wide cistrome studies of NUR77 in PCa, there is strong evidence for it regulating programmed cell death in this cancer [79, 80].

13.3 Mechanisms of NR Cooperation: Bookmarking Functions by Non-steroidal NRs

Mitotic bookmarking functions to retain epigenetic states throughout the cell cycle at gene loci that are poised for immediate reactivation post-mitotically (Fig. 13.1). This involves the retention of histone variants, regulatory proteins and modifications, and some selected TFs. Bookmarking mechanisms prevent the spreading of heterochromatin into genomic regions which are pre-marked for TF future actions. In this manner, these epigenetic mechanisms regulate genes that coordinately control cell growth and lineage maintenance following mitosis. Furthermore, it is clear these mechanisms are corrupted in carcinogenesis and tumor maintenance leading to deregulated proliferation and compromised control of differentiation [81–86].

Several Type II NRs have been reported to have bookmarking properties independent of ligand exposure, again reflecting their predominant nuclear location. NR1I2/PXR remains constitutively associated with mitotic chromatin specifically at the *CYP3A4* promoter during mitosis [88]. A region of PXR contains a 'mitotic chromatin binding-determining region' which exerts these functions. The bookmarking property of PXR is impeded by direct interaction with the orphan NR small heterodimer partner (SHP) perhaps underscoring the importance of this function [89]. Other examples of NRs appearing to play a bookmarking function include NR3B2/ESRBB, which is a major pluripotency TF that remains bound to key regulatory regions during mitosis [90]; it is bound widely with at least 10,000 binding sites and maintains nucleosome positioning during mitosis to ensure the rapid post-mitotic re-establishment of functional regulatory complexes at selected enhancers and promoters [91].

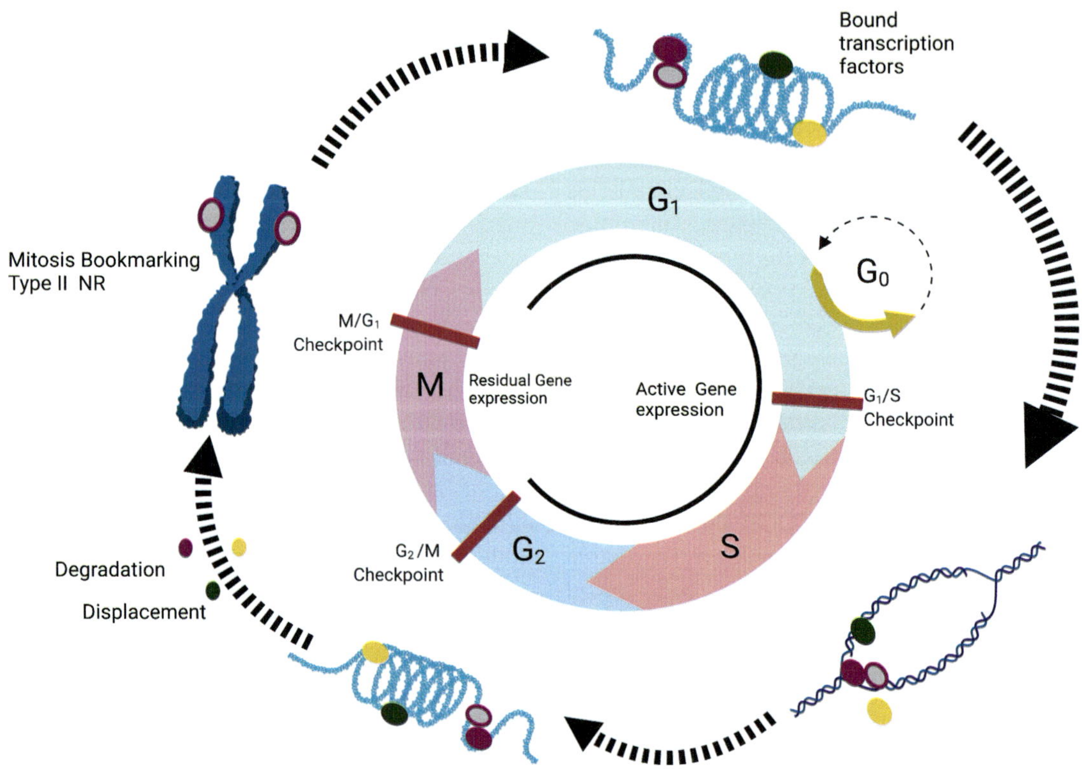

Fig. 13.1 A model for bookmarking function by Type II nuclear receptors for Type I nuclear receptors. As cells go through the cell cycle and division chromatin assumes different conformations, becoming most compacted during metaphase of mitosis. Prior to this, many proteins to chromatin associations are lost as a result of degradation and displacement. However, a number of transcription factors are retained such that transcription, or the marking of sites for transcription, can be activated rapidly in G_1. This function is termed bookmarking and there is evidence that Type II nuclear receptors that are nuclear resident in both the presence and absence of ligand (mitosis, purple open symbols) can serve this function for other nuclear receptors (G_1, purple solid symbol)

Similarly, HNF4α bookmarks specific genomic regions and keeps them competent for future activation during liver development [92].

The raises an interesting question of whether this bookmarking property is a generalized feature of NRs, and specifically those NRs that are nuclear resident independent of ligand exposure maybe retaining enhancer access through mitosis for other NRs. This concept is supported by examples above of Type II and Type I NR co-regulation of gene expression programs. Given that non-steroidal NRs in PCa and BrCa are frequently disrupted for example with decreased expression (e.g., RARγ), this may suggest that Type II NRs bookmark and regulate the actions of AR and ERα. However, there is also evidence of ligand activated (and therefore nuclear resident) AR and ERα being associated with mitotic chromatin although it is unclear if these complexes are the cause or consequence of other NRs/TFs serving as bookmarking factors [87].

More generally, there are clear examples of AR and ERα being genomically relocated to other sites during cancer initiation and progression, and in response to NR-targeted therapies. For example, the AR is reprogrammed specifically to genomic sites that are normally regulated in development only in the transition to metastatic PCa by reactivating latent regulatory elements active in fetal prostate organogenesis [93]. It is a tantalizing prospect that the interactions between Type II and Type I NRs is in part underpinned by Type II NR bookmarking enhancers and regulatory regions that are regulated by Type I NR binding to promote cell fate decisions such as differentiation. Furthermore, disruption of these Type II NR complexes potentially disrupts these functions.

13.4 Genomic Approaches to Defining Type I and II NR Cistromes and Interactions

Methods to map histone and TF genomic interactions emerged in the 1990s with the development of ChIP approaches [94–96], and became genome-wide with the advent of microarray technologies giving rise to so-called ChIP–chip [97] approaches, and then subsequently ChIP-Seq [98]. This key technology has been profoundly improved and diversified to tackle limitations such as protein abundance, cross-linking efficiency and antibody availability and specificity. For example, Cleavage Under Targets & Release Using Nuclease (CUT&RUN) and CUT&Tag has made it easier to study TF binding and histone modifications at genome scale [99, 100]. Similarly, the development of ATAC-Seq (Transposase Accessible Chromatin followed by high-throughput sequencing) [101] has enabled the measurement of chromatin accessibility and has also been refined to address single cells and to improve accuracy. More widely, genomic approaches are advancing rapidly to encompass single cell resolution, which allows ever more complex biological questions to be addressed [102]. In parallel, CRISPR technologies are enabling the tagging of proteins, and DNA and epigenome editing, to more establish conditional cell contexts with which to test NR functions more accurately [103, 104].

Matching these wet-lab advancements has been an equally explosive growth in the dry-lab to develop and refine the analyses of cistromic data and combine it with parallel transcriptomic data. This challenge of integrating cistrome to transcriptome data is surprisingly complex. For example, defining NR:enhancer:gene interactions that are driven by NRs is challenging because of the large number of NR and coregulator interactions, which are altered by diverse and interdependent genetic and epigenetic mechanisms, and are further controlled by the 3-D genome [105–107]. Thus, NR:enhancer:gene relationships are dynamic and non-linear, with each gene regulated by multiple enhancers in a time- and signal-dependent manner [108, 109], and occur over large genomic distances [110].

Defining the statistical significance of NR:gene relationships, or even NR:NR:gene relationships, is a question of whether a NR signal-to-gene-expression relationship is occurring more than predicted by chance, which in turn requires defining the background of NR:gene relationships. Random sampling methods such as bootstrapping can be used to simulate the distribution of NR:gene relationships changes across the genome for statistical comparison [111, 112], and parsimonious annotation of the genome, for example with the ChromHMM algorithm [113] to define epigenetic states, or the ROSE algorithm to define super-enhancers [114] [115] can refine these statistical challenges. Furthermore, testing the overlap of target NR ChIP-Seq data with comprehensive data sets, such as contained in Cistrome DB [116], allows co-enrichment testing of hundreds of TF and histone modification ChIP-Seq datasets to reveal the extent of enrichment with other NRs and their coregulators. RNA-Seq undertaken in parallel treatments can be matched with these highly annotated cistromic data to define cistrome-transcriptome relationships and test their phenotypic associations for example using Kolmogorov–Smirnov tests to examine differences in cumulative distribution plots for cistrome binding sites with respect to nearest gene, and again using bootstrapping approaches to measure how the specific cistrome-relationships associate with gene expression patterns [117].

Thus, there are many routes through testing NR:gene relationships and this most likely underpins the frequently divergent findings in the literature. On top of this there are multiple methods for cistrome [118, 119] or transciptome [120, 121] analyses and as yet there are few commonly accepted protocol standards, in contrast, for example to the MIAME-compliant protocols for microarray analyses [122]. Therefore, it is unsurprising that for a given NR there is little consensus on the number of significant binding sites, what motifs are most enriched, what the genomic distribution is and how it relates to transcription.

13.5 Conclusion

Non-steroidal NRs have been somewhat neglected from a genomic perspective, although it is clear their actions and interactions with steroidal NRs are biologically impactful. In this chapter we attempted to provide a broad overview of the advances in understanding non-steroidal nuclear receptor cistromes and their interaction with other AR and ERα in PCa and BrCa and highlighted the expanding impact of the genome wide studies in NR biology. These NRs are potential therapeutic targets in cancer and may be exploited to augment traditional therapeutic approaches. Cistromic studies are rapidly advancing and revealing unprecedented insights into the interactions between Type I and Type II NRs, even with some methodological ambiguities.Funding*MJC* acknowledge support in part from the Prostate program of the Department of Defense Congressionally Directed Medical Research Programs [W81XWH-20-1-0373; W81XWH-21-1-0850]; the Breast program of the Department of Defense Congressionally Directed Medical Research Programs [W81XWH-21-1-0555]; Prostate Cancer UK [RIA18-ST2–022]. *MJC* also acknowledges National Institute of Health Cancer Center Support Grant (P30CA016058) to the OSUCCC The James.

References

1. De Bosscher K, Desmet SJ, Clarisse D, Estebanez-Perpina E, Brunsveld L (2020) Nuclear receptor crosstalk – defining the mechanisms for therapeutic innovation. Nat Rev Endocrinol 16:363–377
2. Zhao L, Zhou S, Gustafsson JA (2019) Nuclear receptors: recent drug discovery for cancer therapies. Endocr Rev 40:1207–1249
3. Long MD, Campbell MJ (2015) Pan-cancer analyses of the nuclear receptor superfamily. Nucl Receptor Res 2
4. Duez H, Staels B (2010) Nuclear receptors linking circadian rhythms and cardiometabolic control. Arterioscler Thromb Vasc Biol 30:1529–1534
5. Sonoda J, Pei L, Evans RM (2008) Nuclear receptors: decoding metabolic disease. FEBS Lett 582:2–9
6. Takahashi N, Goto T, Hirai S, Uemura T, Kawada T (2009) Genome science of lipid metabolism and obesity. Forum Nutr 61:25–38
7. Ruskovska T et al (2021) Systematic Bioinformatic analyses of nutrigenomic modifications by polyphenols associated with Cardiometabolic health in humans-evidence from targeted nutrigenomic studies. Nutrients 13
8. Bravo-Ruiz I, Medina MA, Martinez-Poveda B (2021) From food to genes: transcriptional regulation of metabolism by lipids and carbohydrates. Nutrients 13
9. Tao LJ, Seo DE, Jackson B, Ivanova NB, Santori FR (2020) Nuclear hormone receptors and their ligands: metabolites in control of transcription. Cell 9
10. Fuller PJ, Yang J, Young MJ (2017) 30 years of the mineralocorticoid receptor: Coregulators as mediators of mineralocorticoid receptor signalling diversity. J Endocrinol 234:T23–T34
11. Schnyder S, Kupr B, Handschin C (2017) Coregulator-mediated control of skeletal muscle plasticity - a mini-review. Biochimie 136:49–54
12. Obeid JP, Zafar N, El Hokayem J (2016) Steroid hormone receptor Coregulators in endocrine cancers. IUBMB Life 68:504–515
13. Dasgupta S, Lonard DM, O'Malley BW (2014) Nuclear receptor coactivators: master regulators of human health and disease. Annu Rev Med 65:279–292
14. Weikum ER, Liu X, Ortlund EA (2018) The nuclear receptor superfamily: a structural perspective. Protein Sci 27:1876–1892
15. Evans RM, Mangelsdorf DJ (2014) Nuclear receptors, RXR, and the big bang. Cell 157:255–266
16. Paul J (1981) Sir George Beatson and the Royal Beatson Memorial Hospital. Med Hist 25:200–201
17. Huggins C, Stevens R, Hodges CV (1941) Studies on prostatic cancer: II. The effects of castration on advanced carcinoma of the prostate gland. Arch Surg 43:209–223
18. Smith KW, Thompson PD, Rodriguez EP, Mackay L, Cobice DF (2019) Effects of vitamin D as a regulator of androgen intracrinology in LNCAP prostate cancer cells. Biochem Biophys Res Commun 519:579–584
19. Olokpa E, Bolden A, Stewart LV (2016) The androgen receptor regulates PPARgamma expression and activity in human prostate cancer cells. J Cell Physiol 231:2664–2672
20. Eskra JN, Kuiper JW, Walden PD, Bosland MC, Ozten N (2017) Interactive effects of 9-cis-retinoic acid and androgen on proliferation, differentiation, and apoptosis of LNCaP prostate cancer cells. Eur J Cancer Prev 26:71–77
21. Wang JH, Tuohimaa P (2008) Calcitriol and TO-901317 interact in human prostate cancer LNCaP cells. Gene Regul Syst Bio 2:97–105
22. Cai Y et al (2003) Cytochrome P450 genes are differentially expressed in female and male hepatocyte retinoid X receptor alpha-deficient mice. Endocrinology 144:2311–2318

23. Lareyre JJ et al (2000) Characterization of an androgen-specific response region within the 5′ flanking region of the murine epididymal retinoic acid binding protein gene. Biol Reprod 63:1881–1892
24. Yetkin D, Balli E, Ayaz F (2021) Antiproliferative activity of tamoxifen, vitamin D3 and their concomitant treatment. EXCLI J 20:1394–1406
25. Yang S et al (2021) Differential effects of estrogen receptor alpha and beta on endogenous ligands of peroxisome proliferator-activated receptor gamma in papillary thyroid cancer. Front Endocrinol (Lausanne) 12:708248
26. Tanaka N et al (2021) Pemafibrate, a novel selective PPARalpha modulator, attenuates tamoxifen-induced fatty liver disease. Clin J Gastroenterol 14:846–851
27. Verma A, Cohen DJ, Jacobs TW, Boyan BD, Schwartz Z (2021) The relative expression of ERalpha isoforms ERalpha66 and ERalpha36 controls the cellular response to 24R,25-Dihydroxyvitamin D3 in breast cancer. Mol Cancer Res 19:99–111
28. Hasan N, Sonnenschein C, Soto AM (2019) Vitamin D3 constrains estrogen's effects and influences mammary epithelial organization in 3D cultures. Sci Rep 9:7423
29. Fan P, Abderrahman B, Chai TS, Yerrum S, Jordan VC (2018) Targeting peroxisome proliferator-activated receptor gamma to increase estrogen-induced apoptosis in estrogen-deprived breast cancer cells. Mol Cancer Ther 17:2732–2745
30. de Vera IMS (2018) Advances in orphan nuclear receptor pharmacology: a new era in drug discovery. ACS Pharmacol Transl Sci 1:134–137
31. Hendrickson WK et al (2011) Vitamin D receptor protein expression in tumor tissue and prostate cancer progression. J Clin Oncol 29:2378–2385
32. Feldman D, Krishnan AV, Swami S, Giovannucci E, Feldman BJ (2014) The role of vitamin D in reducing cancer risk and progression. Nat Rev Cancer 14:342–357
33. Bouillon R et al (2008) Vitamin D and human health: lessons from vitamin D receptor null mice. Endocr Rev 29:726–776
34. Fleet JC et al (2019) Vitamin D signaling suppresses early prostate carcinogenesis in TgAPT121 mice. Cancer Prev Res (Phila) 12:343–356
35. Baumann B, Lugli G, Gao S, Zenner M, Nonn L (2019) High levels of PIWI-interacting RNAs are present in the small RNA landscape of prostate epithelium from vitamin D clinical trial specimens. Prostate 79:840–855
36. Tavera-Mendoza LE et al (2017) Vitamin D receptor regulates autophagy in the normal mammary gland and in luminal breast cancer cells. Proc Natl Acad Sci U S A 114:E2186–E2194
37. Zinser G, Packman K, Welsh J (2002) Vitamin D(3) receptor ablation alters mammary gland morphogenesis. Development 129:3067–3076
38. Vanhevel J, Verlinden L, Doms S, Wildiers H, Verstuyf A (2022) The role of vitamin D in breast cancer risk and progression. Endocr Relat Cancer 29:R33–R55
39. Kittler R et al (2013) A comprehensive nuclear receptor network for breast cancer cells. Cell Rep 3:538–551
40. Hua S, Kittler R, White KP (2009) Genomic antagonism between retinoic acid and estrogen signaling in breast cancer. Cell 137:1259–1271
41. Al-Dhaheri M et al (2011) CARM1 is an important determinant of ERalpha-dependent breast cancer cell differentiation and proliferation in breast cancer cells. Cancer Res 71:2118–2128
42. Agadir A et al (1994) Retinoic acid receptors: involvement in acute promyelocytic leukemia. Cell Mol Biol (Noisy-le-Grand) 40:263–274
43. Lin RJ et al (1998) Role of the histone deacetylase complex in acute promyelocytic leukaemia. Nature 391:811–814
44. de The H (2018) Differentiation therapy revisited. Nat Rev Cancer 18:117–127
45. Spira AI, Carducci MA (2003) Differentiation therapy. Curr Opin Pharmacol 3:338–343
46. Aboseif SR, Dahiya R, Narayan P, Cunha GR (1997) Effect of retinoic acid on prostatic development. Prostate 31:161–167
47. Lohnes D et al (1993) Function of retinoic acid receptor gamma in the mouse. Cell 73:643–658
48. Long MD et al (2019) The miR-96 and RARgamma signaling axis governs androgen signaling and prostate cancer progression. Oncogene 38:421–444
49. de Ruijter TC et al (2020) Prognostic DNA methylation markers for hormone receptor breast cancer: a systematic review. Breast Cancer Res 22:13
50. Bakavicius A et al (2019) Urinary DNA methylation biomarkers for prediction of prostate cancer upgrading and upstaging. Clin Epigenetics 11:115
51. Liu X, Giguere V (2014) Inactivation of RARbeta inhibits Wnt1-induced mammary tumorigenesis by suppressing epithelial-mesenchymal transitions. Nucl Recept Signal 12:e004
52. Jiang D et al (2014) Meta-analyses of methylation markers for prostate cancer. Tumour Biol 35:10449–10455
53. Niewiadomska-Cimicka A et al (2017) Genome-wide analysis of RARbeta transcriptional targets in mouse striatum links retinoic acid signaling with Huntington's disease and other neurodegenerative disorders. Mol Neurobiol 54:3859–3878
54. Wang J et al (2016) ROR-gamma drives androgen receptor expression and represents a therapeutic target in castration-resistant prostate cancer. Nat Med 22:488–496
55. Cai D et al (2019) RORgamma is a targetable master regulator of cholesterol biosynthesis in a cancer subtype. Nat Commun 10:4621
56. Du J, Xu R (2012) RORalpha, a potential tumor suppressor and therapeutic target of breast cancer. Int J Mol Sci 13:15755–15766

57. Mao W et al (2021) RORalpha suppresses cancer-associated inflammation by repressing respiratory complex I-dependent ROS generation. Int J Mol Sci 22
58. Decara J et al (2020) Peroxisome proliferator-activated receptors: experimental targeting for the treatment of inflammatory bowel diseases. Front Pharmacol 11:730
59. Adhikary T et al (2013) Inverse PPARbeta/delta agonists suppress oncogenic signaling to the ANGPTL4 gene and inhibit cancer cell invasion. Oncogene 32:5241–5252
60. Jiao X et al (2021) Ppargamma1 facilitates ErbB2-mammary adenocarcinoma in mice. Cancers (Basel) 13
61. Galbraith LCA et al (2021) PPAR-gamma induced AKT3 expression increases levels of mitochondrial biogenesis driving prostate cancer. Oncogene 40:2355–2366
62. Liu S et al (2014) Differential roles of PPARgamma vs TR4 in prostate cancer and metabolic diseases. Endocr Relat Cancer 21:R279–R300
63. Kaikkonen S, Paakinaho V, Sutinen P, Levonen AL, Palvimo JJ (2013) Prostaglandin 15d-PGJ(2) inhibits androgen receptor signaling in prostate cancer cells. Mol Endocrinol 27:212–223
64. Govindarajan R et al (2007) Thiazolidinediones and the risk of lung, prostate, and colon cancer in patients with diabetes. J Clin Oncol 25:1476–1481
65. Butler R, Mitchell SH, Tindall DJ, Young CY (2000) Nonapoptotic cell death associated with S-phase arrest of prostate cancer cells via the peroxisome proliferator-activated receptor gamma ligand, 15-deoxy-delta12,14-prostaglandin J2. Cell Growth Differ 11:49–61
66. Zheng K, Chen S, Hu X (2022) Peroxisome proliferator activated receptor gamma Coactivator-1 alpha: a double-edged sword in prostate cancer. Curr Cancer Drug Targets
67. Siddappa M et al (2020) Identification of transcription factor co-regulators that drive prostate cancer progression. Sci Rep 10:20332
68. Torrano V et al (2016) The metabolic co-regulator PGC1alpha suppresses prostate cancer metastasis. Nat Cell Biol 18:645–656
69. Shukla S et al (2017) Aberrant activation of a gastrointestinal transcriptional circuit in prostate cancer mediates castration resistance. Cancer Cell 32:792–806 e797
70. Wang Z et al (2020) Nuclear receptor HNF4alpha performs a tumor suppressor function in prostate cancer via its induction of p21-driven cellular senescence. Oncogene 39:1572–1589
71. Pihlajamaa P et al (2014) Tissue-specific pioneer factors associate with androgen receptor cistromes and transcription programs. EMBO J 33:312–326
72. Miyajima N et al (1988) Identification of two novel members of erbA superfamily by molecular cloning: the gene products of the two are highly related to each other. Nucleic Acids Res 16:11057–11074
73. Ladias JA, Karathanasis SK (1991) Regulation of the apolipoprotein AI gene by ARP-1, a novel member of the steroid receptor superfamily. Science 251:561–565
74. Erdos E, Balint BL (2020) NR2F2 orphan nuclear receptor is involved in estrogen receptor alpha-mediated transcriptional regulation in luminal a breast cancer cells. Int J Mol Sci 21
75. Jiang G et al (2019) Cooperativity of co-factor NR2F2 with Pioneer factors GATA3, FOXA1 in promoting ERalpha function. Theranostics 9:6501–6516
76. Sosa MS et al (2015) NR2F1 controls tumour cell dormancy via SOX9- and RARbeta-driven quiescence programmes. Nat Commun 6:6170
77. Khalil BD et al (2022) An NR2F1-specific agonist suppresses metastasis by inducing cancer cell dormancy. J Exp Med 219
78. Guo H et al (2021) NR4A1 regulates expression of immediate early genes, suppressing replication stress in cancer. Mol Cell 81:4041–4058 e4015
79. Agostini-Dreyer A, Jetzt AE, Stires H, Cohick WS (2015) Endogenous IGFBP-3 mediates intrinsic apoptosis through modulation of Nur77 phosphorylation and nuclear export. Endocrinology 156:4141–4151
80. Lee KW et al (2007) Contribution of the orphan nuclear receptor Nur77 to the apoptotic action of IGFBP-3. Carcinogenesis 28:1653–1658
81. Bellec M et al (2022) The control of transcriptional memory by stable mitotic bookmarking. Nat Commun 13:1176
82. Caravaca JM et al (2013) Bookmarking by specific and nonspecific binding of FoxA1 pioneer factor to mitotic chromosomes. Genes Dev 27:251–260
83. Kadauke S et al (2012) Tissue-specific mitotic bookmarking by hematopoietic transcription factor GATA1. Cell 150:725–737
84. Zaidi SK et al (2010) Mitotic bookmarking of genes: a novel dimension to epigenetic control. Nat Rev Genet 11:583–589
85. Kouskouti A, Talianidis I (2005) Histone modifications defining active genes persist after transcriptional and mitotic inactivation. EMBO J 24:347–357
86. Michelotti EF, Sanford S, Levens D (1997) Marking of active genes on mitotic chromosomes. Nature 388:895–899
87. Kumar S, Chaturvedi NK, Kumar S, Tyagi RK (2008) Agonist-mediated docking of androgen receptor onto the mitotic chromatin platform discriminates intrinsic mode of action of prostate cancer drugs. Biochim Biophys Acta 1783:59–73
88. Rana M, Dash AK, Ponnusamy K, Tyagi RK (2018) Nuclear localization signal region in nuclear receptor PXR governs the receptor association with mitotic chromatin. Chromosom Res 26:255–276
89. Kumar S, Vijayan R, Dash AK, Gourinath S, Tyagi RK (2021) Nuclear receptor SHP dampens transcription function and abrogates mitotic chromatin association of PXR and ERalpha via intermolecular

interactions. Biochim Biophys Acta Gene Regul Mech 1864:194683
90. Festuccia N et al (2016) Mitotic binding of Esrrb marks key regulatory regions of the pluripotency network. Nat Cell Biol 18:1139–1148
91. Festuccia N et al (2019) Transcription factor activity and nucleosome organization in mitosis. Genome Res 29:250–260
92. Karagianni P, Moulos P, Schmidt D, Odom DT, Talianidis I (2020) Bookmarking by non-pioneer transcription factors during liver development establishes competence for future gene activation. Cell Rep 30:1319–1328 e1316
93. Pomerantz MM et al (2020) Prostate cancer reactivates developmental epigenomic programs during metastatic progression. Nat Genet 52:790–799
94. Franke A et al (1992) Polycomb and polyhomeotic are constituents of a multimeric protein complex in chromatin of Drosophila melanogaster. EMBO J 11:2941–2950
95. Dedon PC, Soults JA, Allis CD, Gorovsky MA (1991) A simplified formaldehyde fixation and immunoprecipitation technique for studying protein-DNA interactions. Anal Biochem 197:83–90
96. Orlando V, Paro R (1995) Chromatin multiprotein complexes involved in the maintenance of transcription patterns. Curr Opin Genet Dev 5:174–179
97. Buck MJ, Lieb JD (2004) ChIP-chip: considerations for the design, analysis, and application of genome-wide chromatin immunoprecipitation experiments. Genomics 83:349–360
98. Johnson DS, Mortazavi A, Myers RM, Wold B (2007) Genome-wide mapping of in vivo protein-DNA interactions. Science 316:1497–1502
99. Kaya-Okur HS et al (2019) CUT&tag for efficient epigenomic profiling of small samples and single cells. Nat Commun 10:1930
100. Skene PJ, Henikoff S (2017) An efficient targeted nuclease strategy for high-resolution mapping of DNA binding sites. elife 6
101. Buenrostro JD, Giresi PG, Zaba LC, Chang HY, Greenleaf WJ (2013) Transposition of native chromatin for fast and sensitive epigenomic profiling of open chromatin, DNA-binding proteins and nucleosome position. Nat Methods 10:1213–1218
102. King HW et al (2021) Integrated single-cell transcriptomics and epigenomics reveals strong germinal center-associated etiology of autoimmune risk loci. Sci Immunol 6:eabh3768
103. Chakravarti R et al (2022) A review on CRISPR-mediated epigenome editing: a future directive for therapeutic management of cancer. Curr Drug Targets
104. Liu G, Lin Q, Jin S, Gao C (2022) The CRISPR-Cas toolbox and gene editing technologies. Mol Cell 82:333–347
105. Guo Y et al (2018) CRISPR-mediated deletion of prostate cancer risk-associated CTCF loop anchors identifies repressive chromatin loops. Genome Biol 19:160
106. Rhie SK et al (2019) A high-resolution 3D epigenomic map reveals insights into the creation of the prostate cancer transcriptome. Nat Commun 10:4154
107. Taberlay PC et al (2016) Three-dimensional disorganization of the cancer genome occurs coincident with long-range genetic and epigenetic alterations. Genome Res 26:719–731
108. Roadmap Epigenomics C et al (2015) Integrative analysis of 111 reference human epigenomes. Nature 518:317–330
109. Thurman RE et al (2012) The accessible chromatin landscape of the human genome. Nature 489:75–82
110. Jung I et al (2019) A compendium of promoter-centered long-range chromatin interactions in the human genome. Nat Genet 51:1442–1449
111. Franke J, Neumann MH (2000) Bootstrapping neural networks. Neural Comput 12:1929–1949
112. Long MD et al (2015) Integrative genomic analysis in K562 chronic myelogenous leukemia cells reveals that proximal NCOR1 binding positively regulates genes that govern erythroid differentiation and Imatinib sensitivity. Nucleic Acids Res 43:7330–7348
113. Ernst J, Kellis M (2017) Chromatin-state discovery and genome annotation with ChromHMM. Nat Protoc 12:2478–2492
114. Kim DH et al (2021) BET Bromodomain inhibition blocks an AR-repressed, E2F1-activated treatment-emergent neuroendocrine prostate cancer lineage plasticity program. Clin Cancer Res 27:4923–4936
115. Whyte WA et al (2013) Master transcription factors and mediator establish super-enhancers at key cell identity genes. Cell 153:307–319
116. Layer RM et al (2018) GIGGLE: a search engine for large-scale integrated genome analysis. Nat Methods 15:123–126
117. Long MD, van Den B, Singh P, Battaglia S, Campbell MJ (2015) Integrative genomic analysis in K562 chronic myelogenous leukemia cells reveals that proximal NCOR1 binding positively regulates genes that govern erythroid differentiation and Imatinib sensitivity. Nucleic Acids Res 43:7330–7348
118. Feng J, Liu T, Qin B, Zhang Y, Liu XS (2012) Identifying ChIP-seq enrichment using MACS. Nat Protoc 7:1728–1740
119. Lun AT, Smyth GK (2016) Csaw: a Bioconductor package for differential binding analysis of ChIP-seq data using sliding windows. Nucleic Acids Res 44:e45
120. Law CW et al (2016) RNA-seq analysis is easy as 1-2-3 with limma, Glimma and edgeR. F1000Res 5
121. Patro R, Duggal G, Love MI, Irizarry RA, Kingsford C (2017) Salmon provides fast and bias-aware quantification of transcript expression. Nat Methods 14:417–419
122. Brazma A et al (2001) Minimum information about a microarray experiment (MIAME)-toward standards for microarray data. Nat Genet 29:365–371

Part V

New Developments in Transcriptional Control by Nuclear Receptors

Protein Condensation in the Nuclear Receptor Family; Implications for Transcriptional Output

14

Monique D. Appelman, Elle E. Hollaar, Jurian Schuijers, and Saskia W. C. van Mil

Abstract

The nuclear receptor superfamily is a group of transcriptional regulators that orchestrate multiple vital processes such as inflammation, metabolism, and cell proliferation. In recent years, it has become clear that some nuclear receptors form condensates in living cells. These condensates contain high concentrations of proteins and can contain millions of molecules. At these sites, high concentrations of nuclear receptors and co-factors potentially contribute to efficient transcription. While condensate formation has been observed for some nuclear receptors, the majority have unknown condensate formation abilities. Condensate formation abilities for these NRs would implicate an additional layer of regulation for the entire nuclear receptor family. Here, we consider the nuclear receptor superfamily, the current evidence for condensate formation of some of its members and the potential of the whole superfamily to form condensates. Insights into the regulation of assembly or disassembly of nuclear receptor condensates and our considerations for the understudied family members imply that condensate biology might be an important aspect of nuclear receptor-regulated gene transcription.

Keywords

Transcriptional regulation · Nuclear receptors · Condensate formation · intrinsically disordered regions · Biomolecular condensates · Liquid-liquid phase separation

14.1 The Nuclear Receptor Superfamily

The nuclear receptor (NR) superfamily is a group of transcription factors (TFs), of which many members are activated by different ligands such as steroid and thyroid hormones [1, 2]. There is a lot of interest in these receptors because ligand-activated NRs regulate multiple essential processes such as inflammation, metabolism, and cell proliferation. NRs regulate these processes by recruiting co-factors to specific promoter or enhancer sites, which results in transcriptional activation or repression [2–6]. Dysregulation of several NRs is implicated in cancer, atherosclerosis, diabetes, and other pathologies [2–6].

Currently, 48 members of the NR family are known in humans [6, 7]. These differ in their

Authors Jurian Schuijers and Saskia WC van Mil share senior authorship.

M. D. Appelman · E. E. Hollaar · J. Schuijers · S. W. C. van Mil (✉)
Center for Molecular Medicine, UMC Utrecht and Utrecht University, Utrecht, The Netherlands
e-mail: S.W.C.vanMil@umcutrecht.nl

M. J. Campbell, C. L. Bevan (eds.), *Nuclear Receptors in Human Health and Disease*, Advances in Experimental Medicine and Biology 1390, https://doi.org/10.1007/978-3-031-11836-4_14

ligand and DNA binding domains, which is well summarized elsewhere [7, 8]. In general, NRs have a standard protein structure that consists of multiple domains in a specific order: the N-terminal domain (NTD), DNA binding domain (DBD), hinge, ligand binding domain (LBD), and the C-terminal domain [3, 5, 9, 10]. The NTD contains an activator function-1 region (AF1), which is responsible for interactions with co-factors, and this region is also important for transcriptional activation [3, 5, 9, 11]. The DBD is the most conserved region in the NR, comprises two zinc fingers and is responsible for targeting the NR to a specific DNA sequence. The hinge region is not only described to function as a connecter between the DBD and the LBD involved in nuclear translocation, but is in many cases also described as a region that is post-translationally modified, and this influences NR transactivation and ligand sensitivity [3, 5, 7]. The LBD is formed by 12 conserved α-helical regions numbered from H1 to H12 that undergo allosteric changes after ligand binding, leading to the activation of the NR. The activator function-2 region (AF-2) is a part of the LBD, which is responsible for transcriptional activation by recruitment of coregulator proteins and the transcription complex [3, 11–13].

In addition to the similarities in structure between the NRs, the transcriptional activation and repression by NRs are also regulated in a common way. Some NRs behave as transcriptional repressors when their ligand is absent. This repression is mediated by the recruitment of co-repressors such as NCoR1 (nuclear receptor co-repressor) in the unliganded state. This leads, for example, to the mobilization of histone deacetylases, and the resulting deacetylated histones lead to a more condensed chromatin structure. This prohibits RNA polymerase II binding, preventing transcriptional activation [7, 14–16]. Upon ligand binding and the resulting conformational change in the NR, the co-repressor complex is released [7, 15, 17]. Other NRs are localized in the cytosol in the unliganded state. Upon binding of the ligand, the NR translocates to the nucleus and binds to the DNA.

Subsequently, co-activators are recruited to the NR. Over 350 NR co-activators are known, of which some co-activators have histone acetyltransferase (HAT) activity which, in contrast to the histone deacetylases, leads to histone acetylation and decondensation of the chromatin [7, 18, 19]. Co-activators also support initiation of transcription by catalyzing the assembly of the transcription preinitiation complex at promoters [7, 18, 19].

Activated NRs bind as monomer to a specific DNA hexameric sequence, or as a homodimer or heterodimer to a dual hexameric repeat, which can be positioned in an inverted, everted, or direct orientation [3, 5, 7, 9, 12, 17]. Ultimately, NR-cofactor binding to these so-called hormone responsive elements (HRE) results in the recruitment of RNA polymerase II and the activation of transcription.

For transcriptional activation, it is vital that ligand, co-factors, RNA polymerase II and NR find each other in the crowded environment of the cell at the appropriate time and place to regulate transcription successfully [20–23]. As we will highlight in the next paragraphs, it is thought that NR condensation is the means by which this intricate process is achieved.

14.2 A Condensate Model for NR Transcriptional Regulation

Compartmentalization of the necessary proteins into organelles with membranes (e.g., nucleus, endoplasmic reticulum, lysosomes) is one of the methods by which a cell regulates the spatial and temporal localization of proteins required to assert certain functions. The second way this localization is regulated is the formation of membraneless organelles [20, 22, 24].

These membraneless organelles, also called biomolecular condensates, are compartments within a cell in which biomolecules such as proteins and nucleic acids assemble, and are typified by their droplet-like structure [20–25]. Although consensus around the biomolecular condensate term is recent, these structures have been observed since the late 1800s, when E.B. Wilsons described a liquid droplet-like organization in protoplasm using simple light microscopy [20, 26]. Recently, it has been demonstrated that several TFs, including some NRs (e.g., estrogen receptor alpha (ERα), glucocorticoid receptor (GR)), can form condensates within the nucleus [16, 20–25, 27, 28]. Moreover, these condensates have been observed to play a role in efficient transcriptional regulation [16, 20–25]. Condensates are likely to be formed through a biophysical process of phase separation. In phase separation, part of a homogenous solution demixes into two phases, a dense phase and a dilute phase [20, 23, 26, 29]. In cells, a specific type of phase separation can drive condensate formation by forming a liquid compartment in a liquid environment; this process is called liquid-liquid phase separation (LLPS) [20, 23, 26].

Currently, the evidence for LLPS underlying condensate formation is incomplete, mainly because some of the characteristics of LLPS are difficult to demonstrate in a cellular context [23]. However, there are now numerous examples that show that many different proteins can form condensates, which result in cellular compartments with high concentrations of these proteins. Among these examples are several NRs and co-factors, indicating that condensate biology likely plays a role in NR function.

Formation of protein condensates is driven by weak multivalent interactions between proteins, and is dependent on protein concentration [20–25]. These weak multivalent interactions often involve prion-like domains or, more broadly, intrinsically disordered regions (IDRs). Prion-like domains in proteins are defined by the ability to assume multiple conformational states, and one of these states enabling binding to other copies of the same protein, which is favorable for condensate formation [30, 31]. An IDR is characterized by a low number of hydrophobic amino acids and enrichment in polar, charged and aromatic residues [21, 22, 25, 32]. These IDR properties also result in lack of a fixed 3D structure and these amino acids facilitate multivalent interactions that can potentially drive condensate formation (Fig. 14.1) [21, 25, 32].

The importance of these domains in condensate formation was demonstrated using OCT4 (octamer-binding transcription factor 4). OCT4

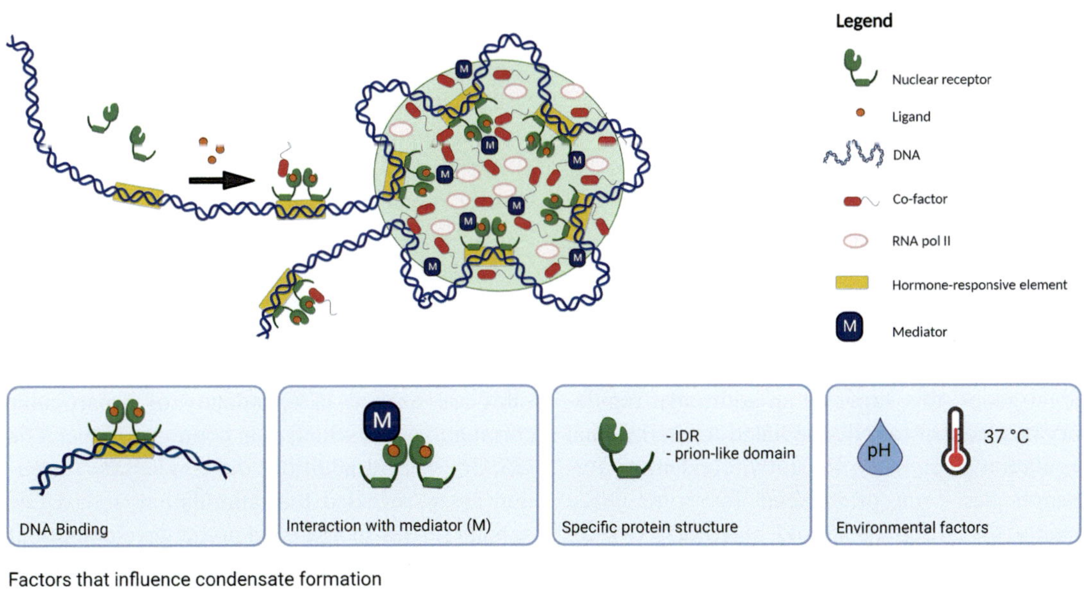

Fig. 14.1 Schematic overview of NR condensate formation. Upon ligand binding, NRs bind to the hormone-responsive elements in DNA. Interactions with Mediator complex, specific protein structural elements (e.g. IDR and prion-like domains), and different environmental factors promote NR condensate formation

induced the formation of condensates that included the essential co-factor complex Mediator [33]. Disrupting the IDR of OCT4 blocked the formation of condensates, demonstrating the dependency of condensate formation on the IDR of OCT4. Interestingly, the lack of condensate formation was accompanied by a lower transcriptional output of OCT4 target genes [20, 33, 34]. Besides showing the importance of certain IDRs for condensate formation, this example suggests that condensate formation is involved in increasing the activation potential of TFs, which may include NRs.

Besides IDRs, important factors that influence condensate formation of TFs are the DNA accessibility and density of TF binding motifs. Digestion of DNA *in situ* disrupts condensate formation and adjusting the density of DNA elements controls condensate nucleation, indicating a role for DNA in nucleating condensates (Fig. 14.1) [22, 35, 36]. In addition to DNA, environmental factors such as temperature, ionic strength, protein and RNA concentrations, osmolarity, and pH levels have been suggested to play a role in condensate formation [20, 25, 36–38] (Fig. 14.1). The capacity of condensates to integrate so many biological signals, together with their effect on transcriptional output, suggests condensate formation is an additional layer of regulation of TFs, including NRs.

14.3 Evidence of NR Condensate Formation

Since condensate formation results in the assembly of TFs, transcription complexes and co-factors, thereby increasing the transcription of target genes, this suggests an additional regulatory mechanism for NR-mediated transcriptional regulation (Fig. 14.1). Recently, condensate formation has been established for some NRs, mainly the group of steroid receptors. One of these steroid receptors is the glucocorticoid receptor (GR), which is activated upon glucocorticoid binding in the cytoplasm. GR subsequently translocates to the nucleus and binds to its HRE [39]. The existence of GR condensates has been shown via expression of a GFP-fusion construct in several cell lines [35, 36, 40, 41].

Several GR domains have been implicated to be essential for its condensation. Deletion of either the DBD or LBD reduced the number of GR condensates in cells [36, 42]. Moreover, mutation of a single amino acid (phenylalanine at position 623) decreased ligand binding, which reduced the number of condensates compared to wild-type GR, suggesting that ligand binding is essential for GR-condensate formation [42, 43]. Interestingly, deletion of the NTD did reduce the number of condensates in vitro, but not in cell culture experiments [36, 41, 42]. This is remarkable because the NTD contains an IDR, which are often found to be crucial in condensate formation [20, 33, 34].

Nevertheless, the NTD has a function in the formation of condensates under certain environmental circumstances [36, 41]. An increase in NaCl, osmolarity, or temperature induces a rise in the number of condensates for wild-type GR [36, 41] and is reversible, suggesting that the formed condensates are not the result of abnormal aggregate development [36]. However, NaCl treatment could not induce an increase in condensate formation of GR lacking the NTD [36]. This suggests that the NTD in GR is involved in condensate formation upon specific environmental cues. Contrarily, the LBD and DBD are essential for condensate formation independent of environmental cues [36, 41, 42]. This shows that specific domains within one protein can have different effects on condensate regulation.

To investigate the potential role of DNA density in the formation of GR condensates, Stortz et al. performed a GR condensate formation assay [36]. This demonstrated that GR condensates are formed independently of a particular chromatin state, which is in contrast to other TFs [35, 36, 41]. In addition, the condensate formation assay revealed that stimulation with a GR agonist results in GR condensate development at specific locations in the nucleus. Moreover, a lack of specific GR DNA binding motifs leads to a decrease in GR condensate formation [41]. Overall, these results indicate that specific GR binding DNA sequences are necessary for condensate formation. In these condensates, cofactors, such as Mediator, G9a, and SRC (steroid receptor co-activator), colocalize with GR [35, 41, 44].

The GR is not the only NR for which condensate formation has been described. Condensates containing ERα [28, 33, 45–48], mineralocorticoid receptor (MR) [49, 50], progesterone receptor (PR) [51, 52], and androgen receptor (AR) have also been demonstrated [46, 53, 54]. Similar to the GR, the LBD and DBD are essential in condensate formation of these NRs [50, 51, 54]. These NRs do not have prion-like domains, but do contain an IDR located in the NTD [11, 31, 41]. Interestingly, only the NTD of the AR has been found to have a crucial role in condensate formation, independent of environmental cues or other domains [50, 51, 54]. The contribution to condensate formation of the NTD is still debated for other NRs [45, 48, 53, 54].

Like GR, the condensates of ERα, PR, and AR colocalize with Mediator, but also with other NR-specific cofactors [45, 51, 53, 54]. For example, condensates of ERα together with MegaTrans components were observed at enhancer clusters together with estrogen responsive genes upon estrogen stimulation [28]. Interestingly, knockdown of Mediator decreased AR condensate formation and transcriptional output, which was not the case for GR (Fig. 14.1) [54]. This indirectly suggests that AR-condensate formation influences transcription. Another way of investigating the effect of NR condensate formation on target gene expression is with the aliphatic alcohol 1,6-hexanediol (HD) [36, 54, 55]. HD disrupts hydrophobic interactions between proteins and is used to target condensate formation. Despite pleiotropic effects of HD, it was shown that HD treatment disrupts ERα and AR condensates, which led to decreased gene activity of their target genes [28, 45, 47, 48, 54]. While these methods to determine transcriptional effects of condensate formation do not demonstrate a direct effect, Wei et al. showed a direct link between TF condensate formation and transcription by establishing that nascent RNA is enriched in the condensates, compared to an even distribution of RNA when these condensates are absent [56]. These results suggest that also for NRs there can be a direct link between condensate formation and transcriptional regulation.

In conclusion, condensate formation has been demonstrated for multiple NRs. The exact role of the different NR structures, such as the IDR, and the precise mechanism behind condensate formation are still unknown or might differ for each NR. However, the expectation is that NR condensate formation can influence transcription.

14.4 Potential Condensate Formation of the NR Superfamily

Currently, condensate formation has been described for five NRs (ERα, PR, AR, GR, and MR). These five NRs have characteristics typically associated with condensate formation, such as an IDR and interaction with Mediator (Fig. 14.1). To estimate the relevance of condensate biology for the NR family, we here predict the potential to form condensates for the other NR family members based on the characteristics of these five NRs.

To gain insight into the ability of the other NRs to form condensates, we have used different phase separation prediction tools. The characteristics of the condensate forming GR [35, 36, 40, 41, 57, 58], AR [54], MR [49, 50], PRs [51, 52], and ERα [47] were used to set a baseline for predicting condensate formation.

Typically, phase-separation prediction tools use one aspect that is important for phase separation, such as the presence of an IDR, a prion-like domain, or charged amino acids. However, dSCOPE (Detecting Sequence Critical fOr Phase sEparation) uses a combination of these factors to predict if a protein has a phase separation domain, a domain that has a combination of factors favorable for phase separation including IDRs, charged amino acids, low complexity, hydropathy, polarity, and a prion-like domains [59]. Therefore, dSCOPE was used to predict the presence of a "phase separation domain" in all 48 NRs (Table 14.1). dSCOPE predicted a phase separation domain in 22 out of the 48 human NRs.

Table 14.1 Overview of the prediction analysis of all 48 NRs on their condensates formation characteristics

Name NR	Abbreviation	Gene symbol	% disordered	Predicted Phase separation domain	Predicted Prion-like domain	Reported Mediator interaction
Mineralocorticoid receptor	MR*	NR3C2	66	Yes	Yes	–
Androgen receptor	AR*	NR3C4	60	No	Yes	+ [64]
Progesterone receptor	PR*	NR3C3	58	Yes	No	–
Glucocorticoid receptor	GR*	NR3C1	54	No	No	+ [64]
Estrogen receptor-α	ERα*	NR3A1	47	Yes	No	+ [64]
Rev-Erbα	Rev-Erbα	NR1D1	67	Yes	No	–
Neuron-derived orphan receptor 1	NOR1	NR4A3	64	Yes	No	+ [64]
Retinoid X receptor-β	RXRβ	NR2B2	64	Yes	No	+ [64]
Retinoic acid receptor-γ	RARγ	NR1B3	61	Yes	No	+ [64]
Neuron-derived clone 77	NUR77	NR4A1	59	Yes	No	+ [64]
Retinoic acid receptor-α	RARα	NR2B1	59	No	No	+ [64]
Retinoic acid receptor-β	RARβ	NR2B2	58	No	No	+ [64]
Photoreceptor cell-specific nuclear receptor	PNR	NR2E3	58	No	No	–
Nuclear receptor related 1	NURR1	NR4A2	55	Yes	Yes	+ [64]
Rev-Erbβ	Rev-Erbβ	NR1D2	55	Yes	Yes	–
Liver X receptor-β	LXRβ	NR1H2	54	Yes	Yes	+ [65]
Retinoid X receptor-γ	RXRγ	NR2B3	54	Yes	No	+ [64]
Estrogen-related receptor-α	ERRα	NR3B1	53	No	No	–
Retinoid X receptor-α	RXRα	NR2B1	52	No	No	+ [64]
RAR-related orphan receptor-γ	RORγ	NR1F3	50	Yes	No	–
Farnesoid X receptor	FXR	NR1H4	49	Yes	Yes	+ [66, 67]
Steroidogenic factor 1	SF1	NR5A1	49	Yes	No	–
Estrogen receptor-β	ERβ	NR3A2	49	No	Yes	+ [64]
Liver X receptor-α	LXRα	NR1H3	49	No	No	+ [65]
Hepatocyte nuclear factor-4-γ	HNF4γ	NR2A2	48	No	No	+ [64]
Liver receptor homolog-1	LRH-1	NR5A2	48	No	No	+ [68]
Thyroid hormone receptor-α	TRα	NR1A1	47	No	No	+ [64]
Germ cell nuclear factor	GCF	NR6A1	47	No	No	–
Hepatocyte nuclear factor-4-α	HNF4α	NR2A1	46	No	No	+ [64]
RAR-related orphan receptor-α	RORα	NR1F1	46	Yes	No	–
Chicken ovalbumin upstream promoter-transcription factor I	COUP-TFI	NR2F1	43	Yes	Yes	–
V-erbA-related	EAR-2	NR2F6	43	Yes	No	–
Estrogen-related receptor-γ	ERRγ	NR3B3	43	Yes	No	–
Testicular receptor 4	TR4	NR2C2	43	No	No	–
Chicken ovalbumin upstream promoter-transcription factor II	COUP-TFII	NR2F2	42	Yes	Yes	–
Estrogen-related receptor-β	ERRβ	NR3B2	41	Yes	No	–
Testicular receptor 2	TR2	NR2C1	40	No	No	–
Pregnane X receptor	PXR	NR1I2	40	No	No	–
Vitamin D receptor	VDR	NR1I1	37	No	No	+ [64]
Homologue of the drosophila tailless gene	TLX	NR2E1	34	No	No	–

(continued)

Table 14.1 (continued)

Name NR	Abbreviation	Gene symbol	% disordered	Predicted Phase separation domain	Predicted Prion-like domain	Reported Mediator interaction
Dosage-sensitive sex reversal, adrenal hypoplasia critical region, on chromosome X, gene 1	DAX	NR0B1	33	No	No	–
RAR-related orphan receptor-β	ROR-β	NR1F2	32	No	No	+ [69]
Thyroid hormone receptor-β	TRβ	NR1A2	32	No	No	+ [64]
Small heterodimer partner	SHP	NR0B2	30	Yes	No	–
Peroxisome proliferator-activated receptor-β/δ	PPAR-β/δ	NR1C2	30	No	No	+ [64]
Peroxisome proliferator-activated receptor-α	PPARα	NR1C1	29	No	No	+ [64]
Constitutive androstane receptor	CAR	NR1I3	29	No	No	+ [64]
Peroxisome proliferator-activated receptor-γ	PPARγ	NR1C3	25	No	No	+ [64]

The * symbol indicates NRs with described condensate formation

However, only three out of the five described condensate-forming NRs (ERα, PR, AR, GR, and MR) were predicted to have a phase separation domain by dSCOPE. To avoid relying on a single algorithm, two other common prediction programs were investigated for their ability to predict condensate formation.

Firstly, PondR (Predictor Of Naturally Disordered Regions) with predictor VSL2 was used (Table 14.1 and Fig. 14.2) [60, 61]. PondR provides a disorder score for each amino acid in a particular protein and when applied to the NR family it showed an IDR in most of the 48 NR family members, including each of the five benchmark NRs.

Secondly, PLAAC (Prion-Like Amino Acid Composition) was used to predict a prion-like domain in NRs. PLAAC predicts a prion-like domain in only two out of the five NRs known to form condensates, so the prion-like domain is unfit to predict condensate formation on its own. These results illustrate the difficulty of accurately predicting condensate formation. However, by comparing the results of the three prediction programs for these five NRs, some features seem to be common and potentially required for the formation of these condensates and could be used as criteria to estimate the likelihood of NR condensate formation.

Firstly, the presence of an IDR and the presence of either phase separation or prion-like domain(s) correlate with a higher likelihood to form condensates. A second, data-informed, factor is interaction with the Mediator complex which has been demonstrated to influence condensate formation not only for some NRs but also for other TFs [47, 51]. We have ranked the 48 NR factors according to likelihood of condensate formation based on the information above (Table 14.1). Based on the different prediction methods, all of the NRs have some hallmarks associated with condensate formation and we estimate that it is likely that a large portion of the family will indeed form condensates *in vivo*.

Future studies will validate whether the NRs indeed form condensates. Overexpression studies should be interpreted with care and at least be validated with endogenous NR expression (for example, by generating knock-in of an endogenously expressed NR-mEGFP fusion). Next, condensate formation upon ligand addition can be determined by confocal microscopy [36, 62]. To exclude aggregate formation, fluorescence recovery after photobleaching (FRAP) should be performed, to establish that the NR condensates are dynamic structures that exchange molecules with their surroundings [36, 62]. Together, such experiments will validate whether NRs can form

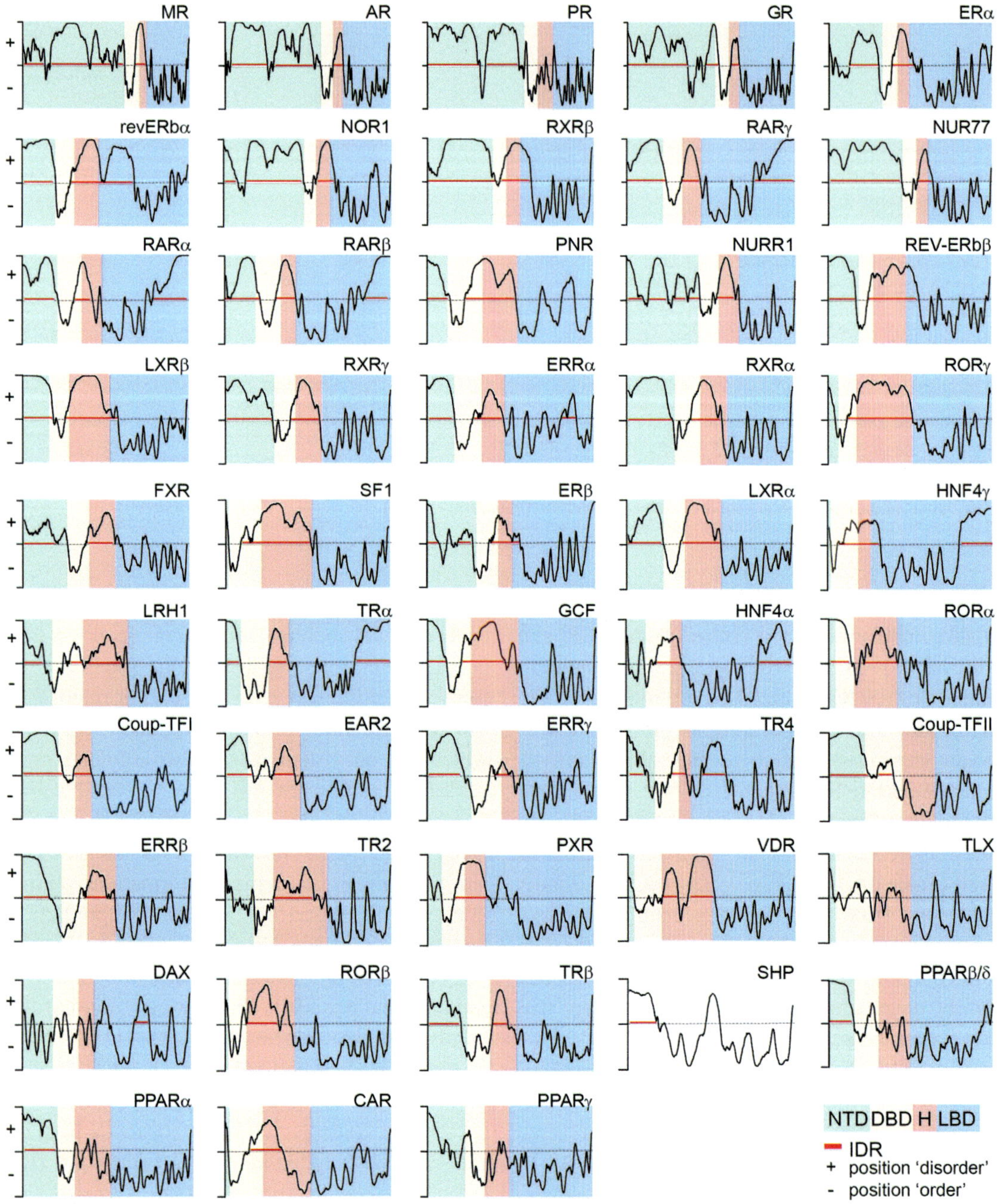

Fig. 14.2 PONDR Predictions for the NR family. See text for more information

liquid-like condensates in cells. Subsequently, essential NR domains can be investigated by means of deletion mutants or inactivating point mutations. Lastly, the effect on transcription should be demonstrated by nascent RNA labeling in the presence and absence of NR condensates [63]. These studies will provide insight in the function of condensate formation as a mechanism of transcriptional regulation for NR target genes.

Detailed understanding of NR condensates will be crucial to identify new methods to manipulate transcriptional output. In conclusion, based on the chosen prediction tools, many more NRs outside the five for which experimental evidence is available likely have the ability to form protein condensates.

14.5 Conclusion and Future Perspectives

The past few years have provided a lot of new insights into NR condensate formation suggesting that nuclear protein condensates partly regulate NR function [70, 71]. We used three different prediction programs to predict NR condensation for the complete NR family based on the characteristics of the five NRs (ERα, PR, AR, GR, MR) for which condensate formation has been established [59, 61, 72]. This showed that NR condensate formation is likely to be much more common than current experimental data has shown, potentially affecting a broad swathe if not all of the NR family.

Current knowledge of NR condensates and the implications for transcription is based on experiments using HD to disrupt interactions or by knocking out co-factors [45, 47, 48, 54], which need to be carefully interpreted because the pleiotropic effects of HD or potential indirect effects of co-factor knock-out. Therefore, detailed studies of NR condensate formation and its influence on transcription are necessary to provide more insight into how NR transcription is regulated.

Other transcriptional regulators, such as TAF15 (TATA-binding protein-associated factor 15) and p300, can form condensates that enhance transcriptional output and gene activation [56, 70]. This demonstrates that these transcriptional regulators influence transcription, implicating that NR condensates can potentially also directly influence transcriptional output. The different prediction tools showed that TAF15 has a predicted percentage disordered of 93%, a phase separation domain and prion-like domain [56, 59, 61, 72]. This supports our suggestion that a high fraction of disordered protein combined with a phase separation domain, and a prion-like domain enhance the chance of condensate formation.

However, there is an important difference between NRs and other TF such as TAF15. TF activation is complex and can involve different intracellular signal transduction pathways, while NRs are directly activated by lipophilic ligands [73]. The five described NRs form condensates only in the presence of their ligand, suggesting a role for ligands in NR condensate formation [39]. NR ligands thereby add to the complexity of condensate regulation.

Further detailed studies on the underlying forces of NR condensate formation and the influence of condensate formation on transcription will provide a better understanding of how NR condensate formation can influence transcription, and could potentially be exploited to manipulate this therapeutically.

References

Robinson-Rechavi M, Garcia HE, Laudet V (2003) The nuclear receptor superfamily. J Cell Sci 116(4):585–586

Sever R, Glass CK (2013) Signaling by nuclear receptors. Cold Spring Harb Perspect Biol 5(3):a016709

Weikum ER, Liu X, Ortlund EA (2018) The nuclear receptor superfamily: a structural perspective. Protein Sci 27(11):1876–1892

Mangelsdorf DJ, Thummel C, Beato M et al (1995) The nuclear receptor superfamily: the second decade. Cell 83(6):835

Olefsky JM (2001) Nuclear receptor minireview series. J Biol Chem 276(40):36863–36864

Kumar R, McEwan IJ (2012) Allosteric modulators of steroid hormone receptors: structural dynamics and gene regulation. Endocr Rev 33(2):271–299

Pawlak M, Lefebvre P, Staels B (2012) General molecular biology and architecture of nuclear receptors. Curr Top Med Chem 12(6):486–504

Moras D, Gronemeyer H (1998) The nuclear receptor ligand-binding domain: structure and function. Curr Opin Cell Biol 10(3):384–391

Mukha A, Kalkhoven E, van Mil SW (2021) Splice variants of metabolic nuclear receptors: Relevance for metabolic disease and therapeutic targeting. Biochim Biophys Acta (BBA) - Mol Basis Dis 2021:166183

Mazaira GI, Zgajnar NR, Lotufo CM et al (2018) The nuclear receptor field: a historical overview and future challenges. Nucl Receptor Res 5:101320

Simons SS Jr, Edwards DP, Kumar R (2014) Minireview: dynamic structures of nuclear hormone receptors: new promises and challenges. Mol Endocrinol 28(2):173–182

Porter BA, Ortiz MA, Bratslavsky G, Kotula L (2019) Structure and function of the nuclear receptor superfamily and current targeted therapies of prostate cancer. Cancers 11(12):1852

Wärnmark A, Treuter E, Wright AP, Gustafsson J (2003) Activation functions 1 and 2 of nuclear receptors: molecular strategies for transcriptional activation. Mol Endocrinol 17(10):1901–1909

Yoon H, Chan DW, Huang Z et al (2003) Purification and functional characterization of the human N-CoR complex: the roles of HDAC3, TBL1 and TBLR1. EMBO J 22(6):1336–1346

Sladek FM (2011) What are nuclear receptor ligands? Mol Cell Endocrinol 334(1–2):3–13

Cho W, Spille J, Hecht M et al (2018) Mediator and RNA polymerase II clusters associate in transcription-dependent condensates. Science 361(6400):412–415

Holzer G, Markov GV, Laudet V (2017) Evolution of nuclear receptors and ligand signaling: toward a soft key–lock model? Curr Top Dev Biol 125:1–38

Han SJ, Lonard DM, O'Malley BW (2009) Multi-modulation of nuclear receptor coactivators through posttranslational modifications. Trends Endocrinol Metab 20(1):8–15

McKenna NJ, Lanz RB, O'Malley BW (1999) Nuclear receptor coregulators: cellular and molecular biology. Endocr Rev 20(3):321–344

Peng L, Li E, Xu L (2020) From start to end: Phase separation and transcriptional regulation. Biochim Biophys Acta, Gene Regul Mech 2020:194641

Banani SF, Lee HO, Hyman AA, Rosen MK (2017) Biomolecular condensates: organizers of cellular biochemistry. Nat Rev Mol Cell Biol 18(5):285–298

Sabari BR, Dall'Agnese A, Young RA (2020) Biomolecular condensates in the nucleus. Trends Biochem Sci

Bhat P, Honson D, Guttman M (2021) Nuclear compartmentalization as a mechanism of quantitative control of gene expression. Nat Rev Mol Cell Biol:1–18

Alberti S, Hyman AA (2021) Biomolecular condensates at the nexus of cellular stress, protein aggregation disease and ageing. Nat Rev Mol Cell Biol 22(3):196–213

Shin Y, Brangwynne CP (2017) Liquid phase condensation in cell physiology and disease. Science 357(6357)

Woringer M, Darzacq X (2018) Protein motion in the nucleus: from anomalous diffusion to weak interactions. Biochem Soc Trans 46(4):945–956

Sabari BR, Dall'Agnese A, Boija A et al (2018) Coactivator condensation at super-enhancers links phase separation and gene control. Science 361(6400)

Nair SJ, Yang L, Meluzzi D et al (2019) Phase separation of ligand-activated enhancers licenses cooperative chromosomal enhancer assembly. Nat Struct Mol Biol 26(3):193–203

Boeynaems S, Alberti S, Fawzi NL et al (2018) Protein phase separation: a new phase in cell biology. Trends Cell Biol 28(6):420–435

Gotor NL, Armaos A, Calloni G et al (2020) RNA-binding and prion domains: the yin and yang of phase separation. Nucleic Acids Res 48(17):9491–9504

Li L, McGinnis JP, Si K (2018) Translational control by prion-like proteins. Trends Cell Biol 28(6):494–505

Dignon GL, Best RB, Mittal J (2020) Biomolecular phase separation: from molecular driving forces to macroscopic properties. Annu Rev Phys Chem 71:53–75

Boija A, Klein IA, Sabari BR et al (2018) Transcription factors activate genes through the phase-separation capacity of their activation domains. *Cell* 175(7):1842-1855. e16

Hnisz D, Shrinivas K, Young RA, Chakraborty AK, Sharp PA (2017) A phase separation model for transcriptional control. Cell 169(1):13–23

Stortz M, Presman DM, Bruno L et al (2017) Mapping the dynamics of the glucocorticoid receptor within the nuclear landscape. Sci Rep 7(1):1–14

Stortz M, Pecci A, Presman DM, Levi V (2020) Unraveling the molecular interactions involved in phase separation of glucocorticoid receptor. BMC Biol 18:1–20

King JT, Shakya A (2021) Phase separation of DNA: from past to present. Biophys J

Strom AR, Brangwynne CP (2019) The liquid nucleome–phase transitions in the nucleus at a glance. J Cell Sci 132(22):jcs235093

Stortz M, Presman DM, Pecci A, Levi V (2021) Phasing the intranuclear organization of steroid hormone receptors. Biochem J 478(2):443–461

Htun H, Barsony J, Renyi I, Gould DL, Hager GL (1996) Visualization of glucocorticoid receptor translocation and intranuclear organization in living cells with a green fluorescent protein chimera. Proc Natl Acad Sci 93(10):4845–4850

Frank F, Liu X, Ortlund EA (2021) Glucocorticoid receptor condensates link DNA-dependent receptor dimerization and transcriptional transactivation. Proc Natl Acad Sci 118(30)

Schaaf MJ, Lewis-Tuffin LJ, Cidlowski JA (2005) Ligand-selective targeting of the glucocorticoid receptor to nuclear subdomains is associated with decreased receptor mobility. Mol Endocrinol 19(6):1501–1515

Kauppi B, Jakob C, Färnegårdh M et al (2003) The three-dimensional structures of antagonistic and agonistic forms of the glucocorticoid receptor ligand-binding domain: RU-486 induces a trans-conformation that leads to active antagonism. J Biol Chem 278(25):22748–22754

Voegel JJ, Heine MJ, Zechel C, Chambon P, Gronemeyer H (1996) TIF2, a 160 kDa transcriptional mediator for the ligand-dependent activation function AF-2 of nuclear receptors. EMBO J 15(14):3667–3675

Stenoien DL, Mancini MG, Patel K, Allegretto EA, Smith CL, Mancini MA (2000) Subnuclear trafficking of estrogen receptor-α and steroid receptor coactivator-1. Mol Endocrinol 14(4):518–534

Ahmed J, Meszaros A, Lazar T, Tompa P (2021) DNA-binding domain as the minimal region driving RNA-dependent liquid–liquid phase separation of androgen receptor. Protein Sci 30(7):1380–1392

Saravanan B, Soota D, Islam Z et al (2020) Ligand dependent gene regulation by transient ERα clustered enhancers. PLoS Genet 16(1):e1008516

Tanida T, Matsuda KI, Yamada S, Hashimoto T, Kawata M (2015) Estrogen-related receptor β reduces the subnuclear mobility of estrogen receptor α and suppresses estrogen-dependent cellular function. J Biol Chem 290(19):12332–12345

Fejes-Tóth G, Pearce D, Náray-Fejes-Tóth A (1998) Subcellular localization of mineralocorticoid receptors in living cells: effects of receptor agonists and antagonists. Proc Natl Acad Sci 95(6):2973–2978

Pearce D, Náray-Fejes-Tóth A, Fejes-Tóth G (2002) Determinants of subnuclear organization of mineralocorticoid receptor characterized through analysis of wild type and mutant receptors. J Biol Chem 277(2):1451–1456

Arnett-Mansfield RL, Graham JD, Hanson AR et al (2007) Focal subnuclear distribution of progesterone receptor is ligand dependent and associated with transcriptional activity. Mol Endocrinol 21(1):14–29

Muñoz-Gil G, Romero C, Mateos N et al (2020) Phase separation of tunable biomolecular condensates predicted by an interacting particle model. bioRxiv

Saitoh M, Takayanagi R, Goto K et al (2002) The presence of both the amino-and carboxyl-terminal domains in the AR is essential for the completion of a transcriptionally active form with coactivators and intranuclear compartmentalization common to the steroid hormone receptors: a three-dimensional imaging study. Mol Endocrinol 16(4):694–706

Zhang F, Wong S, Lee J et al (2021) Dynamic phase separation of the androgen receptor and its coactivators to regulate gene expression. bioRxiv

Duster R, Kaltheuner IH, Schmitz M, Geyer M (2021) 1, 6-hexanediol, commonly used to dissolve liquid–liquid phase separated condensates, directly impairs kinase and phosphatase activities. J Biol Chem 296

Wei M, Chang Y, Shimobayashi SF, Shin Y, Strom AR, Brangwynne CP (2020) Nucleated transcriptional condensates amplify gene expression. Nat Cell Biol 22(10):1187–1196

Weikum ER, Knuesel MT, Ortlund EA, Yamamoto KR (2017) Glucocorticoid receptor control of transcription: precision and plasticity via allostery. Nat Rev Mol Cell Biol 18(3):159–174

van Steensel B, Brink M, van der Meulen K et al (1995) Localization of the glucocorticoid receptor in discrete clusters in the cell nucleus. J Cell Sci 108(9):3003–3011

Li S, Yu K, Zhang Q et al (2021) dSCOPE: A software to detect sequences critical for liquid-liquid phase separation. bioRxiv

Shen B, Chen Z, Yu C, Chen T, Shi M, Li T (2021) Computational screening of biological phase-separating proteins. Genomics Proteomics Bioinformatics

Xue B, Dunbrack RL, Williams RW, Dunker AK, Uversky VN (2010) PONDR-FIT: a meta-predictor of intrinsically disordered amino acids. Biochim Biophys Acta Protein Proteomics 1804(4):996–1010

Alberti S, Gladfelter A, Mittag T (2019) Considerations and challenges in studying liquid-liquid phase separation and biomolecular condensates. Cell 176(3):419–434

McSwiggen DT, Hansen AS, Teves SS et al (2019) Evidence for DNA-mediated nuclear compartmentalization distinct from phase separation. elife 8:e47098

Chen W, Roeder RG (2011) Mediator-dependent nuclear receptor function. Semin Cell Dev Biol 22(7):749–758

Ranjan A, Ansari SA (2018) Therapeutic potential of mediator complex subunits in metabolic diseases. Biochimie 144:41–49

Torra IP, Freedman LP, Garabedian MJ (2004) Identification of DRIP205 as a coactivator for the farnesoid X receptor. J Biol Chem 279(35):36184–36191

Wang S, Lai K, Moy FJ, Bhat A, Hartman HB, Evans MJ (2006) The nuclear hormone receptor farnesoid X receptor (FXR) is activated by androsterone. Endocrinology 147(9):4025–4033

Cornelison JL, Cato ML, Johnson AM et al (2020) Development of a new class of liver receptor homolog-1 (LRH-1) agonists by photoredox conjugate addition. Bioorg Med Chem Lett 30(16):127293

Atkins GB, Hu X, Guenther MG, Rachez C, Freedman LP, Lazar MA (1999) Coactivators for the orphan nuclear receptor RORα. Mol Endocrinol 13(9):1550–1557

Wagh K, Garcia DA, Upadhyaya A (2021) Phase separation in transcription factor dynamics and chromatin organization. Curr Opin Struct Biol 71:148–155

Klosin A, Oltsch F, Harmon T et al (2020) Phase separation provides a mechanism to reduce noise in cells. Science 367(6476):464–468

Lancaster AK, Nutter-Upham A, Lindquist S, King OD (2014) PLAAC: a web and command-line application to identify proteins with prion-like amino acid composition. Bioinformatics 30(17):2501–2502

Barnes PJ, Drazen JM, Rennard SI, Thomson NC (2009) Asthma and COPD: basic mechanisms and clinical management. Elsevier

15 Prostate Cancer Epigenetic Plasticity and Enhancer Heterogeneity: Molecular Causes, Consequences and Clinical Implications

Jeroen Kneppers, Andries M. Bergman, and Wilbert Zwart

Abstract

Prostate cancer (PCa) proliferation is dictated by androgen receptor (AR) signaling, which regulates gene expression through *cis*-regulatory regions including proximal and distal enhancers. The repertoire of AR interactions at enhancers is dependent on tissue and cellular contexts and thus shape a spectrum of phenotypes through such epigenetic heterogeneity. Moreover, PCa is a multifocal disease and displays a high degree of intra- and inter-tumor heterogeneity, adding to the phenotypic complexity. It is increasingly becoming clear that PCa may be considered an epigenetic disease caused by various molecular causes with profound consequences and clinical implications which are underpinned by enhancer interaction heterogeneity.

In this review, we provide a detailed overview of molecular interactors that affect prostate cancer epigenetic heterogeneity, such as coding and non-coding somatic variants, large scale structural variations, pioneer factor binding at enhancers and various contexts that influence enhancer engagement heterogeneity in PCa development and progression. Finally, we explore how the vast heterogeneity in epigenetic profiles identified in recent omics studies results in distinct genomic subtypes which predict disease progression and thus offer opportunities in biomarker discovery and further personalizing cancer treatment. As such, heterogeneous enhancer interactions take center stage in elucidating mechanisms of prostate cancer progression, patient prognostication, therapy discovery and overcoming acquired treatment resistance.

Andries M Bergman and Wilbert Zwart share senior authorship.

J. Kneppers
Division of Oncogenomics, Oncode Institute, Netherlands Cancer Institute, Amsterdam, The Netherlands

A. M. Bergman
Division of Oncogenomics, Oncode Institute, Netherlands Cancer Institute, Amsterdam, The Netherlands

Division of Medical Oncology, Netherlands Cancer Institute, Amsterdam, The Netherlands
e-mail: a.bergman@nki.nl

W. Zwart (✉)
Division of Oncogenomics, Oncode Institute, Netherlands Cancer Institute, Amsterdam, The Netherlands

Laboratory of Chemical Biology and Institute for Complex Molecular Systems, Department of Biomedical Engineering, Eindhoven University of Technology, Eindhoven, The Netherlands
e-mail: w.zwart@nki.nl

M. J. Campbell, C. L. Bevan (eds.), *Nuclear Receptors in Human Health and Disease*, Advances in Experimental Medicine and Biology 1390, https://doi.org/10.1007/978-3-031-11836-4_15

Keywords
Prostate cancer · Enhancers · Androgen receptor · Epigenetics · Heterogeneity · Cistrome plasticity · Personalized medicine

Abbreviations

AR	Androgen receptor
ARBS	Androgen receptor binding site
ARE	Androgen response element
ADT	Androgen deprivation therapy
ChIP-seq	Chromatin immunoprecipitation followed by sequencing
CRPC	Castration resistant prostate cancer
DHT	Dihydrotestosterone
GWAS	Genome wide association study
NR	Nuclear receptor
PCa	Prostate cancer
PDX	Patient derived xenograft
PIN	Prostatic intraepithelial neoplasia
PTM	Post-transcriptional modification
SE	Super enhancer
SNP	Single nucleotide polymorphism
TAD	Topologically associating domains
TF	Transcription factor
TSS	Transcriptional start site

15.1 Introduction

Enhancers were first discovered in 1981 when researchers of two independent groups found simian virus (SV40) DNA sequences 3kb distal to the SV40 promoter capable of stimulating transcriptional output of a linked β-globin gene by 200-fold when transfected in mammalian cells [1, 2]. These experiments generated a more complete understanding of how gene regulation emerges from an interplay between often distally located enhancers and proximal promoter regions. An onset of subsequent studies discovered not only that enhancers are general genomic features in a variety of organisms including mammals [3–7], but also that defects in enhancers can lead to pathogenesis [8–11]. Although the human genome contains approximately 20,000 protein coding genes [12], currently roughly fifty times more non-coding regulatory regions have been described across tissue types [13, 14], prompting a reassessment of non-coding genome functionality. Moreover, genome-wide association studies (GWAS) have shown that variants involved in human disease are enriched at non-coding regulatory elements over coding sequences [15]. Early genome-wide studies identified the total repertoire of promoter and enhancer sequences based on a combination of ChIP-seq and chromatin accessibility assays with specific histone modifications such as high H3K27ac signal [16–18], whereas later a high proportion of H3K4 mono- versus tri-methylation allowed researchers to separate enhancers from promoters [19, 20].

Successive research endeavors characterized enhancer sequences to have the following properties: (1) activated enhancers mediate strong transcriptional activation of the gene it controls [1, 2], (2) activation is independent from the orientation of the enhancer element [1, 2], (3) enhancers function in a tissue specific manner [3], (4) enhancer sequences are bidirectionally transcribed as short enhancer RNA (eRNA) transcripts [21], (5) enhancers possess regulatory multiplicity, in which a single enhancer can activate multiple promoters of linked genes, whereas multiple enhancers can also regulate a single promoter [22, 23], (6) activation can be exerted in *cis* over genomic distances up to megabases away [24], (7) enhancers are scattered throughout 98% of the non-coding human genome [25]. The last property was an unexpected finding of modern genome sequencing and annotation by the Encyclopedia of DNA Elements (ENCODE) project, showing that a large proportion of the non-coding genome has regulatory control over the expression of the coding genome [25]. Interestingly, changes in non-coding regulatory elements are frequently observed in oncogenesis [10, 11, 26].

15.2 Prostate Cancer as Enhancer-Driven Disease

Prostate cancer (PCa) is the second-most commonly diagnosed malignancy in men worldwide [27]. PCa is mainly driven by the nuclear receptor androgen receptor (AR) [28], that acts as a master transcription regulator of cell proliferation when bound to its cognate ligand dihydrotestosterone (DHT) [29, 30]. While blockade of the AR signaling axis using androgen deprivation therapy (ADT) as a first line of treatment is initially successful [31, 32], over time resistance to ADT inevitably occurs and remaining cancer cells rebound as lethal castration resistant prostate cancer (CRPC) [33, 34]. AR signaling persists during CRPC despite castration level circulating testosterone, which highlights the essentiality of AR signaling in PCa cells. Sustained successful PCa treatment is challenged by the heterogeneous nature of PCa, which is present on multiple levels (Fig. 15.1).

PCa is a multifocal disease with ~60–90% of patients presenting multiple independent primary tumor foci at time of diagnosis [35–38]. Such foci exhibit inter-lesion heterogeneity, which

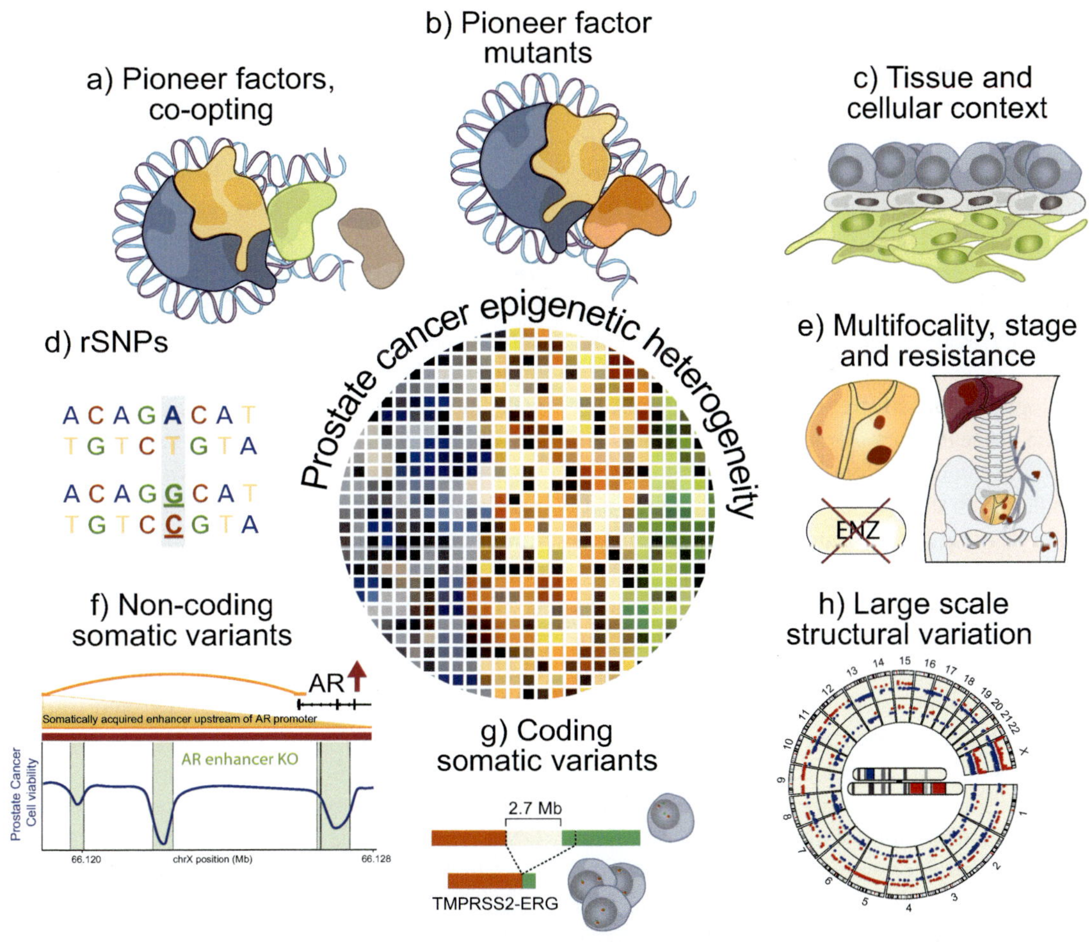

Fig. 15.1 Overview of sources of prostate cancer epigenetic heterogeneity: Epigenetic heterogeneity in AR enhancer interactions varies in different tissues and cellular contexts and can be induced by (**a**) pioneer factor binding at chromatin and (**b**) pioneer factor mutations altering chromatin binding properties, (**c**) tissue and cellular context, (**d**) germline PCa risk single nucleotide polymorphisms, (**e**) PCa multifocality, disease stage and acquirement of therapy resistance, (**f**) non-coding and (**g**) coding somatic variants and (**h**) large scale structural variations that amplify or delete genomic regions

manifests in differences in cell morphology, tumor microenvironment and degrees of aggressiveness [39, 40]. Contrastingly, metastatic PCa lesions were reported to predominantly share a homogeneous, monoclonal background [41]. While primary local interventions, such as radiotherapy and prostatectomy, affect the entire prostate and treat all foci successfully, these treatments are associated with significant adverse effects [42, 43]. An alternative approach revolves around limited local treatments that ablate only the largest tumor focus while sparing the prostate and limiting adverse effects. However, these strategies are complicated by PCa heterogeneity, as remaining lesions may still metastasize at a later stage [44, 45]. Second, intratumoral heterogeneity is observed in genetically diverse cell populations within a single tumor focus and arises from tumor microenvironmental cues, lineage plasticity, as well as genetic and epigenetic defects [46–50]. Genomic inter-tumor heterogeneity manifests itself in the shape of small-scale genetic mutations like single nucleotide variants (SNVs), while copy number alterations (CNAs) and translocations of large-scale genomic elements are even more likely to impact tumor development [46, 51, 52]. Third, such events also impact *cis*-regulatory elements such as enhancers that tightly control expression on the same DNA strand, which disrupts epigenetic regulatory networks leading to profound phenotypic differences and loss of cellular identity [53].

An increasing amount of evidence illuminates a role for heterogeneous epigenetic regulation in PCa through AR [37, 54, 55], but how can intra- and intertumoral heterogeneous enhancer interactions shape a spectrum of phenotypes and outcomes in PCa? As heterogeneity seems to be pervasive in tumors, one can ask the question what the contributions of different sources of heterogeneity in the progression of PCa are. Clearly, research questions and efforts have converged on elucidating the role of AR as oncogenic driver and the emergence of resistance. Can we apply such knowledge of AR chromatin interaction profiles and their dysregulation to attempt overcoming resistance by optimizing and personalizing PCa treatment based on heterogeneity? In this review, we aim to address these questions by providing a comprehensive overview of recent progress that has been made on this subject and indicate which therapeutic avenues future research might illuminate.

15.3 AR Biology and Enhancer Regulation in Prostate Cancer

Historically, nuclear receptors were investigated in the context of their activity at promoter elements. For AR and PCa, prostate specific antigen (PSA; encoded by *KLK3*) represents a highly characterized example of AR promoter binding, with specificity to prostate tissue and high androgen inducibility [56, 57]. However, later studies revealed that AR binding at promoters is an exception and represents a relatively rare event, as compared to AR binding at enhancers [58]. Activated steroidal (Type I), nuclear receptors like AR possess the capacity to regulate transcription of target genes through binding at enhancer elements that are located distally from a target gene's transcriptional start site (TSS) [59, 60]. Such distal regulation offers tight, but also highly modular control of transcription in response to hormonal cues, with many co-regulators involved in transcriptional output [61]. Specifically, AR becomes activated upon binding with its cognate ligand dihydrotestosterone (DHT) in the cytosol, dimerizes and subsequently translocates to the nucleus where it binds to AR binding sites (ARBS) [29, 62]. Although AR's DNA binding domain recognizes and binds androgen response elements (AREs) consisting of dihexameric palindromes on the DNA [63], ARE presence is not a strict requirement for DNA binding, since AR cooperates with interacting TFs bound at AP-1, MYC, KLF and SREBF motifs [64, 65].

Recruitment of co-factors to enhancers is required for DNA looping and subsequent enhancer-promoter interactions. Factors bound at enhancers provide scaffolding for the large mediator complex to bind transiently and further recruit the transcriptional machinery [66–69].

Indeed, mediator's MED1 subunit contains LXXLL binding motifs that strongly interacts with the AR-AF2 domain in a ligand-dependent manner [70] and recently a cryo-EM study reported steroid receptor coactivator 3 (NCOA3/SRC-3) to interact with an FXXLF binding motif in AR's N-terminal domain, enabling p300/CREB-binding protein (CBP) recruitment [71]. Since mediator recruits RNA polymerase II (RNAPII) and activates expression at promoters, enhancers can affect expression over large distances without direct promoter contact, which was demonstrated for PCa and AR by collaboration with ERG [72]. We provide a graphical overview of proteins involved in AR promoter-enhancer interactions in Fig. 15.2.

Additionally, transcription also occurs at enhancer loci when active AR complexes recruit RNAPII polymerases [73]. In contrast to RNAPII activity at gene-coding promoters resulting in mRNAs, bidirectional transcription at RNAPII-occupied enhancers gives rise to small, unstable eRNAs [21]. Ascribing specific functionality to a number of eRNAs has succeeded in the context of gene expression [74, 75] and fine-tuning co-activator function at gene promoters [76]. Although defining general functionality of eRNAs remains challenging, TF activity at enhancers can be inferred through RNAPII stochastic models quantifying co-localization of TF binding motifs and eRNAs [77, 78]. These findings were further corroborated by transgenic embryonic assays, showing that enhancer functionality can be predicted by the level and directionality of eRNA transcription [79]. Finally, combining RNA-seq with chromatin accessibility data through ATAC-seq has been used to map eRNA transcript abundance on a genome-wide scale in neuronal cell populations in different activation states, providing first evidence that eRNA function is dependent on genomic context and partially dependent on sequence [80]. Next to eRNA transcription at enhancers, other studies also revealed the existence of large and dynamic transcriptional hubs at highly active loci of TF binding [81–83]. Such loci containing many active enhancer elements often regulate key differentiation processes in development and tissue

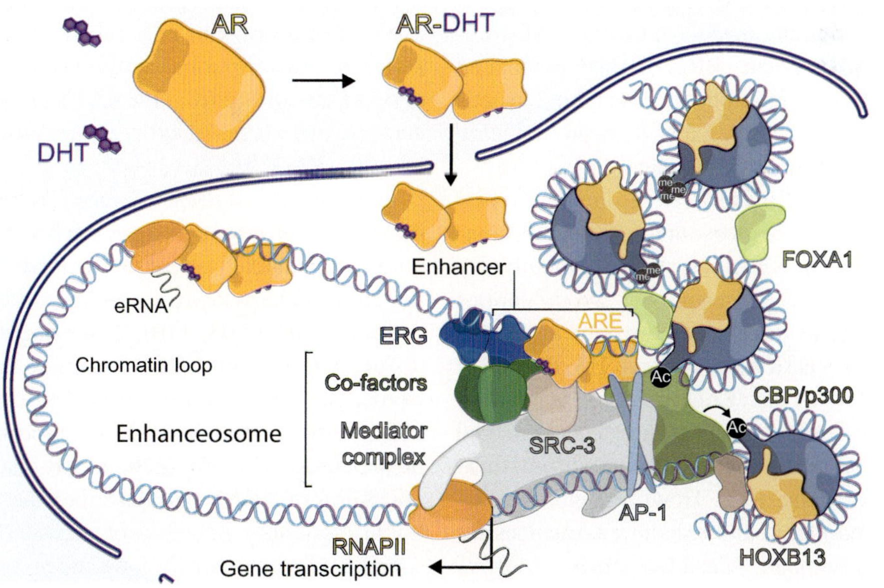

Fig. 15.2 Graphical overview of AR action at enhancers: AR binds DHT and dimerizes in the cytoplasm prior to nuclear translocation. Pioneer factor FOXA1 opens chromatin wrapped tightly around histones, allowing AR dimers to bind the chromatin through AREs and other regulatory elements. Co-factors and transcriptional machinery components such as SRC-3, CBP, p300, AP1, mediator complex and RNAPII are recruited to facilitate gene transcription, while RNAPII activity at AR-bound enhancers results in bidirectional transcription of eRNAs

identity and have been dubbed 'super enhancers' (SEs). Since clusters of enhancers in close proximity recruit many TFs, SEs form phase-separated condensates [84] with a local high-density biomolecule assembly of RNAPII [83], co-activators MED1, BRD4 [82, 83] and KLF4 [85]. However, the true number, distribution, and the proposed synergistic transcriptional activation of SEs is a matter of ongoing research and scientific debate [86].

15.4 PCa-Specific Pioneer Factors as Source of Regulatory Heterogeneity in AR Binding

Transcriptionally silent chromatin is required for maintaining correct cellular identity dictated by a specific subset of genes transcribed from active chromatin, tightly regulating cell fate decisions. Pioneer factors like forkhead box protein A (FOXA1) open up condensed chromatin [87], so that transcription factors (TFs) and ultimately transcriptional coactivator complexes such as CBP and p300 [88, 89] and other coregulators like SRC-3 can bind [90, 91]. Additionally, SWI/SNF chromatin remodelers (or human BAF complex: ATP-dependent BRG1/BRM associated factors) and other co-modulators can bind to activate and repress expression through inducing chromatin conformation changes [61].

Transcriptionally inactive chromatin or heterochromatin is nucleosome-dense and compactly folded DNA characterized mainly by histone tail post-transcriptional modifications (PTMs) of up to three methyl groups at histone H3 lysine 9 (H3K9me1-3) and H3 lysine 27 (H3K27me1-3) [92]. Consequently, gene transcription is silenced as TFs are physically blocked by nucleosomes from binding heterochromatin at enhancer elements [92]. However, pioneer factors open chromatin and enhancer sequences for TF binding [87]. In PCa development, FOXA1 and homeobox B13 (HOXB13) expression levels are increased while their mode of action is reprogrammed, allowing for altered regulation of AR-mediated transcription [61, 93]. Additionally, GATA2 and OCT1 have also been found to cooperate with AR to mediate androgen response in PCa growth [58, 94].

As a result of these functions, pioneer factors facilitate AR binding through nucleosome displacement, thereby inducing an open chromatin conformation which is characterized by 'active' enhancer histone modifications and which is permissive to TF binding [95–98]. AR binding at DNA is mostly pioneered by FOXA1 binding to chromatin, marked by hypomethylated DNA and presence of histone modifications H3K4me1 and H3K4me2 [99–101]. FOXA1 was first identified as an AR interactor when FOXA1 binding motifs were found located adjacent to ARBS in prostate gene regulatory regions for human PSA and rat probasin (PSA orthologue) [102]. Additionally, AR's DNA binding domain interacts directly with FOXA1's forkhead domain [102, 103]. Genome-wide FOXA1-bound sites were shown to be cell-line specific and differentially functional between breast and PCa cell lines [99, 104], with genome-wide FOXA1 binding at the majority of ARBS later confirmed specifically in PCa cell lines LNCaP and VCaP [105, 106]. Interestingly, silencing of FOXA1 triggers a switch in AR binding at ARBS, altering gene expression profiles in PCa cell lines [105–108]. As such, transcriptional activity of diverse gene networks resulting from FOXA1's pioneer factor activity, are tissue-specific and control cellular identity [87, 109].

Interestingly, ARBS are rarely found at promoters, as the vast majority of ARBS are found at putative enhancer sequences located distally of the target gene's locus depending on tissue and cellular context [93, 110]. Taken together, such distal *cis*-regulatory ARBS constitute the AR cistrome; the term cistrome was first coined in a 2008 study on FOXA1 and ERα binding sites in breast cancer [99]. As such, an AR cistrome is a collection of ARBS that describes the transcriptional regulatory potential of activated AR in a specific context, which have been extensively reported in many different contexts such as healthy prostate tissue, PCa cell lines and tissues from varying stages of PCa [61, 93, 110–114]. Additionally, AR cistromes also vary in different cell type contexts like fibroblasts [115], macrophages [116], male breast cancer [111] and female breast cancer [117]. In this review, we focus on AR function in prostate epithelial cells, mostly in the context of PCa. In the following section, we address the question of how context-dependent AR cistromes influence PCa heterogeneity and how shifts in AR cistromes affect tumor progression.

15.5 AR Cistromes are Heterogeneous Between Different Tissue, Cellular and Tumor Contexts

Prostate development is a complex process dependent on the presence of androgens and developmental pathways requiring activation of diverse genes at different stages and tissue identities [118]. As such, various types of prostate tissue are thought to be driven by different AR cistromes during development and tissue maintenance, but also during tumor initiation [119, 120]. One of the first studies to dissect the differences between AR cistromes in prostate tissue compared histologically normal prostate tissues with prostate cancers, which were both enriched in epithelial cell content [93]. A core set of tumor associated ARBS (T-ARBS) was found to co-localize with FOXA1 and HOXB13 binding, which was absent at normal associated ARBS (N-ARBS), providing the first clinical evidence of AR cistrome reprogramming [93].

Furthermore, overexpression of FOXA1 and HOXB13 in benign prostate cells induced a change in AR cistrome reminiscent of reprogramming in PCa cells, showing that in tumorigenesis HOXB13 may act as a pioneer factor and induces different AR cistromic repertoires that influence disease progression [121, 122]. This finding was later confirmed by a study that found somatic structural variants to impact master TF *cis*-regulatory regions, altering binding for various factors including AR, FOXA1, HOXB13 and SOX9, which in turn may influence prostate oncogenesis [123]. Additionally, such a malignancy-associated shift in AR signalling can also be pioneered by GATA2 and c-JUN [58, 124, 125]. GATA2 is a zinc-finger TF that normally regulates developmental gene expression but also influences AR chromatin binding by enabling access to additional putative ARBS prior to androgen stimulation [94]. Newly accessible ARBS include those near the AR locus, resulting in a GATA2 pioneered elevation of AR expression, which can further be enhanced by co-occupancy by FOXA1 at GATA2-pioneered sites [94]. c-JUN dimerizes with FOS to form the AP-1 complex which transactivates gene expression of PCa driver ETV1 [124]. Moreover, c-JUN's expression levels were found to correlate with AR transcriptional activity and knockdown of c-JUN abrogated AR-dependent PCa cell proliferation [64, 126]. Although c-JUN can control AR binding and has been implicated in AR malignancy shift, pioneering activity by c-JUN has not been formally proven. Taken together, an ensemble of TFs modulates AR through enabling chromatin accessibility at newly activated ARBS, thereby expanding the repertoire of possible AR cistromes that are associated with a context-dependent PCa AR signalling malignancy shift.

Acquired cancer therapy resistance is deeply rooted in inter- and intra-tumor heterogeneity, in which a certain cell population manages to overcome and adapt to therapy-induced selection over other populations [49]. In androgen-depleted conditions, PCa cell subpopulations that lose prostate differentiation while gaining resistance to AR signaling inhibition have been shown to survive and acquire an aggressive pathological phenotype [127]. As such, tumor progression can be viewed as an evolutionary dynamic process, in which tumor cells not only reprogram epigenetic control of cell identity or acquire a new phenotype, but also communicate differentially with their tumor microenvironment (TME) [50, 128]. While PCa cell lines -mostly derived from patients with advanced disease- are typically typical studied in the absence of a TME context, recently a push has been made to boost the diversity of clinical stages represented in PCa models in which a TME is present, using patient-derived xenografts (PDXs) [129].

Diverse PCa cell lines and PDX models contain ARBS that are shared, but there are also ARBS that are specifically found in a single cell line, that partly recapitulate the intrinsic inter-patient heterogeneity [113]. Although AR cistromes in prostatic epithelial cells and tissues take center stage, AR cistromes are also heterogeneous between cell types of the prostate TME, which can interact with tumors and influence growth [47]. AR cistromes of PCa stroma constituent cells like fibroblasts and macrophages have been dissected and were found to deviate from AR cistromes reported in epithelial cells [115, 116, 130]. The context dependency of the AR cistrome in these TME-associated cell constituents functionally contributes to PCa progression by affecting PCa migration potential or by supporting PCa invasiveness through AR signaling.

On a final note, diverse AR cistromes are also found in both ER^+ and ER^- (molecular apocrine) breast cancer. AR cistromes in both breast cancer subtypes are also facilitated by FOXA1, yet with opposing forces on tumor driving potential, with AR acting as driver in ER^- but as tumor suppressor in ER^+ breast cancers [111, 117, 131, 132]. Clearly, the topic of cancer cistrome heterogeneity is wide-ranging and has been reviewed previously [133–135]. Therefore, we will focus on which AR cistromic heterogeneity occurs within the different stages of PCa progression from initiation to development of metastatic CRPC.

15.6 AR Cistromic Heterogeneity Progressively Develops from PCa Initiation to Neuroendocrine Differentiation

Early stage primary PCa is confined to the prostate, with lesions initiating in the glandular tissue lesions in the form of prostatic intraepithelial neoplasia (PIN) lesions in which DNA damage caused by oxidative stress and inflammation in the prostate gland plays an important role [136–138]. PCa tumorigenesis is genomically characterized by the occurrence of SNVs, small deletions and gene fusions, while AR activity is highly heterogeneous among tumors [37, 54]. Interestingly, different primary tumor foci in the same prostate rarely share SNVs or structural variation at regulatory elements, further highlighting the multiclonal heterogeneous nature of primary tumors [123]. SNV accumulation in tumor foci was also found to rarely drive pro-oncogenic processes, providing a potential explanation for PCa indolence [51, 139, 140]. However, the myriad of SNVs present at regulatory elements alter the transactivation potential of enhancers, especially of those regulating master TF activity [123]. Moreover, primary tumors do have an enrichment of SNVs in ARBS that are somatically acquired in tumors, thus providing a source of genetic heterogeneity in PCa that may affect epigenetic regulation [123]. To study epigenetic regulation in PCa, we previously undertook epigenetic analyses to dissect AR cistrome heterogeneity in primary tumors by integrating gene expression data with AR cistrome data with enhancer-mapping histone modification marks (H3K27ac, H3K27me3 and H3K4me3) [55].

Three major epigenetic subtypes were revealed in primary PCa tissues, two of which were dominated by TMPRSS2-ERG fusion status, while a third was characterized by low activity and chromatin binding of AR, but with high WNT and FGF signalling [55]. TMPRSS2-ERG fusions lead to a particularly reprogrammed cistrome, as evidenced by a different H3K27ac profile that enables co-opting of ERG of AR, FOXA1 and HOXB13 resulting in AR cistromic heterogeneity [141]. Although AR profiles in primary disease do not appear to have prognostic potential by themselves, AR cistrome reprogramming continuously occurs during disease progression [55]. Somatic structural variants, such as either TMPRSS2-ERG gene fusions or coding mutations in FOXA1 and SPOP are also found associated with AR cistrome plasticity and are discussed in-depth later.

15.7 Metastatic PCa Heterogeneity

PCa mortality is predominantly caused by metastatic disease, in which tumor cells preferentially spread from a primary lesion to locoregional lymph nodes and bones [27, 142, 143]. Somatically acquired large-scale structural enhancer variants are common in cancer [144]; a process which accelerates in metastatic disease [145] and affects TF binding, chromatin organization and gene expression [146]. In metastatic PCa, large scale structural variations at either coding or cis-regulatory sequences represent a class of key oncogenic events often coupled with copy number alterations (CNAs) such as gains at critical oncogenes including AR, MYC, CDK12, or losses at tumor-suppressor genes including TP53 and BRCA2 [147]. Recently, a study was reported that integrated pan-cancer genomics data with clinical information and functional genome-scale CRISPRi screens in metastatic PCa models to discover additional drivers of metastatic PCa, revealing that KIF4A knockdown alters genome-wide chromatin accessibility and acts as a driver of metastatic PCa aggressiveness with concomitant poor prognosis [148].

Prognostication of PCa patients based on pathological and genomic biomarkers could distinguish those patients with high-risk of developing aggressive disease over those with indolent PCa, paving the way for prognostication based on epigenetic status [149–151]. Another study from our group compared genome-wide AR binding, chromatin accessibility and gene expression between primary PCa and ADT-resistant tumors and integrated these with publically available clinical and genomic cancer databases [151]. The resulting gene expression signature could predict outcome in primary PCa patients in independent cohorts, suggesting that an underlying pro-metastatic AR cistrome may already be present in patients with primary patients whose disease eventually progressed [151]. This notion was further supported by a study that epigenetically profiled tissues in the disease progression spectrum from normal prostate epithelium to primary PCa to metastatic disease [121]. Normal prostate epithelium already displays regulatory elements that are prepopulated by FOXA1 and HOXB13, which AR later binds in metastatic PCa to drive fetal prostate developmental programs [121]. These two studies together underline the relevance of studying PCa state transitions epigenetically as a crucial method to understand molecular underpinnings underlying PCa progression and it critically suggests that inter-tumor PCa heterogeneity is strongly associated with cistromic heterogeneity.

Difficult to treat metastatic castration resistant prostate cancer (mCRPC) arises once metastatic PCa growth has been restored through reactivation of AR signaling pathways in an ADT-induced, low testosterone environment [152]. mCRPC is characterized by a distinct AR cistrome that is reprogrammed by CRPC specific TFs such as STAT, MYC and E2F, while such heterogeneity is not captured by cell lines but only found in tissues [113]. Later, a first report on AR, FOXA1 and CTCF binding in multiple metastatic tumors in an individual patient confirmed a robust, metastasis-specific transcriptional program despite few inter-lesion differences in the AR cistrome, showing that the metastatic AR cistrome between different affected organs is surprisingly similar [153]. Potent AR inhibitors such as enzalutamide and darolutamide are administered to suppress the AR signaling axis after CRPC emerges [154, 155]. Under the pressure of such therapies, mCRPC can further differentiate towards lethal neuroendocrine prostate cancer (NEPC) in the last stages of PCa, which rarely arises *de novo* and is characterized by absent AR signaling, neuroendocrine marker expression and loss of TP53 and RB1 [156, 157]. Additionally, neuroendocrine differentiation is characterized by a concomitant aberrant global shift in DNA methylation and altered expression of epigenetic modifiers and TFs [156, 158, 159]. Support for such epigenetic deregulation in NEPC was recently reported in genetically engineered NEPC mouse model by using single cell transcriptomics and chromatin accessibility methods, which revealed that *Ascl1* and *Pou2f3* are differentially regulated in dedifferentiated cell populations marked by shifts in global DNA methylation

[160]. Moreover, the FOXA1 cistrome is extensively reprogrammed during NEPC [161]. Taken together, an image emerges in which enhancer plasticity in each of the different PCa stages leads to adaptation and progression through rewiring of AR cistromes.

15.8 Non-coding and Protein Coding Somatic Mutations Induce AR Cistromic Heterogeneity

Somatic mutations are a prominent feature of metastatic PCa [147, 162–164], in which AR plays a key role. A multitude of studies reported that in the metastatic disease setting CNAs can lead to the amplification of a SE cluster driving AR expression, providing evidence for *de novo* rewiring of the AR cistrome as a powerful oncogenic driver [147, 162, 165]. Moreover, it was recently reported that AR binding sites are highly mutated in PCa, potentially due to faulty base excision repair at abasic sites [166]. Similarly, during NEPC differentiation, the FOXA1 promoter loses regulatory contact with its key enhancer while simultaneously acquiring *de novo* regulation from a further distally located super-enhancer [161]. Therefore, somatic mutations in pioneer factor binding sites represent another distinct class of non-coding somatic mutations causing epigenetic heterogeneity in PCa.

Conversely, FOXA1 protein coding somatic mutations are frequently occurring across disease stages [54, 167], with a substantial subset of primary PCa, mCRPC and NEPC tumors harboring recurrent SNVs in the FOXA1 coding sequence [168–170]. SNVs in FOXA1 that alter its pioneering function are mostly truncations, indels and missense mutations that converge on three mutational hotspots: the Wing2 region, the forkhead DNA binding domain and C-terminal truncations [171, 172]. Firstly, Wing2 hotspot mutants make up roughly half of all FOXA1 coding mutations which are enriched in the primary stage of PCa, suggesting emergence during localized disease. Moreover, Wing2 mutants exhibit greater pioneering activity than the effect of overexpression of wild-type FOXA1 [171, 172]. Secondly, forkhead DNA binding domain mutation R219 affects a highly conserved part of the forkhead domain that contacts the DNA, altering pioneering activity and activating a mesenchymal/neuroendocrine transcriptional program driven by WNT-signaling [171, 172]. Interestingly, $FOXA1^{R219}$ is acquired in PCa transitioning from primary to metastatic disease and its binding motifs differ markedly from canonical FOXA1-binding motifs, shutting down normal luminal differentiation programmes [171, 172]. Finally, 20% of FOXA1 mutations are frameshift truncations that result in loss of FOXA1's C-terminal transactivating domain. Such mutants show markedly higher DNA binding affinity resulting in altered chromatin binding, engaging an expanded total cistrome for FOXA1 [171–174]. Taken together, FOXA1 mutations are powerful drivers of AR cistromic reprogramming and plasticity by co-opting novel ARBS and transcriptional programs.

Another powerful and frequently recurring oncogenic driver in AR cistromic rewiring is the TMPRSS2-ERG fusion event that occurs in ~50% of patients and is a common initiator of prostate tumorigenesis [175–178], while tumor suppressor PTEN loss co-occurs with TMPRSS2-ERG in aggressive metastatic PCa [179–182]. Specifically, the promoter of TMPRSS2 is fused to the proto-oncogenic transcription factor ERG (ETV1, 4 or 5), causing aberrant overexpression of ERG that in turn drives a PCa oncogenic transcriptional program through ERG-mediated AR recruitment at novel and existing ARBS [110, 141, 178, 183]. Moreover, overexpressed ERG was recently reported to co-opt AR and FOXA1 bound sites to drive expression of DLX, a homeobox-containing TF whose elevated expression is linked to aggressive metastatic disease [184]. These findings further highlight the biological role of TMPRSS2-ERG fusions in advanced PCa beyond its better-understood role in primary disease.

Moreover, mutations occurring in speckle-type pox virus and zinc finger protein (SPOP) were proposed to further exacerbate ERG-driven PCa [185], since the E3 ubiquitin ligase SPOP is a tumor suppressor gene and frequently mutated in PCa [168, 186, 187]. Wild-type SPOP promotes ubiquitination and subsequent proteolytic degradation of critical PCa drivers including ERG [185, 188], AR [189, 190], Myc [191], BRD4 [192, 193] and SRC-3 [194], while SPOP's suppressing function is disrupted by binding cleft mutations [90, 189, 194], leading to a reprogrammed AR cistrome [195]. For instance, SRC-3's oncogenic role as steroid receptor coactivator in PCa is supported by its association with poor prognosis and aggressive phenotype [90, 91, 196, 197]. SRC-3 was proven to associate with AR at enhancers under androgen stimulation, increasing PSA expression [198] and later to be involved in expression of many AR-driven genes [199]. Many proliferation pathways are activated by SRC-3, amongst which MAPK/ERK signaling [200, 201] and Akt-mTOR signaling in PCa cells [91], while homozygous SRC-3 knockout in mice leads to PCa tumor growth arrest and prolonged survival [202].

Interestingly, co-occurring SPOP and ERG mutations are mutually exclusive [203] and the initially proposed SPOP-mutant stabilization was later explained as case of synthetic lethality that prevents appearance of this phenotype [204]. Bromodomain histone reader ZMYND11 is stabilized by mutated SPOP which in turn represses ERG function [204], further corroborating earlier observed paradoxal antagonism of ERG on AR signaling through auto-inhibitory PRMT5 methylation of AR [110, 205]. Additionally, an LXXLL AR interacting motif in the ETS domain of ERG was identified with affinity similar to AR coactivating peptides [206] through mutational studies and ERG-stimulated AR activation, suggesting that AR and ERG can directly interact resulting in a reprogrammed AR cistrome [207].

15.9 Risk SNPs and Somatic Mutations are Enriched at AR-Bound Enhancers

Another source of heterogeneity in AR cistromics comes in the form of germline and somatic sequence variation. With 80% of the cancer risk single nucleotide polymorphisms (rSNPs) [208] mapping to intronic and intergenic regions, a relatively large subset of these are enriched in *bona fide* enhancer elements over other non-coding regions when correcting for size [15, 26]. PCa genome-wide association studies (GWASs) and subsequent studies functionally annotated rSNPs as risk enhancers [209], associated rSNPs with higher risk of developing disease [210] and catalogued rSNPs found from a large pool of PCa tumors [211]. All studies report overrepresentation of rSNPs in enhancer elements that are linked to PCa master TFs with potential transcriptionally altering consequences. Further screening using high-throughput measurement of protein-bound oligo retention times, in which TFs in nuclear extracts bound to SNP-containing oligos are pulled down, found that 20 rSNPs were associated with decreased AR binding in LNCaP [212]. Interestingly, one rSNP was located at the center of a cluster of AR, HOXB13 and FOXA1 binding sites, of which specifically FOXA1 binding was decreased which translated to lower regulatory and transcriptional potential of PCa oncogene RGS17 [212].

Similarly, some PCa rSNPs within well-characterized enhancers influence PCa cell viability [123], as exemplified by enhancers that are located in a single topological associating domain regulating MYC [213, 214]: PCAT1 and PCAT2 [215–218]. Another high-throughput epigenomic study provides evidence that rSNPs create or perturb TF binding sites including AR, as exemplified by a rSNP abrogating AR-mediated repression of the putative oncogene CDKN2B-AS1 which influences cell cycle regulation [219]. Generally, heritable PCa risk is associated with a strong enrichment of PCa rSNPs in prostate-lineage specific enhancers

[121]. As such, rSNPs contribute to AR cistromic heterogeneity by perturbing and creating TF binding sites that affect PCa progression.

15.10 Clinical Implications and Biomarker Development of Heterogeneity in Epigenetic Subtypes

It is increasingly becoming more apparent that PCa may be considered an epigenetic disease in which many key cell identity processes are disrupted and different transcriptional programs are initiated through AR cistromic rewiring, orchestrated by reprogrammed FOXA1 and HOXB13 [121, 161, 220]. The future clinical potential of targeting enhancer-gene pairs in cancer is promising, as such interactions have been systematically charted for in the TCGA pan-cancer dataset, with aberrant enhancer activation observed in most cancers [221]. Since aberrant enhancer activation and cistromic heterogeneity appears to be a key feature of PCa, specific epigenetic states and biomarkers ensuing from such states offer great opportunities for informed clinical decisions based on epigenetic subtypes.

Our previous integrative epigenetic profiling study in primary prostate cancer has revealed a PCa subtype independent of TMPRSS2-ERG status, characterized by low mutational burden together with neutral copy number and AR expression but a contrastingly low AR activity and chromatin binding [55]. Since this subtype with heterogeneous TMPRSS2-ERG status is potentially driven by NGF, FGF and WNT signaling and associated with poor outcome [119], therapeutic opportunities may exploit applying small molecule inhibitors (SMIs) targeting these pathways [222–224], particularly for this subpopulation of patients. Further comparing AR chromatin binding patterns between disease states and contexts allows for the dissection of heterogeneous epigenetic subtypes and may accelerate PCa progression biomarker discovery [151, 225], expanding cistromic studies to other proteins such as CTCF [226, 227], ETS [178, 228], FOS [229, 230], HOXB13 [151, 225], KLF9 [151, 231, 232], SP1 [233, 234], SPOP [204, 228] and XBP1 [113, 151, 235].

Another distinct class of SMIs are epigenetic drugs targeting histone deacetylases (HDACs) expressed highly in primary PCa [236] and the enzymatic subunit of the polycomb repressive complex EZH2, which is overexpressed in CRPC [237] and co-occupies reprogrammed AR cistromes in NEPC [238]. Both HDAC and EZH2 promote transcriptional silencing through remodeling chromatin conformation, either deacetylation or methylation of histone tail modifications. Inhibition of EZH2 with SMIs [239] could help overcome ADT resistance and increase effectiveness of AR inhibition in CRPC patients and is suggested to potentiate PCa tumors to PD-1 checkpoint inhibition [240]. Although the HDAC inhibitor vorinostat is an effective inhibitor of PCa proliferation by synergizing with AR antagonists in cells and *in vivo* [241, 242], HDAC inhibition is associated with significant toxicity in patients which currently prevents phase III clinical investigation for PCa [243, 244]. Alternatively, FOXA1 chromatin binding can be indirectly repressed through inhibition of H3K4 demethylation by transcriptional repressor KDM1A (LSD1), which synergizes with AR antagonists *in vivo* and associates with FOXA1 [245]. Contrastingly, direct inhibition of FOXA1 with the SMI JQ1 abrogates FOXA1 binding with co-repressors, which alleviates repression of gene pathways associated with PCa invasion [246].

Finally, PCa's inclination towards inter- and intra-tumor heterogeneity necessitates enhanced minimally-invasive biomarker detection relying on a combination of classic and novel urine- or blood-based prognostic biomarkers [247, 248], which can be highly impactful by preventing the reported systematic overtreatment of patients with indolent disease [139, 211, 249, 250].

15.11 Future Outlook

The dissection of heterogeneity among populations of tumor cells and their TME has recently made exceptional progress through the implementation of single-cell omics technologies [251, 252]. First, a massive transcriptomic heterogeneity was found within tumors, with multiple distinct transcriptional programs and cellular subsets associated with PCa progression [253]. Second, persistent resistant cells without stem cell properties were found to repopulate tumors upon treatment [254], with high cell cycle turnover in resistant cells showing a heterogeneous response towards ADT therapies, such as with enzalutamide [255]. Finally, single cell epigenomics and cistromics studies are yet to be reported for PCa, but such technologies have been applied for identifying heterogeneous chromatin states in breast cancer [256] and were demonstrated to infer single cell heterogeneity in chromatin accessibility [257, 258]. These studies uncover the clinical impact of shifts in heterogeneous cell populations under therapeutic pressure, and underline how single-cell genomics and transcriptomics have improved our understanding of intra-tumor heterogeneity. Clearly, the future application of single cell epigenomics and cistromics technologies would provide a formidable tool to understand the consequences of epigenetic heterogeneity in the context of cancer and facilitate the identification of novel drug targets.

Tracing multiple foci in patients using their genomic profiles allows for dissection of heterogeneous patterns of metastatic spread [259]. It is becoming increasingly clear that PCa metastatic seeding occurs heterogeneously through asynchronous and cross-metastatic seeding [260, 261] with tumor lineages evolving differently [143, 262], which may have direct consequences on the level of epigenetic heterogeneity [153] as well as clinical decision-making [45]. As such, longitudinal sampling might offer the most comprehensive and dynamic view of heterogeneity in AR cistromes during the course of PCa, which to date has only been applied for blood-derived cfDNA methylomes [263]. Although currently unreported, we anticipate longitudinal translational studies with coupled single cell epigenomics and cistromics, so that epigenetic developments become embedded as an intrinsic component of clinical trials, allowing for a precise identification of the dynamics and heterogeneity of epigenetic subtypes to ultimately contribute to improved data-driven clinical decision-making.

Concluding, PCa presents many heterogeneous facets that diverge in AR cistromic reprogramming and contribute to PCa development, progression and therapy response. Taken together, there appear to be distinct and programmatic epigenetic alterations in which normal enhancer binding is altered during PCa initiation and progression, ultimately leading to heterogeneous AR cistromes between tumors, dictating markedly different transcriptional programs with different prognostication between patients. Future technological developments may facilitate a full epigenomic and cistromic characterization of PCa heterogeneity in patient samples, ultimately contributing to personalized medicine. Knowledge gained from such cistromic studies may facilitate the discovery of novel biomarkers for tailored therapeutics and lead to better patient prognostication. As such, AR cistrome heterogeneity in PCa resembles a shifting fingerprint of the tumor: personal and reflective of a specific transcriptional regulatory potential, yet dynamic and subject to change over time.

References

1. Banerji J, Rusconi S, Schaffner W (1981) Expression of a beta-globin gene is enhanced by remote SV40 DNA sequences. Cell 27:299–308
2. Moreau P et al (1981) The SV40 72 base repair repeat has a striking effect on gene expression both in SV40 and other chimeric recombinants. Nucleic Acids Res 9:6047–6068
3. Banerji J, Olson L, Schaffner W (1983) A lymphocyte-specific cellular enhancer is located downstream of the joining region in immunoglobulin heavy chain genes. Cell 33:729–740
4. Gillies SD, Morrison SL, Oi VT, Tonegawa S (1983) A tissue-specific transcription enhancer element is located in the major intron of a rearranged immunoglobulin heavy chain gene. Cell 33:717–728
5. Mercola M, Wang XF, Olsen J, Calame K (1983) Transcriptional enhancer elements in the mouse

immunoglobulin heavy chain locus. Science 221:663–665
6. Struhl K (1984) Genetic properties and chromatin structure of the yeast gal regulatory element: an enhancer-like sequence. Proc Natl Acad Sci U S A 81:7865–7869
7. Shepherd B, Garabedian MJ, Hung MC, Wensink PC (1985) Developmental control of Drosophila yolk protein 1 gene by cis-acting DNA elements. Cold Spring Harb Symp Quant Biol 50:521–526
8. Kioussis D, Vanin E, deLange T, Flavell RA, Grosveld FG (1983) Beta-globin gene inactivation by DNA translocation in gamma beta-thalassaemia. Nature 306:662–666
9. Driscoll MC, Dobkin CS, Alter BP (1989) Gamma delta beta-thalassemia due to a de novo mutation deleting the 5' beta-globin gene activation-region hypersensitive sites. Proc Natl Acad Sci U S A 86:7470–7474
10. Erikson J, ar-Rushdi A, Drwinga HL, Nowell PC, Croce CM (1983) Transcriptional activation of the translocated c-myc oncogene in burkitt lymphoma. Proc Natl Acad Sci U S A 80:820–824
11. Wiman KG et al (1984) Activation of a translocated c-myc gene: role of structural alterations in the upstream region. Proc Natl Acad Sci U S A 81:6798
12. Nurk S et al (2021) The complete sequence of a human genome. 2021.05.26.445798 https://www.biorxiv.org/content/10.1101/2021.05.26.445798v1. https://doi.org/10.1101/2021.05.26.445798
13. Moore JE et al (2020) Expanded encyclopaedias of DNA elements in the human and mouse genomes. Nature 583:699–710
14. Domcke S et al (2020) A human cell atlas of fetal chromatin accessibility. Science 370:eaba7612
15. Maurano MT et al (2012) Systematic localization of common disease-associated variation in regulatory DNA. Science 337:1190–1195
16. Furey TS (2012) ChIP-seq and beyond: new and improved methodologies to detect and characterize protein-DNA interactions. Nat Rev Genet 13:840–852
17. Klemm SL, Shipony Z, Greenleaf WJ (2019) Chromatin accessibility and the regulatory epigenome. Nat Rev Genet 20: 207–220
18. Mp C et al (2010) Histone H3K27ac separates active from poised enhancers and predicts developmental state. Proc Natl Acad Sci U S A 107:21931–21936
19. Heintzman ND et al (2007) Distinct and predictive chromatin signatures of transcriptional promoters and enhancers in the human genome. Nat Genet 39:311–318
20. Robertson AG et al (2008) Genome-wide relationship between histone H3 lysine 4 mono- and tri-methylation and transcription factor binding. Genome Res 18:1906–1917
21. Kim T-K et al (2010) Widespread transcription at neuronal activity-regulated enhancers. Nature 465:182–187
22. Mohrs M et al (2001) Deletion of a coordinate regulator of type 2 cytokine expression in mice. Nat Immunol 2:842–847
23. Fulco CP et al (2019) Activity-by-Contact model of enhancer-promoter regulation from thousands of CRISPR perturbations. Nat Genet 51:1664
24. Fulco CP et al (2016) Systematic mapping of functional enhancer–promoter connections with CRISPR interference. Science 354:769–773
25. ENCODE Project Consortium (2012) An integrated encyclopedia of DNA elements in the human genome. Nature 489:57–74
26. Sur I, Taipale J (2016) The role of enhancers in cancer. Nat Rev Cancer 16:483–493
27. Sung H et al (2021) Global cancer statistics 2020: GLOBOCAN estimates of incidence and mortality worldwide for 36 cancers in 185 countries. CA Cancer J Clin 71:209–249
28. Yang Q, Fung K-M, Day WV, Kropp BP, Lin H-K (2005) Androgen receptor signaling is required for androgen-sensitive human prostate cancer cell proliferation and survival. Cancer Cell Int 5:8
29. Brinkmann AO et al (1999) Mechanisms of androgen receptor activation and function. J Steroid Biochem Mol Biol 69:307–313
30. Heinlein CA, Chang C (2004) Androgen receptor in prostate cancer. Endocr Rev 25:276–308
31. Prostate Cancer Trialists' Collaborative Group (2000) Maximum androgen blockade in advanced prostate cancer: an overview of the randomised trials. Lancet 355:1491–1498
32. Ohlson N, Wikström P, Stattin P, Bergh A (2005) Cell proliferation and apoptosis in prostate tumors and adjacent non-malignant prostate tissue in patients at different time-points after castration treatment. Prostate 62:307–315
33. Oh WK, Kantoff PW (1998) Management of hormone refractory prostate cancer: current standards and future prospects. J Urol 160:1220–1229
34. Chi KN et al (2009) Castration-resistant prostate cancer: from new pathophysiology to new treatment targets. Eur Urol 56:594–605
35. Aihara M, Wheeler TM, Ohori M, Scardino PT (1994) Heterogeneity of prostate cancer inradical prostatectomy specimens. Urology 43:60–66
36. Andreoiu M, Cheng L (2010) Multifocal prostate cancer: biologic, prognostic, and therapeutic implications. Hum Pathol 41:781–793
37. Espiritu SMG et al (2018) The evolutionary landscape of localized prostate cancers drives clinical aggression. Cell 173:1003–1013.e15
38. Carm KT et al (2019) Interfocal heterogeneity challenges the clinical usefulness of molecular classification of primary prostate cancer. Sci Rep 9:13579
39. Algaba F, Montironi R (2010) Impact of prostate cancer multifocality on its biology and treatment. J Endourol 24:799–804
40. Cyll K et al (2017) Tumour heterogeneity poses a significant challenge to cancer biomarker research. Br J Cancer 117:367–375

41. Liu W et al (2009) Copy number analysis indicates monoclonal origin of lethal metastatic prostate cancer. Nat Med 15:559–565
42. Chade DC et al (2012) Cancer control and functional outcomes of salvage radical prostatectomy for radiation-recurrent prostate cancer: a systematic review of the literature. Eur Urol 61:961–971
43. Mottet N et al (2017) EAU-ESTRO-SIOG guidelines on prostate cancer. Part 1: screening, diagnosis, and local treatment with curative intent. Eur Urol 71:618–629
44. van der Poel HG et al (2018) Focal therapy in primary localised prostate cancer: The European Association of Urology position in 2018. Eur Urol 74:84–91. https://doi.org/10.1016/j.eururo.2018.01.001
45. Kneppers J et al (2019) Frequent clonal relations between metastases and non-index prostate cancer lesions. JCI Insight 4: e124756
46. Martinez P et al (2013) Parallel evolution of tumour subclones mimics diversity between tumours. J Pathol 230:356–364
47. Berglund E et al (2018) Spatial maps of prostate cancer transcriptomes reveal an unexplored landscape of heterogeneity. Nat Commun 9:2419
48. Fane M, Weeraratna AT (2020) How the ageing microenvironment influences tumour progression. Nat Rev Cancer 20:89–106
49. Marusyk A, Janiszewska M, Polyak K (2020) Intratumor heterogeneity: the Rosetta stone of therapy resistance. Cancer Cell 37:471
50. Bozic I, Wu CJ (2020) Delineating the evolutionary dynamics of cancer from theory to reality. Nat Can 1:580–588
51. Boutros PC et al (2015) Spatial genomic heterogeneity within localized, multifocal prostate cancer. Nat Genet 47:736–745
52. Bakhoum SF, Cantley LC (2018) The multifaceted role of chromosomal instability in cancer and its microenvironment. Cell 174:1347–1360
53. Flavahan WA, Gaskell E, Bernstein BE (2017) Epigenetic plasticity and the hallmarks of cancer. Science 357: eaal2380
54. (2015) The molecular taxonomy of primary prostate cancer. Cell 163:1011–1025
55. Stelloo S et al (2018) Integrative epigenetic taxonomy of primary prostate cancer. Nat Commun 9:1–12
56. Wang MC, Valenzuela LA, Murphy GP, Chu TM (1979) Purification of a human prostate specific antigen. Investig Urol 17:159–163
57. Riegman PH, Vlietstra RJ, van der Korput JA, Brinkmann AO, Trapman J (1991) The promoter of the prostate-specific antigen gene contains a functional androgen responsive element. Mol Endocrinol 5:1921–1930
58. Wang Q et al (2007) A hierarchical network of transcription factors governs androgen receptor-dependent prostate cancer growth. Mol Cell 27:380
59. Carroll JS et al (2006) Genome-wide analysis of estrogen receptor binding sites. Nat Genet 38:1289–1297
60. Massie CE et al (2011) The androgen receptor fuels prostate cancer by regulating central metabolism and biosynthesis. EMBO J 30:2719–2733
61. Stelloo S et al (2018) Endogenous androgen receptor proteomic profiling reveals genomic subcomplex involved in prostate tumorigenesis. Oncogene 37:313–322
62. Itkonen H, Mills IG (2012) Chromatin binding by the androgen receptor in prostate cancer. Mol Cell Endocrinol 360:44–51
63. Roche PJ, Hoare SA, Parker MG (1992) A consensus DNA-binding site for the androgen receptor. Mol Endocrinol 6:2229–2235
64. Hsu C-C, Hu C-D (2013) Transcriptional activity of c-Jun is critical for the suppression of AR function. Mol Cell Endocrinol 372:12
65. Wilson S, Qi J, Filipp FV (2016) Refinement of the androgen response element based on ChIP-Seq in androgen-insensitive and androgen-responsive prostate cancer cell lines. Sci Rep 6:1–15
66. Kornberg RD (2005) Mediator and the mechanism of transcriptional activation. Trends Biochem Sci 30:235–239
67. Kagey MH et al (2010) Mediator and cohesin connect gene expression and chromatin architecture. Nature 467:430–435
68. Borggrefe T, Yue X (2011) Interactions between subunits of the Mediator complex with gene-specific transcription factors. Semin Cell Dev Biol 22:759–768
69. Zhao H et al (2021) Structure of mammalian Mediator complex reveals Tail module architecture and interaction with a conserved core. Nat Commun 12:1355
70. Chen W, Roeder R (2011) Mediator-dependent nuclear receptor functions. Semin Cell Dev Biol 22:749
71. Yu X et al (2020) Structural insights of transcriptionally active, full-length androgen receptor coactivator complexes. Mol Cell 79:812–823.e4
72. Zhang Z et al (2019) An AR-ERG transcriptional signature defined by long range chromatin interactomes in prostate cancer cells. Genome Res 29:223–235
73. De Santa F et al (2010) A large fraction of extragenic RNA pol II transcription sites overlap enhancers. PLoS Biol 8:e1000384
74. Shii L, Song L, Maurer K, Zhang Z, Sullivan KE (2017) SERPINB2 is regulated by dynamic interactions with pause-release proteins and enhancer RNAs. Mol Immunol 88:20–31
75. Austenaa LMI et al (2015) Transcription of mammalian cis-regulatory elements is restrained by actively enforced early termination. Mol Cell 60:460–474
76. Aguilo F et al (2016) Deposition of 5-methylcytosine on enhancer RNAs enables the coactivator function of PGC-1α. Cell Rep 14:479–492

77. Azofeifa JG, Dowell RD (2017) A generative model for the behavior of RNA polymerase. Bioinformatics 33:227–234
78. Azofeifa JG et al (2018) Enhancer RNA profiling predicts transcription factor activity. Genome Res 28:334–344
79. Mikhaylichenko O et al (2018) The degree of enhancer or promoter activity is reflected by the levels and directionality of eRNA transcription. Genes Dev 32:42–57
80. Carullo NVN et al (2020) Enhancer RNAs predict enhancer–gene regulatory links and are critical for enhancer function in neuronal systems. Nucleic Acids Res 48:9550–9570
81. Chong S et al (2018) Imaging dynamic and selective low-complexity domain interactions that control gene transcription. Science 361:eaar2555
82. Sabari BR et al (2018) Coactivator condensation at super-enhancers links phase separation and gene control. Science 361:eaar3958
83. Cho W-K et al (2018) Mediator and RNA polymerase II clusters associate in transcription-dependent condensates. Science 361:412–415
84. Hnisz D, Shrinivas K, Young RA, Chakraborty AK, Sharp PA (2017) A phase separation model for transcriptional control. Cell 169:13–23
85. Sharma R et al (2021) Liquid condensation of reprogramming factor KLF4 with DNA provides a mechanism for chromatin organization. Nat Commun 12:5579
86. Choi J et al (2021) Evidence for additive and synergistic action of mammalian enhancers during cell fate determination. elife 10:e65381
87. Zaret KS (2020) Pioneer transcription factors initiating gene network changes. Annu Rev Genet 54:367–385
88. Visel A et al (2009) ChIP-seq accurately predicts tissue-specific activity of enhancers. Nature 457:854–858
89. Wang Z et al (2009) Genome-wide mapping of HATs and HDACs reveals distinct functions in active and inactive genes. Cell 138:1019–1031
90. Gnanapragasam VJ, Leung HY, Pulimood AS, Neal DE, Robson CN (2001) Expression of RAC 3, a steroid hormone receptor co-activator in prostate cancer. Br J Cancer 85:1928–1936
91. Zhou H-J et al (2005) SRC-3 is required for prostate cancer cell proliferation and survival. Cancer Res 65:7976–7983
92. Allshire RC, Madhani HD (2018) Ten principles of heterochromatin formation and function. Nat Rev Mol Cell Biol 19:229
93. Pomerantz MM et al (2015) The androgen receptor cistrome is extensively reprogrammed in human prostate tumorigenesis. Nat Genet 47:1346–1351
94. Wu D et al (2014) Three-tiered role of the pioneer factor GATA2 in promoting androgen-dependent gene expression in prostate cancer. Nucleic Acids Res 42:3607–3622
95. Cirillo LA et al (2002) Opening of compacted chromatin by early developmental transcription factors HNF3 (FoxA) and GATA-4. Mol Cell 9:279–289
96. ENCODE Project Consortium (2011) A user's guide to the encyclopedia of DNA elements (ENCODE). PLoS Biol 9:e1001046
97. Kharchenko PV et al (2011) Comprehensive analysis of the chromatin landscape in Drosophila melanogaster. Nature 471:480–485
98. Mayran A, Drouin J (2018) Pioneer transcription factors shape the epigenetic landscape. J Biol Chem 293:13795–13804
99. Lupien M et al (2008) FoxA1 translates epigenetic signatures into enhancer driven lineage-specific transcription. Cell 132:958
100. Wang Q et al (2009) Androgen receptor regulates a distinct transcription program in androgen-independent prostate cancer. Cell 138:245–256
101. Sérandour AA et al (2011) Epigenetic switch involved in activation of pioneer factor FOXA1-dependent enhancers. Genome Res 21:555
102. Gao N et al (2003) The role of hepatocyte nuclear factor-3 alpha (Forkhead Box A1) and androgen receptor in transcriptional regulation of prostatic genes. Mol Endocrinol 17:1484–1507
103. Yu X et al (2005) Foxa1 and Foxa2 interact with the androgen receptor to regulate prostate and epididymal genes differentially. Ann N Y Acad Sci 1061:77–93
104. Eeckhoute J et al (2009) Cell-type selective chromatin remodeling defines the active subset of FOXA1-bound enhancers. Genome Res 19:372–380
105. Wang D et al (2011) Reprogramming transcription by distinct classes of enhancers functionally defined by eRNA. Nature 474:390–394
106. Sahu B et al (2011) Dual role of FoxA1 in androgen receptor binding to chromatin, androgen signalling and prostate cancer. EMBO J 30:3962–3976
107. Sahu B et al (2013) FoxA1 specifies unique androgen and glucocorticoid receptor binding events in prostate cancer cells. Cancer Res 73:1570–1580
108. Jin H-J, Zhao JC, Wu L, Kim J, Yu J (2014) Cooperativity and equilibrium with FOXA1 define the androgen receptor transcriptional program. Nat Commun 5:3972
109. Iwafuchi-Doi M, Zaret KS (2016) Cell fate control by pioneer transcription factors. Development (Cambridge, England) 143:1833
110. Yu J et al (2010) An integrated network of androgen receptor, polycomb, and TMPRSS2-ERG gene fusions in prostate cancer progression. Cancer Cell 17:443–454
111. Severson TM et al (2018) Characterizing steroid hormone receptor chromatin binding landscapes in male and female breast cancer. Nat Commun 9:482
112. Mei S et al (2017) Cistrome Data Browser: a data portal for ChIP-Seq and chromatin accessibility data in human and mouse. Nucleic Acids Res 45:D658–D662

113. Sharma NL et al (2013) The androgen receptor induces a distinct transcriptional program in castration-resistant prostate cancer in man. Cancer Cell 23:35–47
114. Chen Z et al (2015) Agonist and antagonist switch DNA motifs recognized by human androgen receptor in prostate cancer. EMBO J 34:502–516
115. Cioni B et al (2018) Loss of androgen receptor signaling in prostate cancer-associated fibroblasts (CAFs) promotes CCL2- and CXCL8-mediated cancer cell migration. Mol Oncol 12:1308–1323
116. Cioni B et al (2020) Androgen receptor signalling in macrophages promotes TREM-1-mediated prostate cancer cell line migration and invasion. Nat Commun 11:4498
117. Hickey TE et al (2021) The androgen receptor is a tumor suppressor in estrogen receptor–positive breast cancer. Nat Med 27:310–320
118. Toivanen R, Shen MM (2017) Prostate organogenesis: tissue induction, hormonal regulation and cell type specification. Development (Cambridge, England) 144:1382
119. Zhao SG et al (2017) Associations of luminal and basal subtyping of prostate cancer with prognosis and response to androgen deprivation therapy. JAMA Oncol 3:1663–1672
120. Li F et al (2020) ERG orchestrates chromatin interactions to drive prostate cell fate reprogramming. J Clin Invest 130:5924–5941
121. Pomerantz MM et al (2020) Prostate cancer reactivates developmental epigenomic programs during metastatic progression. Nat Genet 52:790–799
122. Brechka H, Bhanvadia RR, VanOpstall C, Vander Griend DJ (2017) HOXB13 mutations and binding partners in prostate development and cancer: function, clinical significance, and future directions. Genes Dis 4:75–87
123. Mazrooei P et al (2019) Cistrome partitioning reveals convergence of somatic mutations and risk variants on master transcription regulators in primary prostate tumors. Cancer Cell 36:674–689.e6
124. Cai C, Hsieh C-L, Shemshedini L (2007) c-Jun has multiple enhancing activities in the novel cross talk between the androgen receptor and Ets variant gene 1 in prostate cancer. Mol Cancer Res 5:725–735
125. Copeland BT, Du J, Pal SK, Jones JO (2019) Factors that influence the androgen receptor cistrome in benign and malignant prostate cells. Mol Oncol 13:2616
126. Chen S-Y et al (2006) c-Jun enhancement of androgen receptor transactivation is associated with prostate cancer cell proliferation. Oncogene 25:7212–7223
127. Nouri M et al (2017) Therapy-induced developmental reprogramming of prostate cancer cells and acquired therapy resistance. Oncotarget 8:18949–18967
128. Faivre EJ et al (2020) Selective inhibition of the BD2 bromodomain of BET proteins in prostate cancer. Nature 578:306–310
129. Navone NM et al (2018) Movember GAP1 PDX project: an international collection of serially transplantable prostate cancer patient-derived xenograft (PDX) models. Prostate 78:1262–1282
130. Leach DA et al (2017) Cell-lineage specificity and role of AP-1 in the prostate fibroblast androgen receptor cistrome. Mol Cell Endocrinol 439:261–272
131. Robinson JLL et al (2011) Androgen receptor driven transcription in molecular apocrine breast cancer is mediated by FoxA1. EMBO J 30:3019–3027
132. Michmerhuizen AR, Spratt DE, Pierce LJ, Speers CW (2020) ARe we there yet? Understanding androgen receptor signaling in breast cancer. NPJ Breast Cancer 6:47
133. Guo M, Peng Y, Gao A, Du C, Herman JG (2019) Epigenetic heterogeneity in cancer. Biomarker Res 7:23
134. Carter B, Zhao K (2021) The epigenetic basis of cellular heterogeneity. Nat Rev Genet 22:235–250
135. Flach KD, Zwart W (2016) The first decade of estrogen receptor cistromics in breast cancer. J Endocrinol 229:R43–R56
136. Bostwick DG (2000) Prostatic intraepithelial neoplasia. Curr Urol Rep 1:65–70
137. Gupta-Elera G, Garrett AR, Robison RA, O'Neill KL (2012) The role of oxidative stress in prostate cancer. Eur J Cancer Prev 21:155–162
138. Papachristodoulou A et al (2021) NKX3.1 localization to mitochondria suppresses prostate cancer initiation. Cancer Discov 11:2316–2333
139. Daskivich TJ et al (2011) Overtreatment of men with low-risk prostate cancer and significant comorbidity. Cancer 117:2058–2066
140. Løvf M et al (2018) Multifocal primary prostate cancer exhibits high degree of genomic heterogeneity. Eur Urol 75: 498–505
141. Kron KJ et al (2017) TMPRSS2–ERG fusion co-opts master transcription factors and activates NOTCH signaling in primary prostate cancer. Nat Genet 49:1336–1345
142. Datta K, Muders M, Zhang H, Tindall DJ (2010) Mechanism of lymph node metastasis in prostate cancer. Future Oncol 6:823–836
143. Mangiola S et al (2016) Comparing nodal versus bony metastatic spread using tumour phylogenies. Sci Rep 6:33918
144. Beroukhim R et al (2010) The landscape of somatic copy-number alteration across human cancers. Nature 463:899–905
145. Stopsack KH et al (2019) Aneuploidy drives lethal progression in prostate cancer. PNAS 116:11390–11395
146. Albert FW, Kruglyak L (2015) The role of regulatory variation in complex traits and disease. Nat Rev Genet 16:197–212
147. Quigley DA et al (2018) Genomic hallmarks and structural variation in metastatic prostate cancer. Cell 174:758–769.e9
148. Das R et al (2021) An integrated functional and clinical genomics approach reveals genes driving

aggressive metastatic prostate cancer. Nat Commun 12:4601
149. Irshad S et al (2013) A molecular signature predictive of indolent prostate cancer. Sci Transl Med 5:202ra122
150. Lalonde E et al (2014) Tumour genomic and microenvironmental heterogeneity for integrated prediction of 5-year biochemical recurrence of prostate cancer: a retrospective cohort study. Lancet Oncol 15:1521–1532
151. Stelloo S et al (2015) Androgen receptor profiling predicts prostate cancer outcome. EMBO Mol Med 7:1450–1464
152. Kirby M, Hirst C, Crawford ED (2011) Characterising the castration-resistant prostate cancer population: a systematic review. Int J Clin Pract 65:1180–1192
153. Severson TM et al (2021) Epigenetic and transcriptional analysis reveals a core transcriptional program conserved in clonal prostate cancer metastases. Mol Oncol 15:1942–1955
154. Scher HI et al (2012) Increased survival with enzalutamide in prostate cancer after chemotherapy. N Engl J Med 367:1187–1197
155. Fizazi K et al (2019) Darolutamide in nonmetastatic, castration-resistant prostate cancer. N Engl J Med 380:1235–1246
156. Beltran H et al (2016) Divergent clonal evolution of castration-resistant neuroendocrine prostate cancer. Nat Med 22:298–305
157. Puca L, Vlachostergios PJ, Beltran H (2019) Neuroendocrine differentiation in prostate cancer: emerging biology, models, and therapies. Cold Spring Harb Perspect Med 9:a030593
158. Ruan L, Wang L, Wang X, He M, Yao X (2017) SIRT1 contributes to neuroendocrine differentiation of prostate cancer. Oncotarget 9:2002–2016
159. Reina-Campos M et al (2019) Increased serine and one-carbon pathway metabolism by PKCλ/ι deficiency promotes neuroendocrine prostate cancer. Cancer Cell 35:385–400.e9
160. Brady NJ et al (2021) Temporal evolution of cellular heterogeneity during the progression to advanced AR-negative prostate cancer. Nat Commun 12:3372
161. Baca SC et al (2021) Reprogramming of the FOXA1 cistrome in treatment-emergent neuroendocrine prostate cancer. Nat Commun 12:1979
162. Viswanathan SR et al (2018) Structural alterations driving castration-resistant prostate cancer revealed by linked-read genome sequencing. Cell 174: 433–447
163. van Dessel LF et al (2019) The genomic landscape of metastatic castration-resistant prostate cancers reveals multiple distinct genotypes with potential clinical impact. Nat Commun 10:5251
164. Mayrhofer M et al (2018) Cell-free DNA profiling of metastatic prostate cancer reveals microsatellite instability, structural rearrangements and clonal hematopoiesis. Genome Med 10:85
165. Takeda DY et al (2018) A somatically acquired enhancer of the androgen receptor is a noncoding driver in advanced prostate cancer. Cell 174:422–432.e13
166. Morova T et al (2020) Androgen receptor-binding sites are highly mutated in prostate cancer. Nat Commun 11:832
167. Robinson D et al (2015) Integrative clinical genomics of advanced prostate cancer. Cell 161:1215–1228
168. Barbieri CE et al (2012) Exome sequencing identifies recurrent SPOP, FOXA1 and MED12 mutations in prostate cancer. Nat Genet 44:685–689
169. Grasso CS et al (2012) The mutational landscape of lethal castration-resistant prostate cancer. Nature 487:239–243
170. Beltran H et al (2020) Circulating tumor DNA profile recognizes transformation to castration-resistant neuroendocrine prostate cancer. J Clin Invest 130:1653–1668
171. Adams EJ et al (2019) FOXA1 mutations alter pioneering activity, differentiation and prostate cancer phenotypes. Nature 571:408–412
172. Parolia A et al (2019) Distinct structural classes of activating FOXA1 alterations in advanced prostate cancer. Nature 571:413–418. https://doi.org/10.1038/s41586-019-1347-4
173. Gao S et al (2019) Forkhead domain mutations in FOXA1 drive prostate cancer progression. Cell Res 29:770
174. Iwafuchi M et al (2020) Gene network transitions in embryos depend upon interactions between a pioneer transcription factor and core histones. Nat Genet 52:418–427
175. Tomlins SA et al (2005) Recurrent fusion of TMPRSS2 and ETS transcription factor genes in prostate cancer. Science 310:644–648
176. Tomlins SA et al (2008) Role of the TMPRSS2-ERG gene fusion in prostate cancer. Neoplasia 10:177–188
177. Klezovitch O et al (2008) A causal role for ERG in neoplastic transformation of prostate epithelium. PNAS 105:2105–2110
178. Chen Y et al (2013) ETS factors reprogram the androgen receptor cistrome and prime prostate tumorigenesis in response to PTEN loss. Nat Med 19:1023–1029
179. Krohn A et al (2012) Genomic deletion of PTEN is associated with tumor progression and early PSA recurrence in ERG fusion-positive and fusion-negative prostate cancer. Am J Pathol 181:401–412
180. Ahearn TU et al (2016) A prospective investigation of PTEN loss and ERG expression in lethal prostate cancer. J Natl Cancer Inst 108:djv346
181. Punnoose EA et al (2015) PTEN loss in circulating tumour cells correlates with PTEN loss in fresh tumour tissue from castration-resistant prostate cancer patients. Br J Cancer 113:1225–1233
182. Salami SS et al (2019) Circulating tumor cells as a predictor of treatment response in clinically localized prostate cancer. JCO Precis Oncol 3:PO.18.00352

183. Tomlins SA et al (2009) ETS gene fusions in prostate cancer: from discovery to daily clinical practice. Eur Urol 56:275–286
184. Goel S et al (2021) Transcriptional network involving ERG and AR orchestrates Distal-less homeobox-1 mediated prostate cancer progression. Nat Commun 12:5325
185. Gan W et al (2015) SPOP promotes ubiquitination and degradation of the ERG oncoprotein to suppress prostate cancer progression. Mol Cell 59:917 930
186. Nagai Y et al (1997) Identification of a novel nuclear speckle-type protein, SPOP. FEBS Lett 418:23–26
187. Zhuang M et al (2009) Structures of SPOP-substrate complexes: insights into molecular architectures of BTB-Cul3 ubiquitin ligases. Mol Cell 36:39–50
188. An J et al (2015) Truncated ERG oncoproteins from TMPRSS2-ERG fusions are resistant to SPOP-mediated proteasome degradation. Mol Cell 59:904–916
189. An J, Wang C, Deng Y, Yu L, Huang H (2014) Destruction of full-length androgen receptor by wild-type SPOP, but not prostate-cancer-associated mutants. Cell Rep 6:657–669
190. Geng C et al (2014) Androgen receptor is the key transcriptional mediator of the tumor suppressor SPOP in prostate cancer. Cancer Res 74:5631–5643
191. Geng C et al (2017) SPOP regulates prostate epithelial cell proliferation and promotes ubiquitination and turnover of c-MYC oncoprotein. Oncogene 36:4767–4777
192. Janouskova H et al (2017) Opposing effects of cancer-type-specific SPOP mutants on BET protein degradation and sensitivity to BET inhibitors. Nat Med 23:1046–1054
193. Dai X et al (2017) Prostate cancer-associated SPOP mutations confer resistance to BET inhibitors through stabilization of BRD4. Nat Med 23:1063–1071
194. Geng C et al (2013) Prostate cancer-associated mutations in speckle-type POZ protein (SPOP) regulate steroid receptor coactivator 3 protein turnover. Proc Natl Acad Sci U S A 110:6997–7002
195. Grbesa I et al (2021) Reshaping of the androgen-driven chromatin landscape in normal prostate cells by early cancer drivers and effect on therapeutic sensitivity. Cell Rep 36:109625
196. Tien JC-Y et al (2013) The steroid receptor coactivator-3 is required for the development of castration-resistant prostate cancer. Cancer Res 73:3997
197. Yan J et al (2008) Steroid receptor coactivator-3/AIB1 promotes cell migration and invasiveness through focal adhesion turnover and matrix metalloproteinase expression. Cancer Res 68:5460
198. Louie MC et al (2003) Androgen-induced recruitment of RNA polymerase II to a nuclear receptor–p160 coactivator complex. PNAS 100:2226–2230
199. Zhou XE et al (2010) Identification of SRC3/AIB1 as a preferred coactivator for hormone-activated androgen receptor. J Biol Chem 285:9161
200. Migliaccio A et al (2000) Steroid-induced androgen receptor–oestradiol receptor β–Src complex triggers prostate cancer cell proliferation. EMBO J 19:5406–5417
201. Migliaccio A et al (2007) Inhibition of the SH3 domain-mediated binding of Src to the androgen receptor and its effect on tumor growth. Oncogene 26:6619–6629
202. Chung AC-K et al (2007) Genetic ablation of the amplified in breast cancer 1 inhibits spontaneous prostate cancer progression in mice. Cancer Res 67:5965–5975
203. Shoag J et al (2018) SPOP mutation drives prostate neoplasia without stabilizing oncogenic transcription factor ERG. J Clin Invest 128:381–386
204. Bernasocchi T et al (2021) Dual functions of SPOP and ERG dictate androgen therapy responses in prostate cancer. Nat Commun 12:734
205. Mounir Z et al (2016) ERG signaling in prostate cancer is driven through PRMT5-dependent methylation of the Androgen Receptor. elife 5:e13964
206. Hsu C-L et al (2014) Identification of a new androgen receptor (AR) co-regulator BUD31 and related peptides to suppress wild-type and mutated AR-mediated prostate cancer growth via peptide screening and X-ray structure analysis. Mol Oncol 8:1575–1587
207. Wasmuth EV et al (2020) Modulation of androgen receptor DNA binding activity through direct interaction with the ETS transcription factor ERG. PNAS 117:8584–8592
208. Welter D et al (2014) The NHGRI GWAS Catalog, a curated resource of SNP-trait associations. Nucleic Acids Res 42:D1001–D1006
209. Hazelett DJ et al (2014) Comprehensive functional annotation of 77 prostate cancer risk loci. PLoS Genet 10:e1004102
210. Chen H et al (2015) Systematic enrichment analysis of potentially functional regions for 103 prostate cancer risk-associated loci. Prostate 75:1264–1276
211. Whitington T et al (2016) Gene regulatory mechanisms underpinning prostate cancer susceptibility. Nat Genet 48:387–397
212. Zhang P et al (2018) High-throughput screening of prostate cancer risk loci by single nucleotide polymorphisms sequencing. Nat Commun 9:1–12
213. Ahmadiyeh N et al (2010) 8q24 prostate, breast, and colon cancer risk loci show tissue-specific long-range interaction with MYC. PNAS 107:9742–9746
214. Wasserman NF, Aneas I, Nobrega MA (2010) An 8q24 gene desert variant associated with prostate cancer risk confers differential in vivo activity to a MYC enhancer. Genome Res 20:1191–1197
215. Guo H et al (2016) Modulation of long noncoding RNAs by risk SNPs underlying genetic predispositions to prostate cancer. Nat Genet 48:1142–1150
216. Han Y et al (2016) Prostate cancer susceptibility in men of African ancestry at 8q24. J Natl Cancer Inst 108: djv431

217. Chung S et al (2011) Association of a novel long non-coding RNA in 8q24 with prostate cancer susceptibility. Cancer Sci 102:245–252
218. Kim T et al (2014) Long-range interaction and correlation between MYC enhancer and oncogenic long noncoding RNA CARLo-5. Proc Natl Acad Sci U S A 111:4173
219. Wang T et al (2021) Integrative epigenome map of the normal human prostate provides insights into prostate cancer predisposition. Front Cell Dev Biol 9:723676
220. Augello MA et al (2019) CHD1 loss alters AR binding at lineage-specific enhancers and modulates distinct transcriptional programs to drive prostate tumorigenesis. Cancer Cell 35:603
221. Chen H et al (2018) A Pan-cancer analysis of enhancer expression in nearly 9000 patient samples. Cell 173:386–399.e12
222. Chen W-Y et al (2021) Nerve growth factor interacts with CHRM4 and promotes neuroendocrine differentiation of prostate cancer and castration resistance. Commun Biol 4:1–14
223. Zhang Z et al (2018) Inhibition of the Wnt/β-catenin pathway overcomes resistance to enzalutamide in castration-resistant prostate cancer. Cancer Res 78:3147–3162
224. Capozzi M et al (2019) Lenvatinib, a molecule with versatile application: from preclinical evidence to future development in anti-cancer treatment. Cancer Manag Res 11:3847
225. Jeong T-O et al (2012) Evaluation of HOXB13 as a molecular marker of recurrent prostate cancer. Mol Med Rep 5:901–904
226. Taslim C et al (2012) Integrated analysis identifies a class of androgen-responsive genes regulated by short combinatorial long-range mechanism facilitated by CTCF. Nucleic Acids Res 40:4754–4764
227. Höflmayer D et al (2020) Expression of CCCTC-binding factor (CTCF) is linked to poor prognosis in prostate cancer. Mol Oncol 14:129–138
228. Liu D et al (2021) Tumor subtype defines distinct pathways of molecular and clinical progression in primary prostate cancer. J Clin Invest 131: e147878
229. Shemshedini L, Knauthe R, Sassone-Corsi P, Pornon A, Gronemeyer H (1991) Cell-specific inhibitory and stimulatory effects of Fos and Jun on transcription activation by nuclear receptors. EMBO J 10:3839–3849
230. Lu H et al (2016) αvβ6 integrin promotes castrate-resistant prostate cancer through JNK1-mediated activation of androgen receptor. Cancer Res 76:5163–5174
231. Shen P et al (2014) KLF9, a transcription factor induced in flutamide-caused cell apoptosis, inhibits AKT activation and suppresses tumor growth of prostate cancer cells. Prostate 74:946–958
232. Shen P et al (2021) KLF9 suppresses cell growth and induces apoptosis via the AR pathway in androgen-dependent prostate cancer cells. Biochem Biophys Rep 28:101151
233. Lu S, Jenster G, Epner DE (2000) Androgen induction of cyclin-dependent kinase inhibitor p21 gene: role of androgen receptor and transcription factor Sp1 complex. Mol Endocrinol 14:753–760
234. Bedolla RG et al (2012) Predictive value of Sp1/Sp3/FLIP signature for prostate cancer recurrence. PLoS One 7:e44917
235. Sheng X et al (2019) IRE1α-XBP1s pathway promotes prostate cancer by activating c-MYC signaling. Nat Commun 10:1–12
236. Weichert W et al (2008) Histone deacetylases 1, 2 and 3 are highly expressed in prostate cancer and HDAC2 expression is associated with shorter PSA relapse time after radical prostatectomy. Br J Cancer 98:604–610
237. Varambally S et al (2002) The polycomb group protein EZH2 is involved in progression of prostate cancer. Nature 419:624–629
238. Davies A et al (2021) An androgen receptor switch underlies lineage infidelity in treatment-resistant prostate cancer. Nat Cell Biol 23:1023–1034
239. Bai Y et al (2019) Inhibition of enhancer of zeste homolog 2 (EZH2) overcomes enzalutamide resistance in castration-resistant prostate cancer. J Biol Chem 294:9911–9923
240. Morel KL et al (2021) EZH2 inhibition activates a dsRNA–STING–interferon stress axis that potentiates response to PD-1 checkpoint blockade in prostate cancer. Nat Cancer 2:444–456
241. Marrocco DL et al (2007) Suberoylanilide hydroxamic acid (vorinostat) represses androgen receptor expression and acts synergistically with an androgen receptor antagonist to inhibit prostate cancer cell proliferation. Mol Cancer Ther 6:51–60
242. Butler LM et al (2000) Suberoylanilide hydroxamic acid, an inhibitor of histone deacetylase, suppresses the growth of prostate cancer cells in vitro and in vivo. Cancer Res 60:5165–5170
243. Bradley D et al (2009) Vorinostat in advanced prostate cancer patients progressing on prior chemotherapy (National Cancer Institute Trial 6862). Cancer 115:5541–5549
244. Rana Z, Diermeier S, Hanif M, Rosengren RJ (2020) Understanding failure and improving treatment using HDAC inhibitors for prostate cancer. Biomedicines 8
245. Gao S et al (2020) Chromatin binding of FOXA1 is promoted by LSD1-mediated demethylation in prostate cancer. Nat Genet 52:1011–1017
246. Wang L, Xu M, Kao C-Y, Tsai SY, Tsai M-J (2020) Small molecule JQ1 promotes prostate cancer invasion via BET-independent inactivation of FOXA1. J Clin Invest 130:1782–1792
247. Narayan VM (2020) A critical appraisal of biomarkers in prostate cancer. World J Urol 38:547–554
248. Koo KM, Mainwaring PN, Tomlins SA, Trau M (2019) Merging new-age biomarkers and nanodiagnostics for precision prostate cancer management. Nat Rev Urol 16:302–317

249. Antonelli A et al (2018) Biological effect of neo-adjuvant androgen-deprivation therapy assessed on specimens from radical prostatectomy: a systematic review. Minerva Urol Nefrol 70:370–379
250. Loeb S et al (2014) Overdiagnosis and overtreatment of prostate cancer. Eur Urol 65:1046–1055
251. Shalek AK et al (2013) Single-cell transcriptomics reveals bimodality in expression and splicing in immune cells. Nature 498:236
252. Buenrostro JD et al (2015) Single-cell chromatin accessibility reveals principles of regulatory variation. Nature 523:486–490
253. Chen S et al (2021) Single-cell analysis reveals transcriptomic remodellings in distinct cell types that contribute to human prostate cancer progression. Nat Cell Biol 23:87–98
254. Karthaus WR et al (2020) Regenerative potential of prostate luminal cells revealed by single-cell analysis. Science 368:497–505
255. Taavitsainen S et al (2021) Single-cell ATAC and RNA sequencing reveal pre-existing and persistent cells associated with prostate cancer relapse. Nat Commun 12:5307
256. Grosselin K et al (2019) High-throughput single-cell ChIP-seq identifies heterogeneity of chromatin states in breast cancer. Nat Genet 51:1060–1066
257. Ramani V et al (2017) Massively multiplex single-cell Hi-C. Nat Methods 14:263
258. Zhang R, Zhou T, Ma J (2022) Multiscale and integrative single-cell Hi-C analysis with Higashi. Nat Biotechnol 40: 254–261
259. ICGC Prostate UK Group et al (2015) The evolu tionary history of lethal metastatic prostate cancer. Nature 520:353–357
260. Hong MKH et al (2015) Tracking the origins and drivers of subclonal metastatic expansion in prostate cancer. Nat Commun 6: 6605
261. Macintyre G et al (2017) How subclonal modeling is changing the metastatic paradigm. Clin Cancer Res 23:630–635
262. Haffner MC et al (2013) Tracking the clonal origin of lethal prostate cancer. J Clin Invest 123:4918–4922
263. Silva R et al (2021) Longitudinal analysis of individual cfDNA methylome patterns in metastatic prostate cancer. Clin Epigenetics 13:168

Epigenetic Coregulation of Androgen Receptor Signaling

16

Rayzel C. Fernandes, Damien A. Leach, and Charlotte L. Bevan

Abstract

The androgen receptor (AR) is a ligand-activated transcription factor belonging to the nuclear receptor (NR) superfamily. As with other members of the NR family, transcriptional activity of the AR is regulated by interactions with coregulatory proteins, which either enhance (coactivators) or repress (corepressors) its transcriptional activity. AR associated coregulators are functionally diverse, but a large fraction are epigenetic histone and chromatin modifiers. Epigenetic coregulators are recruited to gene regulatory regions as part of multi-protein complexes, often acting in a dynamic and inter-dependent manner to remodel chromatin, thereby allowing or inhibiting the access of AR-associated transcriptional machinery to target genes; functional consequences being regulation of transcriptional output. Epigenetic modifiers, including those that function as AR coregulators, are frequently mutated or aberrantly expressed in prostate cancer and are implicated in disease progression. Some of these modifiers are being investigated as therapeutic targets in several cancer types and could potentially be used to modulate aberrant AR activity in prostate cancer. In this chapter we will summarise the functional role of epigenetic coregulators in AR signalling, their dysregulation during prostate cancer progression and the current status of drugs targeting these enzymes.

Keywords

Epigenetics · Androgen receptor coregulators · Transcriptional regulation · Histone modifiers · Remodellers · Prostate cancer · Therapy

16.1 Introduction

The androgen receptor (AR) is a ligand-activated, DNA-binding transcription factor (TF), belonging to the nuclear receptor (NR) superfamily, that mediates responses to the androgenic ("male") steroid hormones, most prominent of which are dihydrotestosterone and testosterone [1]. The AR-driven transcription program is a key determinant of organ morphogenesis during development and regulates functioning of the normal adult prostate, but is also the main driver of prostate carcinogenesis and disease progression [2, 3].

To activate its full transcriptional program, the AR must be bound and activated by ligands such as testosterone and dihydrotestosterone. AR

R. C. Fernandes · D. A. Leach · C. L. Bevan (✉)
Department of Surgery & Cancer, Imperial College London, London, UK
e-mail: charlotte.bevan@imperial.ac.uk

M. J. Campbell, C. L. Bevan (eds.), *Nuclear Receptors in Human Health and Disease*, Advances in Experimental Medicine and Biology 1390, https://doi.org/10.1007/978-3-031-11836-4_16

binds to sequence-specific regulatory regions in the genome, where it interacts with accessory proteins called coregulators and transcriptional machinery to drive target gene expression [4]. Coregulators can be broadly defined as members of multi-protein complexes that associate directly or indirectly with transcription factors (TFs) and affect their output. These proteins are indispensable for TF functioning since they are rate-limiting factors for transcriptional activity that can either promote (coactivators) or supress (corepressors) target gene expression, without affecting basal transcriptional levels [5–7]. Coregulators also dictate target gene specificity, with each coregulator associated with transcription of specific subsets of TF target genes [8]. The AR is associated with coregulators that encompass a wide variety of functional diversity and modes of action which can be broadly classified into the categories (not mutually exclusive) of epigenetic regulators, chaperones, transcriptional regulators, DNA repair proteins, cytoskeletal proteins and signal transducers, among others [9].

Epigenetic proteins are a key subset of coregulatory partners of AR and many other TFs as they are essential for transcriptional processes, regulating chromatin structure as well as accessibility. Alterations in epigenetic machinery proteins are frequent in prostate cancer and have been suggested to drive carcinogenesis and evolution of treatment resistance, as well as contribute to inter and/or intra tumoral heterogeneity [10, 11]. A number of these outcomes are the result of altered epigenetic coregulators disrupting AR signalling and such coregulators thus represent potential targets for therapeutics.

16.2 AR Structure and Coregulator Binding Interactions

The androgen receptor is a 919 amino acid protein (although size can vary due to the presence of a polymorphic polyglutamine tract) that can structurally be divided into four distinct domains: an N-terminal activation domain, (AF1), a DNA binding domain (DBD), a hinge region, and a ligand binding domain (LBD). The LBD contains the ligand binding pocket, a ligand-dependent activation domain (AF2) and binding function 3 (BF3) site [12–14]. A nuclear localization signal (NLS) is present in the DBD and hinge region [15]. In its unliganded form, the AR is largely localized in the cytoplasm in a heterocomplex with chaperones and immunophilins, which maintains it in a conformation conducive to ligand binding [16]. Upon binding to ligands, the AR dissociates from the chaperone complex and translocates to the nucleus where it homodimerizes and binds to palindromic dihexameric recognition sequences, termed androgen response elements (AREs), within regulatory regions (enhancers or promoters) of target genes [17, 18]. At AREs, the AR recruits and cooperates with other factors including coregulators; interactions between AR and partner protein domains leading to the assembly of large multiprotein complexes that are necessary for transcriptional regulation (Fig. 16.1A).

Coregulator recruitment by the AR is determined by type of ligand. Binding of an agonist within the LBD causes a conformational change, inducing formation of the AF2 hydrophobic binding cleft that recruits coactivators for receptor transactivation [19]. Interaction with AF2 is

Fig. 16.1 (continued) elements (AREs) and associates with coregulatory proteins and transcriptional machinery (**B**) Pioneer factors open compacted chromatin, allowing AR to access DNA (**C**) AR-associated ATP-dependent chromatin remodellers enhance chromatin accessibility (**D**) Histone modifiers are recruited by the AR and modify surrounding histones (**E**) Epigenetic readers "read" histone marks and act as a link to recruit other protein complexes (**f**) Chromatin looping is required to link enhancers elements to gene promoters

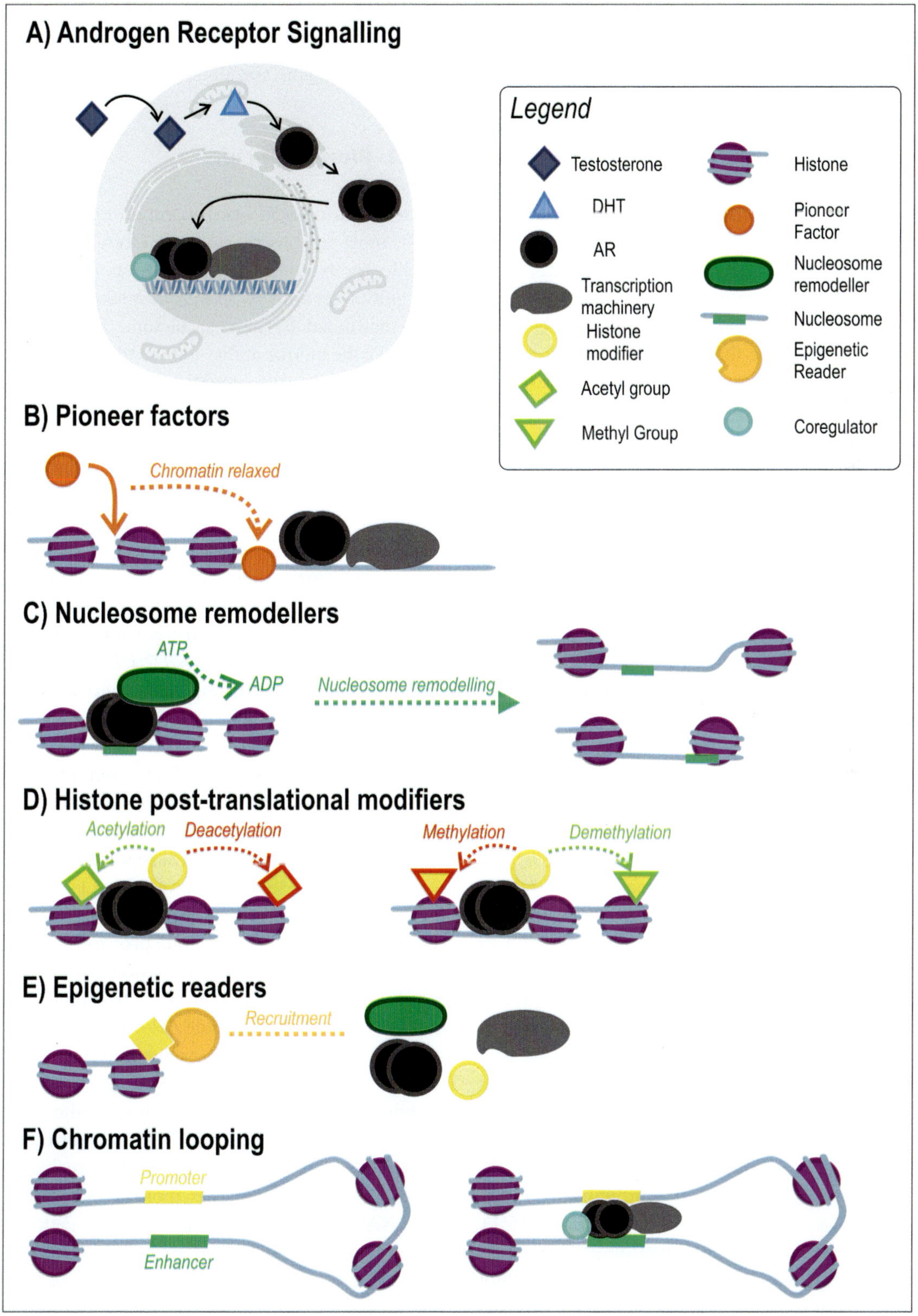

Fig. 16.1 Epigenetic regulation of androgen receptor signalling (**A**) Schematic diagram of AR mediated transcription in prostate cells. Testosterone is converted to dihydrotestosterone (DHT) in prostate cells. DHT binds to the AR, promotes its dimerization and translocation to the nucleus. In the nucleus AR binds to androgen response

mediated by short, alpha-helical LXXLL or FXXLF motifs (L = leucine, F = phenylalanine, X = any amino acid) in coactivators, although the AR AF2 appears to preferentially bind to the FXXLF motif [20, 21]. Corepressors use extended LXXLL-like motifs, called corepressor nuclear receptor boxes (CoRNR boxes); these can interact with the AR following antagonist binding, which promotes conformational changes that accommodate this bulkier motif [22]. The AF2 coactivator binding site is blocked by this conformation of antagonist-bound AR [23]. Although these are the best-characterised modes of interaction, coregulator recruitment can also occur via interactions between the AR BF3 site, N-terminal AF1 or DBD and with regions other than the LXXLL/FXXLF motif in coregulators [13, 14, 19, 24].

16.3 AR-Coregulator Mediated Alteration of the Chromatin Landscape

Genomic DNA in eukaryotic nuclei is complexed around histone octamers to form nucleosomes, arrays of which are further coiled into heterochromatin. Chromatin is further compacted into higher order fibres, i.e. chromosomes [25, 26]. This compact organization of genomic DNA, by hindering indiscriminate access of transcription factors to binding sequences, facilitates tight regulation of gene expression. Key steps in gene transcription include dynamic reorganization of chromatin by transcription factor complexes, recruitment of basal transcription machinery, assembly of the preinitiation complex (PIC) at promoters and RNA polymerase activity [27].

The AR transcription complex assembled at target gene regulatory regions contains several coregulators that are modifiers of chromatin structure. This includes nucleosome remodellers, histone modifying/interacting proteins and mediators of chromatin looping (Fig. 16.1). Coregulators are initially recruited by interactions with the AR but subsequent coregulator recruitment is also dependent on targeting by and interplay with coregulators that are already part of the complex [28]. The coregulator composition within an AR complex is likely to be both cell and target gene specific.

16.3.1 Pioneer Factors

Due to the aforementioned supercoiling of chromatin, most regions of genomic DNA are inaccessible for binding by transcription factors so their initial binding to recognition sequences is often facilitated by pioneer factors. These proteins have the unique ability to bind to and "relax" compacted chromatin, enabling access for other transcription factors and regulatory proteins (Fig. 16.1B) [29]. Pioneer factors have various mechanisms for de-compacting chromatin which include disrupting histone-DNA contacts to destabilize chromatin, evicting histones and recruiting chromatin modifiers [30, 31].

Pioneer factors collaborate with nuclear receptors to regulate distinct tissue-specific transcriptional programs [32]. The AR-associated pioneer factors FOXA1, HNF4α and AP-2α, for instance, regulate distinct AR cistromes (i.e., genome-wide AR binding sites) in the prostate, kidney and epididymis, respectively [33]. Besides FOXA1, other pioneer factors such as GATA2 and HOXB13 regulate the AR cistrome in normal as well as prostate cancer cells [34]. Indeed, these factors are critical for prostate cancer transformation and progression. Overexpression of FOXA1, for example, has been shown to increase AR chromatin binding to facilitate prostate cancer growth [35], while ectopic expression of FOXA1 and HOXB13 in a normal prostate epithelial cell line was shown to redistribute AR binding sites to resemble the pattern in prostate tumours [36]. Moreover, silencing of FOXA1 reprograms AR binding in prostate cancer [37]. Besides the full-length AR, both FOXA1 and GATA2 contribute to androgen deprivation therapy (ADT) resistant prostate cancer by acting as pioneer factors for the DNA binding of AR variants [38, 39]. In addition to enabling chromatin access directly, pioneer factors also facilitate recruitment of histone modifiers and remodellers for further chromatin decompaction [40, 41].

16.3.2 Nucleosome Remodellers

Besides binding to regions of the genome that are already open, AR also has the ability to further influence chromatin accessibility by regulating nucleosome occupancy at target enhancers [42–44]. To mediate these changes, the AR recruits a class of regulators that remodel chromatin using the energy from ATP hydrolysis to remove or reposition nucleosomes (Fig. 16.1C) [45]. AR activity is primarily coactivated by two subfamilies of ATP-dependent chromatin remodellers – the SWItch/Sucrose Non-Fermentable (SWI/SNF) and Chromodomain Helicase DNA-binding (CHD) proteins (Table 16.1).

The SWI/SNF remodelling complex is a large complex comprised of 11–15 subunits (variable by context) including ATPases (BRG1, BRM) and core or associated factors that confer specificity [87, 88, 91]. Of these, BAF60a, BAF57 and SRG3 (a mouse homolog of human BAF155) interact with/coregulate the AR [87, 88, 91]. The ATPase present may also confer specificity, as BRG1 appears to regulate chromatin accessibility for a subset of AR target genes [92]. Most tumour types have mutations in one or more subunits of the SWI/SNF complex [93]. Unlike in other malignancies however, mutations of SWI/SNF subunits are uncommon in PCa but expression levels are often altered during disease progression [89, 94]. In benign and malignant prostate tissues BRG1 and BRM are reciprocally expressed, with increased BRG1/decreased BRM expression associated with cancer progression and metastasis [89]. Loss of BRM is also associated with prostatic hyperplasia and castration resistance in murine prostatic epithelia, but BRM-containing SWI/SNF complexes appears to be preferred for AR activity in cell line models, which is likely to be sustained by BRG1 or other remodelling complexes upon loss of BRM [95, 96]. SWI/SNF subunits can coactivate AR independently of the remodeller complex's ATPase function – SRG3, a core subunit of the mouse SWI/SNF complex, enhances AR transactivation even in the absence of both BRG1 and BRM [91].

The CHD family of remodellers consists of nine members (CHD 1-9) characterized by an N-terminal chromodomains and a central ATPase domain [97]. While some CHD proteins function as monomers, others are part of multiprotein complexes [98]. Members of the CHD family, have divergent functions, for example CHD8 acts as an AR coactivator and is upregulated in PCa whereas CHD1, which is associated with AR transcription at specific enhancers, is frequently deleted in PCa [90, 99]. Interestingly, although AR and CHD1 associate on chromatin and have significant overlap in their chromatin-bound interactome, they do not appear not to interact directly but may be bridged by overlapping interacting coregulators [99, 100].

16.3.3 Histone Post-translational Modifiers

Besides pioneer factors and nucleosome remodelling complexes, histone post-translational modifiers are also major regulators of chromatin accessibility (Fig. 16.1D). Histone octamers within nucleosomes consist of two copies of each of the canonical histones H3, H4, H2A and H2B [101]. Variant versions also exist, which can substitute for canonical histones and play essential roles during replication, gene regulation and repair [102]. Structurally, histone proteins contain a histone fold region and a tail region, with the fold regions responsible for formation of the octamer. Histone tails protrude out of the nucleosome and are targets for modifications that regulate chromatin structure (Fig. 16.2). Histone modifications occur on multiple residues within these N-terminal tails and also in the histone body [103]. Currently, at least 80 histone post translational modifications (PTMs) have been identified and include acetylation, methylation, phosphorylation, ubiquitylation, cronoylation, succinylation, and sumoylation events. These modifications are frequently regulated in a coordinated manner with combined modifications governing regulatory events. Mechanisms by which histone PTMs modulate DNA accessibility differ: acetylation and phosphorylation for

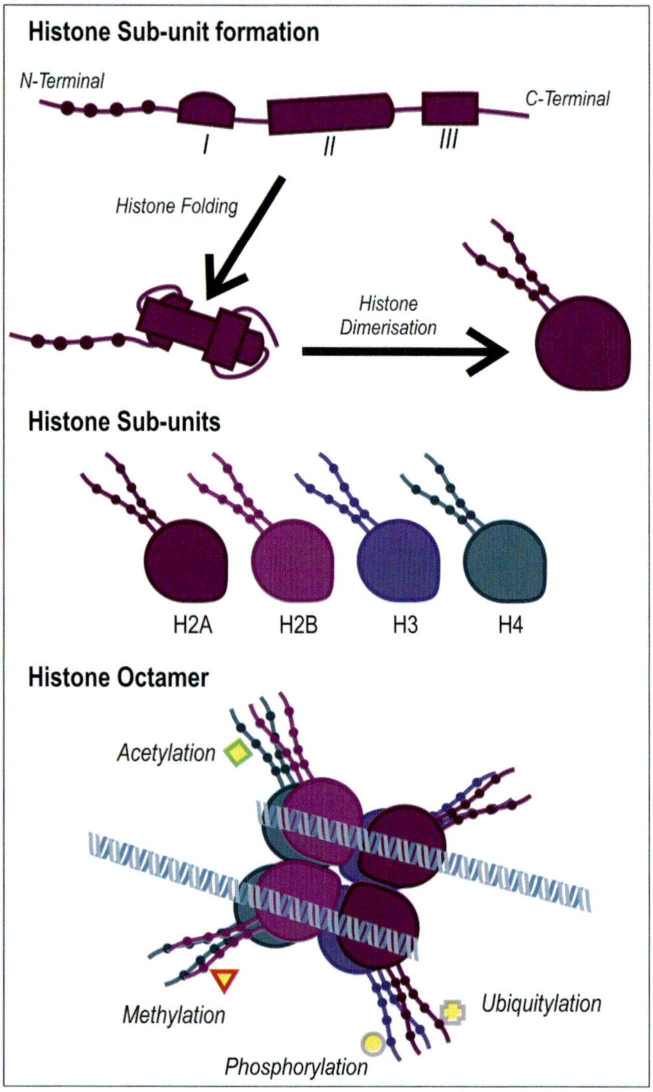

Fig. 16.2 Histone structure and formation of octamers. Histone proteins are characterized by a tail region and a histone fold motif made up of a loop (I), central helix (II) and a short helix (III). The fold domain facilitates histone heterodimerization: two H2A-H2B dimers and a H3-H4 tetramer combine to form the octamer around which DNA is wound to form nucleosomes. Histone tails protruding from nucleosomes are targeted for post translational modifications

instance, alter the charge on histones thereby disrupting electrostatic interactions with DNA, whereas methylation enhances or disrupts interactions with chromatin binding factors [104].

Histone modifying/interacting proteins can be broadly classed as: writers, which deposit marks; those that remove marks, termed erasers; and readers that sense the modification and effect changes. AR associated histone writers and erasers modulate gene expression largely via changes in acetylation at lysine residues and methylation at lysine/arginine residues in histone tails (Table 16.1).

16.3.3.1 Histone Acetylases/Deacetylases

Histone acetylation is generally permissive of gene activation, and deacetylation is generally restrictive, with marks such as H3K27ac, H4K16ac, H3K9Ac and H3K14Ac enriched at active enhancers and/or promoters [105, 106]. Histone acetyl transferases (HATs), such as members of the NCOA/p160/Steroid Receptor Coactivator (SRC) family, p300/CBP and PCAF, are some of the earliest coregulators recruited by agonist activated AR [107]. Although p160/SRC proteins have weak histone acetylase activity

Table 16.1 AR-associated epigenetic coregulators and their published roles in prostate cancer

Coregulator	Function	Coregulator type	Role in PCa	References
Acetylases/Deacetylases				
CBP	Acetylates various residues on H1, H2, H3, H4	Coactivator	Upregulated in CRPC	[46, 47]
p300	Acetylates various residues on H1,H2,H3,H4; Acetylates AR	Coactivator	Upregulated in CRPC	[46, 47]
PCAF	H3K14, H3K9 acetylase	Coactivator	Oncogenic	[48]
KAT5/TIP60	Acetylates various residues on H2A, H3, H4; Acetylates AR	Coactivator	Upregulated in aggressive PCa	[49]
KAT7	H3K14/K23, H4K5/K8/K12 acetylase	Corepressor	Unknown	[50]
KAT8	H4K16 acetylase	Coactivator	Oncogenic	[51]
NCOA1/ SRC1	H3, H4 acetylase	Coactivator	Oncogenic; Upregulated in PCa metastases	[52]
NCOA2/ SRC2/TIF2	Weak histone acetylase activity	Coactivator	Overexpressed in recurrent PCa	[53]
NCOA3/ SRC3/AIB1	H3, H4 acetylase	Coactivator	Oncogenic	[54]
HDAC1	Deacetylates H2A, H2B, H3 and H4	Corepressor	Upregulated in PCa	[55, 56]
HDAC2	Deacetylates H2A, H2B, H3 and H4	Corepressor	Upregulated in PCa	[56, 57]
HDAC3	Deacetylates H2A, H2B, H3 and H4	Corepressor	Upregulated in PCa	[56, 57]
HDAC7	Deacetylates H2A, H2B, H3 and H4	Corepressor	Unknown	[58]
SIRT1	Deacetylates H1 H2A, H2B, H3 and H4; Deacetylates AR	Corepressor	Upregulated in PCa	[59, 60]
NCOR1	Required for recruitment and/or activity of HDACs	Corepressor	Downregulated in PCa	[61, 62]
NCOR2/ SMRT	Required for recruitment and/or activity of HDACs	Corepressor	Reduced expression associated with shorter disease-free survival	[62, 63]
Methylases/Demethylases				
CARM1	Asymmetric H3R16, R26 methylase	Coactivator	Upregulated in PCa	[64]
EZH2	H3 methylase	Coactivator	Oncogenic	[65]
PRMT1	H4R3 methylation	Coactivator	Oncogenic	[66]
NSD1	H3K36me2 methylase	Coactivator	Upregulated in metastatic PCa	[67, 68]
NSD2	H3K36me1/2 methylase	Coactivator	Pro-metastatic	[69, 70]
SET1	H3K4 methyltransferase	Coactivator	Oncogenic	[71]
SET9	H3K4me1,2 methylase, methylates AR	Coactivator	Oncogenic	[72, 73]
G9A	H3K9 methylation	Coactivator	Upregulated in PCa	[74]
KDM1A/ LSD1	H3, H4 demethylase	Coactivator	Oncogenic	[75]
KDM3A/ JMJD1A	H3K9me1/2 demethylase	Coactivator	Oncogenic	[76]
KDM4A/ JMJD2A	H3K9me3 demethylase	Coactivator	Oncogenic, Upregulated in PCa	[24]
KDM4B/ JMJD2B	H3K9 demethylase	Coactivator	Oncogenic, Upregulated in PCa	[77]

(continued)

Table 16.1 (continued)

Coregulator	Function	Coregulator type	Role in PCa	References
KDM4C/ JMJD2C	H3K9 demethylase	Coactivator	Oncogenic	[78]
KDM4D/ JMJD2D	H3K9me3 demethylase	Coactivator	Upregulated in PCa	[24]
KDM5B/ JARID1B	H3K4 demethylase	Coactivator	Upregulated in PCa	[79]
KDM7A	H3, H4 demethylase	Coactivator	Oncogenic; Upregulated in PCa	[80]
KMD8/ JMJD5	H3K36me2 demethylase	Coactivator	Upregulated in PCa	[81]
Epigenetic readers				
ING1b	H3K4me3 reader	Corepressor	Tumour suppressive; Downregulated in CRPC	[82]
ING2	H3K4me3 reader	Corepressor	Tumour suppressive	[83]
ING3	H3K4me3 reader	Coactivator	Oncogenic	[84]
TRIM24	H3K4, H3K23Ac reader	Coactivator	Oncogenic; Upregulated in CRPC	[85]
TDRD3	Asymmetric H3R17me2 and H4R3me2 reader	Coactivator	Unknown	[86]
Chromatin remodellers				
BAF57	SWI/SNF subunit	Coactivator	Oncogenic	[87]
BAF60a	SWI/SNF subunit	Coactivator	Unknown	[88]
BRG1	SWI/SNF subunit	Coactivator	Oncogenic	[89]
BRM	SWI/SNF subunit	Coactivator	Oncogenic	[89]
CHD8	CHD remodeller	Coactivator	Oncogenic	[90]

they have been proposed to act as a bridge, recruiting the more potent p300/CBP and PCAF HATs as well as the methyltransferase, coactivator-associated arginine methyltransferase 1 (CARM1) [107–109]. Members of the p160/SRC family are required for optimal expression of AR targets, with disruption of the interaction between AR and SRC-1 shown to selectively inhibit AR activity [110].

p300 and CBP are paralogous proteins that serve as critical coactivators of NR activity and are associated with the H3K18Ac/H3K27Ac active marks [111]. Both proteins interact with the AR, are recruited to regulatory regions of AR targets such as PSA and promote AR transcriptional activation [112]. p300, however appears to be dominant in the context of AR signalling, regulating many more AR targets than CBP [113, 114]. p300/CBP and the acetylation marks it deposited also appear to be necessary for recruitment of the SWI/SNF complex [28]. Several members of the evolutionarily conserved MYST family of HATs also serve as AR coregulators: while Tip60/Kat5 and KAT8 are coactivators, KAT7 has been shown to repress AR activity [50, 115, 116].

Histone deacetylases (HDACs), which catalyse the removal of acetyl groups from histone and other proteins, are often recruited in cooperation with corepressors by antagonist bound AR: HDAC1 and 2, for instance, are recruited to AR target promoters along with the NCOR and SMRT corepressors in the presence of the AR antagonist bicalutamide [107]. Bicalutamide also represses AR transcriptional activity by recruiting the HDAC sirtuin1 (SIRT1), which in turn likely contributes to gene repression by deacetylating histone H3 at target promoters and enhancers [117]. Additionally, SIRT1 is able to reduce AR coactivation by p300 [59]. Some HDACs, such as HDAC1 and HDAC3, however, are required for transcription of AR activated genes as they facilitate coactivator and RNA PolII recruitment, suggesting that hyperacetylated chromatin may in some circumstances inhibit recruitment of these essential factors [118].

Both HATs and HDACs can also influence AR transcriptional activity independently of their histone modifying properties, by modifying the AR itself. The AR can be post-translationally modified by acetylation, phosphorylation, methylation and sumoylation; these affect protein stability, interactions with other proteins, localization and structure [119]. AR acetylation, carried out by p300, PCAF and TIP60/KAT5, occurs at lysine residues within a conserved KLKK motif in the AR hinge region and is critical for hormone induced activation, augmentation of AR activity, corepressor detachment and coactivator recruitment [55, 120, 121]. Conversely, HDACs serve to inhibit AR activity. This can happen either directly, such as when HDAC1 deacetylates AR to downregulate its activity, or indirectly by HDAC4 through a SUMOylation dependent mechanism [55, 122]. Interestingly, both TIP60/KAT5 and HDAC1 can co-exist in the same complex along with AR, potentially antagonizing each other's actions to control AR activity [55].

16.3.3.2 Histone Methylases/Demethylases

The effects of histone methylation on gene activity are nuanced, and depend on the residue being modified and the number of methyl groups added [123]. Histone methylation is regulated by histone methyl transferases (HMTs) or demethylases (HDMs) which modify either arginine or lysine residues. Generally, methylation at H3K4, H3K36, H3K79 and H3R17 is associated with gene activation, while methylation at H3K9, H3K27 and H4K20 is associated with repression [124, 125].

As part of AR transcriptional complexes, HMTs/HDMs work in concert with other coregulators to affect modification at multiple histone residues, with different outcomes on AR activity. Examples of AR associated histone lysine modifiers include SET9, which activates transcription by methylating H3K4 at enhancers and TSS regions, but prevents deposition of repressive dimethylation marks on H3K9 and JARID1B, which inhibits transcription via the removal of two and three methyl groups from H3K4 [72, 79]. Some methyltransferases are recruited to the AR transcriptional complex by HATs: e.g. CARM1, which methylates H3, requires the presence of the acetylases NCOA1/SRC1 or TIF2 to enhance AR activity [126]. On the other hand, methyltransferase activity can facilitate histone acetylation – the PRMT1 methyltransferase influences AR activity by methylating H4R3, which consequently results in H4 acetylation by p300, while SET9 is necessary for androgen induced recruitment of P/CAF [66, 72]. These interactions between different histone modification events also underscore the complexity of these regulatory events to finely tune and conditionally regulate transcription.

Other instances of interplay between histone modifiers include the demethylases KDM1A/LSD1 and JMJD2C, which act cooperatively to demethylate H3K9 resulting in activation of AR transcriptional activity [78]. KDM1A/LSD1, however, can also form a complex with the RCOR1/CoREST corepressor to demethylate H3K4, thereby turning off AR responsive enhancers [127, 128]. Histone acetylation and methylation at AREs can also involve crosstalk with phosphorylation. Protein kinase C-related kinase 1 (PRK1) acts in an androgen dependent manner to phosphorylate H3T11, which subsequently enhances demethylation of H3K9 by JMJD2C or LSD1and acetylation of H3K9/K14, resulting in upregulation of AR activity [129]. PRK1 also promotes phosphorylation of H3T6 via protein kinase C beta I (PKCβI), which prevents KDM1A/LSD1 from demethylating H3K4 during AR-dependent gene activation [130].

Like acetylation, methylation of the AR protein at the KLKK (K = Lysine, L = Leucine) motif also affects its activity. Thus far, the HMT SET9 has been shown to directly methylate AR and enhance transcriptional activity. The demethylase KDM4B however can indirectly stabilize AR by interacting with it and potentially masking ubiquitin acceptor sites [73, 77].

16.3.4 Epigenetic Readers

Epigenetic marks established by modifiers are recognized and interpreted by effector proteins, called epigenetic readers, to modify chromatin structure. Reader proteins contain domains such as the plant homeodomain (PHD), Bromodomain and extra terminal (BET), Chromodomain (CHD), WD40 repeat (WDR) or Tudor domains which determine binding specificity, with the PHD, CHD and Tudor domains recognizing methylated lysine/arginine residues while BET proteins bind to acetylated lysines [131]. Reader proteins/domains link histone marks to other histone modifiers, or to remodelling, transcription, repair or other complexes (Fig. 16.1E). Of the PHD domain proteins the ING family, which binds to the H3K4 methylation mark and subsequently recruit HATs and HDACs, are AR coregulators [104]. While ING1 and 2 are corepressors, ING3 promotes AR transcriptional activity [82–84]. ING1 and ING2 are potentially recruited to AR transcriptional complexes through their reader activity. Their corepressor activity may relate to their role in recruiting the mSIN3A/HDAC repressor complex [82, 83]. While ING3 can target the AR coactivator HAT TIP60/Kat5 to H3K4me3 marks through its PHD reader activity, this mechanism does not appear to be contribute to ING3 mediated AR transactivation. In this instance, ING3 has a cytoplasmic role scaffolding and increasing cytoplasmic TIP60 and AR interaction, subsequently enhancing receptor acetylation and nuclear translocation [84]. TDRD3, a Tudor domain reader of H3/H4 arginine marks, is an AR transcriptional coactivator likely functioning as a scaffolding molecule for assembly of protein complexes [86].

16.3.5 Chromatin Looping

Most AR enhancers, like enhancers generally, are located distal to promoters of target genes, which necessitates long-range interactions if they are to regulate gene expression. Regulatory regions for AR target genes were initially defined as located within 20–50 kb of the gene but an AR-bound enhancer-target gene interaction spanning 650 kb has recently been reported [132, 133]. These interactions are mediated by AR and other proteins (including coregulators) bound at both sites and lead to the formation of chromatin loops (Fig. 16.1F). Several well-known AR targets, including PSA and TMPRSS2, are regulated by chromatin looping [134, 135]. The Mediator multi-subunit complex is a key regulator of gene expression through the formation of enhancer-promoter chromatin loops. The mediator complex bridges TFs at the enhancer with RNA pol II and preinitiation complex at promoters [136]. The MED1/TRAP220 subunit of this complex is a coactivator for the AR and other NRs, co-recruited along with the AR to AREs upon androgen stimulation [137]. MED1 depletion, or inhibition of its interaction with AR, leads to a reversal of androgen induced transcriptional changes in prostate cancer cell lines [137].

16.4 Dysregulated Expression and Function of Coregulators Promotes PCa Progression by Multiple Mechanisms

The AR plays a central role in in prostate carcinogenesis and targeting it by ADT, with drugs such as enzalutamide, remains the standard of care for recurrent, advanced, and metastatic disease. ADT aims to block the action of the AR by either reducing levels of AR agonists (androgens) or by inhibiting the AR with antagonists (antiandrogens). While this is initially successful, patients usually progress to ADT-resistant prostate cancer (ADT-R-PC) within a few years [138]. ADT-R-PC is characterized by disease progression despite ADT, but the AR signalling axis remaining active in the majority of cases. Resistance to ADT via persistent AR signalling can occur via a number of AR signalling alterations, including AR amplifications, mutations, variants and coregulator associated mechanisms [138].

Dysregulated coregulator function and expression is a frequent feature of ADT-R-PC, suggesting an important role in disease progression and therapy resistance (Table 16.1) [8]. Indeed, mechanisms by which epigenetic coregulators can induce aberrant AR signalling include (i) activating AR under low hormone conditions, (ii) post-translational modification of the AR or associated proteins (iii) facilitating interactions between AR and other factors and (iv) inducing expression of AR target genes in the absence of AR.

Taking the first such mechanism, coregulators can enable activation of the AR in the absence or low levels of agonists, thus escaping ADT-induced androgen blockade. TRIM24, for example, is a bromodomain containing histone acetyl reader that displays increasing expression in recurrent disease and as PCa progresses from primary to CRPC. Under low hormone conditions, TRIM24 can promote proliferation of PCa cells. This is attributed to its ability to regulate more AR responsive and cell cycle associated genes under hormone-starved compared to hormone-stimulated conditions [85]. Additionally, AR and TRIM24 coactivated genes are upregulated in CRPC and are predictive of recurrence [85]. This has been proposed to be a result of TRIM24 concomitantly binding to acetylated histones and AR, thus anchoring AR to chromatin, under androgen-depleted conditions [85]. As another example, increased levels of NCOA2/SRC-2/TIF2 in post-ADT recurrent PCa are proposed to activate AR signalling by increasing responsiveness of AR to lower affinity androgens [139]. Changes in SWI/SNF remodelling components are also able to contribute to disease progression and hormone independent disease. BAF57, which is upregulated with increasing tumour grade, enhances AR transactivation under androgen-depleted conditions [140].

Epigenetic coregulators that modify histones often also possess the ability to modify and stabilize AR and other proteins, thus contributing to androgen independence through this mechanism. For instance, overexpression of TIP60 in CRPC increases levels of acetylated AR, stabilizing it and consequently leading to increased localization in the nucleus despite the absence of androgens [141]. Increased expression of p300 has been demonstrated to be directly correlated with PCa proliferation, and to be a potential marker predictive of aggressiveness and acquired ADT resistance. One mechanism for this might be its ability to acetylate and stabilize the histone demethylase JMJD1A, which results in enhanced AR activity and resistance to enzalutamide [142, 143]. In the case of the MED1 mediator subunit, phosphorylation by ERK or CDK7 is required for its coactivator activity [137, 144]. In enzalutamide resistant PCa cells, increased levels of phosphorylated MED1 are suggested to contribute to restored AR signalling [137].

Additionally, the ability of some coregulators to scaffold interactions between AR and other factors can indirectly promote aberrant AR activity. ING3 promotes activation of the AR by serving as a scaffold to increase interaction between AR and TIP60 [84]. This consequently leads to increased AR stability through acetylation, and activation of target gene transcription [84]. ING3 is potentially important for androgen independent growth since knockdown of this protein prevented cell growth under conditions that mimic ADT [84].

Finally, coregulators have been shown to compensate for loss of AR signalling by inducing expression of genes that drive PCa growth. Phosphorylated MED1 can induce expression of the AR target UBE2C through chromatin looping, in both AR positive and negative ADT-R-PC, to drive cell growth [145]. In another example, p300 promotes androgen independent expression from the canonical AR target PSA promoter following long-term exposure of cells to IL-6, a cytokine elevated in patients with androgen-independent disease [146].

16.5 Therapeutic Targeting of AR Epigenetic Coregulators

Epigenetic enzymes have been of interest as therapeutic targets for the last few decades for several reasons, including: the reversible nature of epigenetic modifications; the tendency of epigenetic proteins to be differentially expressed in disease conditions; and the ability to inhibit these proteins using small molecule inhibitors [147]. Several small molecule inhibitors are available for epigenetic coregulators that coactivate AR function, and have been tested preclinically or in early clinical trials for prostate cancer (Table 16.2). For the p300/CBP HAT, perhaps the most promising candidate so far is CCS1477 (Inobrodib), a potent and selective bromodomain inhibitor currently in Phase1/2 trials for metastatic PCa and other solid tumours [46]. In vitro, this molecule inhibits PCa cell growth as well as signalling by AR or AR splice variants, and demonstrates anti-tumour activity in vivo [46]. Of the MYST family of HATs, inhibitors exist for KAT5,7 and 8 coactivators, although none have progressed beyond testing in cell line models. The NU9056 inhibitor for KAT5/TIP60 can affect AR levels and expression of PSA in PCa cell lines, potentially via inhibiting acetylase activity [49]. Furthermore, ADT-R-PC cell line models were more sensitive to NU9056 compared to androgen responsive lines, suggesting therapeutic potential [49]. Likewise, histone deacetylase inhibitors (HDACIs) are antiproliferative in preclinical models of prostate cancer. Inhibiting HDACs in PCa may seem counterintuitive given that HAT activity activates AR signalling, however HDACs are frequently upregulated in PCa and their inhibition has been found to supress AR signalling. As mentioned earlier, HDAC 1 and 3 can activate AR transcription by coactivator and PolII recruitment, an effect that is abrogated by HDACIs [118]. As another example, the LAQ824 HDACI represses AR activity by inducing acetylation of the HSP90 chaperone protein, which leads to its dissociation from the AR and subsequently AR degradation [153]. HDACIs also reduce AR mRNA and protein at the transcriptional level [118]. HDACIs are more effective in AR-positive prostate cell lines supporting the concept of these drugs acting in part through AR signalling Indeed, synergistic effects have been observed in vitro on combining HDAC inhibitors with the anti-androgen bicalutamide and such combinations have been assessed in clinical trials [154, 155]. Bicalutamide has been shown to repress AR gene expression by recruiting HDACs [117], so synergistic effects observed on cotreatment with HDACIs are potentially mediated by other pathways. Histone methyltransferase and demethylase inhibitors are similarly promising candidates for therapeutic use with some, like the EZH2 inhibitor GSK126, showing synergistic effects with enzalutamide [151] (Table 16.2). Unlike acetylation, transcriptional effects of methylation tend to be residue specific, hence either methyltransferase or demethylase inhibitors may be required for inhibiting growth.

16.6 Conclusion

Androgen receptor coregulation by epigenetic enzymes is integral to its transcriptional activity. Epigenetic coregulators modulate AR transcriptional activity by diverse mechanisms, some of which have been adapted by prostate cancer cells

Table 16.2 Inhibitors targeting androgen receptor epigenetic coregulators in prostate cancer

Coregulator	Inhibitors	Testing status in prostate cancer	References
p300/CBP	CCS1477	Phase 1/2 clinical trial	[46]
HDACs	Panobinostat, Vorinostat	Phase 1/2 clinical trial (with Bicalutamide)	NCT00878436 NCT00589472
TIP60/KAT5	NU9056	Cell lines	[49]
CARM1	EZM2302	Cell lines	[148]
NSD2	MCTP-39	Cell lines, mouse xenografts	[69]
KDM7A	TC-E 5002	Cell lines	[80]
KDM1A/LSD1	INCB059872	Cell lines	[149]
KDM4A/B	NSC636819	Cell lines	[150]
EZH2	GSK126 PF-06821497	Cell lines Phase 1 clinical trial (for CRPC)	[151] NCT03460977
TRIM24	TRIM24-C34	Cell lines	[152]

to drive disease progression and/or therapy resistance. Epigenetic targets represent promising targets for PCa therapy, but only a few have currently made it to clinical testing. Further investigations into the role of these proteins in AR signalling have the potential for developing new therapies, particularly those that can work in combination with androgen pathway targeting therapy.

Acknowledgements We gratefully acknowledge support from Prostate Cancer UK (ref RA18-ST2-022) and the Prostate Cancer Foundation during the writing of this review.

References

1. Brinkmann A et al (1999) Mechanisms of androgen receptor activation and function. J Steroid Biochem Mol Biol 69(1–6):307–313
2. Marker PC et al (2003) Hormonal, cellular, and molecular control of prostatic development. Dev Biol 253(2):165–174
3. Dehm SM, Tindall DJ (2006) Molecular regulation of androgen action in prostate cancer. J Cell Biochem 99(2):333–344
4. Heinlein CA, Chang C (2002) Androgen receptor (AR) coregulators: an overview. Endocr Rev 23(2):175–200
5. McKenna NJ, Lanz RB, O'Malley BW (1999) Nuclear receptor coregulators: cellular and molecular biology. Endocr Rev 20(3):321–344
6. Giudici M et al (2015) Nuclear receptor coregulators in metabolism and disease. Handbook Exp Pharmacol 233:95–135
7. Cai C et al (2011) Androgen receptor gene expression in prostate cancer is directly suppressed by the androgen receptor through recruitment of lysine-specific demethylase 1. Cancer Cell 20(4):457–471
8. Liu S et al (2017) A comprehensive analysis of coregulator recruitment, androgen receptor function and gene expression in prostate cancer. elife 6:e28482
9. Heemers HV, Tindall DJ (2007) Androgen receptor (AR) coregulators: a diversity of functions converging on and regulating the AR transcriptional complex. Endocr Rev 28(7):778–808
10. Yegnasubramanian S, De Marzo AM, Nelson WG (2019) Prostate cancer epigenetics: from basic mechanisms to clinical implications. Cold Spring Harb Perspect Med 9(4):a030445
11. Kukkonen K et al (2021) Chromatin and epigenetic dysregulation of prostate cancer development, progression, and therapeutic response. Cancers 13(13):3325
12. Brinkmann A et al (1989) The human androgen receptor: domain structure, genomic organization and regulation of expression. J Steroid Biochem 34(1–6):307–310
13. Bevan CL et al (1999) The AF1 and AF2 domains of the androgen receptor interact with distinct regions of SRC1. Mol Cell Biol 19(12):8383–8392
14. Jehle K et al (2014) Coregulator control of androgen receptor action by a novel nuclear receptor-binding motif. J Biol Chem 289(13):8839–8851
15. Zhou Z et al (1994) A ligand-dependent bipartite nuclear targeting signal in the human androgen receptor. Requirement for the DNA-binding domain and modulation by NH2-terminal and carboxyl-terminal sequences. J Biol Chem 269(18):13115–13123
16. Dehm SM, Tindall DJ (2007) Androgen receptor structural and functional elements: role and regulation in prostate cancer. Mol Endocrinol 21(12):2855–2863
17. Pratt WB, Toft DO (1997) Steroid receptor interactions with heat shock protein and immunophilin chaperones. Endocr Rev 18(3):306–360
18. Lonergan PE, Tindall DJ (2011) Androgen receptor signaling in prostate cancer development and progression. J Carcinog 10:20
19. He B et al (1999) Activation function 2 in the human androgen receptor ligand binding domain mediates interdomain communication with the NH2-terminal domain. J Biol Chem 274(52):37219–37225
20. He B et al (2002) The FXXLF motif mediates androgen receptor-specific interactions with coregulators. J Biol Chem 277(12):10226–10235
21. Dubbink HJ et al (2004) Distinct recognition modes of FXXLF and LXXLL motifs by the androgen receptor. Mol Endocrinol 18(9):2132–2150
22. Hodgson MC et al (2008) Structural basis for nuclear receptor corepressor recruitment by antagonist-liganded androgen receptor. Mol Cancer Ther 7(10):3187–3194
23. Tan M et al (2015) Androgen receptor: structure, role in prostate cancer and drug discovery. Acta Pharmacol Sin 36(1):3–23
24. Shin S, Janknecht R (2007) Activation of androgen receptor by histone demethylases JMJD2A and JMJD2D. Biochem Biophys Res Commun 359(3):742–746
25. Tremethick DJ (2007) Higher-order structures of chromatin: the elusive 30 nm fiber. Cell 128(4):651–654
26. Bednar J et al (1998) Nucleosomes, linker DNA, and linker histone form a unique structural motif that directs the higher-order folding and compaction of chromatin. Proc Natl Acad Sci 95(24):14173–14178
27. Hager GL, McNally JG, Misteli T (2009) Transcription dynamics. Mol Cell 35(6):741–753
28. Huang ZQ et al (2003) A role for cofactor–cofactor and cofactor–histone interactions in targeting p300, SWI/SNF and Mediator for transcription. EMBO J 22(9):2146–2155

29. Zaret KS, Carroll JS (2011) Pioneer transcription factors: establishing competence for gene expression. Genes Dev 25(21):2227–2241
30. Cirillo LA et al (2002) Opening of compacted chromatin by early developmental transcription factors HNF3 (FoxA) and GATA-4. Mol Cell 9(2):279–289
31. Zaret KS (2020) Pioneer transcription factors initiating gene network changes. Annu Rev Genet 54:367–385
32. Evans RM, Mangelsdorf DJ (2014) Nuclear receptors, RXR, and the big bang. Cell 157(1):255–266
33. Pihlajamaa P et al (2014) Tissue-specific pioneer factors associate with androgen receptor cistromes and transcription programs. EMBO J 33(4):312–326
34. Hankey W, Chen Z, Wang Q (2020) Shaping chromatin states in prostate cancer by pioneer transcription factors. Cancer Res 80(12):2427–2436
35. Robinson JL et al (2014) Elevated levels of FOXA1 facilitate androgen receptor chromatin binding resulting in a CRPC-like phenotype. Oncogene 33(50):5666–5674
36. Pomerantz MM et al (2015) The androgen receptor cistrome is extensively reprogrammed in human prostate tumorigenesis. Nat Genet 47(11):1346
37. Sahu B et al (2011) Dual role of FoxA1 in androgen receptor binding to chromatin, androgen signalling and prostate cancer. EMBO J 30(19):3962–3976
38. Chaytor L et al (2019) The pioneering role of GATA2 in androgen receptor variant regulation is controlled by bromodomain and extraterminal proteins in castrate-resistant prostate cancer. Mol Cancer Res 17(6):1264–1278
39. Jones D et al (2015) FOXA1 regulates androgen receptor variant activity in models of castrate-resistant prostate cancer. Oncotarget 6(30):29782
40. He B et al (2014) GATA2 facilitates steroid receptor coactivator recruitment to the androgen receptor complex. Proc Natl Acad Sci 111(51):18261–18266
41. Jozwik KM et al (2016) FOXA1 directs H3K4 monomethylation at enhancers via recruitment of the methyltransferase MLL3. Cell Rep 17(10):2715–2723
42. Andreu-Vieyra C et al (2011) Dynamic nucleosome-depleted regions at androgen receptor enhancers in the absence of ligand in prostate cancer cells. Mol Cell Biol 31(23):4648–4662
43. He HH et al (2010) Nucleosome dynamics define transcriptional enhancers. Nat Genet 42(4):343
44. Tewari AK et al (2012) Chromatin accessibility reveals insights into androgen receptor activation and transcriptional specificity. Genome Biol 13(10):1–17
45. Narlikar GJ, Fan H-Y, Kingston RE (2002) Cooperation between complexes that regulate chromatin structure and transcription. Cell 108(4):475–487
46. Welti J et al (2021) Targeting the p300/CBP axis in lethal prostate cancer. Cancer Discov 11(5):1118–1137
47. Waddell AR, Huang H, Liao D (2021) CBP/p300: critical co-activators for nuclear steroid hormone receptors and emerging therapeutic targets in prostate and breast cancers. Cancers 13(12):2872
48. Gong A-Y et al (2012) miR-17-5p targets the p300/CBP-associated factor and modulates androgen receptor transcriptional activity in cultured prostate cancer cells. BMC Cancer 12(1):1–10
49. Coffey K et al (2012) Characterisation of a Tip60 specific inhibitor, NU9056, in prostate cancer 7(10) :1–12
50. Sharma M et al (2000) Androgen receptor interacts with a novel MYST protein, HBO1. J Biol Chem 275(45):35200–35208
51. Jaganathan A et al (2014) Coactivator MYST1 regulates nuclear factor-κB and androgen receptor functions during proliferation of prostate cancer cells. Mol Endocrinol 28(6):872–885
52. Agoulnik IU et al (2005) Role of SRC-1 in the promotion of prostate cancer cell growth and tumor progression. Cancer Res 65(17):7959–7967
53. Agoulnik IU et al (2006) Androgens modulate expression of transcription intermediary factor 2, an androgen receptor coactivator whose expression level correlates with early biochemical recurrence in prostate cancer. Cancer Res 66(21):10594–10602
54. Zhou H-J et al (2005) SRC-3 is required for prostate cancer cell proliferation and survival. Cancer Res 65(17):7976–7983
55. Gaughan L et al (2002) Tip60 and histone deacetylase 1 regulate androgen receptor activity through changes to the acetylation status of the receptor. J Biol Chem 277(29):25904–25913
56. Weichert W et al (2008) Histone deacetylases 1, 2 and 3 are highly expressed in prostate cancer and HDAC2 expression is associated with shorter PSA relapse time after radical prostatectomy. Br J Cancer 98(3):604–610
57. Chng KR et al (2012) A transcriptional repressor co-regulatory network governing androgen response in prostate cancers. EMBO J 31(12):2810–2823
58. Karvonen U, Jänne OA, Palvimo JJ (2006) Androgen receptor regulates nuclear trafficking and nuclear domain residency of corepressor HDAC7 in a ligand-dependent fashion. Exp Cell Res 312(16):3165–3183
59. Fu M et al (2006) Hormonal control of androgen receptor function through SIRT1. Mol Cell Biol 26(21):8122–8135
60. Jung-Hynes B et al (2009) Role of sirtuin histone deacetylase SIRT1 in prostate cancer: a target for prostate cancer management via its inhibition? J Biol Chem 284(6):3823–3832
61. Lopez SM et al (2016) Nuclear receptor corepressor 1 expression and output declines with prostate cancer progression. Clin Cancer Res 22(15):3937–3949
62. Burd CJ, Morey LM, Knudsen KE (2006) Androgen receptor corepressors and prostate cancer. Endocr Relat Cancer 13(4):979–994
63. Long MD et al (2021) Reduced NCOR2 expression accelerates androgen deprivation therapy failure in prostate cancer. Cell Rep 37(11):110109

64. Hong H et al (2004) Aberrant expression of CARM1, a transcriptional coactivator of androgen receptor, in the development of prostate carcinoma and androgen-independent status. Cancer Interdiscip Int J Am Cancer Soc 101(1):83–89
65. Liu Q et al (2019) Polycomb group proteins EZH2 and EED directly regulate androgen receptor in advanced prostate cancer. Int J Cancer 145(2):415–426
66. Wang H et al (2001) Methylation of histone H4 at arginine 3 facilitating transcriptional activation by nuclear hormone receptor. Science 293(5531):853–857
67. Bianco-Miotto T et al (2010) Global levels of specific histone modifications and an epigenetic gene signature predict prostate cancer progression and development. Cancer Epidemiol Prev Biomark 19(10):2611–2622
68. Wang X et al (2001) Identification and characterization of a novel androgen receptor coregulator ARA267-α in prostate cancer cells. J Biol Chem 276(44):40417–40423
69. Aytes A et al (2018) NSD2 is a conserved driver of metastatic prostate cancer progression. Nat Commun 9(1):1–14
70. Kang H-B et al (2009) The histone methyltransferase, NSD2, enhances androgen receptor-mediated transcription. FEBS Lett 583(12):1880–1886
71. Lee K-H et al (2020) MLL5, a histone modifying enzyme, regulates androgen receptor activity in prostate cancer cells by recruiting co-regulators, HCF1 and SET1. BMB Rep 53(12):634
72. Ko S et al (2011) Lysine methylation and functional modulation of androgen receptor by Set9 methyltransferase. Mol Endocrinol 25(3):433–444
73. Gaughan L et al (2011) Regulation of the androgen receptor by SET9-mediated methylation. Nucleic Acids Res 39(4):1266–1279
74. Lee DY et al (2006) Histone H3 lysine 9 methyl transferase G9a is a transcriptional coactivator for nuclear receptors. J Biol Chem 281(13):8476–8485
75. Metzger E et al (2005) LSD1 demethylates repressive histone marks to promote androgen-receptor-dependent transcription. Nature 437(7057):436–439
76. Wilson S et al (2017) The histone demethylase KDM3A regulates the transcriptional program of the androgen receptor in prostate cancer cells. Oncotarget 8(18):30328
77. Coffey K et al (2013) The lysine demethylase, KDM4B, is a key molecule in androgen receptor signalling and turnover. Nucleic Acids Res 41(8):4433–4446
78. Wissmann M et al (2007) Cooperative demethylation by JMJD2C and LSD1 promotes androgen receptor-dependent gene expression. Nat Cell Biol 9(3):347–353
79. Xiang Y et al (2007) JARID1B is a histone H3 lysine 4 demethylase up-regulated in prostate cancer. Proc Natl Acad Sci 104(49):19226–19231
80. Lee KH et al (2018) Histone demethylase KDM7A controls androgen receptor activity and tumor growth in prostate cancer. Int J Cancer 143(11):2849–2861
81. Wang H-J et al (2019) KDM8/JMJD5 as a dual coactivator of AR and PKM2 integrates AR/EZH2 network and tumor metabolism in CRPC. Oncogene 38(1):17–32
82. Esmaeili M et al (2016) The tumor suppressor ING1b is a novel corepressor for the androgen receptor and induces cellular senescence in prostate cancer cells. J Mol Cell Biol 8(3):207–220
83. Esmaeili M et al (2016) A novel crosstalk between the tumor suppressors ING1 and ING2 regulates androgen receptor signaling. J Mol Med 94(10):1167–1179
84. Nabbi A et al (2017) ING3 promotes prostate cancer growth by activating the androgen receptor. BMC Med 15(1):1–14
85. Groner AC et al (2016) TRIM24 is an oncogenic transcriptional activator in prostate cancer. Cancer Cell 29(6):846–858
86. Yang Y et al (2010) TDRD3 is an effector molecule for arginine-methylated histone marks. Mol Cell 40(6):1016–1023
87. Link KA et al (2005) BAF57 governs androgen receptor action and androgen-dependent proliferation through SWI/SNF. Mol Cell Biol 25(6):2200–2215
88. Van De Wijngaart DJ et al (2009) Functional screening of FxxLF-like peptide motifs identifies SMARCD1/BAF60a as an androgen receptor cofactor that modulates TMPRSS2 expression. Mol Endocrinol 23(11):1776–1786
89. Sun A et al (2007) Aberrant expression of SWI/SNF catalytic subunits BRG1/BRM is associated with tumor development and increased invasiveness in prostate cancers. Prostate 67(2):203–213
90. Menon T, Yates JA, Bochar DA (2010) Regulation of androgen-responsive transcription by the chromatin remodeling factor CHD8. Mol Endocrinol 24(6):1165–1174
91. Hong CY et al (2005) Modulation of androgen receptor transactivation by the SWI3-related gene product (SRG3) in multiple ways. Mol Cell Biol 25(12):4841–4852
92. Launonen K-M et al (2021) Chromatin-directed proteomics-identified network of endogenous androgen receptor in prostate cancer cells. Oncogene:1–13
93. Mittal P, Roberts CW (2020) The SWI/SNF complex in cancer—biology, biomarkers and therapy. Nat Rev Clin Oncol 17(7):435–448
94. Reisman D, Glaros S, Thompson E (2009) The SWI/SNF complex and cancer. Oncogene 28(14):1653–1668
95. Shen H et al (2008) The SWI/SNF ATPase Brm is a gatekeeper of proliferative control in prostate cancer. Cancer Res 68(24):10154–10162
96. Marshall TW et al (2003) Differential requirement of SWI/SNF for androgen receptor activity. J Biol Chem 278(33):30605–30613

97. Marfella CG, Imbalzano AN (2007) The Chd family of chromatin remodelers. Mutat Res Fundam Mol Mech Mutagen 618(1–2):30–40
98. Murawska M, Brehm A (2011) CHD chromatin remodelers and the transcription cycle. Transcription 2(6):244–253
99. Augello MA et al (2019) CHD1 loss alters AR binding at lineage-specific enhancers and modulates distinct transcriptional programs to drive prostate tumorigenesis. Cancer Cell 35(4):603–617. e8
100. Burkhardt L et al (2013) CHD1 is a 5q21 tumor suppressor required for ERG rearrangement in prostate cancer. Cancer Res 73(9):2795–2805
101. Kornberg RD (1974) Chromatin structure: a repeating unit of histones and DNA. Science 184(4139):868–871
102. Martire S, Banaszynski LA (2020) The roles of histone variants in fine-tuning chromatin organization and function. Nat Rev Mol Cell Biol 21(9):522–541
103. Mersfelder EL, Parthun MR (2006) The tale beyond the tail: histone core domain modifications and the regulation of chromatin structure. Nucleic Acids Res 34(9):2653–2662
104. Bannister AJ, Kouzarides T (2011) Regulation of chromatin by histone modifications. Cell Res 21(3):381–395
105. Taylor GC et al (2013) H4K16 acetylation marks active genes and enhancers of embryonic stem cells, but does not alter chromatin compaction. Genome Res 23(12):2053–2065
106. Karmodiya K et al (2012) H3K9 and H3K14 acetylation co-occur at many gene regulatory elements, while H3K14ac marks a subset of inactive inducible promoters in mouse embryonic stem cells. BMC Genomics 13(1):1–18
107. Shang Y, Myers M, Brown M (2002) Formation of the androgen receptor transcription complex. Mol Cell 9(3):601–610
108. Spencer TE et al (1997) Steroid receptor coactivator-1 is a histone acetyltransferase. Nature 389(6647):194–198
109. Xu J, Wu R-C, O'Malley BW (2009) Normal and cancer-related functions of the p160 steroid receptor co-activator (SRC) family. Nat Rev Cancer 9(9):615–630
110. Nakka M, Agoulnik IU, Weigel NL (2013) Targeted disruption of the p160 coactivator interface of androgen receptor (AR) selectively inhibits AR activity in both androgen-dependent and castration-resistant AR-expressing prostate cancer cells. Int J Biochem Cell Biol 45(4):763–772
111. Jin Q et al (2011) Distinct roles of GCN5/PCAF-mediated H3K9ac and CBP/p300-mediated H3K18/27ac in nuclear receptor transactivation. EMBO J 30(2):249–262
112. Louie MC et al (2003) Androgen-induced recruitment of RNA polymerase II to a nuclear receptor–p160 coactivator complex. Proc Natl Acad Sci 100(5):2226–2230
113. Ianculescu I et al (2012) Selective roles for cAMP response element-binding protein binding protein and p300 protein as coregulators for androgen-regulated gene expression in advanced prostate cancer cells. J Biol Chem 287(6):4000–4013
114. Raisner R et al (2018) Enhancer activity requires CBP/P300 bromodomain-dependent histone H3K27 acetylation. Cell Rep 24(7):1722–1729
115. Halkidou K et al (2003) Expression of Tip60, an androgen receptor coactivator, and its role in prostate cancer development. Oncogene 22(16):2466–2477
116. Kim J-Y et al (2016) KAT8 regulates androgen signaling in prostate cancer cells. Mol Endocrinol 30(8):925–936
117. Dai Y et al (2007) Sirtuin 1 is required for antagonist-induced transcriptional repression of androgen-responsive genes by the androgen receptor. Mol Endocrinol 21(8):1807–1821
118. Welsbie DS et al (2009) Histone deacetylases are required for androgen receptor function in hormone-sensitive and castrate-resistant prostate cancer. Cancer Res 69(3):958–966
119. Coffey K, Robson CN (2012) Regulation of the androgen receptor by post-translational modifications. J Endocrinol 215(2):221–237
120. Fu M et al (2000) p300 and P/CAF acetylate the androgen receptor at sites governing hormone-dependent transactivation. J Biol Chem 275:20853–20860
121. Fu M et al (2003) Acetylation of androgen receptor enhances coactivator binding and promotes prostate cancer cell growth. Mol Cell Biol 23(23):8563–8575
122. Yang Y et al (2011) Inhibition of androgen receptor activity by histone deacetylase 4 through receptor SUMOylation. Oncogene 30(19):2207–2218
123. Greer EL, Shi Y (2012) Histone methylation: a dynamic mark in health, disease and inheritance. Nat Rev Genet 13(5):343–357
124. Hyun K et al (2017) Writing, erasing and reading histone lysine methylations. Exp Mol Med 49(4):e324–e324
125. Majumder S et al (2006) Involvement of arginine methyltransferase CARM1 in androgen receptor function and prostate cancer cell viability. Prostate 66(12):1292–1301
126. Chen D et al (1999) Regulation of transcription by a protein methyltransferase. Science 284(5423):2174–2177
127. Cai C et al (2014) Lysine-specific demethylase 1 has dual functions as a major regulator of androgen receptor transcriptional activity. Cell Rep 9(5):1618–1627
128. Deng X et al (2017) Protein arginine methyltransferase 5 functions as an epigenetic activator of the androgen receptor to promote prostate cancer cell growth. Oncogene 36(9):1223–1231
129. Metzger E et al (2008) Phosphorylation of histone H3 at threonine 11 establishes a novel chromatin mark for transcriptional regulation. Nat Cell Biol 10(1):53–60

130. Metzger E et al (2010) Phosphorylation of histone H3T6 by PKCβ I controls demethylation at histone H3K4. Nature 464(7289):792–796
131. Arrowsmith CH et al (2012) Epigenetic protein families: a new frontier for drug discovery. Nat Rev Drug Discov 11(5):384–400
132. Takeda DY et al (2018) A somatically acquired enhancer of the androgen receptor is a noncoding driver in advanced prostate cancer. Cell 174(2):422–432. e13
133. Stelloo S, Bergman AM, Zwart W (2019) Androgen receptor enhancer usage and the chromatin regulatory landscape in human prostate cancers. Endocr Relat Cancer 26(5):R267–R285
134. Wang Q, Carroll JS, Brown M (2005) Spatial and temporal recruitment of androgen receptor and its coactivators involves chromosomal looping and polymerase tracking. Mol Cell 19(5):631–642
135. Wang Q et al (2007) A hierarchical network of transcription factors governs androgen receptor-dependent prostate cancer growth. Mol Cell 27(3):380–392
136. Russo JW, Nouri M, Balk SP (2019) Androgen receptor interaction with mediator complex is enhanced in castration-resistant prostate cancer by CDK7 phosphorylation of MED1. Cancer Discov 9(11):1490–1492
137. Ur Rasool R et al (2019) CDK7 inhibition suppresses castration-resistant prostate cancer through MED1 inactivation. Cancer Discov 9(11):1538–1555
138. Pienta KJ, Bradley D (2006) Mechanisms underlying the development of androgen-independent prostate cancer. Clin Cancer Res 12(6):1665–1671
139. Gregory CW et al (2001) A mechanism for androgen receptor-mediated prostate cancer recurrence after androgen deprivation therapy. Cancer Res 61(11):4315–4319
140. Balasubramaniam S et al (2013) Aberrant BAF57 signaling facilitates prometastatic phenotypes. Clin Cancer Res 19(10):2657–2667
141. Shiota M et al (2010) Tip60 promotes prostate cancer cell proliferation by translocation of androgen receptor into the nucleus. Prostate 70(5):540–554
142. Xu S et al (2020) p300-mediated acetylation of histone demethylase JMJD1A prevents its degradation by ubiquitin ligase STUB1 and enhances its activity in prostate cancer. Cancer Res 80(15):3074–3087
143. Debes JD et al (2003) p300 in prostate cancer proliferation and progression. Cancer Res 63(22):7638–7640
144. Belakavadi M et al (2008) MED1 phosphorylation promotes its association with mediator: implications for nuclear receptor signaling. Mol Cell Biol 28(12):3932–3942
145. Chen Z et al (2011) Phospho-MED1-enhanced UBE2C locus looping drives castration-resistant prostate cancer growth. EMBO J 30(12):2405–2419
146. Debes JD et al (2005) p300 regulates androgen receptor–independent expression of prostate-specific antigen in prostate cancer cells treated chronically with interleukin-6. Cancer Res 65(13):5965–5973
147. Ganesan A et al (2019) The timeline of epigenetic drug discovery: from reality to dreams. Clin Epigenetics 11(1):1–17
148. Drew AE et al (2017) Identification of a CARM1 inhibitor with potent in vitro and in vivo activity in preclinical models of multiple myeloma. Sci Rep 7(1):1–13
149. Civenni G et al (2018) INCB059872, a novel FAD-directed LSD1 Inhibitor, is active in prostate cancer models and impacts prostate cancer stem-like cells. AACR
150. Chu C-H et al (2014) KDM4B as a target for prostate cancer: structural analysis and selective inhibition by a novel inhibitor. J Med Chem 57(14):5975–5985
151. Shankar E et al (2020) Dual targeting of EZH2 and androgen receptor as a novel therapy for castration-resistant prostate cancer. Toxicol Appl Pharmacol 404:115200
152. Fong K-w et al (2018) TRIM28 protects TRIM24 from SPOP-mediated degradation and promotes prostate cancer progression. Nat Commun 9(1):1–15
153. Chen L et al (2005) Chemical ablation of androgen receptor in prostate cancer cells by the histone deacetylase inhibitor LAQ824. Mol Cancer Ther 4(9):1311–1319
154. Marrocco DL et al (2007) Suberoylanilide hydroxamic acid (vorinostat) represses androgen receptor expression and acts synergistically with an androgen receptor antagonist to inhibit prostate cancer cell proliferation. Mol Cancer Ther 6(1):51–60
155. Suraweera A, O'Byrne KJ, Richard DJ (2018) Combination therapy with histone deacetylase inhibitors (HDACi) for the treatment of cancer: achieving the full therapeutic potential of HDACi. Front Oncol 8:92

Part VI

Clinical Translation

17 Clinical Translation: Targeting the Estrogen Receptor

Ciara Metcalfe and Jennifer O. Lauchle

Abstract

Estrogen Receptor alpha (ERα) stands as one of the most successfully prosecuted drug targets in oncology, beginning with the approval of tamoxifen for women with ERα positive (ER+) breast cancer over 40 years ago. The field continued to advance with the development of aromatase inhibitors and the pure antiestrogen fulvestrant. With multiple endocrine therapies approved for the treatment of ER+ breast cancer, efforts to generate novel ERα-targeted therapeutics somewhat diminished in the early 2000s. Today however, there are at least eight new molecular entities targeting ERα under active clinical investigation, each with the aim of bringing further benefit to patients. This remarkable re-energizing of the field was spurred in part by the discovery of highly prevalent ERα mutations as a mechanism of resistance to standard-of-care therapies, which provided unequivocal evidence of the continued, and broad, dependence of tumors on ERα, despite relapsing after earlier lines of endocrine therapy. Re-engagement of the pharmaceutical and biotechnology industries with ERα as a drug target has been further underpinned by the impressive advances made in medicinal chemistry, enabling desirable mechanistic features – high potency full ERα antagonism – to be combined with improved drug-like properties – oral bioavailability and optimized pharmacokinetics. In this chapter, we describe the rich history and science behind the currently evolving landscape of ERα targeting in breast cancer.

Keywords

Breast cancer · Estrogen receptor · Endocrine therapy · Tamoxifen (SERM) · Aromtase inhibitors · Fulvestrant (SERD)

17.1 Looking Back: The Beginnings of Endocrine Targeting

Though tamoxifen stands as the first approved small molecule agent for the treatment of ER+ breast cancer, modulation of the endocrine axis has been exploited therapeutically since the late 1890s. Dr. George Beatson, ascribing to the philosophy that an understanding of normal physiology can inform our understanding of the diseased state, leveraged observations from the mammary

C. Metcalfe (✉)
Department of Discovery Oncology, Genentech, South San Francisco, CA, USA
e-mail: metcalfe.ciara@gene.com

J. O. Lauchle
Early Clinical Development, Genentech, South San Francisco, CA, USA

M. J. Campbell, C. L. Bevan (eds.), *Nuclear Receptors in Human Health and Disease*, Advances in Experimental Medicine and Biology 1390, https://doi.org/10.1007/978-3-031-11836-4_17

glands of lactating sheep and rabbits to hypothesize that the ovaries may provide key proliferative signals supporting breast cancer growth [1]. Motivated by this hypothesis, he was the first to perform a bilateral oophorectomy, and reported meaningful clinical benefit. This surgical strategy became standard-of-care for advanced breast cancer in the following years, and was subsequently replaced by oophorectomy by radiation, and then in the mid 1990s by chemical ovarian suppression, through treatment of premenopausal women with synthetic analogs of luteinizing hormone releasing hormone (LHRH agonists) [2].

17.2 The Advent of Small Molecule Modulators of ER Function

The identification and isolation of the "*primary ovarian hormone*" now known as estrogen, at the beginning of the twentieth century, ultimately unlocked our ability to manipulate relevant hormonal signaling events with small molecules [3]. This discovery by Edgar Allen and Edward Doisy paved the way for the creation of synthetic estrogens, as well as anti-estrogenic derivatives. One such derivative was ICI 46,474 (later known as tamoxifen), synthesized by Dora Richardson at Imperial Chemical Industries (ICI, which was to become Zeneca, and subsequently AstraZeneca). Initially designed as an anti-estrogen in the context of ICI's oral contraceptive program, tamoxifen was found to stimulate, rather than suppress ovulation in women, scuppering its potential as a contraceptive agent. Fortunately, Arthur Walpole, the biologist who led the contraceptive research team at ICI had a parallel interest in leveraging anti-estrogens for the treatment of breast cancer. When the contraceptive program stalled, Walpole, working together with V. Craig Jordan and others, kept tamoxifen alive as an anti-cancer agent. The non-linear development path and series of serendipitous events leading to the approval of tamoxifen, first as a treatment for advanced breast cancer, and then as the first chemopreventive/adjuvant treatment for any cancer, is well described and documented by V. Craig Jordan [3, 4] and also by Viviane Quirke [5].

The surprising observation that tamoxifen stimulated, rather than suppressed, ovulation in women was an early indicator of the complex pharmacology for which tamoxifen is now well-recognized. Today, tamoxifen is designated as a selective ERα modulator, or SERM, describing its function in *modulating*, rather than fully and consistently antagonizing, ERα activity. This nuance is tied to the molecular make-up of the ERα protein, which harbors three major functional domains: (1) the N-terminal activation function 1 (AF1) domain (2) a central DNA-binding domain and (3) a C-terminal ligand binding domain (LBD), that receives the endogenous activating ligand, estrogen, as well as the suppressive therapeutic ligands. Both the AF1 and LBD regions have the potential to regulate expression of ERα target genes. Upon binding to the LBD, tamoxifen [more specifically, its active metabolite 4-hydroxytamoxifen (4-OHT)] induces a conformation that prevents recruitment of co-activator proteins to this domain. The ability of tamoxifen to outcompete estrogen for binding to the LBD, together with "deactivating" this domain, provides a molecular basis for the reduction of ERα activity - meaning lower expression of ERα target genes - relative to what is achieved by estrogen. Importantly however, despite disabling the LBD, tamoxifen triggers ERα dimerization and increased binding of ERα to chromatin, allowing for activity of the AF1 domain, which can drive some degree of ERα target gene expression, so-called "partial agonist" activity [6, 7]. Since the AF1 domain is sensitive to input from other pathways and co-activator proteins, the relative strength of such partial agonist activity is highly context dependent [8]. Specifically, tamoxifen's partial agonist activity is deemed to be relatively weak in therapy-naïve breast cancer, and stronger in endometrial cells, in line with its ability to suppress estrogen-dependent proliferation of breast cancer cells, while driving estrogen-like signaling and morphological phenotypes in the uterus.

The partial agonist property of tamoxifen is relevant both from a safety perspective, causing an increased risk of endometrial cancer and thromboembolism, and also from the perspective of therapeutic resistance. Continuous *in vivo* passaging of MCF-7 breast cancer cells in the presence of tamoxifen results in cellular adaptations and a "dialing-up" of its partial agonist properties, ultimately generating tumors in which tamoxifen acts to stimulate ERα signaling and proliferation [9, 10]. Such cellular adaptation, whereby the partial agonist effect of tamoxifen is amplified, is likewise thought to contribute to tamoxifen resistance, and sub-optimal activity, in patients.

17.3 Alternative Strategies for Endocrine Suppression: Targeting the Ligand

During the time that the tamoxifen work was unfolding, Angela and Harry Brodie proposed that targeting the production of estrogen may be a more effective strategy for inhibiting ERα activity than targeting the receptor, and may be associated with fewer safety concerns. Estrogen is produced primarily in the ovaries, but can additionally be synthesized from the peripheral conversion of androgens via aromatization. The aromatase enzyme, encoded by the *CYP19A1* gene and expressed in a number of non-ovarian tissues including adipose and bone, is thus the key regulator of estrogen production in women with ovarian suppression, encompassing postmenopausal women.

The Brodie team embarked on a systematic series of structure/function studies, examining nearly 100 steroidal aromatase inhibitors (AIs), which led to the identification of 4-hydroxyandrostenedione (4-OH-A) [11]. A collaborative group including Angela Brodie, Charles Coombes, Paul Goss and Mitch Dowsett launched the first clinical trial of 4-OH-A as a candidate selective AI for the treatment of breast cancer, and demonstrated efficacy of 4-OH-A (subsequently named formestane), including in women who had progressed on tamoxifen. The full historical account of the development of aromatase inhibitors has been described by the Brodies and colleagues in two excellent reviews [12, 13].

There are currently three highly potent and selective AIs approved for the treatment of ER+ breast cancer: exemestane, a type I steroidal inhibitor, and letrozole and anastrozole, type II non-steroidal inhibitors. Exemestane is an analog of the natural aromatase substrate androstenedione, and is converted by aromatase into a reactive intermediate that covalently interacts with the substrate binding domain, permanently deactivating it, and leading to destabilization of the enzyme. It was hypothesized that irreversible inhibition of aromatase by exemestane may lead to superior outcomes versus the irreversible inhibition achieved by the non-steroidal AIs letrozole and anastrozole. The mild androgenic effect of exemestane was additionally speculated to be therapeutically advantageous relative to the non-steroidal inhibitors. However, a large adjuvant study of exemestane versus anastrozole in over 7500 postmenopausal women with early breast cancer (NCIC CTG MA.27) showed neither to be superior [14]. The results of this long-term study – showing that the steroidal and non-steroidal AIs were associated with similar efficacy – were consistent with what was observed in a Phase III neoadjuvant study, in which postmenopausal women with early breast cancer were exposed to 16 weeks of either exemestane, letrozole, or anastrozole prior to surgery (ACOSOG Z1031) [15].

Critically, clinical trials comparing the activity of the AIs to tamoxifen bore out the early hypothesis that targeting the synthesis of estrogen may be more therapeutically effective than modulating the activity of its receptor with tamoxifen. Specifically, in the ATAC trial comparing anastrozole to tamoxifen in postmenopausal women with localised breast cancer, anastrozole significantly prolonged disease-free survival, time-to-recurrence, and significantly reduced distant metastases. Anastrozole was also associated with fewer side-effects than tamoxifen, in particular gynecological and vascular events, but arthralgia (bone pain) and bone fractures were increased [16]. The Breast International Group (BIG) 1-98 Collaborative Group likewise demonstrated that in postmenopausal women with endocrine-sensitive breast cancer, adjuvant treatment with letrozole reduced the risk of recurrent disease, especially at distant sites, relative to what was achieved with tamoxifen [17]. Notably, the superiority of AIs over tamoxifen in these large adjuvant trials, each of which included the study of over 8000 women, were in line with observations generated in short-term neoadjuvant studies. In particular, the IMPACT study was specifically designed to test the hypothesis that the clinical and/or biologic effects of neoadjuvant anastrozole and tamoxifen might predict the outcome of ATAC adjuvant therapy trial. While there was no significant difference in tumor objective response between the treatment arms, as assessed by both caliper and ultrasound, there was a significant difference in suppression of the proliferation marker Ki67, after both 2 and 12 weeks of treatment, with anastrozole being superior to tamoxifen [18, 19]. Interestingly, apoptosis was not increased in any of the treatment arms, providing the first biological demonstration in patient tumors that ERα signaling supports the progression of cells through the cell cycle, but is not required for cell survival. The P024 trial similarly compared both clinical and biological effects of neoadjuvant AI, in this case letrozole, to tamoxifen, and likewise demonstrated that letrozole was significantly more effective than tamoxifen in reducing tumor proliferation, as measured by immunohistochemistry of Ki67 [20]. Together, these studies demonstrated the potential of neoadjuvant trials to predict long-term outcomes comparing different endocrine therapies, and additionally highlighted the value of such studies in providing material for exploratory correlative science.

The AIs have additionally been demonstrated to be superior to tamoxifen in the metastatic setting (reviewed in [21]). Despite making improvements relative to tamoxifen, resistance to AIs still emerges relatively rapidly in metastatic disease, with median time to progression in those pivotal studies being in the range of 8 to 11 months. Sequencing of metastatic tumor DNA and circulating tumor DNA (ctDNA) has revealed that a major source of therapeutic resistance to the AIs is through acquisition of hotspot mutations in the LBD of ERα. Such mutations have been demonstrated to support estrogen-independence of ERα, allowing for re-activation of ERα signaling and its downstream proliferative program despite the action of AIs in suppressing synthesis of the activating ligand. The discovery of the mutations in *NR3A1*/ESR1 (encoding ERα), their biological features, and how they impact patient outcomes is described in more detail in Chap. 12.

17.4 In Pursuit of a Pure Antiestrogen

While the Brodie team had proposed aromatase inhibition as a means to overcome the partial agonist effect of tamoxifen, a team at ICI, principally Jean Bowler and Alan Wakeling, proposed seeking ERα binding ligands that could achieve "*complete blockade of all stimulatory actions of estrogens*", which they defined as pure antiestrogens. They speculated that treatment of breast cancer patients with such ligands "*may result in a more rapid, complete and longer-lasting tumor remission*" relative to what could be achieved with the SERMs [22]. The ICI team drew inspiration from work conducted by a group of French researchers, Robert Bucourt and colleagues, who were developing an affinity chromatography system for the purification of ERα protein. Bucourt et al. generated a series of estrogen derivatives intended to capture ERα protein and showed that chemical "spacer chains" added to the 7α-position of estradiol maintained binding to ERα [23]. This observation led to a systematic medicinal chemistry effort to identify novel pure antiestrogens via modification of long-chain alkyl substitutes in the 7α-position of estradiol. The uterus of rat and mouse was used as a key *in vivo* screening tool, leveraged to identify derivatives that were capable of completely blocking the trophic actions of estrogens AND that were devoid of any estrogenic activity themselves, and were thus unlike the SERMs that had been described thus far. This effort led to the synthesis and characterization of the first steroidal pure antiestrogen ICI 164384, which was further optimized for potency to generate ICI 182780, now known as fulvestrant, named to reflect full estrogen receptor antagonism [24].

After the identification of fulvestrant as a "pure antiestrogen", mechanistic studies conducted in ovariectomized adult female mice showed that fulvestrant treatment led to an acute loss of ERα protein in the uterus, without decreasing the expression of the ESR1 gene. This data led to the hypothesis that fulvestrant "*may cause its antagonistic effect by producing a rapid disappearance of the ER from the target tissue, resulting in an insufficient amount of ERα to bind the native ligand and elicit agonist responses*" [25]. Subsequent preclinical studies, including in ER+ breast cancer cell lines, supported that original observation and further showed fulvestrant-induced ER turnover to be mediated by the ubiquitin-proteasome system [25–29]. A clinical comparison of fulvestrant and tamoxifen additionally showed that a single 250 mg intramuscular dose of fulvestrant led to a greater decrease in ERα protein levels than was achieved by 14–21 days of daily oral tamoxifen dosing (59% decrease by fulvestrant versus 39% by tamoxifen [30];). This same study additionally evaluated levels of the progesterone receptor (PR), as an ERα target gene, and showed fulvestrant to decrease PR protein (−67%), while tamoxifen treatment elevated PR levels (63%), consistent with distinct transcriptional effects of these two agents. The term SERD (selective ERα downregulator) was coined to reflect the distinct behavior of fulvestrant in increasing ERα turnover, and has over time displaced the original descriptors of "pure antiestrogen" and "full ERα antagonist", which were based on the intended signaling effects sought in the campaign leading to the identification of fulvestrant.

While fulvestrant's capacity to deplete ERα protein provided a seemingly compelling mechanistic explanation for its full ERα antagonism, work conducted by Suzanne Wardell and Donald McDonnell raised questions about this proposed mechanism. In interrogating the contribution of ERα degradation to the overall pharmacology of fulvestrant, they demonstrated that ERα degradation is a saturable process that is separable from its antagonist efficacy [31]. They argued that it is fulvestrant's ability to (1) competitively displace estrogen from the LBD and (2) induce a conformational change in ERα that is incompatible with transcriptional activation, that are likely to be the most important pharmacological characteristics of this pure antiestrogen. Indeed, prior to that work, the Mancini team, working with Bert O'Malley, made the observation using live-cell imaging, that tamoxifen and fulvestrant differentially impacted the mobility of ERα, with fulvestrant dramatically and acutely slowing the dynamic movement of ERα within the nucleus [32]. These data, together with the observations from the McDonnell laboratory, suggested that there was perhaps more to the pharmacology of fulvestrant than increased ERα turnover. The subsequent discovery of chemically distinct pure antiestrogens later provided an opportunity to revisit these important mechanistic questions, and is further described below.

As a pure antiestrogen, fulvestrant harbors the highly desirable mechanistic property of fully suppressing the effects of estrogen without exhibiting any estrogenic action on its own; precisely the features ICI was seeking to improve upon the SERMs. However, the clinical development of fulvestrant, and the ability to formally test the hypothesis that full ERα antagonism would lead to superior clinical outcomes, has been challenged by fulvestrant's poor drug like properties. Specifically, fulvestrant's lack of oral bioavailability has necessitated delivery by large-volume intramuscular injection, which created difficulties in determining an optimal dose. Fulvestrant was initially explored in the clinic at 125 mg, but a planned interim analysis found no evidence for clinical efficacy, leading to halting that particular study (reviewed in [33]). Pivotal Phase III studies (Trials 0020 and 0021) comparing fulvestrant to anastrozole in second line metastatic breast cancer were subsequently conducted at a once per month 250 mg intramuscular dosing schedule, and showed fulvestrant to be at least as effective as anastrozole with respect to time to progression, leading to the first approval for fulvestrant in advanced breast cancer [34]. Pharmacokinetic (PK) analyses conducted in the course of these studies demonstrated that it takes approximately 3–6 months for fulvestrant to reach steady-state levels [34, 35]. Based on those PK observations, the 250 mg once per month schedule was next evaluated with an additional "loading dose" at day 14 in the first month of treatment, though eventually, the recommended dose was increased to 500 mg once per month, plus a day 14 loading dose, supported by results from the CONFIRM trial [36]. Critically, this 500 mg dosing regime of fulvestrant proved to be superior to the AI anastrozole, providing proof of concept that direct ERα antagonism by a pure antiestrogen may be superior to indirect targeting of ERα via suppression of estrogen synthesis, for patients with locally advanced or metastatic breast cancer that were endocrine therapy-naïve [37]. A subgroup analysis from this study suggested that there may be enhanced treatment effects with fulvestrant versus anastrozole in patients with non-visceral disease in particular. Interestingly, a meta-analysis of visceral versus non-visceral metastatic ER+ breast cancer, that also included a focused assessment of liver metastases, suggested that the efficacy of endocrine therapies may be dependent on both the particular agent and also the disease-site [38]. Importantly though, beyond ESR1 mutations, the full set of clinical and biological features that might predict for better outcomes for fulvestrant (and SERDs in general) versus AIs remain to be more deeply explored and understood.

17.4.1 Prospective Optimization of ERα Degradation: Contemporary SERD/SERM Hybrids

The challenges faced by fulvestrant, related to lack of oral bioavailability and its route of administration, suggested that there may still be room for improvement for antiestrogens with this particular mechanism of action. Indeed, clinical imaging studies leveraging labeled estradiol to measure ERα availability in metastatic breast cancer tumors showed that fulvestrant, at the currently recommended dose, does not achieve target saturation in all patients i.e. full ERα occupancy [39]. Importantly, residual ERα availability was associated with early progression. Such data suggests that gains in target saturation (meaning increased engagement of antagonists with ERα protein), ideally with an orally bioavailable drug, may drive improvements in patient outcomes. This hypothesis, together with a growing appreciation for the high prevalence of the estrogen-independent ESR1 mutations, that promote resistance to the AIs, has triggered a huge wave of investment in the identification of orally bioavailable SERDs.

Given that a key feature of fulvestrant is its ability to deplete ERα protein, ERα degradation particularly in the context of MCF-7 cells, which have served as a workhorse model, took center stage in strategies to identify orally bioavailable SERD molecules. The first two new molecular entities to emerge from prospective optimization of ER degradation, and that were clinically investigated, were GDC-0810 (originally developed by Seragon and subsequently acquired by Genentech), and AZD9496, developed by AstraZeneca. Intriguingly, while both molecules robustly degraded ERα in MCF-7 cells *in vitro* and *in vivo*, consistent with a fulvestrant-like mechanism of action, they both exhibited partial agonist-like effects in the rodent uterus, albeit to a lesser extent than tamoxifen [40, 41].

Such context-dependent SERD vs SERM activity was initially believed to be tissue-dependent, with partial agonism being restricted to the endometrial context. However, further investigation showed that these molecules do exhibit partial agonist activity in a subset of ER+ breast cancer cell lines, which negatively impacts their anti-proliferative potential relative to the pure antiestrogens that show no estrogenic activity in the uterus, nor in breast cancer cell lines [42]. Thus, neither GDC-0810 nor AZD9496 met the definition of a "pure antiestrogen" originally articulated by Bowler, Wakeling and colleagues i.e. those molecules that would *"complete effectively with oestradiol without inducing any oestrogen-like actions"* [22, 42]. Indeed, though oral bioavailability was achieved with AZD9496, it failed to demonstrate superiority over fulvestrant in a pre-surgical study assessing various pharmacodynamic biomarkers related to on-target pathway suppression and proliferation ([43], described further below), perhaps highlighting the importance of maintaining full ERα antagonism for maximal pathway suppression and anti-proliferative effect.

17.4.2 Latest Generation ER Antagonists: The -Esterants

GDC-0927 followed GDC-0810 in development, as a more potent ERα antagonist that lacked estrogenic activity in the uterus and in breast cancer cell lines, more clearly fitting the criteria of a "pure antiestrogen" [44]. While GDC-0927 made potency and mechanistic gains over GDC-0810, it suffered from suboptimal PK properties, leading to a significant pill burden that precluded progression to pivotal clinical trials. Importantly though, the discovery of GDC-0927 as an additional pure antiestrogen, which has a nonsteroidal chemical structure unrelated to that of fulvestrant, provided an opportunity to revisit the mechanism by which this class of agents avoids ERα agonism. Guided by prior observations from the study of fulvestrant [31, 32], we dissected molecular events following engagement of ERα with GDC-0927, compared to a partial ERα agonist from the same chemical series [42]. These studies showed that the impact of the pure antiestrogens and the partial ERα antagonists diverged prior to depletion of ERα protein. In particular, acute treatment with partial agonists establishes accessibility at ERα binding motifs, while the pure antiestrogens maintain reduced chromatin access at those same sites, despite largely similar ERα chromatin binding. These data were consistent with the hypothesis from the McDonnell team, that features independent of ERα degradation may drive functional antagonism of ERα. Further, the phenotype of nuclear ERα immobilization described for fulvestrant was shown to extend to GDC-0927, but not to the partial agonist from the same chemical series. A set of point mutations in helix 12 of the LBD, including those identified by Benita Katzellenbogen and others as altering the full antagonist profile of fulvestrant [45–47], were shown to prevent ERα immobilization caused by GDC-0927, and impart partial agonist activity to this molecule. Together these data suggest that immobilization of ERα by the pure antiestrogens is a key feature leading to functional suppression of ERα, likely through disabling the N-terminal intrinsically disordered transactivation domain as well as the LBD, with proteasome-mediated turnover of ERα following its immobilization. These observations follow a growing appreciation that the dynamic movement of transcription factors in general, is critical for their function [48, 49].

Independent, highly intensive medicinal chemistry campaigns to generate further-optimized candidates for clinical investigation led to the identification of orally bioavailable pure antiestrogens/SERDs that include giredestrant (GDC-9545), amcenestrant (SAR439859), camizestrant (AZD9833), imlunestrant (LY3484356), rintodestrant (G1T48) and OP-1250 (see Table 17.1) [50–54]. Notably, giredestrant, amcenestrant, camizestrant each immobilize nuclear ERa at saturating concentrations in vitro, consistent with ERa immobilization being a general phenotype of the pure antiestrogens ([55], and data not published). This observation additionally suggests that the mechanism of these molecules is likely highly similar, suggest-

Table 17.1 Classification of ER antagonists

Therapeutic Class	Name
Currently approved agents (as 2022)	Tamoxifen (SERM), Exemestane (AI, type I), Anastrozole (AI, type II), Letrozole (AI, type II), Fulvestrant (SERD)
Investigational full ER antagonists/ SERDs	Amcenestrant, Camizestrant, Giredestrant, Imlunestrant, OP-1250
Investigational SERD/SERM hybrids	Bazedoxifene, Elacestrant, Lasofoxifene
Investigational covalent ER antagonist (SERCA)	H3B-6545
Investigational ER PROTAC	ARV-471

ing that differentiation in the clinic will most likely be driven by a combination of features that include potency, DMPK, safety, and in the context of combination studies (with CDK4/6 inhibitors for example), drug-drug interactions. A series of potentially registrational clinical trials are currently underway, assessing the efficacy and safety of a number of these molecules, compared to standard-of-care therapies, in both metastatic and early ER+ breast cancer. For a detailed comparative description of these molecules see review from Chen et al. [56]. In addition to the new molecular entities described above, molecules originally identified as SERMs but retrospectively observed to have context-dependent ER degradation potential (perhaps best classified as SERD/SERM hybrids) are additionally being revisited in the context of ER+ breast cancer, including elacestrant (RAD1901) [57, 58],, lasofoxifene [59] and bazedoxifene [60, 61].

17.5 Novel Approaches to ER Antagonists: Heterobifunctional Degraders and Covalent Binders

While much of the recent activity in developing next generation ERα therapeutics has focused on monomeric pure antiestrogen/SERD molecules, with fulvestrant serving as inspiration, there have been parallel efforts to pursue novel classes of ERα antagonists. In particular, Arvinas have identified a heterobifunctional proteolysis targeting chimera (PROTAC®), ARV-471, for development in ER+ breast cancer [62]. PROTACs rely on the chemical induction of proximity between a target of interest, in this case ERα, with an E3 ligase, in order to drive ubiquitination of the target, and its subsequent degradation by the proteasome. The complexity of these molecules, requiring optimization of two ligand/target interfaces, together with their increased size relative to traditional small molecule therapeutics, is such that achieving optimal drug-like properties, including oral bioavailability, has been a major challenge. Importantly however, the entry of ARV-471 into the clinic, and the demonstration of on-target pharmacodynamics as well as early signs of anti-tumor activity upon oral delivery [63], has provided important proof-of-principle for the PROTAC approach, setting the stage for its application not only to ERα, but also to additional therapeutic targets. Critically, the mechanism by which ARV-471 degrades ERα, via direct recruitment of an E3 ligase and independent of ERα immobilization, is entirely distinct from ERα degradation driven by the -esterants, which is immobilization dependent. As such, it seems possible that mechanisms of resistance to these classes of agents will also differ, pointing to the possibility of sequencing them, should they progress to approval.

H3B Biosciences have taken yet another approach, leveraging a reactive cysteine in the LBD to generate a covalent (meaning irreversible) inhibitor, defined as a SERCA, selective ER covalent antagonist. The SERCA H3B-6545 is currently being evaluated in a Phase I/II study with preliminary evidence of clinical benefit in patients with mBC who have received prior endocrine therapy [64]. Preclinical data from an earlier generation SERCA, H3B-5942, showed this molecule to exhibit partial agonism similar to that of tamoxifen in rats, as might be anticipated from a potent LBD binder/inhibitor that fails to disable the N-terminal domain through either ER immobilization or degradation [65]. Intriguingly though, H3B-5942 was unlike tamoxifen in its lack of agonism in the Ishikawa cell line in vitro and also in its biochemical co-activator binding profile. Together, these data suggest that while the SERCA class can exhibit ERα agonism, its activity may be distinct from tamoxifen and other SERMs. The full preclinical and clinical profile of H3B-6545, and how those features relate to efficacy and differentiation from oral SERDs and other ER-targeted therapies currently under evaluation in clinical trials, remain to be described.

17.6 Bringing It Back to the Biology: Neoadjuvant Studies as Potentially Valuable Testing Grounds

Neoadjuvant studies of ER+ breast cancer have been utilized to compare the biological effects of ERα therapies. Though pathologic complete response (pCR) is rare in ER+ breast cancer, tumor samples can be leveraged to measure post-treatment changes in ERα protein levels, pathway activity (via levels of target genes) and Ki67. Reduction in Ki67 following neoadjuvant endocrine therapy (often for 2 weeks) is understood to be a valid marker of suppression of cellular proliferation, and correlates with recurrence free survival outcomes for adjuvant therapy (reviewed by Dr. Arteaga and colleagues [66]). Results comparing endocrine therapies in neoadjuvant trials have resulted in similar conclusions as the outcomes of large adjuvant trials, which make it an attractive setting in which to generate data early in the development of new therapies for breast cancer. In addition to comparing effects of different endocrine therapies, for example tamoxifen versus AI, these studies have also been used to evaluate multiple doses of a particular agent. Two fulvestrant regimens (500 mg day 1 and day 14 then every 4 weeks versus 250 mg every 4 weeks) were compared in the NEWEST trial with 16 weeks of dosing prior to surgery [67]. After 4 weeks of treatment, reductions in Ki67 mean label index (mean percent change from baseline −78.8% versus −47.7% $p < 0.0001$) and both ERα and PR were significantly greater with the 500 mg dose of fulvestrant than the 250 mg dose (ER 50% vs. 14%, $p < 0.0001$, PR 81% vs. 46%, $p = 0.0018$). The pattern of greater reduction in the 500 mg cohort compared with 250 mg persisted in the tumor samples evaluated after 16 weeks but were not statistically significant. As the fulvestrant dose of 500 mg led to improved outcomes compared to the 250 mg dose in randomized trials in metastatic breast cancer [36], the possibility that incorporating pharmacodynamic data from neoadjuvant studies may have influenced dose and clinical development decisions, potentially minimizing the number of patients treated with lower doses of fulvestrant, should be considered as we evaluate the clinical activity of the latest collection of novel ERα targeted therapies relative to standard-of-care.

Recently discovered ERα therapeutics have employed neoadjuvant studies for both exploration of dose and comparison with current endocrine therapies to guide subsequent clinical trials. The first presurgical study to evaluate target inhibition of an oral SERD/SERM hybrid was a comparison of 250 mg of AZD9496 for 5–14 days with Fulvestrant 500 mg. AZD9496 exposure resulted in reduction in ERα and PR H-scores, and Ki-67 levels from baseline, but AZD9496 was not superior to fulvestrant in any of these measurements [43].

Giredestrant was initially evaluated in the metastatic setting at doses from 10 mg to 250 mg with several patients demonstrating clinical benefit at the lowest dose levels. Given that all dose levels were well tolerated, this raised questions about the most appropriate dose to further develop. To address this challenge a window of opportunity neoadjuvant study was conducted to evaluate 3 doses of GDC-9545 (10 mg, 30 mg and 100 mg) following 2 weeks treatment. Biological readouts of giredestrant activity included an assessment of ERα and PR protein levels, a predefined ERα activity signature by RNAseq, and reduction in Ki67. The study concluded that the 30 mg dose achieved maximal on-target activity and supported the decision to further develop giredestrant at this dose level [68]. Similarly, in the SERENA-3 neoadjuvant breast cancer trial, patients will be randomized to receive one of two doses of camizestrant for approximately 1 week and will compare ERα protein levels post treatment as primary outcome. Based on this initial dataset, the study has the potential to open a second stage and compare several doses of camizestrant and possibly fulvestrant [69].

The results of the first study comparing an oral SERD/pure antiestrogen versus an aromatase inhibitor in the neoadjuvant setting were reported in December 2021. The coopERA study was designed to compare the Ki67 reduction from baseline following 2 weeks of treatment with either giredestrant or anastrazole as the primary analysis [70]. Following 2 weeks of treatment with single agent endocrine therapy, patients had the CDK4/6 inhibitor palbociclib added to their treatment regimen, for 16 weeks, prior to surgery. The coopERA study met its primary endpoint of superior Ki67 suppression of giredestrant compared to anastrazole at 2 weeks. The relative reduction of Ki67 from baseline to week 2 was 75% in the giredestrant arm and 67% in anastrazole arm. Furthermore, a greater number of tumors exhibited CCCA (complete cell cycle arrest, defined as a Ki67 score less than or equal to 2.7%) in the giredestrant arm compared to the anastrazole arm, for both Ki67 high and low subgroups. Additional neoadjuvant studies and adjuvant studies are comparing novel endocrine therapies with aromatase inhibitors. The results of these trials will determine if novel ERα agents result in improved outcomes in adjuvant therapy and will further our understanding of the ability of neoadjuvant studies to predict benefit in adjuvant treatment regimens.

Many of the neoadjuvant studies extend pharmacodynamic assessments beyond ERα and PR protein levels by IHC by, for example, leveraging RNA sequencing to assess ERα transcriptional activity, and including an assessment of potential blood-based biomarkers such as circulating tumor DNA. The community is gathering data and outcomes of trials to assess the potential of these novel biomarkers to inform prognosis and therapeutic benefit of endocrine therapies. Neoadjuvant studies additionally have the potential to inform rational combinations of endocrine and targeted therapeutics and may also be leveraged to explore potential mechanisms of resistance. In particular, tumor samples collected from patients enrolled in the FELINE trial comparing letrozole and letrozole plus ribociclib allowed for the application of single cell transcriptomics to study tumor response. This creative approach to sample interrogation revealed that resistance patterns were distinct in patients treated with single agent endocrine therapy (AI) versus combination treatment of endocrine therapy (AI) plus CDK4/6 inhibition, with combination therapy selecting for apparent estrogen independence and upregulation of growth factor receptors [71].

Neoadjuvant studies may have some limitations in predicting the maximal clinical benefit of a therapeutic in the adjuvant setting given the short duration of dosing for comparisons. For example, *ESR1* mutations are rare in newly diagnosed HR+ BC, and estrogen-independence, a major limitation of AIs, may evolve only after long-term treatment. However, these short-duration biomarker-focused studies do provide a unique opportunity early in development of new agents for pre- and on-treatment biopsies to generate a dose response relationship for target inhibition and cellular activity, as well as insights into potential mechanisms (or limitations) of therapeutic benefit. These rich biological datasets can be used in combination with data from clinical trials in patients with metastatic breast cancer to assist in optimization of dose and design of adjuvant trials.

17.7 Outlook

The optimized nature of the ER therapeutics currently being clinically evaluated, relative to earlier generations, is such that the community is now in a position to explore more fully the hypothesis first articulated by ICI in the 1990's, that "*complete blockade of all stimulatory actions of estrogens*" will bring further benefit to patients. Advances in technologies that support high resolution profiling of patient samples, for example single cell RNA-sequencing and high-sensitivity ctDNA analyses, are further aiding our ability to interrogate the impact of these emerging therapeutics on tumor biology. The simultaneous pursuit of numerous next-generation ERα-targeted agents by a variety of pharmaceutical and biotechnology companies has created a highly competitive landscape. The intense competition in ER+ breast cancer has accelerated and elevated resource investments in these clinical programs, and has also spurred creative trial design and sophisticated measures of tumor response and evolution, centered around demonstrating and maximizing the potential benefit of these agents for patients. While it is impossible for us to predict the outcome of the multitude of studies that are currently ongoing, the much hoped-for scenario is that at least one of these agents will demonstrate meaningful gains over the current standard of care, providing important additional options for women with ER+ breast cancer.

References

1. Beatson GT (1896) On the treatment of inoperable cases of carcinoma of the mamma: suggestions for a new method of treatment, with illustrative cases. Trans Med Chir Soc Edinb 15
2. Tan SH, Wolff AC (2007) Luteinizing hormone-releasing hormone agonists in premenopausal hormone receptor-positive breast cancer. Clin Breast Cancer 7
3. Allen E, Doisy EA (1923) An ovarian hormone: preliminary report on its localization, extraction and partial purification, and action in test animals. JAMA 81:819–821
4. Jordan VC (2021) Turning scientific serendipity into discoveries in breast cancer research and treatment: a tale of PhD students and a 50-year roaming tamoxifen team. Breast Cancer Res Treat 190
5. Quirke VM (2017) Tamoxifen from failed contraceptive pill to best-selling breast cancer medicine: a case-study in pharmaceutical innovation. Front Pharmacol 8
6. Liu H, Lee ES, Deb Los Reyes A, Zapf JW, Jordan VC (2001) Silencing and reactivation of the selective estrogen receptor modulator-estrogen receptor alpha complex. Cancer Res 61:3632–3639
7. McInerney EM, Katzenellenbogen BS (1996) Different regions in activation function-1 of the human estrogen receptor required for antiestrogen- and estradiol-dependent transcription activation. J Biol Chem 271:24172–24178
8. Shang Y, Brown M (2002) Molecular determinants for the tissue specificity of SERMs. Science 295
9. Wolf DM, Jordan VC (1994) Characterization of tamoxifen stimulated MCF-7 tumor variants grown in athymic mice. Breast Cancer Res Treat 31:117–127
10. Gottardis MM, Jordan VC (1988) Development of tamoxifen-stimulated growth of MCF-7 tumors in athymic mice after long-term antiestrogen administration. Cancer Res 48:5183–5187
11. Brodie AM, Schwarzel WC, Shaikh AA, Brodie HJ (1977) The effect of an aromatase inhibitor, 4-hydroxy-4-androstene-3,17-dione, on estrogen-dependent processes in reproduction and breast cancer. Endocrinology 100
12. Santen RJ, Brodie H, Simpson ER, Siiteri PK, Brodie A (2009) History of aromatase: saga of an important biological mediator and therapeutic target. Endocr Rev 30:343–375
13. Chumsri S, Howes T, Bao T, Sabnis G, Brodie A (2011) Aromatase, aromatase inhibitors, and breast cancer. J Steroid Biochem Mol Biol 125:13–22
14. Goss PE et al (2013) Exemestane versus anastrozole in postmenopausal women with early breast cancer: NCIC CTG MA.27 – a randomized controlled phase III trial. J Clin Oncol 31
15. Ellis MJ et al (2010) ACOSOG Z1031: a randomized phase II trial comparing exemestane, letrozole, and anastrozole in postmenopausal women with clinical stage II/III estrogen receptor-positive breast cancer. J Clin Oncol. https://doi.org/10.1200/jco.2010.28.18_suppl.lba513
16. Howell A et al (2005) Results of the ATAC (Arimidex, Tamoxifen, Alone or in Combination) trial after completion of 5 years' adjuvant treatment for breast cancer. Lancet 365
17. Thürlimann B et al (2005) A comparison of letrozole and tamoxifen in postmenopausal women with early breast cancer. N Engl J Med 353
18. Smith IE et al (2005) Neoadjuvant treatment of postmenopausal breast cancer with anastrozole, tamoxifen, or both in combination: the Immediate Preoperative Anastrozole, Tamoxifen, or Combined with Tamoxifen (IMPACT) multicenter double-blind randomized trial. J Clin Oncol 23:5108–5116
19. Dowsett M et al (2009) Short-term changes in Ki-67 during neoadjuvant treatment of primary breast cancer with anastrozole or tamoxifen alone or combined correlate with recurrence-free survival. Clin Cancer Res 11:951s–958s
20. Ellis MJ, Ma C (2007) Letrozole in the neoadjuvant setting: the P024 trial. Breast Cancer Res Treat 105:33
21. Smith IE, Dowsett M (2003) Aromatase inhibitors in breast cancer. N Engl J Med 348
22. Wakeling AE (1990) Therapeutic potential of pure antioestrogens in the treatment of breast cancer. J Steroid Biochem Mol Biol 37
23. Bucourt R, Vignau M, Torelli V (1978) New biospecific adsorbents for the purification of estradiol receptor. J Biol Chem 253:8221–8228
24. Wakeling AE, Bowler J (1992) ICI 182,780, a new antioestrogen with clinical potential. J Steroid Biochem Mol Biol 43
25. Gibson MK et al (1991) The mechanism of ICI 164,384 antiestrogenicity involves rapid loss of estrogen receptor in uterine tissue. Endocrinology 129
26. Borrás M et al (1996) Estrogenic and antiestrogenic regulation of the half-life of covalently labeled estrogen receptor in MCF-7 breast cancer cells. J Steroid Biochem Mol Biol 57:203–213
27. Dauvois S, Danielian PS, White R, Parker MG (1992) Antiestrogen ICI 164,384 reduces cellular estrogen receptor content by increasing its turnover. Proc Natl Acad Sci U S A 89:4037–4041
28. Nicholson RI et al (1995) Responses to pure antiestrogens (ICI 164384, ICI 182780) in estrogen-sensitive

and -resistant experimental and clinical breast cancer. Ann N Y Acad Sci 761:148–163
29. Pink JJ, Jordan VC (1996) Models of estrogen receptor regulation by estrogens and antiestrogens in breast cancer cell lines. Cancer Res 56:2321–2330
30. Robertson JF et al (2001) Comparison of the short-term biological effects of 7alpha-[9-(4,4,5,5,5-pentafluoropentylsulfinyl)-nonyl]estra-1,3,5,(10)-triene-3,17beta-diol (Faslodex) versus tamoxifen in postmenopausal women with primary breast cancer. Cancer Res 61
31. Wardell SE, Marks JR, McDonnell DP (2011) The turnover of estrogen receptor α by the selective estrogen receptor degrader (SERD) fulvestrant is a saturable process that is not required for antagonist efficacy. Biochem Pharmacol 82
32. Stenoien DL et al (2001) FRAP reveals that mobility of oestrogen receptor-alpha is ligand- and proteasome-dependent. Nat Cell Biol 3
33. Robertson JF et al (2014) A good drug made better: the fulvestrant dose-response story. Clin Breast Cancer 14
34. Robertson JF et al (2003) Fulvestrant versus anastrozole for the treatment of advanced breast carcinoma in postmenopausal women: a prospective combined analysis of two multicenter trials. Cancer 98
35. Robertson JF et al (2004) Pharmacokinetic profile of intramuscular fulvestrant in advanced breast cancer. Clin Pharmacokinet 43
36. Di Leo A et al (2010) Results of the CONFIRM phase III trial comparing fulvestrant 250 mg with fulvestrant 500 mg in postmenopausal women with estrogen receptor-positive advanced breast cancer. J Clin Oncol 28
37. Robertson JFR et al (2016) Fulvestrant 500 mg versus anastrozole 1 mg for hormone receptor-positive advanced breast cancer (FALCON): an international, randomised, double-blind, phase 3 trial. Lancet 388
38. Robertson JFR et al (2021) Meta-analyses of visceral versus non-visceral metastatic hormone receptor-positive breast cancer treated by endocrine monotherapies. NPJ Breast Cancer 7
39. van Kruchten M et al (2015) Measuring residual estrogen receptor availability during fulvestrant therapy in patients with metastatic breast cancer. Cancer Discov 5
40. Weir HM et al (2016) AZD9496: an oral estrogen receptor inhibitor that blocks the growth of ER-positive and ESR1-mutant breast tumors in preclinical models. Cancer Res 76:3307–3318
41. Joseph JD et al (2016) The selective estrogen receptor downregulator GDC-0810 is efficacious in diverse models of ER+ breast cancer. elife 5
42. Guan J et al (2019) Therapeutic Ligands Antagonize Estrogen Receptor Function by Impairing Its Mobility. Cell 178
43. Robertson JFR et al (2020) A Randomized, Open-label, Presurgical, Window-of-Opportunity Study Comparing the Pharmacodynamic Effects of the Novel Oral SERD AZD9496 with Fulvestrant in Patients with Newly Diagnosed ER + HER2 - Primary Breast Cancer. Clin Cancer Res 26
44. Dickler MN et al (2018) Abstract PD5-10: a first-in-human phase I study to evaluate the oral selective estrogen receptor degrader (SERD), GDC-0927, in postmenopausal women with estrogen receptor positive (ER+) HER2-negative metastatic breast cancer (BC). Cancer Res 78:PD5–P10
45. Arao Y et al (2011) Estrogen receptor α AF-2 mutation results in antagonist reversal and reveals tissue selective function of estrogen receptor modulators. Proc Natl Acad Sci U S A 108:14986–14991
46. Wrenn CK, Katzenellenbogen BS (1993) Structure-function analysis of the hormone binding domain of the human estrogen receptor by region-specific mutagenesis and phenotypic screening in yeast. J Biol Chem 268:24089–24098
47. Montano MM, Ekena K, Krueger KD, Keller AL, Katzenellenbogen BS (1996) Human estrogen receptor ligand activity inversion mutants: receptors that interpret antiestrogens as estrogens and estrogens as antiestrogens and discriminate among different antiestrogens. Mol Endocrinol 10:230–242
48. Hager GL, McNally JG, Misteli T (2009) Transcription dynamics. Mol Cell 35:741–753
49. Liu Z, Tjian R (2018) Visualizing transcription factor dynamics in living cells. J Cell Biol 217:1181–1191
50. Liang J et al (2021) GDC-9545 (Giredestrant): a potent and orally bioavailable selective estrogen receptor antagonist and degrader with an exceptional preclinical profile for ER+ breast cancer. J Med Chem 64:11841–11856
51. Scott JS et al (2020) Discovery of AZD9833, a Potent and Orally Bioavailable Selective Estrogen Receptor Degrader and Antagonist. J Med Chem 63:14530–14559
52. El-Ahmad Y et al (2020) Discovery of 6-(2,4-Dichlorophenyl)-5-[4-[(3S)-1-(3-fluoropropyl)pyrrolidin-3-yl]oxyphenyl]-8,9-dihydro-7H-benzo[7]annulene-2-carboxylic acid (SAR439859), a Potent and Selective Estrogen Receptor Degrader (SERD) for the Treatment of Estrogen-Receptor-Positive Breast Cancer. J Med Chem 63:512–528
53. Hodges-Gallagher L, Harmon CL, Sun R, Myles DC, Kushner P (2020) Abstract 4376: OP-1250, a complete estrogen receptor antagonist (CERAN) that shrinks estrogen receptor positive tumors and exhibits favorable pharmacokinetics. Cancer Res 80:4376–4376
54. Andreano KJ et al (2020) G1T48, an oral selective estrogen receptor degrader, and the CDK4/6 inhibitor lerociclib inhibit tumor growth in animal models of endocrine-resistant breast cancer. Breast Cancer Res Treat 180:635–646
55. Metcalfe C et al (2020) Abstract 3406: GDC-9545: A pure antiestrogen clinical candidate that immobilizes the estrogen receptor and profoundly alters chromatin accessibility in vivo. Cancer Res 80:3406–3406
56. Chen YC et al (2021) Latest generation estrogen receptor degraders for the treatment of hormone receptor-

positive breast cancer. Expert Opin Investig Drugs. https://doi.org/10.1080/13543784.2021.1983542
57. Wardell SE, Nelson ER, Chao CA, Alley HM, McDonnell DP (2015) Evaluation of the pharmacological activities of RAD1901, a selective estrogen receptor degrader. Endocr Relat Cancer 22:713–724
58. Garner F, Shomali M, Paquin D, Lyttle CR, Hattersley G (2015) RAD1901: a novel, orally bioavailable selective estrogen receptor degrader that demonstrates antitumor activity in breast cancer xenograft models. Anti-Cancer Drugs 26:948–956
59. Lainé M et al (2021) Lasofoxifene as a potential treatment for therapy-resistant ER-positive metastatic breast cancer. Breast Cancer Res 23
60. Fanning SW et al (2018) The SERM/SERD bazedoxifene disrupts ESR1 helix 12 to overcome acquired hormone resistance in breast cancer cells. elife 7
61. Wardell SE, Nelson ER, Chao CA, McDonnell DP (2013) Bazedoxifene exhibits antiestrogenic activity in animal models of tamoxifen-resistant breast cancer: implications for treatment of advanced disease. Clin Cancer Res 19:2420–2431
62. Snyder LB et al (2021) Abstract 44: the discovery of ARV-471, an orally bioavailable estrogen receptor degrading PROTAC for the treatment of patients with breast cancer. Cancer Res 81:44–44
63. Hamilton E et al (2022) Abstract PD13-08: first-in-human safety and activity of ARV-471, a novel PROTAC® estrogen receptor degrader, in ER+/HER2-locally advanced or metastatic breast cancer. Cancer Res 82:PD13–PD08
64. Hamilton EP et al (2021) Abstract PD8-06: phase I/II trial of H3B-6545, a novel selective estrogen receptor covalent antagonist (SERCA), in estrogen receptor positive (ER+), human epidermal growth factor receptor 2 negative (HER2-) advanced breast cancer. Cancer Res 81:PD8–P06
65. Puyang X et al (2018) Discovery of selective estrogen receptor covalent antagonists for the treatment of ERα WT and ERα MUT breast cancer. Cancer Discov 8
66. Guerrero-Zotano AL, Arteaga CL (2017) Neoadjuvant trials in ER + breast cancer: a tool for acceleration of drug development and discovery. Cancer Discov 7
67. Kuter I et al (2012) Dose-dependent change in biomarkers during neoadjuvant endocrine therapy with fulvestrant: results from NEWEST, a randomized Phase II study. Breast Cancer Res Treat 133
68. Moore HM et al (2021) Evaluation of pharmacodynamic (PD) and biologic activity in a preoperative window-of-opportunity (WOO) study of giredestrant (GDC-9545) in postmenopausal patients (pts) with estrogen receptor-positive, HER2-negative (ER+/HER2–) operable breast cancer (BC). J Clin Oncol. https://doi.org/10.1200/JCO.2021.39.15_suppl.577
69. Robertson JFR et al (2021) Abstract OT-09-05: a randomized, pre-surgical study to investigate the biological effects of AZD9833 doses in women with ER-positive HER2-negative primary breast cancer (SERENA-3). Cancer Res 81:OT–09–05–OT–09–05
70. Hurvitz SA et al (2022) Abstract PD13-06: neoadjuvant giredestrant (GDC-9545) + palbociclib versus anastrozole + palbociclib in postmenopausal women with estrogen receptor-positive, HER2-negative, untreated early breast cancer: Primary analysis of the randomized, open-label, phase II coopERA breast cancer study. Cancer Res 82:PD13–PD06
71. Griffiths JI et al (2021) Serial single-cell genomics reveals convergent subclonal evolution of resistance as patients with early-stage breast cancer progress on endocrine plus CDK4/6 therapy. Nat Can 2:658–671

18 Drugging the Undruggable: Targeting the N-Terminal Domain of Nuclear Hormone Receptors

Marianne D. Sadar

Abstract

This chapter focuses on the development of drugs targeting the N-terminal domain of nuclear hormone receptors, using progress with the androgen receptor as an example. Historically, development of therapies targeting nuclear hormone receptors has focused on the folded C-terminal ligand-binding domain. Therapies were traditionally not developed to target the intrinsically disordered N-terminal domain as it was considered "undruggable". Recent developments have now shown it is possible to direct therapies to the N-terminal domain. This chapter will provide an introduction of the structure and function of the domains of nuclear hormone receptors, followed by a discussion of the rationale supporting the development of N-terminal domain inhibitors. Chemistry and mechanisms of action of small molecule inhibitors will be described with emphasis on N-terminal domain inhibitors developed to the androgen receptor including those in clinical trials.

M. D. Sadar (✉)
Canada's Michael Smith Genome Sciences Centre, British Columbia Cancer, Vancouver, BC, Canada

Department of Pathology and Laboratory Science, University of British Columbia, Vancouver, BC, Canada
e-mail: msadar@bcgsc.ca

Keywords

Androgen receptor · Intrinsically disordered protein · Drugs · Ralaniten · EPI-002 · Sintokamide

18.1 Introduction

Nuclear hormone receptors share a common modular structural organization that includes a variable N-terminal domain (NTD or A/B domain), a DNA-binding domain (DBD or C domain), a non-conserved hinge region (D domain), and a C-terminal ligand-binding domain (LBD or E domain) [33]. Here we focus on members of nuclear receptor subgroup 3, that include the androgen receptor (AR), two closely related estrogen receptors (ERα and ERβ), glucocorticoid receptor (GR), mineralocorticoid receptor (MR), and progesterone receptor (PR). These are soluble proteins that mediate the effects of lipophilic steroids to regulate the expression of thousands of genes to control the growth and function of cells and tissues [36, 39, 70, 119]. Steroidal hormones diffuse across the cell membrane to bind to the LBDs of hormone receptors, which sets off a series of events that are necessary for transactivation or repression of target genes. First there is a conformational change of the receptor that involves the shedding of interacting chaperones

M. J. Campbell, C. L. Bevan (eds.), *Nuclear Receptors in Human Health and Disease*, Advances in Experimental Medicine and Biology 1390, https://doi.org/10.1007/978-3-031-11836-4_18

followed by translocation of the receptor to the nucleus. The DBD directs binding of the receptor to specific genomic regions on the DNA and dimerization, followed by recruitment of coregulatory proteins, chromatin remodelers, and the general transcriptional machinery necessary for regulating the transcription of target genes [12, 33, 36, 63, 70, 119]. There are two regions within the NTD and LBD called activation functions 1 and 2 (AF-1 and AF-2) respectively that provide the surfaces for interaction with coregulators and the transcriptional machinery [22, 34, 56, 62, 85]. To date, all clinically approved therapies directed against these hormone receptors target AF-2 in their LBDs. However, the recent breakthrough of the discovery of small molecule inhibitors that directly interact with AF-1 of AR has yielded the first ever small molecules that directly interact with the previously-considered "undruggable" NTD of a hormone receptor. The success of drug development against the intrinsically disordered NTD of AR is a precedent in the field of hormone receptors, but it is also worth noting that these molecules were the first drugs that directly bind to any intrinsically disordered target to reach human clinical trials (NCT02606123). Since success in drugging the "undruggable" NTDs of hormone receptors is currently restricted to AR, this review focuses on AR to provide insight into drug development that may have application for other nuclear hormone receptors.

18.2 Modular Structure of Nuclear Hormone Receptors

Nuclear hormone receptors are modular proteins. The steroid hormone receptors vary in size from less than 600 amino acid residues to over 900 residues. Their modular structure includes an intrinsically disordered N-terminal domain (NTD), DNA-binding domain (DBD), hinge region, and C-terminus ligand-binding domain (LBD). There can be substantial amino acid sequence similarity depending on the domain and hormone receptors being compared. For example, the AR-LBD shares 54% sequence similarity with PR-LBD and hence some antiandrogens can inhibit the transcriptional activity of PR [10, 84]; the AR DBD is 76% identical to that of the GR-DBD and not surprisingly they share some common regulatory DNA sequences within the same chromatin loci [20, 102]. This is an important consideration in drug development since the specificity of these hormone receptors involves multiple mechanisms including receptor-specific residues within their ligand-binding pockets but also importantly tissue-specific expression of a hormone receptor which, if not appreciated, could lead to unexpected toxicity in other tissues (for a review see [19]). For example, benign prostate tissue expresses AR but does not express GR, yet in advanced prostate cancer both GR and AR are expressed [52].

18.2.1 Intrinsically Disordered N-terminal Domain (NTD)

The NTDs of these hormone receptors have little sequence conservation (<15%) and vary enormously in size from only 182 amino acid residues for ERα to over 600 residues for MR. The NTD contains activation function 1 (AF-1), which interacts with an abundance of coregulatory proteins [42, 57, 64, 68]. AF-1 of the majority of hormone receptors contains most or all of the transcriptional activity, with the exception of ERα which has most of its transcriptional activity within AF-2 in its LBD [8, 22, 48, 56, 62]. AF-1 and AF-2 can act independently, as demonstrated with deletion and mutational experiments, but generally maximum activity is obtained when AF-1 and AF-2 cooperate in concert ([64, 81]). AF-1 is generally considered ligand-independent. However, AR AF-1 has two transactivation units -1 and 5 (tau-1 and tau-5, respectively) (Fig. 18.1). Tau-1 is considered to be dependent on ligand binding to the receptor and encompasses amino acid residues 101-370, the majority of which are acidic. Amino acid residues 360-485 comprise tau-5, which is considered to be ligand-independent. Deletions of small regions of approximately 100 residues of tau-1 do not eliminate AR transcriptional activity, suggesting that the activity of tau-1 is not attributable to a single

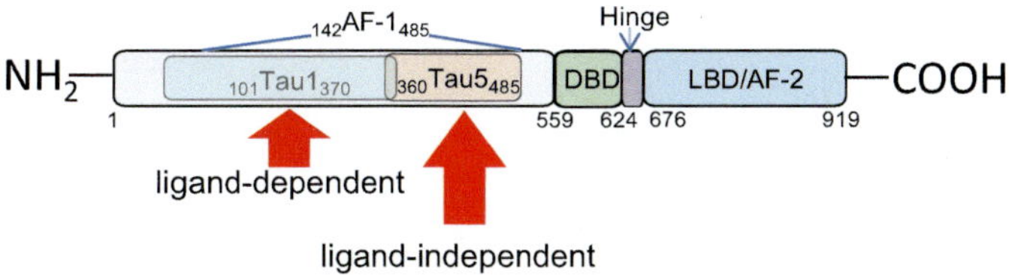

Fig. 18.1 Modular structure of the androgen receptor. AF-1 is within the N-terminal domain (NTD) and contains the ligand-dependent tau1 and ligand-independent tau5. The DNA-binding domain (DBD) contains 65 amino acid residues. The hinge region connects the DBD to the LBD and contains a nuclear localization signal. AF-2 is within the ligand-binding domain (LBD) and contains 249 amino acid residues

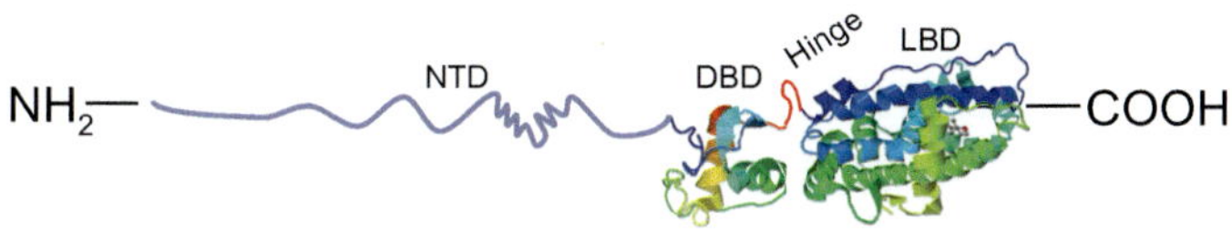

Fig. 18.2 Illustration of the AR displaying the intrinsically disordered NTD and hinge region compared to the folded DBD and LBD

small structural element [55]. This point is of relevance in drug development because it could mean that any small molecule inhibitor, or antibody, directed to tau-1 would need to impact the conformation broadly and not merely a small discrete region. Theoretically this would imply that the recently developed bispecific antibody 3E10-AR441, that binds within tau-1 at residues 299-315 [37], would not be efficacious in blocking all AR transcriptional activity. Additionally, loss of AR-LBD shifts the transcriptional activity of AR from tau-1 to tau-5 [55], which has implications for finding a small molecule inhibitor that blocks both full-length AR in response to ligand (activity mediated through tau-1) and truncated splice variants of AR (AR-Vs) that lack LBD (activity mediated through tau-5).

NTDs of hormone receptors are not amenable to structural analysis by X-ray crystallography due to their intrinsic disorder, thereby impeding drug development (Fig. 18.2). Amino acid residues dictate the disordered state and thereby are "intrinsic" to the coding sequence. Generally, intrinsically disordered proteins or regions are enriched in amino acid residues that have a high net charge, low hydrophobicity, and abundance of proline residues [30, 117, 118]. Cysteine residues can form disulfide bridges that stabilize the protein structure in an oxidizing environment, but under a reducing environment, the disulfide bridges are broken, resulting in the protein becoming less ordered. Figure 18.3 shows a Ronn plot that predicts regions of disorder within the AR-NTD based upon its amino acid sequence. Post-translational modification such as phosphorylation also impacts intra- and intermolecular interactions [7], which in turn impact the conformation of the structure and binding partner preference, plus the protein half-life [23, 40, 124]. Aromatic residues may reveal a molecular recognition region (MoRF) within a region of intrinsic disorder. These MoRFs are of high interest in drug development due to the potential to undergo a disorder-to-order transition with specific interactions or binding. Looking at the amino acid sequence of AR-NTD, there are multiple repeat regions that vary in length that include the polyproline tract (average 9 repeats), polyglycine tract (average 16 repeats), and polyglutamine tract (average 21 CAG repeats). Importantly the AR NTD has several potential MoRFs such as aromatic residues W433, Y445 and F437. An example that emphasizes the qualities that impact structure and function of intrinsically disordered proteins is the interaction of RAP74 with the AR NTD. RAP74 interacts within amino acids residues 423-446 that contain these MoRFs and has weak affinity in the millimolar range (K_D = 1749 μM) that substantially improves

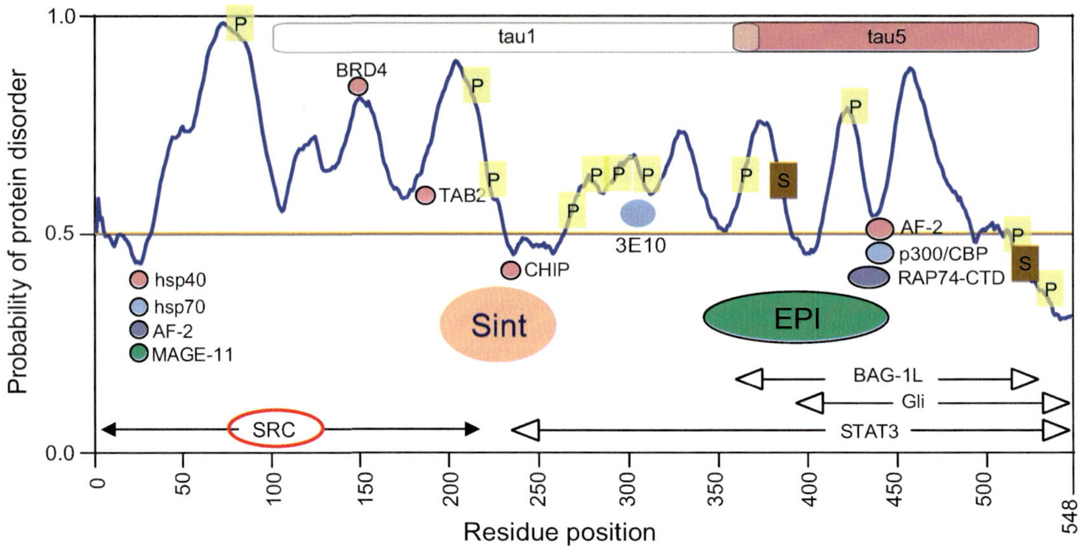

Fig. 18.3 Ronn plot of AR-NTD showing some mapped protein-protein interactions, post-translational modifications, and binding sites for small molecules that directly interact with this domain. The probability of protein disorder across the amino acid residues of the NTD. A probability score below 0.5 is considered ordered (folded) whereas a score above 0.5 is considered disordered. Transactivation units (tau) 1 and 5 are shown within the NTD. The binding sites for EPI-002, sintokamide, and 3E10-AR441 are shown. Regions of posttranslational modification and interactions with some other proteins on the N-terminal domain are shown. Phosphorylation (P) and sumoylation (S). Protein interactions shown include chaperones hsp40 and hsp90; AF-2, activation function-2 in the AR LBD for N/C interaction; RAP74 of the basal transcriptional machinery; p300/CBP, BRD4, TAB2, CHIP, MAGE-11, BAG1L, Gli, STAT3, and SRC

with phosphorylation of S424 of AR to a K_D of 702 μM [26, 109] (Fig. 18.4). Thus, an inhibitor of AR NTD even with an IC_{50} in the very high μM to millimolar range may still have therapeutic value in blocking this weak interaction with RAP74, if such blood levels are achievable without toxicity to other tissues. This difference in molecular mechanism of protein-protein interactions in the NTD compared to ligand-binding to LBD is critical to understand when considering differences between an AR NTD inhibitor and LBD inhibitors, as in the latter case an antiandrogen such as enzalutamide has to compete with the physiological ligand such as dihydrotestosterone (DHT) that has affinity in the low nM range.

The structural plasticity of the NTDs of hormone receptors allows this domain to exist as multiple and changing conformations depending on the environment and interacting partner, but also makes this domain a difficult drug target [17, 29, 35, 82]. The lack of a stable binding site together with shallow clefts for interactions with other proteins creates a challenge in drug development that is unique from the classic "lock-and-key" model for folded proteins. Intrinsically disordered proteins or regions tend to have high specificity and low affinity thereby allowing a rapid interchange of binding partners. Examples for the AR NTD are RAP74 (as described above) and Hsp70, which have binding affinities in the μM range [26, 31]. The degree of helical secondary structure of hormone receptor NTDs increases with binding to interacting proteins to conform to a molten-globule-like conformation referred to as 'collapsed disordered' [61, 71, 97]. NTD-interacting proteins that that are known to increase α-helical content include TATA-binding protein (TBP) [34, 65, 66, 116], CREB-binding protein (CBP) [61], RAP74 subunit of human transcription factor IIF [61, 97] and Jun dimerization protein 2 (JDP2) [116]. These protein-protein interactions induce α-helical structure and lead to additional protein-protein interactions to impact transcriptional activity [106, 121]. Exchange of binding partners may involve unfolding of structured regions in AF-1 with

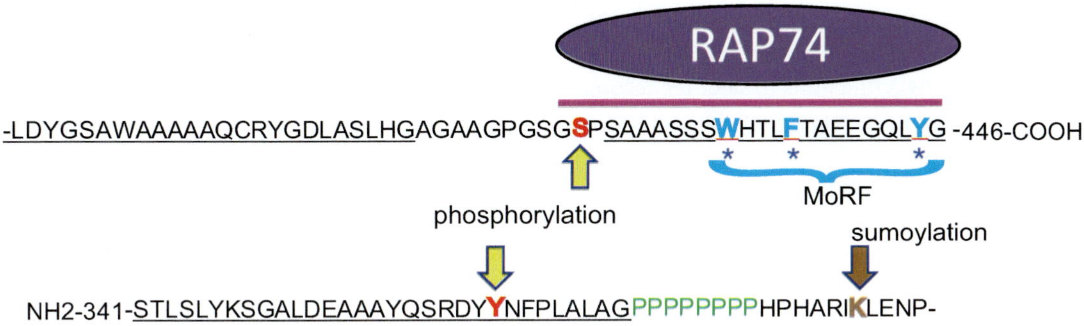

Fig. 18.4 EPI binding sites within tau5 of AR-NTD in the context of RAP74's binding site. Residues in the three EPI-002 binding sites (underlined) and the flanking residues are shown. Phosphorylation of S424 impacts the affinity of RAP74 binding to tau5. MoRFs within the RAP74 site of interaction. Other post-translational modifications that affect AR transcriptional activity and could possibly impact the binding of EPI-002 to tau5 that include Y363 that is phosphorylated by Ack and sumoylation of K386

increased structure within an adjacent region [89]. It is this plasticity that permits hormone receptors to act as a hub of interactions with an extremely large repertoire of binding partners [43, 59, 122]. Recently, low-resolution cryoelectron microscopy revealed that the structure of transcriptionally active full-length AR is unique from ERα in its direct interaction with steroid receptor coactivators (SRCs) and its orientation of dimerization [123]. AR homodimerizes in a head-to-head and tail-to-tail manner and consists of two different conformations of NTD [123]. One AR NTD conformation interacts with a single SRC-3 molecule close to its $_{23}$FQNLF$_{27}$ motif [123], consistent with earlier coimmunoprecipitation studies that showed SRC interacts within amino acid residues 1-233 of the AR-NTD [111]. Conversely, a single p300 molecule interacts with both conformations of NTD [123]. Interaction of a hormone receptor with DNA can also induce tertiary structure and α-helical content of the NTD/AF-1 to encourage protein-protein interactions with cofactors and bridging factors to ultimately impact transcriptional activity [8, 79, 80].

18.2.2 DNA-Binding Domain (DBD) and Hinge Region

The crystal structures of DBDs of hormone receptors have been resolved [105]. This domain is the most conserved in sequence compared to the other domains at greater than 75% for MR, GR, PR, and AR DBDs and 57% between ERα and AR DBDs. Hormone receptor DBDs have three α-helices that are comprised of two zinc finger motifs and a C-terminal extension (CTE). Each zinc finger has four cysteine residues that bind a zinc ion. The first zinc finger subdomain interacts with the major groove of base-specific regions of DNA and is called the P-box. The second zinc finger subdomain stabilizes receptor-DNA interaction through non-specific contacts with the DNA backbone and also contains the distal box (D box) that is involved in receptor dimerization [98]. The 22Rv1 human prostate cancer cell line is commonly used to analyze the effects of drugs on the transcriptional activity of AR including its constitutively active AR-Vs, but this cell line is unique in that its AR carries duplication of exon 3: this encodes an additional zinc finger within its DBD thereby impacting its properties such as protein half-life [108].

The CTE mediates the specificity of AR to recognize androgen response elements (AREs). The majority of AREs have been mapped to enhancers in the regulatory regions of genes regulated by androgens and consist of a repeat of two hexamers separated by a 3 base-pair spacer. It is important to note that there are general response elements that are recognized by all steroid hormone receptors with the exception of ER [112], as well those that are specific to a receptor [18, 27, 38, 54, 119]. Due to this high degree of similarity in sequence and structure across the steroid hormone receptors' DBDs, this domain has been generally considered to be a poor drug target due to challenges to achieve specificity. Although the DBD functions to steer the receptor to specific regulatory regions of the genome, as mentioned above it also primes the NTD for interactions with specific coregulatory proteins. The genomic sequence of a particular response element can influence the transcriptional response of a gene [45].

DBDs are linked to LBDs by the hinge region, which is unstructured. The hinge region is important in the nuclear translocation of the receptor and is sequestered in the absence of ligand [93]. It contains part of the CTE involved in interactions between the DNA and receptor as well as having other functions that are regulated by post-translational modifications within this region [21, 44]. Upon binding DNA there can be a change in conformation of the CTE which stabilizes intramolecular interactions [44] and creates a binding site for coregulatory proteins [13, 95]. Hormone receptors can alter the conformation of DNA to facilitate the assembly of multi-protein complexes within the enhancer or promoter regions of target genes [44].

18.2.3 Ligand-Binding Domain (LBD)

LBDs of hormone receptors function to mediate the effects of steroids and have been the primary target for drug development. The AR-LBD is the direct or indirect target for all currently FDA-approved drugs against the androgen axis. For example, indirect drug targets for AR-LBD are those therapeutics that reduce the levels of androgen that bind to the AR-LBD and include LHRH analogues and CYP17 inhibitors (e.g. abiraterone) that block steroidogenesis. Drugs that directly target AR-LBD include both agonists that are called selective AR modifiers or "SARMs" as well as antagonists that are called, "antiandrogens". Antiandrogens can be steroidal or non-steroidal; non-steroidal antiandrogens have the stem name "lutamide" and include flutamide, nilutamide, bicalutamide, enzalutamide, apalutamide, and darolutamide. Antiandrogens are competitive inhibitors with androgens for the AR-LBD and induce an AR conformation that is not transcriptionally active. The "lutamides" have evolved since the first in class drug flutamide to be more bulky and thereby more effective in disrupting protein-protein interactions. Sequence similarity in the LBDs of hormone receptors manifests with some steroids able to bind to other receptor LBDs in addition to their cognate receptor, based upon the concentrations of steroid and also the existence of point mutations within this domain. For example, AR LBD shares 54% sequence similarity with PR-LBD and not surprisingly steroidal progestins (e.g., cyproterone acetate/6-chloro-17-hydroxy-la,2a-methylenepregna-4,6-diene-3,20-dione acetate) were the first inhibitors discovered against AR [92]. Similarly, the non-steroidal antiandrogens such bicalutamide and enzalutamide bind to PR-LBD to inhibit its transcriptional activity [10, 51].

The crystal structures of LBDs of all hormone receptors have been resolved, in complex with various ligands, with only an agonist-bound conformation available for the AR. The lack of success in obtaining a crystal structure of the AR-LBD in an antagonist conformation has impeded drug development against this important drug target. Crystal structure analyses have revealed that LBDs of hormone receptors are folded into 3 layers that form an anti-parallel α-helical sandwich with up to 12 α-helices (H1-12) and up to 4 short β-strands that may form β-sheets [14, 44, 67, 107, 120]. Hormone receptors lack helix 2 so have 11 helices, with the exception of ER that has all 12 helices [44]. Generally, the binding of an agonist induces conformational changes such that helix 12 stabilizes and covers the ligand-binding pocket to form a hydrophobic cleft and expose the AF2 region. This conformational change provides a binding interface for AF-2 to interact with LxxLL motifs of coactivators such as SRCs [14, 41, 44, 67].

In the absence of ligand, LBDs repress AF-1 transcriptional activities as shown for PR, GR and AR. For these receptors, when their LBDs are deleted the results are constitutively active receptors [8, 22, 48, 55]. Here the ER stands out from the other receptors and emphasizes that its transcriptional activity is largely through AF-2.

Deletion of ER LBD results in a 95% decrease of its transcriptional activity [68], compared to the truncated AR lacking LBD becoming constitutively activated [55]. As mentioned above, AR-LBD also dictates the contribution of transcriptional activity from tau-1 versus tau-5. For the truncated constitutively active AR-Vs that lack LBD, tau-5 would be the dominant tau driving transcriptional activity and thereby a critical drug target.

In diseases such as prostate cancer and some breast cancers, there are some structural alterations in the AR-LBD that are considered to drive the disease and confer resistance to therapies that target the AR-LBD. These structural alterations include deletion or truncation of AR LBD, resulting in constitutively active AR-Vs that are independent of androgens [55]; gain-of-function mutations in the AR-LBD underlying antiandrogen withdrawal syndromes [74]; as well as point mutations that result in promiscuous binding to other steroids [32].

18.3 Androgen Receptor

Full-length AR molecular mass is calculated as 98.9 kDa but when run on SDS-PAGE it migrates as a band of approximately 110 kDa. AR NTD has several polymorphic tracts (see above) that result in its variability in length (generally 547–556 residues), a folded DBD (65 residues), a disordered hinge region (49 residues) and folded LBD (249 residues). Full-length AR is encoded from 8 canonical exons and 7 cryptic exons in the *AR* gene. This gene resides on the X chromosome (*AR* locus: Xq11-Xq12); both males and females have only one functional copy of AR due to X-inactivation. The *AR* gene has binding sites for SP1, NFkB, and c-MYC but lacks elements for TATA and CCAAT in its regulatory region (for reviews see [15, 50]). Tissue-specific activity of AR is modulated by regulation of expression in response to androgen [50] and tissue-specific expression of its coregulators [87]. This results in tissue-specific expression of AR target genes such as prostate-specific antigen (PSA), which is a biomarker for prostate cancer.

Full-length AR mediates the effects of androgens such as testosterone and dihydrotestosterone (DHT) that are required in males for sexual differentiation, maintenance of spermatogenesis, and male gonadotropin regulation. Male reproductive tissue such as the prostate is dependent upon functional AR signaling. The dependency of prostate tissue on androgens provides the rationale for targeting full-length AR for the treatment of prostate cancer using androgen deprivation therapy (ADT) and antiandrogens. In addition to prostate cancer, the androgen axis plays a role in other pathologies such as alopecia, polycystic ovarian syndrome, spinal bulbar muscular dystrophy, androgen insensitivity syndrome, and some breast cancers, thereby emphasizing the need for therapeutic inhibitors of AR transcriptional activity (for a review see [75]).

18.4 Rationale for Developing Inhibitors to the NTD

Interest in developing drugs to the intrinsically disordered AR-NTD predominantly comes from the discovery of constitutively active AR-Vs lacking LBD that are associated with resistance mechanisms in lethal castration-resistance prostate cancer (CRPC) [4, 5, 103, 104] and have been discovered in breast cancer tissues [1, 24, 43, 49]. The fact that all of the transcriptional activity of AR resides in its NTD, also means that inhibitors of the AR-NTD would be effective against full-length AR, gain-of-function mutations in AR-LBD, and other mechanisms of maintained AR transcriptional activities. In other words, an inhibitor to the AR-NTD should block the transcriptional activities of all AR species. Also beneficial is that AR-NTD has little sequence similarity (<15%) to its most closely related hormone receptors and is thereby predicted to be a highly specific drug target.

Due to challenges in discovery of small molecules that directly bind to an intrinsically disordered target, an approach has been to target folded proteins that interact with AR-NTD. The first in vivo proof-of-concept that this could yield a therapeutic response for CRPC was provided using decoys that sequestered AR-NTD interacting proteins [91, 96]. There have also been a number of studies that target an individual interacting binding partner of AR-NTD such as: hsp40/70 to induce degradation of AR-Vs and reduce aggregation of full-length AR with extended polyQ tracts [28, 31, 78, 88]; BRD4 [6]; BAG1L [16, 69, 73]; and SRC-1 and 3 [114] (Fig. 18.3). The approach of targeting an interacting protein rather than the AR-NTD directly has the inherent risk of lack of specificity to blocking AR function since most interacting partners are not unique for AR and interact with many other proteins. Thus, discovery of drugs that directly bind to AR-NTD is of high interest.

18.5 Small Molecule Inhibitors of AR-NTD

Currently all small molecule inhibitors proven to directly bind to the AR-NTD were originally isolated from natural compound libraries [100]. These were libraries of marine sponges that were screened to discover: sintokamides [11, 101]; naphatenones [10, 86]; and EPI-001/ralaniten [3, 90] (Fig. 18.5). Of these three unique chemical scaffolds, the EPI-001/ralaniten analogues were the first drugs against AR-NTD to reach human clinical trials. Importantly, ralaniten was also the first drug that directly binds to any intrinsically disordered target to reach clinical testing, marking a breakthrough in drug discovery for intrinsically disordered proteins. Recently, a second-generation analogue of ralaniten began clinical trials in heavily pretreated men with CRPC (NCT04421222) and in combination with enzalutamide ([46]; NCT05075577). Combinations of EPI compounds have shown improved therapeutic responses for CRPC when combined with antiandrogens [46], radiation [9], PIN1 inhibitors [76], palbociclib [110], mTOR inhibitors [58], taxanes [83], and sintokamides [11]. The sintokamides are still under development for the clinic as potential imaging agents and to use in combination with tau5 inhibitors. Drug development for naphatenones, that were first isolated from the marine sponge Niphates digitalis, was stopped due to their reactivity and alkylation of glutathione [10, 86].

Sintokamides A to E were isolated from the marine sponge Dysidea sp. These were the first small molecules that inhibited the AR-NTD to be published [101]. Sintokamide A (SINT1) directly binds AR AF-1 region to specifically inhibit transactivation of AR NTD and block transcriptional activities of full-length AR and AR-Vs [11]. In vivo studies with sintokamide using human prostate cancer xenografts grown in castrated mice revealed regression of tumours and reduced expression of the AR-regulated gene, PSA [11]. Interestingly, additive inhibition was evident when SINT1 was combined with ralaniten, which suggested SINT1 binds to a site on AF-1 that is unique from that bound by ralaniten [11]. Through studying well-characterized protein-protein interactions with the AR-NTD, differences in blocking interaction of STAT3 with AR-NTD between these compounds revealed that SINT1 probably interacts more N-terminally within tau-1 whereas ralaniten interacts with tau-5 [11, 25]. The inability of the sintokamides to impact IL-6 transactivation of AR and STAT3 interaction with AR-NTD predicts that these compounds would be ineffective against prostatic bone lesions that have elevated levels of IL-6 and are prevalent in men with advanced prostate cancer.

The first EPI compound, EPI-067, was isolated from the marine sponge Geodia lindgreni [2, 99, 100]. Structure activity relationship studies of several hundred analogues yielded EPI-002, a single stereoisomer of the mixture called EPI-001, that was developed for first-in-human clinical trials. These compounds have a chlorohydrin and consistent with the literature were demonstrated to not be reactive as shown at physiological pH in vitro [11] and in vivo using a radioactive imaging agent [51], as well as from patient clinical samples [94, 113]. EPI-002 was established as a first-in-class compound called ralaniten by the USAN Council with a stem name of "aniten" based upon its unique mechanism of action that distinguishes these compounds from the "lutamide" antiandrogens such as enzalu-

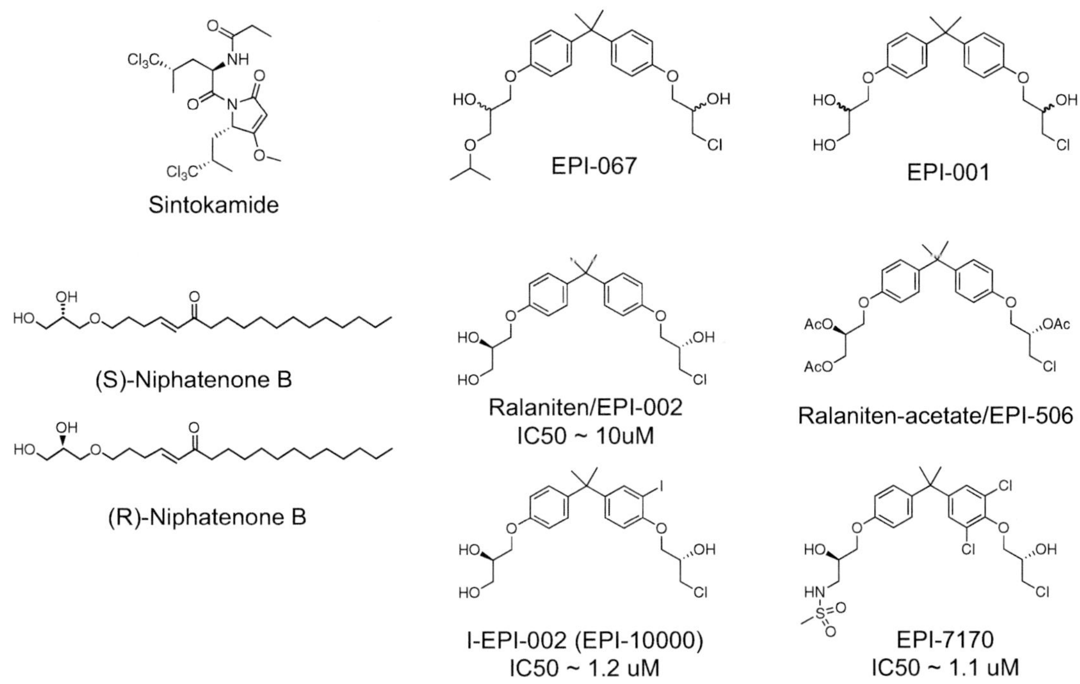

Fig. 18.5 Chemical structures of small molecules validated to directly bind to AF-1 within the AR-NTD. Sintokamides do not block IL-6 transactivation of the AR-NTD or STAT-3 interaction with AR-NTD so are being developed as imaging agents rather than as therapeutics. Niphatenones were reactive and formed glutathione adducts and have been dropped from clinical development. EPI-compounds showing the discovery compound, EPI-067, isolated from a marine sponge [2]. EPI-001 is a mixture of 4 stereoisomers including the active compound ralaniten (EPI-002) which was delivered as a prodrug (ralaniten-acetate or EPI-506) to prostate cancer patients [100]. Addition of a halogen (iodine) to a phenyl ring, I-EPI-002 (EPI-10000), improved the potency by 10-fold compared to EPI-002 [51]. Removal of the primary alcohol as with the second-generation compound EPI-7170 improves the in vivo efficacy presumably due to reduced metabolism [9, 94]

tamide. Ralaniten predominantly binds residues 341-446 of tau-5, including the core unit $_{435}$WHTLF$_{439}$, plus some overlap into tau-1 (101-370) [25] (Figs. 18.3 and 18.4). As predicted, ralaniten inhibits the transcriptional activities of full-length AR, AR-Vs, gain-of-function AR-LBD mutations, AR with altered polyQ tracts, and AR-transcriptional activities with aberrant expression of coactivators and amplified levels of AR [3, 90, 121]. Inhibition of AR transcriptional activity by EPI was specific, with no impact on related human hormone receptors. EPI analogues inhibited AR-NTD interaction with CREB-binding protein (CBP) and RAP74 [3]. They do not induce AR nuclear translocation in the absence of androgen [3]. Importantly EPI analogues block AR DNA binding in the promoters and enhancers of target genes to decrease expression of these genes in response to androgens [3, 90]. In vivo efficacy of EPI as a therapeutic for prostate cancer was demonstrated using xenografts of human prostate cancer cell lines, patient-derived xenografts, and the Herschberger assay [3].

The technical hurdle of aggregation of recombinant intrinsically disordered proteins makes it difficult to provide evidence of direct binding using cell-free assays. In spite of this, evidence of direct interaction of the EPI analogues with the AF-1 region of AR was shown by application of recombinant AF-1 protein in a cell-free assay by fluorescence emission spectroscopy [3] and Click-chemistry probes [90]. Importantly, cell-free assays may be prone to producing artifacts because of the sensitivity of the conformation(s) of an intrinsically disordered protein on its environment and protein-protein interactions. The first evidence of direct binding for any intrinsically disordered protein in cells was provided when EPI bound the endogenous AR in LNCaP human prostate cancer cells. A number of approaches were employed that included both click-chemistry probes and radiolabelled analogues [51, 90]. In vivo data from castrated mice injected with a radiolabelled EPI analogue also provided strong evidence of specificity of EPI compounds for the AR as well as proof-of-concept of the potential of these compounds to image tumours that express AR-NTD [51]. Later, Dr. Salvatella and his team provided NMR data confirming that EPI compounds bind specifically to AF-1 within tau-5 and identified the amino acid residues required for this interaction [25]. There were three regions within AR AF-1 that were required for EPI to bind, implying that EPI binds within a pocket rather than to linear amino acid sequence (Fig. 18.4). Of note, the EPI-binding site on AF-1 is also where RAP74 interacts [26] thereby supporting earlier studies showing EPI blocked this interaction [3]. Post-translational modifications within this region that may alter the binding of EPI include phosphorylation of Y363 and S424 as well as sumoylation of K386 within the flanking region. Studies to address the impact of post-translational modifications on EPI binding to AR-NTD will be important to predict potential resistance mechanisms.

18.6 First-in-Human Clinical Trials

In November 2015, the prodrug of ralaniten, called ralaniten-acetate and also known as EPI-506, was administered to the first CRPC patient enrolled in first-in-human clinical trials (NCT02606123). This marked an important event of ralaniten being the first drug that directly binds to any intrinsically disordered protein to reach a clinical trial - in oncology and even more notably in any disease. This Phase I clinical trial was a dose-escalation study in 28 heavily pre-treated CRPC patients in whom abiraterone and/or enzalutamide had previously failed. The drug did show signs of efficacy in some patients as evidenced by a reduction of serum PSA and stable disease, especially in those patients receiving higher doses. A few patients remained on ralaniten for more than 1 year with stable disease. These indications of efficacy were in spite of patients having steady-state C_{min} blood concentrations that were 50× lower than what would be optimal based upon in vitro data of 25 μM [100]. Notably, the most highly dosed patients who received 3600 mg/daily had blood trough levels of only 200 ng/mL which is equivalent to 0.5 μM. These blood levels are also 48- to 58-fold lower than steady-state C_{min} for enzalutamide and its active metabolite respectively [53]. This extremely poor pharmacokinetic profile for ralaniten resulted in excessive pill burden and ultimately stopping its clinical development, in spite of it being well-tolerated. Subsequent analyses of samples from these patients revealed that ralaniten was oxidized and glucuronidated predominantly at the alcohol groups [94]. A second generation set of analogues have been designed to improve the metabolic stability of this class of drugs and these include EPI-7170 [9] and the clinical compound EPI-7386, which entered clinical trials in June 2020 for men with metastatic CRPC (NCT04421222). Early data released at ASCO-GU in February 2021 stated, "Despite the suboptimal 200 mg dose, one out of three patients who completed 12 weeks of therapy experienced a prostate specific antigen ("PSA") decline of more than 50% after three cycles of EPI-7386 therapy (12 weeks) with ongoing continued PSA declines continuing through six cycles of therapy, despite previously having failed enzalutamide and abiraterone acetate" [72, 77].

18.7 Conclusions

It is estimated that 33–50% of the proteome is intrinsically disordered or has intrinsically disordered regions [115]. The plasticity of intrinsically disordered proteins allows for multiple and changing conformations to enable the exchange of numerous binding partners. Proteins that possess intrinsic disorder tend to have functions within signaling networks and include transcription factors such as nuclear hormone receptors and regulators of the cell cycle. Thus, it is not surprising that intrinsically disordered proteins are associated with many diseases such as cancer, diabetes, cardiovascular disease, and amyloidosis (for a review article see [60]) and are a rich potential source of drug targets. Progress on developing inhibitors that directly bind to the intrinsically disordered AR NTD, that are the first to reach clinical trials in humans for this class of proteins, may help lead future successes against other intrinsically disordered drug targets.

References

1. Aceto N, Bardia A, Wittner BS, Donaldson MC, O'Keefe R, Engstrom A, Bersani F, Zheng Y, Comaills V, Niederhoffer K, Zhu H, Mackenzie O, Shioda T, Sgroi D, Kapur R, Ting DT, Moy B, Ramaswamy S, Toner M, Haber DA, Maheswaran S (2018) AR expression in breast cancer CTCs associates with bone metastases. Mol Cancer Res 16:720–727
2. Andersen RJ (2017) Sponging off nature for new drug leads. Biochem Pharmacol 139:3–14
3. Andersen RJ, Mawji NR, Wang J, Wang G, Haile S, Myung JK, Watt K, Tam T, Yang YC, Banuelos CA, Williams DE, Mcewan IJ, Wang Y, Sadar MD (2010) Regression of castrate-recurrent prostate cancer by a small-molecule inhibitor of the amino-terminus domain of the androgen receptor. Cancer Cell 17:535–546
4. Antonarakis ES, Lu C, Wang H, Luber B, Nakazawa M, Roeser JC, Chen Y, Mohammad TA, Chen Y, Fedor HL, Lotan TL, Zheng Q, De Marzo AM, Isaacs JT, Isaacs WB, Nadal R, Paller CJ, Denmeade SR, Carducci MA, Eisenberger MA, Luo J (2014) AR-V7 and resistance to enzalutamide and abiraterone in prostate cancer. N Engl J Med 371:1028–1038
5. Armstrong AJ, Halabi S, Luo J, Nanus DM, Giannakakou P, Szmulewitz RZ, Danila DC, Healy P, Anand M, Rothwell CJ, Rasmussen J, Thornburg B, Berry WR, Wilder RS, Lu C, Chen Y, Silberstein JL, Kemeny G, Galletti G, Somarelli JA, Gupta S, Gregory SG, Scher HI, Dittamore R, Tagawa ST, Antonarakis ES, George DJ (2019) Prospective multicenter validation of androgen receptor splice variant 7 and hormone therapy resistance in high-risk castration-resistant prostate cancer: the Prophecy study. J Clin Oncol 37:1120–1129
6. Asangani IA, Dommeti VL, Wang X, Malik R, Cieslik M, Yang R, Escara-Wilke J, Wilder-Romans K, Dhanireddy S, Engelke C, Iyer MK, Jing X, Wu YM, Cao X, Qin ZS, Wang S, Feng FY, Chinnaiyan AM (2014) Therapeutic targeting of BET bromodomain proteins in castration-resistant prostate cancer. Nature 510:278–282
7. Bah A, Forman-Kay JD (2016) Modulation of intrinsically disordered protein function by post-translational modifications. J Biol Chem 291:6696–6705
8. Bain DL, Franden MA, Mcmanaman JL, Takimoto GS, Horwitz KB (2000) The N-terminal region of the human progesterone A-receptor. Structural analysis and the influence of the DNA binding domain. J Biol Chem 275:7313–7320
9. Banuelos CA, Ito Y, Obst JK, Mawji NR, Wang J, Hirayama Y, Leung JK, Tam T, Tien AH, Andersen RJ, Sadar MD (2020) Ralaniten sensitizes enzalutamide-resistant prostate cancer to ionizing radiation in prostate cancer cells that express androgen receptor splice variants. Cancers (Basel) 12:1991
10. Banuelos CA, Lal A, Tien AH, Shah N, Yang YC, Mawji NR, Meimetis LG, Park J, Kunzhong J, Andersen RJ, Sadar MD (2014) Characterization of niphatenones that inhibit androgen receptor N-terminal domain. PLoS One 9:e107991
11. Banuelos CA, Tavakoli I, Tien AH, Caley DP, Mawji NR, Li Z, Wang J, Yang YC, Imamura Y, Yan L, Wen JG, Andersen RJ, Sadar MD (2016) Sintokamide A is a novel antagonist of androgen receptor that uniquely binds activation function-1 in its amino-terminal domain. J Biol Chem 291:22231–22243
12. Beato M (1989) Gene regulation by steroid hormones. Cell 56:335–344
13. Blanco JC, Minucci S, Lu J, Yang XJ, Walker KK, Chen H, Evans RM, Nakatani Y, Ozato K (1998) The histone acetylase PCAF is a nuclear receptor coactivator. Genes Dev 12:1638–1651
14. Bourguet W, Germain P, Gronemeyer H (2000) Nuclear receptor ligand-binding domains: three-dimensional structures, molecular interactions and pharmacological implications. Trends Pharmacol Sci 21:381–388
15. Burnstein KL (2005) Regulation of androgen receptor levels: implications for prostate cancer progression and therapy. J Cell Biochem 95:657–669
16. Cato L, Neeb A, Sharp A, Buzon V, Ficarro SB, Yang L, Muhle-Goll C, Kuznik NC, Riisnaes R, Nava Rodrigues D, Armant O, Gourain V, Adelmant G, Ntim EA, Westerling T, Dolling D, Rescigno P, Figueiredo I, Fauser F, Wu J, Rottenberg JT, Shatkina L, Ester C, Luy B, Puchta H, Troppmair J, Jung N, Brase S, Strahle U, Marto JA, Nienhaus GU, Al-Lazikani B, Salvatella X, De Bono JS, Cato AC, Brown M (2017) Development of Bag-1L as a therapeutic target in androgen receptor-dependent prostate cancer. elife 6:e27159
17. Choi UB, Sanabria H, Smirnova T, Bowen ME, Weninger KR (2019) Spontaneous switching among conformational ensembles in intrinsically disordered proteins. Biomolecules 9:114
18. Claessens F, Alen P, Devos A, Peeters B, Verhoeven G, Rombauts W (1996) The androgen-specific pro-

basin response element 2 interacts differentially with androgen and glucocorticoid receptors. J Biol Chem 271:19013–19016
19. Claessens F, Joniau S, Helsen C (2017) Comparing the rules of engagement of androgen and glucocorticoid receptors. Cell Mol Life Sci 74:2217–2228
20. Cleutjens CB, Steketee K, Van Eekelen CC, Van Der Korput JA, Brinkmann AO, Trapman J (1997) Both androgen receptor and glucocorticoid receptor are able to induce prostate-specific antigen expression, but differ in their growth-stimulating properties of LNCaP cells. Endocrinology 138:5293–5300
21. Clinckemalie L, Vanderschueren D, Boonen S, Claessens F (2012) The hinge region in androgen receptor control. Mol Cell Endocrinol 358:1–8
22. Danielsen M, Northrop JP, Jonklaas J, Ringold GM (1987) Domains of the glucocorticoid receptor involved in specific and nonspecific deoxyribonucleic acid binding, hormone activation, and transcriptional enhancement. Mol Endocrinol 1:816–822
23. Darling AL, Uversky VN (2018) Intrinsic disorder and posttranslational modifications: the darker side of the biological dark matter. Front Genet 9:158
24. De Kruijff IE, Sieuwerts AM, Onstenk W, Jager A, Hamberg P, De Jongh FE, Smid M, Kraan J, Timmermans MA, Martens JWM, Sleijfer S (2019) Androgen receptor expression in circulating tumor cells of patients with metastatic breast cancer. Int J Cancer 145:1083–1089
25. De Mol E, Fenwick RB, Phang CT, Buzon V, Szulc E, De La Fuente A, Escobedo A, Garcia J, Bertoncini CW, Estebanez-Perpina E, Mcewan IJ, Riera A, Salvatella X (2016) EPI-001, a compound active against castration-resistant prostate cancer, targets transactivation unit 5 of the androgen receptor. ACS Chem Biol 11:2499–2505
26. De Mol E, Szulc E, Di Sanza C, Martinez-Cristobal P, Bertoncini CW, Fenwick RB, Frigole-Vivas M, Masin M, Hunter I, Buzon V, Brun-Heath I, Garcia J, De Fabritiis G, Estebanez-Perpina E, Mcewan IJ, Nebreda AR, Salvatella X (2018) Regulation of androgen receptor activity by transient interactions of its transactivation domain with general transcription regulators. Structure 26:145–152 e3
27. Devos A, Claessens F, Alen P, Winderickx J, Heyns W, Rombauts W, Peeters B (1997) Identification of a functional androgen-response element in the exon 1-coding sequence of the cystatin-related protein gene crp2. Mol Endocrinol 11:1033–1043
28. Dong J, Wu Z, Wang D, Pascal LE, Nelson JB, Wipf P, Wang Z (2019) Hsp70 binds to the androgen receptor N-terminal domain and modulates the receptor function in prostate cancer cells. Mol Cancer Ther 18:39–50
29. Dunker AK, Brown CJ, Lawson JD, Iakoucheva LM, Obradovic Z (2002) Intrinsic disorder and protein function. Biochemistry 41:6573–6582
30. Dyson HJ, Wright PE (2005) Intrinsically unstructured proteins and their functions. Nat Rev Mol Cell Biol 6:197–208
31. Eftekharzadeh B, Banduseela VC, Chiesa G, Martinez-Cristobal P, Rauch JN, Nath SR, Schwarz DMC, Shao H, Marin-Argany M, Di Sanza C, Giorgetti E, Yu Z, Pierattelli R, Felli IC, Brun-Heath I, Garcia J, Nebreda AR, Gestwicki JE, Lieberman AP, Salvatella X (2019) Hsp70 and Hsp40 inhibit an inter-domain interaction necessary for transcriptional activity in the androgen receptor. Nat Commun 10:3562
32. Eisermann K, Wang D, Jing Y, Pascal LE, Wang Z (2013) Androgen receptor gene mutation, rearrangement, polymorphism. Transl Androl Urol 2:137–147
33. Evans RM (1988) The steroid and thyroid hormone receptor superfamily. Science 240:889–895
34. Fischer K, Kelly SM, Watt K, Price NC, Mcewan IJ (2010) Conformation of the mineralocorticoid receptor N-terminal domain: evidence for induced and stable structure. Mol Endocrinol 24:1935–1948
35. Fisher CK, Stultz CM (2011) Constructing ensembles for intrinsically disordered proteins. Curr Opin Struct Biol 21:426–431
36. Gelmann EP (2002) Molecular biology of the androgen receptor. J Clin Oncol 20:3001–3015
37. Goicochea NL, Garnovskaya M, Blanton MG, Chan G, Weisbart R, Lilly MB (2017) Development of cell-penetrating bispecific antibodies targeting the N-terminal domain of androgen receptor for prostate cancer therapy. Protein Eng Des Sel 30:785–793
38. Green S, Kumar V, Theulaz I, Wahli W, Chambon P (1988a) The N-terminal DNA-binding 'zinc finger' of the oestrogen and glucocorticoid receptors determines target gene specificity. EMBO J 7:3037–3044
39. Griekspoor A, Zwart W, Neefjes J, Michalides R (2007) Visualizing the action of steroid hormone receptors in living cells. Nucl Recept Signal 5:e003
40. He Y, Chen Y, Mooney SM, Rajagopalan K, Bhargava A, Sacho E, Weninger K, Bryan PN, Kulkarni P, Orban J (2015) Phosphorylation-induced conformational ensemble switching in an intrinsically disordered cancer/testis antigen. J Biol Chem 290:25090–25102
41. Heery DM, Kalkhoven E, Hoare S, Parker MG (1997) A signature motif in transcriptional coactivators mediates binding to nuclear receptors. Nature 387:733–736
42. Hermanson O, Glass CK, Rosenfeld MG (2002) Nuclear receptor coregulators: multiple modes of modification. Trends Endocrinol Metab 13:55–60
43. Hickey TE, Irvine CM, Dvinge H, Tarulli GA, Hanson AR, Ryan NK, Pickering MA, Birrell SN, Hu DG, Mackenzie PI, Russell R, Caldas C, Raj GV, Dehm SM, Plymate SR, Bradley RK, Tilley WD, Selth LA (2015) Expression of androgen receptor splice variants in clinical breast cancers. Oncotarget 6:44728–44744
44. Hill KK, Roemer SC, Churchill ME, Edwards DP (2012) Structural and functional analysis of domains of the progesterone receptor. Mol Cell Endocrinol 348:418–429

45. Hilser VJ, Thompson EB (2011) Structural dynamics, intrinsic disorder, and allostery in nuclear receptors as transcription factors. J Biol Chem 286:39675–39682
46. Hirayama Y, Tam T, Jian K, Andersen RJ, Sadar MD (2020) Combination therapy with androgen receptor N-terminal domain antagonist EPI-7170 and enzalutamide yields synergistic activity in AR-V7-positive prostate cancer. Mol Oncol 14:2455–2470
47. Holden NS, George T, Rider CF, Chandrasekhar A, Shah S, Kaur M, Johnson M, Siderovski DP, Leigh R, Giembycz MA, Newton R (2014) Induction of regulator of G-protein signaling 2 expression by long-acting beta2-adrenoceptor agonists and glucocorticoids in human airway epithelial cells. J Pharmacol Exp Ther 348:12–24
48. Hollenberg SM, Giguere V, Segui P, Evans RM (1987) Colocalization of DNA-binding and transcriptional activation functions in the human glucocorticoid receptor. Cell 49:39–46
49. Hu DG, Hickey TE, Irvine C, Wijayakumara DD, Lu L, Tilley WD, Selth LA, Mackenzie PI (2014) Identification of androgen receptor splice variant transcripts in breast cancer cell lines and human tissues. Horm Cancer 5:61–71
50. Hunter I, Hay CW, Esswein B, Watt K, Mcewan IJ (2018) Tissue control of androgen action: the ups and downs of androgen receptor expression. Mol Cell Endocrinol 465:27–35
51. Imamura Y, Tien AH, Pan J, Leung JK, Banuelos CA, Jian K, Wang J, Mawji NR, Fernandez JG, Lin KS, Andersen RJ, Sadar MD (2016) An imaging agent to detect androgen receptor and its active splice variants in prostate cancer. JCI Insight 1:e87850
52. Isikbay M, Otto K, Kregel S, Kach J, Cai Y, Vander Griend DJ, Conzen SD, Szmulewitz RZ (2014) Glucocorticoid receptor activity contributes to resistance to androgen-targeted therapy in prostate cancer. Horm Cancer 5:72–89
53. Ito Y, Sadar MD (2018) Enzalutamide and blocking androgen receptor in advanced prostate cancer: lessons learnt from the history of drug development of antiandrogens. Res Rep Urol 10:23–32
54. Jantzen HM, Strahle U, Gloss B, Stewart F, Schmid W, Boshart M, Miksicek R, Schutz G (1987) Cooperativity of glucocorticoid response elements located far upstream of the tyrosine aminotransferase gene. Cell 49:29–38
55. Jenster G, Van Der Korput HA, Trapman J, Brinkmann AO (1995) Identification of two transcription activation units in the N-terminal domain of the human androgen receptor. J Biol Chem 270:7341–7346
56. Jenster G, Van Der Korput HA, Van Vroonhoven C, Van Der Kwast TH, Trapman J, Brinkmann AO (1991) Domains of the human androgen receptor involved in steroid binding, transcriptional activation, and subcellular localization. Mol Endocrinol 5:1396–1404
57. Johnson AB, O'Malley BW (2012) Steroid receptor coactivators 1, 2, and 3: critical regulators of nuclear receptor activity and steroid receptor modulator (SRM)-based cancer therapy. Mol Cell Endocrinol 348:430–439
58. Kato M, Banuelos CA, Imamura Y, Leung JK, Caley DP, Wang J, Mawji NR, Sadar MD (2016) Cotargeting androgen receptor splice variants and mTOR signaling pathway for the treatment of castration resistant prostate cancer. Clin Cancer Res 22:2744–2754
59. Krasowski MD, Reschly EJ, Ekins S (2008) Intrinsic disorder in nuclear hormone receptors. J Proteome Res 7:4359–4372
60. Kulkarni P, Uversky VN (2019) Intrinsically disordered proteins in chronic diseases. Biomolecules 9:147
61. Kumar R, Betney R, Li J, Thompson EB, Mcewan IJ (2004a) Induced alpha-helix structure in AF1 of the androgen receptor upon binding transcription factor TFIIF. Biochemistry 43:3008–3013
62. Kumar R, Moure CM, Khan SH, Callaway C, Grimm SL, Goswami D, Griffin PR, Edwards DP (2013) Regulation of the structurally dynamic N-terminal domain of progesterone receptor by protein-induced folding. J Biol Chem 288:30285–30299
63. Kumar R, Thompson EB (1999) The structure of the nuclear hormone receptors. Steroids 64:310–319
64. Kumar R, Thompson EB (2003) Transactivation functions of the N-terminal domains of nuclear hormone receptors: protein folding and coactivator interactions. Mol Endocrinol 17:1–10
65. Kumar R, Thompson EB (2019) Role of phosphorylation in the modulation of the glucocorticoid receptor's intrinsically disordered domain. Biomolecules 9:95
66. Kumar R, Volk DE, Li J, Lee JC, Gorenstein DG, Thompson EB (2004b) TATA box binding protein induces structure in the recombinant glucocorticoid receptor AF1 domain. Proc Natl Acad Sci U S A 101:16425–16430
67. Kumar R, Zakharov MN, Khan SH, Miki R, Jang H, Toraldo G, Singh R, Bhasin S, Jasuja R (2011) The dynamic structure of the estrogen receptor. J Amino Acids 2011:812540
68. Kumar V, Green S, Stack G, Berry M, Jin JR, Chambon P (1987) Functional domains of the human estrogen receptor. Cell 51:941–951
69. Kuznik NC, Solozobova V, Jung N, Grassle S, Lei Q, Lewandowski EM, Munuganti R, Zoubeidi A, Chen Y, Brase S, Cato ACB (2021) Development of a benzothiazole scaffold-based androgen receptor N-terminal inhibitor for treating androgen-responsive prostate cancer. ACS Chem Biol 16:2103–2108
70. Laudet V, Hanni C, Coll J, Catzeflis F, Stehelin D (1992) Evolution of the nuclear receptor gene superfamily. EMBO J 11:1003–1013
71. Lavery DN, Mcewan IJ (2008) Structural characterization of the native NH2-terminal transactivation

domain of the human androgen receptor: a collapsed disordered conformation underlies structural plasticity and protein-induced folding. Biochemistry 47:3360–3369
72. Le Moigne R, Pearson P, Lauriault V, Hong NH, Virsik P, Zhou HJ, Cesano A (2021) Preclinical and clinical pharmacology of EPI-7386, an androgen receptor N-terminal domain inhibitor for castration-resistant prostate cancer. J Clin Oncol 39:119
73. Lee I, Kuznik NC, Rottenberg JT, Brown M, Cato ACB (2019) BAG1L: a promising therapeutic target for androgen receptor-dependent prostate cancer. J Mol Endocrinol 62:R289–R299
74. Leone G, Tucci M, Buttigliero C, Zichi C, Pignataro D, Bironzo P, Vignani F, Scagliotti GV, Di Maio M (2018) Antiandrogen withdrawal syndrome (AAWS) in the treatment of patients with prostate cancer. Endocr Relat Cancer 25:R1–R9
75. Leung JK, Tien AH, Sadar MD (2021a) Androgen receptors in the pathology of disease. In: Badr MZ (ed) Nuclear receptors: the art and science of modulator design and discovery. Springer, Cham
76. Leung JK, Imamura Y, Kato M, Wang J, Mawji NR, Sadar MD (2021b) Pin1 inhibition improves the efficacy of ralaniten compounds that bind to the N-terminal domain of androgen receptor. Commun Biol 4:v381
77. Lifesciences (2021) ESSA pharma presents favorable initial phase 1 clinical pharmacology data of EPI-7386 for advanced forms of prostate cancer at the 2021 ASCO genitourinary cancers symposium [Online]. Available: https://lifesciencesbc.ca/members/essa-pharma-presents-favorable-initial-phase-1-clinical-pharmacology-data-of-epi-7386-for-advanced-forms-of-prostate-cancer-at-the-2021-asco-genitourinary-cancers-symposium [Accessed]
78. Liu C, Lou W, Yang JC, Liu L, Armstrong CM, Lombard AP, Zhao R, Noel ODV, Tepper CG, Chen HW, Dall'Era M, Evans CP, Gao AC (2018) Proteostasis by STUB1/HSP70 complex controls sensitivity to androgen receptor targeted therapy in advanced prostate cancer. Nat Commun 9:4700
79. Loven MA, Davis RE, Curtis CD, Muster N, Yates JR, Nardulli AM (2004) A novel estrogen receptor alpha-associated protein alters receptor-deoxyribonucleic acid interactions and represses receptor-mediated transcription. Mol Endocrinol 18:2649–2659
80. Loven MA, Likhite VS, Choi I, Nardulli AM (2001) Estrogen response elements alter coactivator recruitment through allosteric modulation of estrogen receptor beta conformation. J Biol Chem 276:45282–45288
81. Mangelsdorf DJ, Thummel C, Beato M, Herrlich P, Schutz G, Umesono K, Blumberg B, Kastner P, Mark M, Chambon P, Evans RM (1995) The nuclear receptor superfamily: the second decade. Cell 83:835–839
82. Marsh JA, Forman-Kay JD (2012) Ensemble modeling of protein disordered states: experimental restraint contributions and validation. Proteins 80:556–572
83. Martin SK, Banuelos CA, Sadar MD, Kyprianou N (2014) N-terminal targeting of androgen receptor variant enhances response of castration resistant prostate cancer to taxane chemotherapy. Mol Oncol 9:628–639
84. Matias PM, Donner P, Coelho R, Thomaz M, Peixoto C, Macedo S, Otto N, Joschko S, Scholz P, Wegg A, Basler S, Schafer M, Egner U, Carrondo MA (2000) Structural evidence for ligand specificity in the binding domain of the human androgen receptor. Implications for pathogenic gene mutations. J Biol Chem 275:26164–26171
85. Mcinerney EM, Katzenellenbogen BS (1996) Different regions in activation function-1 of the human estrogen receptor required for antiestrogen-and estradiol-dependent transcription activation. J Biol Chem 271:24172–24178
86. Meimetis LG, Williams DE, Mawji NR, Banuelos CA, Lal AA, Park JJ, Tien AH, Fernandez JG, De Voogd NJ, Sadar MD, Andersen RJ (2012) Niphatenones, glycerol ethers from the sponge Niphates digitalis block androgen receptor transcriptional activity in prostate cancer cells: structure elucidation, synthesis, and biological activity. J Med Chem 55:503–514
87. Miller CP, Shomali M, Lyttle CR, O'Dea LS, Herendeen H, Gallacher K, Paquin D, Compton DR, Sahoo B, Kerrigan SA, Burge MS, Nickels M, Green JL, Katzenellenbogen JA, Tchesnokov A, Hattersley G (2011) Design, synthesis, and preclinical characterization of the selective androgen receptor modulator (SARM) RAD140. ACS Med Chem Lett 2:124–129
88. Moses MA, Kim YS, Rivera-Marquez GM, Oshima N, Watson MJ, Beebe KE, Wells C, Lee S, Zuehlke AD, Shao H, Bingman WE 3rd, Kumar V, Malhotra SV, Weigel NL, Gestwicki JE, Trepel JB, Neckers LM (2018) Targeting the Hsp40/Hsp70 chaperone axis as a novel strategy to treat castration-resistant prostate cancer. Cancer Res 78:4022–4035
89. Motlagh HN, Hilser VJ (2012) Agonism/antagonism switching in allosteric ensembles. Proc Natl Acad Sci U S A 109:4134–4139
90. Myung JK, Banuelos CA, Fernandez JG, Mawji NR, Wang J, Tien AH, Yang YC, Tavakoli I, Haile S, Watt K, Mcewan IJ, Plymate S, Andersen RJ, Sadar MD (2013) An androgen receptor N-terminal domain antagonist for treating prostate cancer. J Clin Invest 123:2948–2960
91. Myung JK, Wang G, Chiu HH, Wang J, Mawji NR, Sadar MD (2017) Inhibition of androgen receptor by decoy molecules delays progression to castration-recurrent prostate cancer. PLoS One 12:e0174134
92. Neumann F, Elger W (1966) The effect of a new anti-androgenic steroid, 6-chloro-17-Hydroxy-1alpha, 2alpha-methylenepregna-4,6-diene-3,20-dione acetate (cyproterone acetate) on the sebaceous glands of mice. J Invest Dermatol 46:561–572

93. Ni L, Llewellyn R, Kesler CT, Kelley JB, Spencer A, Snow CJ, Shank L, Paschal BM (2013) Androgen induces a switch from cytoplasmic retention to nuclear import of the androgen receptor. Mol Cell Biol 33:4766–4778
94. Obst JK, Wang J, Jian K, Williams DE, Tien AH, Mawji N, Tam T, Yang YC, Andersen RJ, Chi KN, Montgomery B, Sadar MD (2019) Revealing metabolic liabilities of ralaniten to enhance novel androgen receptor targeted therapies. ACS Pharmacol Transl Sci 2:453–467
95. Poukka H, Karvonen U, Janne OA, Palvimo JJ (2000) Covalent modification of the androgen receptor by small ubiquitin-like modifier 1 (SUMO-1). Proc Natl Acad Sci U S A 97:14145–14150
96. Quayle SN, Mawji NR, Wang J, Sadar MD (2007) Androgen receptor decoy molecules block the growth of prostate cancer. Proc Natl Acad Sci U S A 104:1331–1336
97. Reid J, Kelly SM, Watt K, Price NC, Mcewan IJ (2002) Conformational analysis of the androgen receptor amino-terminal domain involved in transactivation. Influence of structure-stabilizing solutes and protein-protein interactions. J Biol Chem 277:20079–20086
98. Roemer SC, Donham DC, Sherman L, Pon VH, Edwards DP, Churchill ME (2006) Structure of the progesterone receptor-deoxyribonucleic acid complex: novel interactions required for binding to half-site response elements. Mol Endocrinol 20:3042–3052
99. Sadar MD (2011) Small molecule inhibitors targeting the "achilles' heel" of androgen receptor activity. Cancer Res 71:1208–1213
100. Sadar MD (2020) Discovery of drugs that directly target the intrinsically disordered region of the androgen receptor. Expert Opin Drug Discovery 15:551–560
101. Sadar MD, Williams DE, Mawji NR, Patrick BO, Wikanta T, Chasanah E, Irianto HE, Soest RV, Andersen RJ (2008) Sintokamides A to E, chlorinated peptides from the sponge Dysidea sp. that inhibit transactivation of the N-terminus of the androgen receptor in prostate cancer cells. Org Lett 10:4947–4950
102. Sahu B, Laakso M, Pihlajamaa P, Ovaska K, Sinielnikov I, Hautaniemi S, Janne OA (2013) FoxA1 specifies unique androgen and glucocorticoid receptor binding events in prostate cancer cells. Cancer Res 73:1570–1580
103. Scher HI, Graf RP, Schreiber NA, Jayaram A, Winquist E, Mclaughlin B, Lu D, Fleisher M, Orr S, Lowes L, Anderson A, Wang Y, Dittamore R, Allan AL, Attard G, Heller G (2018) Assessment of the validity of nuclear-localized androgen receptor splice variant 7 in circulating tumor cells as a predictive biomarker for castration-resistant prostate cancer. JAMA Oncol 4:1179–1186
104. Scher HI, Lu D, Schreiber NA, Louw J, Graf RP, Vargas HA, Johnson A, Jendrisak A, Bambury R, Danila D, Mclaughlin B, Wahl J, Greene SB, Heller G, Marrinucci D, Fleisher M, Dittamore R (2016) Association of AR-V7 on circulating tumor cells as a treatment-specific biomarker with outcomes and survival in castration-resistant prostate cancer. JAMA Oncol 2:1441–1449
105. Shaffer PL, Jivan A, Dollins DE, Claessens F, Gewirth DT (2004) Structural basis of androgen receptor binding to selective androgen response elements. Proc Natl Acad Sci U S A 101:4758–4763
106. Simons SS (2010) Glucocorticoid receptor cofactors as therapeutic targets. Curr Opin Pharmacol 10:613–619
107. Tan MH, Li J, Xu HE, Melcher K, Yong EL (2015) Androgen receptor: structure, role in prostate cancer and drug discovery. Acta Pharmacol Sin 36:3–23
108. Tepper CG, Boucher DL, Ryan PE, Ma AH, Xia L, Lee LF, Pretlow TG, Kung HJ (2002) Characterization of a novel androgen receptor mutation in a relapsed CWR22 prostate cancer xenograft and cell line. Cancer Res 62:6606–6614
109. Tien AH, Sadar MD (2018) Order within a disordered structure. Structure 26:4–6
110. Tien AH, Sadar MD (2021) Cyclin-dependent kinase 4/6 inhibitor palbociclib in combination with ralaniten analogues for the treatment of androgen receptor-positive prostate and breast cancers. Mol Cancer Ther. molcanther.0411.2021
111. Ueda T, Mawji NR, Bruchovsky N, Sadar MD (2002) Ligand-independent activation of the androgen receptor by interleukin-6 and the role of steroid receptor coactivator-1 in prostate cancer cells. J Biol Chem 277:38087–38094
112. Umesono K, Evans RM (1989) Determinants of target gene specificity for steroid/thyroid hormone receptors. Cell 57:1139–1146
113. Wang L, Wu Y, Zhang W, Kannan K (2012) Widespread occurrence and distribution of bisphenol A diglycidyl ether (BADGE) and its derivatives in human urine from the United States and China. Environ Sci Technol 46:12968–12976
114. Wang Y, Lonard DM, Yu Y, Chow DC, Palzkill TG, O'Malley BW (2011) Small molecule inhibition of the steroid receptor coactivators, SRC-3 and SRC-1. Mol Endocrinol 25:2041–2053
115. Ward JJ, Sodhi JS, Mcguffin LJ, Buxton BF, Jones DT (2004) Prediction and functional analysis of native disorder in proteins from the three kingdoms of life. J Mol Biol 337:635–645
116. Wardell SE, Kwok SC, Sherman L, Hodges RS, Edwards DP (2005) Regulation of the amino-terminal transcription activation domain of progesterone receptor by a cofactor-induced protein folding mechanism. Mol Cell Biol 25:8792–8808
117. Wright PE, Dyson HJ (2015) Intrinsically disordered proteins in cellular signalling and regulation. Nat Rev Mol Cell Biol 16:18–29
118. Xie H, Vucetic S, Iakoucheva LM, Oldfield CJ, Dunker AK, Uversky VN, Obradovic Z (2007) Functional anthology of intrinsic disorder. 1. Biological pro-

cesses and functions of proteins with long disordered regions. J Proteome Res 6:1882–1898
119. Yamamoto KR (1985) Steroid receptor regulated transcription of specific genes and gene networks. Annu Rev Genet 19:209–252
120. Yang J, Young MJ (2009) The mineralocorticoid receptor and its coregulators. J Mol Endocrinol 43:53–64
121. Yang YC, Banuelos CA, Mawji NR, Wang J, Kato M, Haile S, Mcewan IJ, Plymate S, Sadar MD (2016) Targeting androgen receptor activation function-1 with EPI to overcome resistance mechanisms in castration-resistant prostate cancer. Clin Cancer Res 22:4466–4477
122. York B, O'Malley BW (2010) Steroid receptor coactivator (SRC) family: masters of systems biology. J Biol Chem 285:38743–38750
123. Yu X, Yi P, Hamilton RA, Shen H, Chen M, Foulds CE, Mancini MA, Ludtke SJ, Wang Z, O'Malley BW (2020) Structural insights of transcriptionally active, full-length androgen receptor coactivator complexes. Mol Cell 79(812-823):e4
124. Zhou J, Zhao S, Dunker AK (2018) Intrinsically disordered proteins link alternative splicing and post-translational modifications to complex cell signaling and regulation. J Mol Biol 430:2342–2359

Genetic Variation and Mendelian Randomization Approaches

19

Mojgan Yazdanpanah, Nahid Yazdanpanah, and Despoina Manousaki

Abstract

While genome-wide association studies (GWAS) on levels of nuclear receptors are sparse, the genetics of ligands of these receptors (steroid hormones, thyroid hormones, and liposoluble vitamins) have been extensively studied in GWAS of predominantly European populations. Hundreds of genetic variants across the genome have been associated with serum levels of nuclear receptor ligands, shedding light on the physiology of hormone metabolism. These GWAS findings have been used to explore causal associations of these hormones with complex human traits and diseases in Mendelian randomization (MR) studies, and in studies using polygenic risk scores to quantify the genetic predisposition to higher/lower hormone levels. As such, besides providing insights into hormonal pathophysiology and its causal relationship with clinical complications, GWAS-identified genetic markers could ultimately play an important role in the daily clinical management of patients. As large trans-ethnic GWAS on levels of nuclear receptor ligands emerge, and with the fast advances in genotyping techniques and constant decrease of the genotyping costs, studying an individual's genetically predicted hormonal profile could be the next step in personalizing the management of patients with pathologies related to nuclear receptors and their ligands.

Keywords

Nuclear receptors · Ligands · GWAS · Mendelian randomization · Polygenic risk scores

19.1 Introduction

In the decade following 2010, genome-wide association studies (GWAS) have improved our understanding of the polygenic architecture of ligands of nuclear receptors (NR), such as steroid hormones, thyroid hormones, and vitamins A and D. This was done by assessing the genetic factors linked to the amounts of these ligands in the circulation in cohorts of thousands of individuals, of predominantly European ancestry. Although GWAS data on circulating levels of NR themselves are scarce, a wealth of genetic variants

M. Yazdanpanah · N. Yazdanpanah
Research Center of the Sainte-Justine University Hospital, University of Montreal, QC, Canada

D. Manousaki (✉)
Research Center of the Sainte-Justine University Hospital, University of Montreal, QC, Canada

Departments of Pediatrics, Biochemistry and Molecular Medicine, University of Montreal, QC, Canada
e-mail: despina.manousaki@umontreal.ca

M. J. Campbell, C. L. Bevan (eds.), *Nuclear Receptors in Human Health and Disease*, Advances in Experimental Medicine and Biology 1390, https://doi.org/10.1007/978-3-031-11836-4_19

across the genome have been related to levels of their direct and indirect ligands in large-scale GWAS. These findings have shed light on the physiology of hormone metabolism. Also, they have been leveraged to explore causal associations of those hormones with complex human traits and diseases (i.e., common diseases and traits of polygenic etiology) in Mendelian randomization (MR) studies, and in studies using polygenic risk scores (PRS) to quantify the genetic predisposition to higher/lower hormone levels.

Earlier studies have explored the genetics of complex traits, for instance levels of vitamins and steroid hormones, through a candidate gene approach, where genetic variants or genes, supported by prior knowledge on their biological functions, were assessed for association with a target disease or trait. Contrarily, GWAS follow a hypothesis-free approach within which usually 500,000–800,000 genetic variants across the complete genome (known as single nucleotide polymorphisms or SNPs) are genotyped in each study participant using SNP-arrays. Then, using genomic imputation to reference panels (libraries of whole-genome sequenced individuals), up to 10–20 million additional SNPs are imputed. Subsequently, each SNP is tested against the phenotype of interest. To account for multiple testing, a corrected p-value threshold of 5×10^{-8} is employed to attenuate the risk of false-positive findings. Effect sizes of the individual SNPs are typically small, as often expected for common variants (SNPs with minor allele frequency-MAF >5%). These studies therefore require large-scale collaborations to achieve a sufficient number of subjects for adequate statistical power to detect significant associations. Within the recent years, Whole Genome Sequencing (WGS) studies became more accessible and less costly. These studies result in far more information than GWAS on imputed genotype-wide genotyping data, and have enabled identification of rare variants (ie SNPs with a MAF below 1%) with larger effect sizes [1].

MR represents a study design in genetic epidemiology allowing for causal inference [2]. In MR studies, rather than measuring directly the amount of a circulating biomarker (such as a hormone or a vitamin) and testing its causal association with an outcome (trait or disease), SNPs associated with a given biomarker at a genome-wide level are used as instruments to infer its levels. Since the assignment of genetic variants at conception is random (according to Mendel's second law), these SNPs are not influenced by environmental factors which could confound the association between a biomarker and a disease outcome in traditional epidemiological studies [3]. As such, MR limits bias by confounding and moreover addresses reverse causation, which happens when a disease outcome influences the levels of a measured biomarker. In the past 10 years, the two-sample MR study design (where genetic variants related to an exposure and an outcome are assessed in distinct populations) [2] has enabled the use of large-scale GWAS data for biomarker levels (among which serum levels of vitamins and hormones) to explore their causal associations with human diseases and traits.

As mentioned above, GWAS have identified SNPs that are linked to a range of traits and complex diseases. Despite the small individual contributions of those SNPs [4], a major portion of the variance of a disease may be explained by adding the individual effects of all identified SNPs across the genome. To put it another way, a strategy to utilize the knowledge gained from GWAS is to summarize the risk from multiple disease-causing SNPs in PRS which can be computed from individuals' genotype data [5]. These scores are either simple counts (unweighted) or weighted sums of the disease-causing SNP-alleles. By assigning a unique score per individual and placing it in a specific percentile of a normal distribution, PRS allow disease risk stratification [5]. Also, genomic-based prediction of risk could enable risk stratification for a variety of complex diseases and traits (for which PRS exist) in individuals who have previously undergone genome-wide genotyping [6–8]. Further, such genotyping must be undertaken only once over the lifetime and its cost is decreasing rapidly (currently <$50 USD in a research context). Therefore health care systems are increasingly investing in genome-wide genotyping of their populations [9, 10].

In this chapter, we are discussing the results of the most recent GWAS on levels of NR ligands, and the clinical applications of those findings, specifically in MR studies assessing the causal role of NR ligands in human traits and diseases and in PRS predicting disease risk.

19.1.1 GWAS on NR Levels

The absence of commercialized assays to directly measure circulating levels of NR in the blood and of established norms, and also the increased cost related to measuring such levels in large populations explain partly the sparsity of GWAS in this field. To date, a single GWAS has assessed circulating levels of the estrogen receptor alpha (ERα) [11]. One common SNP within the *HRG* (histidine rich glycoprotein gene) was identified to be associated at a genome-wide significant levels with levels of the ERα. This GWAS assessed levels of up to 1124 circulating proteins using the SOMAscan platform in 997 individuals of European ancestry. Genome-wide significant associations were identified for 539 proteins and replicated in a multi-ethnic sample of 338 individuals [11]. No MR studies on NR levels are published so far using the aforementioned GWAS data. As an alternative to those approaches, it is potentially attractive to assess the impact of circulating ligands for NR.

19.1.2 25 Hydroxyvitamin D (Calcidiol)

Calcidiol or 25 hydroxyvitamin D (25OHD) is the precursor of the 1,25 dihydroxyvitamin D (calcitriol), which is the direct ligand of the vitamin D receptor. Because of the stability of its serum levels (in contrast to that of calcitriol levels), 25OHD is the established biomarker of vitamin D status in humans. 25OHD levels in humans are highly heritable, with estimates of heritability in twin and family studies of up to 40% [12, 13]. In the past decade, at least 6 GWAS [14–19] studies were published on levels of 25OHD in Europeans, and two small GWAS were published in non-European populations [20, 21]. No large GWAS on calcitriol levels are available up to now.

Earlier GWAS on 25OHD levels leveraging data from almost ~80,000 individuals have identified SNPs in 6 loci across the genome, among which four genes with direct role in vitamin D synthesis and metabolism (7-Dehydrocholesterol Reductase (*DHCR7*), Cytochrome P450 Family 2 Subfamily R Member 1(*CYP2R1*), *GC* and Cytochrome P450 Family 24 Subfamily A Member 1(*CYP24A1*)) [15]. SNPs in these four genes explained a limited portion of the variance in 25OHD levels (2.4%). Recently, two large GWAS [16, 19] including up to ~430,000 Europeans have substantially increased our understanding on the polygenic architecture of 25OHD levels, indicating over 140 25OHD-related loci. The heritability of 25OHD explained by all SNPs across the genome identified in these two studies was estimated to 16%.

Improved understanding of the genetic determinants of 25OHD has helped re-assess the role of vitamin D in the pathogenesis of skeletal and extraskeletal outcomes through MR. Taken together, the evidence from over 60 vitamin D MR studies published to date [22] does not support a causal role for the overwhelming majority of studied outcomes, including osteoporosis, different kinds of cancer, cardiovascular disease (CVD), autoimmune diseases and all-cause mortality. Despite this, in certain cases where the evidence from MR supported a causal role of vitamin D, for example in multiple sclerosis [23–26], these results had important clinical implications. This is for instance reflected in recent clinical care guidelines for the use of cholecalciferol in preventing multiple sclerosis in those at risk, published by the MS Society of Canada. The recent identification of over 140 25OHD-associated genetic variants [16, 19], allowing a deeper understanding of the genetic determinants contributing to variation in circulating 25OHD levels, will likely potentiate the utilization of these variants as instruments for 25OHD levels in MR studies. Indeed, the most recent MR studies have used up to 250 SNPs as instruments for 25OHD levels to study various outcomes [19, 27].

Vitamin D PRS are being explored as tools to stratify individuals for risk of vitamin D deficiency [28]. In some studies, PRS were applied to study associations of genetically determined vitamin D levels and disease outcomes, using a variable number of SNPs [29, 30]. The variance explained of 25OHD levels was relatively low, varying from 0.3 to 13%. Recently, new vitamin D PRS explained a substantially larger portion of the variance of 25OHD (up to 22%) [31, 32], which facilitates their clinical implementation.

19.1.3 Thyroid Hormones (Thyroxin and Triiodothyronine)

It is estimated that 40–65% of the inter-individual variation in markers of thyroid function is controlled by genetic factors [33]. To characterize these factors, various linkage and candidate gene studies have been performed, which have identified only a limited number of genes. Within the last 10 years, GWAS in the field of thyroid function have identified numerous new genes, while whole-genome sequencing efforts have also yielded interesting findings. While most GWAS focused on levels of thyroid-stimulating hormone (TSH) or diseases causing thyroid dysfunction (Graves disease, Hashimoto thyroiditis), some studies [34–42] explored directly the genetic architecture of thyroxin (free tetraiodothyronine or FT4) and of total triiodothyronine (T3), the results of which are discussed below.

To date, a minimum of 63 SNPs in 48 loci have been associated at a genome-wide significant level with FT4 and/or T3 levels. Among these loci, certain have a known role in TSH synthesis (ex LIM Homeobox 3 [*LHX3*]); the TSH signaling cascade (ex Phosphodiesterase 8B [*PDE8B*]); transcription factors expressed in the thyroid gland (ex Forkhead Box E1 (*FOXE1*)); thyroid angiogenesis (ex Vascular Endothelial Growth Factor A [*VEGFA*]); deiodination steps which convert T4 to the bioactive T3 or its inactive form (ex Iodothyronine Deiodinase 1[*DIO1*] and Iodothyronine Deiodinase 2[*DIO2*]); thyroid hormone transportation (ex Solute Carrier Family 17 Member 4(*SLC17A4*)); or participate in the hepatic metabolism of thyroid hormones (ex Aminoadipate Aminotransferase [*AADAT*]). Other associations are driven by exogenous thyroxine administration. As an example, *PTCSC2* or (Papillary Thyroid Carcinoma Susceptibility Candidate 2) variants have been described in subjects with a history of thyroid cancer, who are typically on high doses of L-thyroxin post thyroidectomy.

In the first WGS GWAS of serum TSH and FT4 levels in a total of 16,335 individuals [36], a new variant at Beta-1,4-Galactosyltransferase 6(*B4GALT6*) with a MAF of 3.2% was identified for FT3. *B4GALT6* is a galactosyltransferase, which is known to inhibit production of cAMP in TSH-stimulated cells.

GWAS-identified variants in a recent study explained 5.6% and 2.3% of the variance in serum TSH and FT4 levels respectively [35]. Applying a PRS based on these variants, it was demonstrated that subjects with scores within the highest PRS quartile had a 6.7 times increased risk of (sub)clinical hypothyroidism compared to subjects with a score within the lowest quartile [35].

19.1.4 Estradiol

Estradiol is the ligand for the ERα sex hormone NR, and has widespread biological effects, being the primary estrogen during reproductive years. Estradiol levels have a well-established role in disease susceptibility, particularly cancer in reproductive tissues in both men and women [43–46].

Family aggregation and twin studies among women estimate the heritability of estradiol levels to up to 45% [47, 48]. Earlier candidate gene studies established associations of estradiol levels with variants within the Cytochrome P450 Family 19 Subfamily A Member 1(*CYP19A1*) gene in post-menopausal women [49–52]. Notably, *CYP19A1* variants have been linked to both estradiol levels and endometrial carcinoma

[50–63], but no association was shown with breast cancer [51, 64]. Also, in postmenopausal women Estrogen Receptor 1(*ESR1*) gene (encodes ERα) haplotypes are related to higher plasma estradiol levels [65, 66], Finally, evidence from candidate gene studies have identified the *vWF* (vin Willebrand Factor) to be associated with Estradiol levels [67].

GWAS have identified only a low number of loci associated with sex hormone levels, with several reproductive hormones yet to be assessed. As an example, a locus in an intron of the Anoctamin 2 (*ANO2)* gene on chromosome 12 was shown to predict estradiol level [68]. A previous GWAS of estradiol levels in postmenopausal women failed to identify any genetic variants reaching genome-wide significance in this region, though this GWAS might have been underpowered to detect this signal. In total, 11 GWAS loci associated with estradiol explained 6.5% of the variance in age-adjusted log-estradiol levels [52]. For several of these loci our knowledge on their regulatory mechanisms remains incomplete.

Observational studies have shown a correlation between estradiol concentration and disease risk, yet not been able to distinguish between a causal relationship or an association driven by confounding. MR studies have therefore provided good evidence that higher post-menopausal estradiol levels are indeed a causal risk factor for endometrial carcinoma, and have a causal effect on estimated bone mineral density (BMD) and fracture risk [69]. A recent MR study, suggested a causal effect of estradiol on BMD in men and confirmed that *CYP19A1* was an important genetic regulator of bone health in men [70], together with two other independent loci on the X-chromosome linked to estradiol levels, and two loci (Tripartite Motif Containing 4(*TRIM4)* and Cytochrome P450 Family 11 Subfamily B Member 1/ Cytochrome P450 Family 11 Subfamily B Member 2(*CYP11B1/B2*)) associated with levels of estrone, the main sex steroid in women in menopause [70]. Using as instruments variants within the *CYP19A1* gene, estradiol levels were inversely associated with risk of thromboembolism in a MR study [71].

Results from MR on causal effects of genetically predicted estradiol on systemic inflammation among women are controversial [72]. Adding to the prevailing body of observational evidence [73, 74], the results of MR on endogenous estrogen exposure and colorectal cancer risk were inconsistent [75].

Altered reproductive hormone levels have been shown to be involved in the pathophysiology of depressive disorders [76–79], and a negative association was found between PRS for estradiol levels and whole hippocampal volume. However, the PRS failed to predict the occurrence of depressive disorders [80].

19.1.5 Testosterone and its Precursors

Testosterone is the main androgen in men produced by the testicles, and is predominantly bound to sex hormone-binding globulin (SHBG) while a smaller fraction is loosely bound to albumin [81]. There is growing evidence that serum testosterone could be a valuable biomarker of men's overall health status. Evidence from several epidemiological studies indicating that low serum testosterone concentrations are related to higher risk of common complex diseases including cardiovascular morbidity, metabolic syndrome, dyslipidemia, hypertension, type 2 diabetes, stroke, atherosclerosis, osteoporosis [82–88] and increased risk of mortality in men [89, 90].

Although testosterone is often thought of as a male hormone, females also produce testosterone, albeit at lower levels. Elevated testosterone levels in women have been described in the context of polycystic ovarian syndrome, insulin resistance, dyslipidemia, and hypertension [82, 91]. Genetics studies on testosterone levels in females failed to yield significant results [52].

Evidence from twin studies suggests that the heritability of serum testosterone levels in men varies between 57% [92] and 60% [93]. Another study reported comparable heritability estimates in younger males (64% in 9-year- olds and 70% in 12-year- olds) [94].

GWAS have determined multiple variants that influence testosterone regulation in healthy adults [52, 68, 95–98], also including variants in *SHBG*. GWAS studies have explained only 5% of variance in testosterone [68]. In earlier GWAS, two autosomal gene loci (Jumonji Domain Containing 1C *(JMJD1C)* and *SHBG)* have shown genome-wide association with serum testosterone [99]. However, the genetic determinants of serum testosterone and therefore the genetic risk factors for low concentrations remain poorly understood. More recently, a GWAS meta-analysis on total testosterone levels in males identified three loci, including two within the *SHBG* [95]. A GWAS in post-menopausal women did not detect genetic variants associated with testosterone levels [52], in accordance with previous studies [52].

MR studies have examined the association of genetic predictors of testosterone with CVD and associated risk factors in men using genetic variants from the Family With Sequence Similarity 9 Member B *(FAM9B), SHBG, CYP19A1* and Estrogen Receptor 2*(ESR2)* genes. Interestingly, none of the observed loci supported a causally protective effect of testosterone [89, 100–102] but instead indicated potential harms for blood pressure [89, 103], lipids [100] and cardiac function [102]. Notably, an intrinsic limitation for MR of sex hormones is that genetic predictors of testosterone on *SHBG* exert pleiotropic effects [95, 97, 99].

MR found a causal link between higher testosterone levels and CVD using Jumonji Domain Containing 1C *(JMJD1C)* variants amongst men in the UK Biobank [104]. Further MR studies in men observed a causal relationship of testosterone with cognition [105] and BMI [106]. While there is an extensive focus on men, few testosterone MR studies included women [107, 108]. The latter studies did not observe evidence of causal links between testosterone and a range of cardiovascular risk factors in women [108]. Evidence from MR also supports an association of elevated testosterone levels with both BMI and waist circumference [109] in both sexes, and of a decreasing effect of testosterone predominantly in men's height [108, 110].

Flynn et al. constructed sex-specific PRS which showed an improved predictive performance for testosterone levels over a sex-combined model [109]. Interestingly, PRS have been applied to access the relationship between sex hormone and mental traits. For instance, PRS for testosterone level was associated with fluid intelligence in middle-aged females [111].

19.1.6 Vitamin A

Vitamin A (VA) is present in the blood mainly in the form of retinol and provitamin. VA is crucial to human health and is involved in many metabolic and physiological processes, for example in vision [112–115], cell differentiation [116, 117], embryonic development [118, 119] and immunity [120]. Retinol is among the foremost biologically active forms of vitamin A and is hypothesized to influence a large range of human diseases including asthma, CVD, infectious diseases and cancer [121].

There is evidence that genetic variants influence circulating retinol level. Family studies have estimated that 30% of the variation in serum retinol is heritable [122]. One case study demonstrated that a mutation in the gene encoding retinol-binding protein 4 (*RBP4*), a major transport protein for retinol in circulation, resulted in abnormally low retinol concentrations [123], while inactivation of the transthyretin (*TTR*) gene, the other major retinol transport protein, also resulted in hypovitaminosis A in mice [124].

GWAS recently examined the genetics of serum retinol concentration in a healthy population, but did not identify variants associated with serum retinol at a genome-wide significance level [125]. Thus, it remains unclear whether common SNPs can explain variations in retinol concentration within the normal range. Another GWAS found two distinct regions that influence circulating retinol levels. The SNPs showing the strongest signal localize to regions that include the biologically plausible candidate genes, Retinol Binding Protein 4 (*RBP4)* and *TTR* [126].

Previously, a GWAS meta-analysis of α-carotene concentrations, which combined results across three diverse study populations, found one locus to be associated with pro-vitamin A carotenoids at a genome-wide level. Specifically, SNPs within the *BCO1* (beta-carotene oxygenase 1, alias *BCMO1*) locus were associated with significantly higher circulating concentrations of β-carotene and the same locus was associated to a lesser extent with α-carotene concentrations [125]. *BCO1* catalyzes the primary step of conversion of pro-vitamin A carotenoids to vitamin A (retinol) in the small intestine. The association between SNPs in *BCO1* and α-carotene was weak, relative to its association with β-carotene [127]. Three novel loci related to serum α-carotene concentrations in a GWAS on a population that consumed a controlled diet. In this study, the Calpain 2/ Calpain 8(*CAPN2/ CAPN8)* locus provided compelling evidence for association with serum α-carotene concentrations [128].

The evidence from genetics was utilized to investigate the causal relationships between vitamin A levels and disease. SNPs near the Retinol Binding Protein 4 (*RBP4)* gene have been linked to vitamin A–related diseases, such as retinitis pigmentosa [129]. A pooled analysis of 9226 cases and 10,420 controls found no association between *BCMO1* individual SNPs or weighted multi-SNP scores and breast cancer risk [130].

Observational studies have suggested associations between greater levels of dietary-derived antioxidants and a lower risk of coronary heart disease (CHD), while randomized clinical trials showed no reduction in CHD risk following antioxidant supplementation. Evidence from MR did not support a protective effect of genetic predisposition to higher antioxidant levels on coronary heart disease risk [131]. Similarly, MR did not support a causal association between antioxidants such as β-carotene and retinol, and ischemic stroke [131]. Furthermore, evidence from MR suggested that higher exposure to β-carotene, or retinol, does not lower the risk of Alzheimer disease [132].

Common variants linked to schizophrenia in a large GWAS aggregated in retinoid genes and were used to formulate a PRS, which was significantly correlated with risk of schizophrenia. The same study detected through whole-genome sequencing a rare variant in the retinoic acid receptor beta gene (*RARB*) in individuals with severe cognitive deficits [133].

19.1.7 Cortisol

Cortisol secretion is regulated by the hypothalamic-pituitary-adrenal axis in response to a variety of biological and environmental factors [134], and psychological stressors [135], and is marked by a circadian rhythm [136], which lead to substantial variation of cortisol levels throughout the day. Cortisol is transported in the circulation by the corticosteroid-binding globulin (CBG), which binds ~90% of the total plasma cortisol; nevertheless, only the free cortisol can access tissues and have biological effects [137].

Plasma and saliva cortisol levels have been utilized in twin studies to study the heritability of cortisol in humans. The heritability of diurnal cortisol secretion has been estimated at 62% [138]. For acute plasma cortisol measures, the estimates range from low (14%) to moderate heritability (45%) [139–141]. Two studies [142, 143] detected no heritability for morning saliva cortisol levels and total day-time secretion, but observed an important contribution of shared environment (>40%).

A mixed candidate gene and GWAS approach found variation within the FKBP Prolyl Isomerase 5 (*FKBP5)* gene to be related to both cortisol levels and depression risk [144]. In a GWAS meta-analysis for plasma cortisol in 12,597 individuals from the CORtisol NETwork (CORNET) consortium, three common SNPs explained approximately 0.5% variation in morning plasma cortisol levels [145]. These SNPs located within the Serpin Family A Member 6 (*SERPINA6)* gene which codes CBG, and the Serpin Family A Member 1 (*SERPINA1)* gene which encodes α1-antitrypsin (which inhibits cleavage of the reactive centre loop that releases cortisol from CBG) [145].

Several population-based cross-sectional studies have reported morning plasma levels of cortisol to be positively linked to plasma glucose, blood pressure and other cardiovascular risk factors [146–150]. A cross-sectional study of the association between circulating cortisol levels and CVD events was inconclusive [151], while a prospective nested case-control study showed a positive association between morning plasma cortisol and incident CVD [151]. A recent study found an enrichment of cortisol and testosterone associations among known CVD loci [152], while two plausible candidate genes (Cytochrome P450 Family 11 Subfamily B Member 1 *(CYP11B1)* and Cytochrome P450 Family 11 Subfamily B Member 2 *(CYP11B2)*) were detected among these loci, which catalyze the conversion of deoxycortisol into cortisol. However, an MR study showed that genetically predicted cortisol level (using 6 SNPs from the CORNET GWAS) was not related to ischemic heart disease, ischemic stroke, type 2 diabetes, or CVD [153]. In another MR study, a genetic instrument for morning plasma cortisol comprising 3 SNPs in the *SERPINA6* explaining 0.5% of the variation in morning plasma cortisol levels was causally related to CVD [154]. A recent study identified new SNPs in the *SERPINA6/A1* locus to be associated with morning plasma cortisol level [155], which were also shown to influence expression of *SERPINA6* in the liver, and expression of adipose tissue genes, suggesting that variations in CBG level can influence the delivery of cortisol to peripheral tissues. The same study used an MR design to demonstrate that genetically determined CBG level is causally linked to ischemic heart disease and myocardial infarction [155].

Some studies have reported an improvement in neuropsychiatric disorders after resolution of hypercortisolism, suggesting a causal effect of cortisol on depression [156]. In line with observational studies, an MR study provided evidence that genetic predisposition to higher serum morning cortisol level was associated with an increased depression score [157]. In this MR, 18 SNP-alleles conferred a 0.07 standard deviation increase in the serum morning cortisol level, while 2 of these alleles were associated with a 0.12 standard deviation increase in salivary cortisol level [157]. In another MR study, genetically determined cortisol and glucocorticoid receptor expression level, using as instruments SNPs in *SERPINA6* and *NR3C1* (the gene encoding the glucocorticoid receptor) increased risk of psychosis [158]. In females, an MR study using a variant in *NR3C1* showed an independent causal effect of glucocorticoid receptor expression on psychosis risk [158].

PRS constructed from 6 SNPs in *SERPINA6, Serpin Family A Member 2 (SERPINA2)*, and *SERPINA1* loci, previously associated with increased morning plasma cortisol levels in adults, also correlated with higher diurnal and stress-induced salivary cortisol patterns in a cohort of 8-year-old children [159].

19.2 Discussion

In the previous sections, we have provided an overview of the present state of knowledge on the genetic basis of NR ligands, including vitamins A and D, steroid hormones, and thyroid hormones. In this final section of this chapter, we are discussing the constraints of the existing studies, strategies to uncover their missing heritability, in addition to directions for future research.

19.2.1 Missing Heritability

The total variation in levels of NR ligands estimated to be determined by genetic factors in family and twin studies is often significantly higher than the variance explained by GWAS findings. This "missing heritability" may be uncovered in several ways. First, increasing sample size could be a reasonable step, as GWAS are statistically limited by the fact that many genetic variants are tested, for which a stringent multiple testing correction is required. The contribution of an increasing sample size in optimizing power for discovery in GWAS has been demonstrated by the increasing number of genes related to NR ligands in larger-scale GWAS. Second, it has been known that the circulating amount of many NR ligands, notably steroid and thyroid hormones, are sex-dependent. Despite this, most published GWAS did not perform gender- specific analyses. Future GWAS should make sure to also include variants on the X chromosome. Third, rare variants with possibly larger effect sizes, not yet identified by GWAS, are expected to substantially contribute to this total variance. Larger GWAS, together with exome and WGS studies are needed to identify and study the effects of these rarer variants. Fourth, next to the GWAS-identified SNPs, part of the variance in NR ligand levels can be attributed to copy-number variations (CNV), as well as epigenetic factors, including DNA methylation and histone modifications, which are themselves likely controlled by SNPs in the regulatory enzymes and targeting complexes. While these epigenetic factors are shown to play an important role in various complex traits, only few efforts have been made to investigate these factors in relation to NR ligands in epigenome-wide association studies [160].

19.2.2 Ethnic Representation

It has been estimated that ~80% of all published GWAS have been performed in populations of European descent. This is also true in GWAS for NR ligands, leading to a strong need for genetic studies in ethnically diverse populations, including Asians and Africans. This can allow implementation of PRS and MR studies using data of these GWAS in populations of African or Asian descent.

19.2.3 Limited Access to Measurements of Direct Ligands of NR

Due to the complexity of measuring intracellular levels of NR ligands (which bind directly to the NR), there are no GWAS assessing directly these levels. As such, the presence or absence of MR evidence on causal association of the circulating levels of ligand and a disease does not preclude those intracellular levels of ligands for NR could have the same causal influence on disease outcomes. GWAS studies looking specifically at intracellular levels of ligands for NR are practically unfeasible, since they require measurement of these levels in large samples. An alternative could be to perform GWAS on circulating free levels of sex steroid hormones, or levels of the active compounds of vitamins (such as the 1,25 dihydroxyvitamin D), which are the direct NR ligands and exert the intracellular effects.

19.2.4 Functional Follow-Up

The identification of genetic variants associated with levels of hormones and vitamins is important for various reasons. First of all, they provide insights into the biological mechanisms regulating hormonal and metabolic functions in humans. In this direction, unveiling the missing heritability could uncover new pathways in steroid hormone physiology. Functional studies for the newly identified genetic variants might enhance our understanding on the role of these genes in regulating levels of these hormones.

Second, while variation in levels of NR ligands has been associated with various adverse outcomes in observational studies, there is still much debate about the presence of true causal effects on clinical endpoints, due to conflicting results between studies, probably resulting from confounding factors and reverse causality. In the past years, MR studies have helped assessing causal associations between NR ligands and the above outcomes by circumventing these factors. A major limitation in MR is the limited variance in the levels of the NR ligands explained by the GWAS SNPs. The identification of more genetic determinants of levels of NR ligands and NR themselves will thus potentiate future MR studies to test causality of these biomarkers on human complex disease. Another limitation of the MR is the presence of pleiotropic effects of the GWAS SNPs used as instruments to infer levels of steroid hormones, which can bias the MR results. In the absence of evidence on the role of these SNPs in steroid hormone physiology, it is impossible to fully address pleiotropy in MR. In this direction, functional follow-up of the GWAS findings could facilitate the detection of invalid instrument in future MR studies.

19.2.5 Future Clinical Applications

Besides providing insights into hormonal pathophysiology and its causal relationship with clinical complications, genetic markers could ultimately have an important role in the daily clinical management of patients. For instance, genetic variation in the deiodinase genes has been proposed as a predictor of response to thyroid hormone replacement therapy in hypothyroid patients [34]. Moreover, the very polygenic architecture of most NR ligands opens new venues for studies on the potential clinical application of PRS. Specifically, these PRS could play a role in predicting individuals with levels of these molecules at the extremes of normal variation. While in other fields of medicine large-scale studies have been performed to study disease risk prediction using PRS, few such efforts have been made in field of steroid hormones and vitamins [31], and the predictive power of such scores is currently still too limited to be used in clinical practice. Despite this, these studies have provided the first insights into the potential clinical use of PRS for NR ligand levels. Larger GWAS on levels of NR ligands will enhance the performance of future PRS to predict individuals with extreme high/low levels of these hormones. With the fast advances in genotyping techniques, thousands of these variants can currently be determined from a single blood sample, for less than 50 USD per patient. Studying the clinical use of these genetic markers will therefore be the next translational step towards personalizing the management of patients with pathologies related to NR and their ligands.

References

1. Hoglund J, Rafati N, Rask-Andersen M, Enroth S, Karlsson T, Ek WE et al (2019) Improved power and precision with whole genome sequencing data in genome-wide association studies of inflammatory biomarkers. Sci Rep 9(1):16844
2. Burgess S, Butterworth A, Thompson SG (2013) Mendelian randomization analysis with multiple genetic variants using summarized data. Genet Epidemiol 37(7):658–665
3. Lawlor DA, Harbord RM, Sterne JA, Timpson N, Davey SG (2008) Mendelian randomization: using genes as instruments for making causal inferences in epidemiology. Stat Med 27(8):1133–1163
4. Pigeyre M, Yazdi FT, Kaur Y, Meyre D (2016) Recent progress in genetics, epigenetics and metagenomics unveils the pathophysiology of human obesity. Clin Sci (Lond) 130(12):943–986
5. Sugrue LP, Desikan RS (2019) What are polygenic scores and why are they important? JAMA 321(18):1820–1821
6. Forgetta VK-BJ, Forest M, Durand A, Bhatnagar S, Kemp J, Morris JA, Kanis JA, Kiel DP, McCloskey EV, Rivadeneira F, Johannson H, Harvey N, Cooper C, Evans DM, Pineau J, Leslie WD, Greenwood CMT, Richards JB (2018) Machine learning to predict osteoporotic fracture risk from genotypes [Internet]. Available from https://www.biorxiv.org/content/early/2018/09/12/413716
7. Khera AV, Chaffin M, Aragam KG, Haas ME, Roselli C, Choi SH et al (2018) Genome-wide polygenic scores for common diseases identify individuals with risk equivalent to monogenic mutations. Nat Genet 50(9):1219–1224
8. Pare G, Mao S, Deng WQ (2017) A machine-learning heuristic to improve gene score prediction of polygenic traits. Sci Rep 7(1):12665
9. Jensen PB, Jensen LJ, Brunak S (2012) Mining electronic health records: towards better research applications and clinical care. Nat Rev Genet 13(6):395–405
10. Joseph J, Grzymski MJC, Metcalf J, Galanopoulos C, Rowan C, Henderson M, Read RW, Reed H, Lipp B, Miceli D, Rybarski S, Slonim A (2018) The healthy nevada project: rapid recruitment for population health study. bioRxiv
11. Suhre K, Arnold M, Bhagwat AM, Cotton RJ, Engelke R, Raffler J et al (2017) Connecting genetic risk to disease end points through the human blood plasma proteome. Nat Commun 8:14357
12. Shea MK, Benjamin EJ, Dupuis J, Massaro JM, Jacques PF, D'Agostino RB Sr et al (2009) Genetic and non-genetic correlates of vitamins K and D. Eur J Clin Nutr 63(4):458–464
13. Livshits G, Karasik D, Seibel MJ (1999) Statistical genetic analysis of plasma levels of vitamin D: familial study. Ann Hum Genet 63(Pt 5):429–439
14. Manousaki D, Dudding T, Haworth S, Hsu YH, Liu CT, Medina-Gomez C et al (2017) Low-frequency synonymous coding variation in CYP2R1 has large effects on vitamin D levels and risk of multiple sclerosis. Am J Hum Genet 101(2):227–238
15. Jiang X, O'Reilly PF, Aschard H, Hsu YH, Richards JB, Dupuis J et al (2018) Genome-wide association study in 79,366 European-ancestry

individuals informs the genetic architecture of 25-hydroxyvitamin D levels. Nat Commun 9(1):260
16. Manousaki D, Mitchell R, Dudding T, Haworth S, Harroud A, Forgetta V et al (2020) Genome-wide association study for vitamin D levels reveals 69 independent loci. Am J Hum Genet 106(3):327–337
17. Wang TJ, Zhang F, Richards JB, Kestenbaum B, van Meurs JB, Berry D et al (2010) Common genetic determinants of vitamin D insufficiency: a genome-wide association study. Lancet 376(9736):180–188
18. Ahn J, Yu K, Stolzenberg-Solomon R, Simon KC, McCullough ML, Gallicchio L et al (2010) Genome-wide association study of circulating vitamin D levels. Hum Mol Genet 19(13):2739–2745
19. Revez JA, Lin T, Qiao Z, Xue A, Holtz Y, Zhu Z et al (2020) Genome-wide association study identifies 143 loci associated with 25 hydroxyvitamin D concentration. Nat Commun 11(1):1647
20. Sapkota BR, Hopkins R, Bjonnes A, Ralhan S, Wander GS, Mehra NK et al (2016) Genome-wide association study of 25(OH) vitamin D concentrations in Punjabi Sikhs: results of the Asian Indian diabetic heart study. J Steroid Biochem Mol Biol 158:149–156
21. Engelman CD, Meyers KJ, Ziegler JT, Taylor KD, Palmer ND, Haffner SM et al (2010) Genome-wide association study of vitamin D concentrations in Hispanic Americans: the IRAS family study. J Steroid Biochem Mol Biol 122(4):186–192
22. Bouillon R, Manousaki D, Rosen C, Trajanoska K, Rivadeneira F, Richards JB (2022) The health effects of vitamin D supplementation: evidence from human studies. Nat Rev Endocrinol 18(2):96–110
23. Gianfrancesco MA, Stridh P, Rhead B, Shao X, Xu E, Graves JS et al (2017) Evidence for a causal relationship between low vitamin D, high BMI, and pediatric-onset MS. Neurology 88(17):1623–1629
24. Mokry LE, Ross S, Ahmad OS, Forgetta V, Smith GD, Goltzman D et al (2015) Vitamin D and risk of multiple sclerosis: a Mendelian randomization study. PLoS Med 12(8):e1001866
25. Rhead B, Baarnhielm M, Gianfrancesco M, Mok A, Shao X, Quach H et al (2016) Mendelian randomization shows a causal effect of low vitamin D on multiple sclerosis risk. Neurol Genet 2(5):e97
26. Jacobs BM, Noyce AJ, Giovannoni G, Dobson R (2020) BMI and low vitamin D are causal factors for multiple sclerosis: a mendelian randomization study. Neurol Neurophysiol Neurosci 7(2)
27. Revez JA, Lin T, Qiao Z, Xue A, Holtz Y, Zhu Z et al (2019) Genome-wide association study identifies 143 loci associated with 25 hydroxyvitamin D concentration. bioRxiv
28. Hatchell KE, Lu Q, Hebbring SJ, Michos ED, Wood AC, Engelman CD (2019) Ancestry-specific polygenic scores and SNP heritability of 25(OH)D in African- and European-ancestry populations. Hum Genet 138(10):1155–1169
29. Chandler PD, Tobias DK, Wang L, Smith-Warner SA, Chasman DI, Rose L et al (2018) Association between vitamin D genetic risk score and cancer risk in a large cohort of U.S. women. Nutrients 10(1)
30. Avinun R, Romer AL, Israel S (2020) Vitamin D polygenic score is associated with neuroticism and the general psychopathology factor. Prog Neuro-Psychopharmacol Biol Psychiatry 100:109912
31. Sinnott-Armstrong N, Tanigawa Y, Amar D, Mars N, Benner C, Aguirre M et al (2021) Genetics of 35 blood and urine biomarkers in the UK biobank. Nat Genet 53(2):185–194
32. Cai M, Xiao J, Zhang S, Wan X, Zhao H, Chen G et al (2021) A unified framework for cross-population trait prediction by leveraging the genetic correlation of polygenic traits. Am J Hum Genet 108(4):632–655
33. Medici M, Visser TJ, Peeters RP (2017) Genetics of thyroid function. Best Pract Res Clin Endocrinol Metab 31(2):129–142
34. Panicker V, Wilson SG, Walsh JP, Richards JB, Brown SJ, Beilby JP et al (2010) A locus on chromosome 1p36 is associated with thyrotropin and thyroid function as identified by genome-wide association study. Am J Hum Genet 87(3):430–435
35. Porcu E, Medici M, Pistis G, Volpato CB, Wilson SG, Cappola AR et al (2013) A meta-analysis of thyroid-related traits reveals novel loci and gender-specific differences in the regulation of thyroid function. PLoS Genet 9(2):e1003266
36. Taylor PN, Porcu E, Chew S, Campbell PJ, Traglia M, Brown SJ et al (2015) Whole-genome sequence-based analysis of thyroid function. Nat Commun 6:5681
37. Gunjaca I, Matana A, Boutin T, Torlak V, Punda A, Polasek O et al (2019) Genome-wide association meta-analysis for total thyroid hormone levels in Croatian population. J Hum Genet 64(5):473–480
38. Popovic M, Matana A, Torlak V, Boutin T, Brdar D, Gunjaca I et al (2019) Genome-wide meta-analysis identifies novel loci associated with free triiodothyronine and thyroid-stimulating hormone. J Endocrinol Investig 42(10):1171–1180
39. Nielsen TR, Appel EV, Svendstrup M, Ohrt JD, Dahl M, Fonvig CE et al (2017) A genome-wide association study of thyroid stimulating hormone and free thyroxine in Danish children and adolescents. PLoS One 12(3):e0174204
40. Teumer A, Chaker L, Groeneweg S, Li Y, Di Munno C, Barbieri C et al (2018) Genome-wide analyses identify a role for SLC17A4 and AADAT in thyroid hormone regulation. Nat Commun 9(1):4455
41. Kwak SH, Park YJ, Go MJ, Lee KE, Kim SJ, Choi HS et al (2014) A genome-wide association study on thyroid function and anti-thyroid peroxidase antibodies in Koreans. Hum Mol Genet 23(16):4433–4442
42. Rhee EP, Ho JE, Chen MH, Shen D, Cheng S, Larson MG et al (2013) A genome-wide association study of the human metabolome in a community-based cohort. Cell Metab 18(1):130–143
43. Slater S, Oliver RT (2000) Testosterone: its role in development of prostate cancer and potential risk

from use as hormone replacement therapy. Drugs Aging 17(6):431–439
44. Bernstein L, Ross RK (1993) Endogenous hormones and breast cancer risk. Epidemiol Rev 15(1):48–65
45. Persson I (2000) Estrogens in the causation of breast, endometrial and ovarian cancers - evidence and hypotheses from epidemiological findings. J Steroid Biochem Mol Biol 74(5):357–364
46. Hsing AW, Reichardt JK, Stanczyk FZ (2002) Hormones and prostate cancer: current perspectives and future directions. Prostate 52(3):213–235
47. Stone J, Folkerd E, Doody D, Schroen C, Treloar SA, Giles GG et al (2009) Familial correlations in postmenopausal serum concentrations of sex steroid hormones and other mitogens: a twins and sisters study. J Clin Endocrinol Metab 94(12):4793–4800
48. Travison TG, Zhuang WV, Lunetta KL, Karasik D, Bhasin S, Kiel DP et al (2014) The heritability of circulating testosterone, oestradiol, oestrone and sex hormone binding globulin concentrations in men: the Framingham heart study. Clin Endocrinol 80(2):277–282
49. Beckmann L, Husing A, Setiawan VW, Amiano P, Clavel-Chapelon F, Chanock SJ et al (2011) Comprehensive analysis of hormone and genetic variation in 36 genes related to steroid hormone metabolism in pre- and postmenopausal women from the breast and prostate cancer cohort consortium (BPC3). J Clin Endocrinol Metab 96(2):E360–E367
50. Dunning AM, Dowsett M, Healey CS, Tee L, Luben RN, Folkerd E et al (2004) Polymorphisms associated with circulating sex hormone levels in postmenopausal women. J Natl Cancer Inst 96(12):936–945
51. Haiman CA, Dossus L, Setiawan VW, Stram DO, Dunning AM, Thomas G et al (2007) Genetic variation at the CYP19A1 locus predicts circulating estrogen levels but not breast cancer risk in postmenopausal women. Cancer Res 67(5):1893–1897
52. Prescott J, Thompson DJ, Kraft P, Chanock SJ, Audley T, Brown J et al (2012) Genome-wide association study of circulating estradiol, testosterone, and sex hormone-binding globulin in postmenopausal women. PLoS One 7(6):e37815
53. Paynter RA, Hankinson SE, Colditz GA, Kraft P, Hunter DJ, De Vivo I (2005) CYP19 (aromatase) haplotypes and endometrial cancer risk. Int J Cancer 116(2):267–274
54. Ahn J, Schumacher FR, Berndt SI, Pfeiffer R, Albanes D, Andriole GL et al (2009) Quantitative trait loci predicting circulating sex steroid hormones in men from the NCI-breast and prostate cancer cohort consortium (BPC3). Hum Mol Genet 18(19):3749–3757
55. Eriksson AL, Lorentzon M, Vandenput L, Labrie F, Lindersson M, Syvanen AC et al (2009) Genetic variations in sex steroid-related genes as predictors of serum estrogen levels in men. J Clin Endocrinol Metab 94(3):1033–1041
56. Kidokoro K, Ino K, Hirose K, Kajiyama H, Hosono S, Suzuki T et al (2009) Association between CYP19A1 polymorphisms and sex hormones in postmenopausal Japanese women. J Hum Genet 54(2):78–85
57. Travis RC, Schumacher F, Hirschhorn JN, Kraft P, Allen NE, Albanes D et al (2009) CYP19A1 genetic variation in relation to prostate cancer risk and circulating sex hormone concentrations in men from the breast and prostate cancer cohort consortium. Cancer Epidemiol Biomark Prev 18(10):2734–2744
58. Beckmann L, Hüsing A, Setiawan VW, Amiano P, Clavel-Chapelon F, Chanock SJ et al (2011) Comprehensive analysis of hormone and genetic variation in 36 genes related to steroid hormone metabolism in pre- and postmenopausal women from the breast and prostate cancer cohort consortium (BPC3). J Clin Endocrinol Metabol 96(2):E360–E3E7
59. Lundin E, Wirgin I, Lukanova A, Afanasyeva Y, Krogh V, Axelsson T et al (2012) Selected polymorphisms in sex hormone-related genes, circulating sex hormones and risk of endometrial cancer. Cancer Epidemiol 36(5):445–452
60. Flote VG, Furberg AS, McTiernan A, Frydenberg H, Ursin G, Iversen A et al (2014) Gene variations in oestrogen pathways, CYP19A1, daily 17beta-estradiol and mammographic density phenotypes in premenopausal women. Breast Cancer Res 16(6):499
61. Tao MH, Cai Q, Zhang ZF, Xu WH, Kataoka N, Wen W et al (2007) Polymorphisms in the CYP19A1 (aromatase) gene and endometrial cancer risk in Chinese women. Cancer Epidemiol Biomark Prev 16(5):943–949
62. Setiawan VW, Doherty JA, Shu XO, Akbari MR, Chen C, De Vivo I et al (2009) Two estrogen-related variants in CYP19A1 and endometrial cancer risk: a pooled analysis in the epidemiology of endometrial cancer consortium. Cancer Epidemiol Biomark Prev 18(1):242–247
63. Low YL, Li Y, Humphreys K, Thalamuthu A, Li Y, Darabi H et al (2010) Multi-variant pathway association analysis reveals the importance of genetic determinants of estrogen metabolism in breast and endometrial cancer susceptibility. PLoS Genet 6(7):e1001012
64. Michailidou K, Hall P, Gonzalez-Neira A, Ghoussaini M, Dennis J, Milne RL et al (2013) Large-scale genotyping identifies 41 new loci associated with breast cancer risk. Nat Genet 45(4):353-61, 61e1-2
65. Schuit SC, de Jong FH, Stolk L, Koek WN, van Meurs JB, Schoofs MW et al (2005) Estrogen receptor alpha gene polymorphisms are associated with estradiol levels in postmenopausal women. Eur J Endocrinol 153(2):327–334
66. Herrington DM, Howard TD, Hawkins GA, Reboussin DM, Xu J, Zheng SL et al (2002) Estrogen-receptor polymorphisms and effects of

estrogen replacement on high-density lipoprotein cholesterol in women with coronary disease. N Engl J Med 346(13):967–974
67. Harrison RL, McKee PA (1984) Estrogen stimulates von Willebrand factor production by cultured endothelial cells. Blood 63(3):657–664
68. Ruth KS, Campbell PJ, Chew S, Lim EM, Hadlow N, Stuckey BG et al (2016) Genome-wide association study with 1000 genomes imputation identifies signals for nine sex hormone related phenotypes. Eur J Hum Genet 24(2):284–290
69. Nethander M, Vandenput L, Eriksson AL, Windahl S, Funck-Brentano T, Ohlsson C (2019) Evidence of a causal effect of estradiol on fracture risk in men. J Clin Endocrinol Metab 104(2):433–442
70. Eriksson AL, Perry JRB, Coviello AD, Delgado GE, Ferrucci L, Hoffman AR et al (2018) Genetic determinants of circulating estrogen levels and evidence of a causal effect of estradiol on bone density in men. J Clin Endocrinol Metab 103(3):991–1004
71. Nethander M, Quester J, Vandenput L, Ohlsson C (2021) Association of genetically predicted serum estradiol with risk of thromboembolism in men: a mendelian randomization study. J Clin Endocrinol Metab
72. Zhao J, Jiang CQ, Lam TH, Liu B, Cheng KK, Kavikondala S et al (2014) Genetically predicted 17beta-estradiol and systemic inflammation in women: a separate-sample Mendelian randomisation analysis in the Guangzhou biobank cohort study. J Epidemiol Community Health 68(8):780–785
73. Elbanna HG, Ebrahim MA, Abbas AM, Zalata K, Hashim MA (2012) Potential value of estrogen receptor beta expression in colorectal carcinoma: interaction with apoptotic index. J Gastrointest Cancer 43(1):56–62
74. Kennelly R, Kavanagh DO, Hogan AM, Winter DC (2008) Oestrogen and the colon: potential mechanisms for cancer prevention. Lancet Oncol 9(4):385–391
75. Neumeyer S, Banbury BL, Arndt V, Berndt SI, Bezieau S, Bien SA et al (2018) Mendelian randomisation study of age at menarche and age at menopause and the risk of colorectal cancer. Br J Cancer 118(12):1639–1647
76. Campbell S, Marriott M, Nahmias C, MacQueen GM (2004) Lower hippocampal volume in patients suffering from depression: a meta-analysis. Am J Psychiatry 161(4):598–607
77. Schmaal L, Veltman DJ, van Erp TG, Samann PG, Frodl T, Jahanshad N et al (2016) Subcortical brain alterations in major depressive disorder: findings from the ENIGMA Major depressive disorder working group. Mol Psychiatry 21(6):806–812
78. Balzer BW, Duke SA, Hawke CI, Steinbeck KS (2015) The effects of estradiol on mood and behavior in human female adolescents: a systematic review. Eur J Pediatr 174(3):289–298
79. Barth C, Steele CJ, Mueller K, Rekkas VP, Arelin K, Pampel A et al (2016) In-vivo dynamics of the human hippocampus across the menstrual cycle. Sci Rep 6:32833
80. Smeeth DM, Dima D, Jones L, Jones I, Craddock N, Owen MJ et al (2019) Polygenic risk for circulating reproductive hormone levels and their influence on hippocampal volume and depression susceptibility. Psychoneuroendocrinology 106:284–292
81. Goldman AL, Bhasin S, Wu FCW, Krishna M, Matsumoto AM, Jasuja R (2017) A reappraisal of Testosterone's binding in circulation: physiological and clinical implications. Endocr Rev 38(4):302–324
82. Haring R, Baumeister SE, Volzke H, Dorr M, Felix SB, Kroemer HK et al (2011) Prospective association of low total testosterone concentrations with an adverse lipid profile and increased incident dyslipidemia. Eur J Cardiovasc Prev Rehabil 18(1):86–96
83. Haring R, Volzke H, Felix SB, Schipf S, Dorr M, Rosskopf D et al (2009) Prediction of metabolic syndrome by low serum testosterone levels in men: results from the study of health in Pomerania. Diabetes 58(9):2027–2031
84. Kupelian V, Page ST, Araujo AB, Travison TG, Bremner WJ, McKinlay JB (2006) Low sex hormone-binding globulin, total testosterone, and symptomatic androgen deficiency are associated with development of the metabolic syndrome in nonobese men. J Clin Endocrinol Metab 91(3):843–850
85. Shores MM, Arnold AM, Biggs ML, Longstreth WT Jr, Smith NL, Kizer JR et al (2014) Testosterone and dihydrotestosterone and incident ischaemic stroke in men in the cardiovascular health study. Clin Endocrinol 81(5):746–753
86. Svartberg J, von Muhlen D, Mathiesen E, Joakimsen O, Bonaa KH, Stensland-Bugge E (2006) Low testosterone levels are associated with carotid atherosclerosis in men. J Intern Med 259(6):576–582
87. Torkler S, Wallaschofski H, Baumeister SE, Volzke H, Dorr M, Felix S et al (2011) Inverse association between total testosterone concentrations, incident hypertension and blood pressure. Aging Male 14(3):176–182
88. Vikan T, Schirmer H, Njolstad I, Svartberg J (2010) Low testosterone and sex hormone-binding globulin levels and high estradiol levels are independent predictors of type 2 diabetes in men. Eur J Endocrinol 162(4):747–754
89. Haring R, Volzke H, Steveling A, Krebs A, Felix SB, Schofl C et al (2010) Low serum testosterone levels are associated with increased risk of mortality in a population-based cohort of men aged 20-79. Eur Heart J 31(12):1494–1501
90. Laughlin GA, Barrett-Connor E, Bergstrom J (2008) Low serum testosterone and mortality in older men. J Clin Endocrinol Metab 93(1):68–75
91. Mody A, White D, Kanwal F, Garcia JM (2015) Relevance of low testosterone to non-alcoholic fatty liver disease. Cardiovasc Endocrinol 4(3):83–89
92. Hoekstra RA, Bartels M, Boomsma DI (2006) Heritability of testosterone levels in 12-year-old

twins and its relation to pubertal development. Twin Res Hum Genet 9(4):558–565
93. Harris JA, Vernon PA, Boomsma DI (1998) The heritability of testosterone: a study of Dutch adolescent twins and their parents. Behav Genet 28(3):165–171
94. Koenis MM, Brouwer RM, van Baal GC, van Soelen IL, Peper JS, van Leeuwen M et al (2013) Longitudinal study of hormonal and physical development in young twins. J Clin Endocrinol Metab 98(3):E518–E527
95. Ohlsson C, Wallaschofski H, Lunetta KL, Stolk L, Perry JR, Koster A et al (2011) Genetic determinants of serum testosterone concentrations in men. PLoS Genet 7(10):e1002313
96. Zhai G, Teumer A, Stolk L, Perry JR, Vandenput L, Coviello AD et al (2011) Eight common genetic variants associated with serum DHEAS levels suggest a key role in ageing mechanisms. PLoS Genet 7(4):e1002025
97. Coviello AD, Haring R, Wellons M, Vaidya D, Lehtimaki T, Keildson S et al (2012) A genome-wide association meta-analysis of circulating sex hormone-binding globulin reveals multiple loci implicated in sex steroid hormone regulation. PLoS Genet 8(7):e1002805
98. Chen CP, Huang JP, Chen YY, Chern SR, Wu PS, Su JW et al (2013) Chromosome 22q11.2 deletion syndrome: prenatal diagnosis, array comparative genomic hybridization characterization using uncultured amniocytes and literature review. Gene 527(1):405–409
99. Jin G, Sun J, Kim ST, Feng J, Wang Z, Tao S et al (2012) Genome-wide association study identifies a new locus JMJD1C at 10q21 that may influence serum androgen levels in men. Hum Mol Genet 21(23):5222–5228
100. Zhao J, Jiang C, Lam TH, Liu B, Cheng KK, Xu L et al (2014) Genetically predicted testosterone and cardiovascular risk factors in men: a Mendelian randomization analysis in the Guangzhou biobank cohort study. Int J Epidemiol 43(1):140–148
101. Svartberg J, Schirmer H, Wilsgaard T, Mathiesen EB, Njolstad I, Lochen ML et al (2014) Single-nucleotide polymorphism, rs1799941 in the Sex Hormone-Binding Globulin (SHBG) gene, related to both serum testosterone and SHBG levels and the risk of myocardial infarction, type 2 diabetes, cancer and mortality in men: the Tromso study. Andrology 2(2):212–218
102. Zhao J, Jiang C, Lam TH, Liu B, Cheng KK, Xu L et al (2015) Genetically predicted testosterone and electrocardiographic QT interval duration in Chinese: a Mendelian randomization analysis in the Guangzhou Biobank Cohort Study. Int J Epidemiol 44(2):613–620
103. Haring R, Teumer A, Volker U, Dorr M, Nauck M, Biffar R et al (2013) Mendelian randomization suggests non-causal associations of testosterone with cardiometabolic risk factors and mortality. Andrology 1(1):17–23
104. Castro MI, Koritnik DR, Rose JC (1986) Fetal plasma insulin and thyroid hormone levels during acute in utero ethanol exposure in a maternal-fetal sheep model. Endocrinology 118(5):1735–1742
105. Zhao JV, Lam TH, Jiang C, Cherny SS, Liu B, Cheng KK et al (2016) A Mendelian randomization study of testosterone and cognition in men. Sci Rep 6:21306
106. Eriksson J, Haring R, Grarup N, Vandenput L, Wallaschofski H, Lorentzen E et al (2017) Causal relationship between obesity and serum testosterone status in men: a bi-directional mendelian randomization analysis. PLoS One 12(4):e0176277
107. Luo S, Au Yeung SL, Zhao JV, Burgess S, Schooling CM (2019) Association of genetically predicted testosterone with thromboembolism, heart failure, and myocardial infarction: mendelian randomisation study in UK biobank. BMJ 364:l476
108. Schooling CM, Luo S, Au Yeung SL, Thompson DJ, Karthikeyan S, Bolton TR et al (2018) Genetic predictors of testosterone and their associations with cardiovascular disease and risk factors: a Mendelian randomization investigation. Int J Cardiol 267:171–176
109. Flynn E, Tanigawa Y, Rodriguez F, Altman RB, Sinnott-Armstrong N, Rivas MA (2021) Sex-specific genetic effects across biomarkers. Eur J Hum Genet 29(1):154–163
110. Handelsman DJ, Yeap B, Flicker L, Martin S, Wittert GA, Ly LP (2015) Age-specific population centiles for androgen status in men. Eur J Endocrinol 173(6):809–817
111. Liang X, Cheng S, Ye J, Chu X, Wen Y, Liu L et al (2021) Evaluating the genetic effects of sex hormone traits on the development of mental traits: a polygenic score analysis and gene-environment-wide interaction study in UK biobank cohort. Mol Brain 14(1):3
112. Dowling JE (1997) Obituary: George Wald(1906–97). Nature 387(6631):356
113. Rando RR (1990) The chemistry of vitamin a and vision. Angew Chem Int Ed Engl 29(5):461–480
114. Maumenee AE (1993) The history of vitamin a and its ophthalmic implications. A personal viewpoint. Arch Ophthalmol 111(4):547–550
115. Wright CB, Redmond TM, Nickerson JM (2015) A history of the classical visual cycle. Prog Mol Biol Transl Sci 134:433–448
116. Ross AC, Gardner EM (1994) The function of vitamin a in cellular growth and differentiation, and its roles during pregnancy and lactation. Adv Exp Med Biol 352:187–200
117. Love JM, Gudas LJ (1994) Vitamin a, differentiation and cancer. Curr Opin Cell Biol 6(6):825–831
118. Zile MH (2001) Function of vitamin a in vertebrate embryonic development. J Nutr 131(3):705–708
119. Clagett-Dame M, DeLuca HF (2002) The role of vitamin a in mammalian reproduction and embryonic development. Annu Rev Nutr 22:347–381
120. Goodman DS (1984) Vitamin A and retinoids in health and disease. N Engl J Med 310(16):1023–1031

121. Krinsky NI, Johnson EJ (2005) Carotenoid actions and their relation to health and disease. Mol Asp Med 26(6):459–516
122. Gueguen S, Leroy P, Gueguen R, Siest G, Visvikis S, Herbeth B (2005) Genetic and environmental contributions to serum retinol and alpha-tocopherol concentrations: the Stanislas family study. Am J Clin Nutr 81(5):1034–1044
123. Biesalski HK, Frank J, Beck SC, Heinrich F, Illek B, Reifen R et al (1999) Biochemical but not clinical vitamin a deficiency results from mutations in the gene for retinol binding protein. Am J Clin Nutr 69(5):931–936
124. Gressner AM, Greiling H (1977) The influence of glycosaminoglycans on the synthesis of polyphenylalanine by rat liver ribosomes. Hoppe Seylers Z Physiol Chem 358(1):69–78
125. Ferrucci L, Perry JR, Matteini A, Perola M, Tanaka T, Silander K et al (2009) Common variation in the beta-carotene 15,15′-monooxygenase 1 gene affects circulating levels of carotenoids: a genome-wide association study. Am J Hum Genet 84(2):123–133
126. Mondul AM, Yu K, Wheeler W, Zhang H, Weinstein SJ, Major JM et al (2011) Genome-wide association study of circulating retinol levels. Hum Mol Genet 20(23):4724–4731
127. Gong X, Marisiddaiah R, Rubin LP (2017) Inhibition of pulmonary beta-carotene 15, 15′-oxygenase expression by glucocorticoid involves PPARalpha. PLoS One 12(7):e0181466
128. D'Adamo CR, Dawson VJ, Ryan KA, Yerges-Armstrong LM, Semba RD, Steinle NI et al (2016) The CAPN2/CAPN8 locus on chromosome 1q is associated with variation in serum alpha-carotene concentrations. J Nutrigenet Nutrigenomics 9(5–6):254–264
129. Cukras C, Gaasterland T, Lee P, Gudiseva HV, Chavali VR, Pullakhandam R et al (2012) Exome analysis identified a novel mutation in the RBP4 gene in a consanguineous pedigree with retinal dystrophy and developmental abnormalities. PLoS One 7(11):e50205
130. Hendrickson SJ, Lindstrom S, Eliassen AH, Rosner BA, Chen C, Barrdahl M et al (2013) Plasma carotenoid- and retinol-weighted multi-SNP scores and risk of breast cancer in the National Cancer Institute breast and prostate cancer cohort consortium. Cancer Epidemiol Biomark Prev 22(5):927–936
131. Luo J, le Cessie S, van Heemst D, Noordam R (2021) Diet-derived circulating antioxidants and risk of coronary heart disease: a Mendelian randomization study. J Am Coll Cardiol 77(1):45–54
132. Williams DM, Hagg S, Pedersen NL (2019) Circulating antioxidants and Alzheimer disease prevention: a Mendelian randomization study. Am J Clin Nutr 109(1):90–98
133. Reay WR, Atkins JR, Quide Y, Carr VJ, Green MJ, Cairns MJ (2020) Polygenic disruption of retinoid signalling in schizophrenia and a severe cognitive deficit subtype. Mol Psychiatry 25(4):719–731
134. West DW, Phillips SM (2012) Associations of exercise-induced hormone profiles and gains in strength and hypertrophy in a large cohort after weight training. Eur J Appl Physiol 112(7):2693–2702
135. Kudielka BM, Hellhammer DH, Wust S (2009) Why do we respond so differently? Reviewing determinants of human salivary cortisol responses to challenge. Psychoneuroendocrinology 34(1):2–18
136. Adam EK, Hawkley LC, Kudielka BM, Cacioppo JT (2006) Day-to-day dynamics of experience--cortisol associations in a population-based sample of older adults. Proc Natl Acad Sci U S A 103(45):17058–17063
137. Henley D, Lightman S, Carrell R (2016) Cortisol and CBG - getting cortisol to the right place at the right time. Pharmacol Ther 166:128–135
138. Bartels M, Van den Berg M, Sluyter F, Boomsma DI, de Geus EJ (2003) Heritability of cortisol levels: review and simultaneous analysis of twin studies. Psychoneuroendocrinology 28(2):121–137
139. Froehlich JC, Zink RW, Li TK, Christian JC (2000) Analysis of heritability of hormonal responses to alcohol in twins: beta-endorphin as a potential biomarker of genetic risk for alcoholism. Alcohol Clin Exp Res 24(3):265–277
140. Inglis GC, Ingram MC, Holloway CD, Swan L, Birnie D, Hillis WS et al (1999) Familial pattern of corticosteroids and their metabolism in adult human subjects--the Scottish adult twin study. J Clin Endocrinol Metab 84(11):4132–4137
141. Meikle AW, Stringham JD, Woodward MG, Bishop DT (1988) Heritability of variation of plasma cortisol levels. Metabolism 37(6):514–517
142. Wust S, Federenko IS, van Rossum EF, Koper JW, Hellhammer DH (2005) Habituation of cortisol responses to repeated psychosocial stress-further characterization and impact of genetic factors. Psychoneuroendocrinology 30(2):199–211
143. Wust S, Federenko I, Hellhammer DH, Kirschbaum C (2000) Genetic factors, perceived chronic stress, and the free cortisol response to awakening. Psychoneuroendocrinology 25(7):707–720
144. Velders FP, Kuningas M, Kumari M, Dekker MJ, Uitterlinden AG, Kirschbaum C et al (2011) Genetics of cortisol secretion and depressive symptoms: a candidate gene and genome wide association approach. Psychoneuroendocrinology 36(7):1053–1061
145. Bolton JL, Hayward C, Direk N, Lewis JG, Hammond GL, Hill LA et al (2014) Genome wide association identifies common variants at the SERPINA6/SERPINA1 locus influencing plasma cortisol and corticosteroid binding globulin. PLoS Genet 10(7):e1004474
146. Walker BR (1996) Abnormal glucocorticoid activity in subjects with risk factors for cardiovascular disease. Endocr Res 22(4):701–708
147. Filipovsky J, Ducimetiere P, Eschwege E, Richard JL, Rosselin G, Claude JR (1996) The relationship of blood pressure with glucose, insulin, heart rate, free fatty acids and plasma cortisol levels according to

degree of obesity in middle-aged men. J Hypertens 14(2):229–235

148. Fraser R, Ingram MC, Anderson NH, Morrison C, Davies E, Connell JM (1999) Cortisol effects on body mass, blood pressure, and cholesterol in the general population. Hypertension 33(6):1364–1368
149. Reynolds RM, Walker BR, Syddall HE, Andrew R, Wood PJ, Phillips DI (2005) Is there a gender difference in the associations of birthweight and adult hypothalamic-pituitary-adrenal axis activity? Eur J Endocrinol 152(2):249–253
150. Hamer M, O'Donnell K, Lahiri A, Steptoe A (2010) Salivary cortisol responses to mental stress are associated with coronary artery calcification in healthy men and women. Eur Heart J 31(4):424–429
151. Reynolds RM, Ilyas B, Price JF, Fowkes FG, Newby DE, Webb DJ et al (2009) Circulating plasma cortisol concentrations are not associated with coronary artery disease or peripheral vascular disease. QJM 102(7):469–475
152. Pott J, Bae YJ, Horn K, Teren A, Kuhnapfel A, Kirsten H et al (2019) Genetic association study of eight steroid hormones and implications for sexual dimorphism of coronary artery disease. J Clin Endocrinol Metab 104(11):5008–5023
153. Kwok MK, Kawachi I, Rehkopf D, Schooling CM (2020) The role of cortisol in ischemic heart disease, ischemic stroke, type 2 diabetes, and cardiovascular disease risk factors: a bi-directional Mendelian randomization study. BMC Med 18(1):363
154. Crawford AA, Soderberg S, Kirschbaum C, Murphy L, Eliasson M, Ebrahim S et al (2019) Morning plasma cortisol as a cardiovascular risk factor: findings from prospective cohort and Mendelian randomization studies. Eur J Endocrinol 181(4):429–438
155. Crawford AA, Bankier S, Altmaier E, Barnes CLK, Clark DW, Ermel R et al (2021) Variation in the SERPINA6/SERPINA1 locus alters morning plasma cortisol, hepatic corticosteroid binding globulin expression, gene expression in peripheral tissues, and risk of cardiovascular disease. J Hum Genet
156. Pivonello R, Simeoli C, De Martino MC, Cozzolino A, De Leo M, Iacuaniello D et al (2015) Neuropsychiatric disorders in Cushing's syndrome. Front Neurosci 9:129
157. Zhou X, Qiao N (2019) Association of cortisol levels with neuropsychiatric functions: a mendelian randomization analysis. Front Endocrinol (Lausanne) 10:564
158. Iftimovici A, Kebir O, He Q, Jay TM, Group IS, Rouleau GA et al (2020) Stress, cortisol and NR3C1 in at-risk individuals for psychosis: a mendelian randomization study. Front Psych 11:680
159. Utge S, Raikkonen K, Kajantie E, Lipsanen J, Andersson S, Strandberg T et al (2018) Polygenic risk score of SERPINA6/SERPINA1 associates with diurnal and stress-induced HPA axis activity in children. Psychoneuroendocrinology 93:1–7
160. Menard V, Eap O, Harvey M, Guillemette C, Levesque E (2009) Copy-number variations (CNVs) of the human sex steroid metabolizing genes UGT2B17 and UGT2B28 and their associations with a UGT2B15 functional polymorphism. Hum Mutat 30(9):1310–1319

Index

M. J. Campbell, C. L. Bevan (eds.), *Nuclear Receptors in Human Health and Disease*, Advances in Experimental Medicine and Biology 1390, https://doi.org/10.1007/978-3-031-11836-4

Printed by Printforce, the Netherlands